CP VIOLATION

Why didn't the matter in our Universe annihilate with antimatter immediately after its creation? The study of **CP** violation may help to answer this fundamental question. Reflecting the explosion of new results over the last decade, this second edition has been substantially expanded. From basic principles to the front-line of research, this account presents the information and theoretical tools necessary to understand this phenomenon.

Charge conjugation, parity and time reversal are introduced, before describing the Kobayashi–Maskawa (KM) theory for **CP** violation and examining our understanding of **CP** violation in kaon decays. Following chapters reveal how the discovery of B mesons provided a new laboratory to study **CP** violation with KM theory predicting large asymmetries, and discuss how these predictions have been confirmed since the first edition of this book. This led to M. Kobayashi and T. Maskawa receiving the 2008 Nobel Prize for Physics. Later chapters describe the search for a new theory of nature's fundamental dynamics. The observation of neutrino oscillations provides opportunities to reveal **CP** violation in the lepton sector, which might drive baryogenesis in a Big Bang Universe. The importance of close links with experiment is stressed, and numerous problems are included. This book is suitable for researchers in high energy, atomic and nuclear physics and in the history and philosophy of science.

IKAROS BIGI was born in Munich, Germany. Following undergraduate and postgraduate studies at the Universities of Munich, Oxford and Stanford, he has taught and researched at the Max-Planck Institute for Physics, CERN, RWTH Aachen, UCLA, the University of Oregon, SLAC and the University of Notre Dame. He is a former scholarship student of the Mazimilianeum Foundation and Scholarship Foundation of the German People and has been appointed both a Heisenberg Fellow and a Max-Kade Fellow.

ICHIRO SANDA was born in Tokyo, and at the age of 14 accompanied his father who was transferred to the United States on business. After a bachelor's degree in physics from the University of Illinois and a Ph.D. from Princeton University, he taught and researched at Columbia University, Fermilab and Rockefeller University. In 1992, after 34 years in the US, he went to Japan as a Professor of physics at Nagoya University. He is now the Chairman of the physics department at Kanagawa University. He is a winner of the 10th Inoue Prize (1993) and the 43rd Nishina Memorial Prize (1997). Both prizes have been awarded for his work in **CP** violation, and on B physics.

Since 1980 the authors have written 14 papers together. In their first paper they explained the special role for **CP** violation played by certain B meson decays; among them was the channel $B \rightarrow \psi K_S$, where the first **CP** asymmetry outside K decays was established in 2001. In 2004 they were jointly awarded the J. J. Sakurai Prize by the American Physical Society 'for pioneering theoretical insights that pointed the way to the very fruitful experimental study of **CP** violation in B decays, and for continuing contributions to the field of **CP** and heavy flavor physics'.

CAMBRIDGE MONOGRAPHS ON PARTICLE PHYSICS NUCLEAR PHYSICS AND COSMOLOGY 28

General Editors: T. Ericson, P. V. Landshoff

CP VIOLATION

I. I. BIGI

Physics Department, University of Notre Dame du Lac

A. I. SANDA

Physics Department, Kanagawa University

CAMBRIDGE
UNIVERSITY PRESS

University Printing House, Cambridge CB2 8BS, United Kingdom

Cambridge University Press is part of the University of Cambridge.

It furthers the University's mission by disseminating knowledge in the pursuit of
education, learning and research at the highest international levels of excellence.

www.cambridge.org
Information on this title: www.cambridge.org/9781107424302

© I. Bigi and A. Sanda 2009

First edition published 1999
Second edition published 2009
First paperback edition 2016

A catalogue record for this publication is available from the British Library

Library of Congress Cataloguing in Publication data
Bigi, I. I.
CP violation / I.I. Bigi, A.I. Sanda. – 2nd ed.
p.   cm. – (Cambridge monographs on particle physics,
nuclear physics, and cosmology ; 28)
ISBN 978-0-521-84794-0
1. CP violation (Nuclear physics)   I. Sanda, A. I. (A. Ichiro)   II. Title.
III. Series.
QC793.3.V5B54   2009
539.7'25–dc22
2009007510

ISBN 978-0-521-84794-0 Hardback
ISBN 978-1-107-42430-2 Paperback

Dedicated to

Colette and Hiroko

Contents

Preface to the second edition

Although the preface to the first edition was written 10 years ago, we see no reason to take anything back from it. Instead we feel confirmed in both our general outlook and in many of our more specific predictions.

- The first direct manifestations of **CP** violation outside the $K^0 - \overline{K}^0$ complex have indeed been found in the decays of neutral B mesons.
- There are already several modes where **CP** asymmetries have been established, and these were actually the expected ones.
- Their size is measured in units of 10%.
- $B_d - \overline{B}_d$ oscillations play an essential role in most cases.
- Last, but not least – the effects are in full agreement with the predictions of the theory of Kobayashi–Maskawa (KM).

The 'battle for supremacy' among theories for the observed **CP** violation that was still hanging in the balance at the end of the last millenium has been decided in favour of KM theory. Now the argument is over the issue of completeness, namely whether there are additional sources of **CP** violation. We know that the answer is most likely affirmative. For understanding the observed baryon number of the Universe as a dynamically generated quantity rather than an arbitrary initial condition requires **CP** violation (in the quark or lepton sector), and KM dynamics cannot fill this role. Probing **CP** invariance more precisely and comprehensively is called for also due to another reason and a most topical one. There are persuasive arguments for the observed electroweak symmetry breaking being driven or at least stabilized by New Physics characterized by the 1 TeV mass scale. *Generic* versions of such New Physics models should already have revealed themselves in flavour-changing neutral currents in B transitions. This has not happened (yet), which has led to the suggestion that the anticipated 'nearby' New Physics must be of the minimal flavour violation variety. We find it much more likely that minimal flavour violation represents an approximate rather than an absolute rule, which will lead to observable deviations from standard model predictions, albeit on a numerically delicate level. Therefore we think it is mandatory to extend the high sensitivity programme of heavy flavour studies. Such considerations become timely with the LHC beginning to operate in 2008.

After we decided to write a second edition we realized how much updating and therefore work was needed. Yet our initial shock gave way to a better appreciation of how much progress has been achieved in the field of heavy flavour studies since the first edition. This in turn led to a sense of deep gratitude to our colleagues on the experimental as well as theoretical side who have made that exciting progress possible.

As indicated above, we see no reason to change the three main goals expressed for the first edition or the intended readership. On the other hand we have updated and even rearranged the material to reflect the greatly changed and expanded experimental and theoretical landscapes. The latter involves both 'streamlining' the previous discussion of theoretical models, since they are no longer viewed as alternatives to KM theory and emphasizing new directions in model building, in particular in the context of supersymmetry and models with extra (space) dimensions.

The first few years of the new millennium have seen discoveries in our field that by any measure were extraordinary. We are confident that even more profound progress will be made in the next two decades.

We want to thank D. Hefferman, S. Mishima, Y. Nakayama, T. Shindou, K. Ukai for bringing typos to our attention and W. Bernreuther, K. Kleinknecht and K. Schubert for pointers to the literature. One of us (I. B.) benefitted greatly from the unique environment at the Aspen Center of Physics while working on this book.

I. I. Bigi
ibigi@nd.edu

A. I. Sanda
sanda@kanagawa-u.ac.jp

Preface to the first edition

Some discoveries in the sciences profoundly change how we view nature. The discovery of parity violation in the weak interactions in 1956 certainly falls into this illustrious category. Yet it just started the shift to a new perspective; it was the discovery of **CP** violation in 1964 by Christenson, Cronin, Fitch and Turlay at Brookhaven National Lab – completely unexpected to almost all despite the experience of 8 years earlier – that established the new paradigm that even in the microscopic regime symmetries should not be assumed to hold a priori, but have to be subjected to determined experimental scrutiny.

It would seem that after the initial period of discoveries little progress has been achieved, since despite dedicated efforts **CP** violation has not been observed outside the decays of K_L mesons, nor can we claim to have come to a real understanding of this fundamental phenomenon.

We have, however, ample reason to expect imminent dramatic changes. Firstly, direct **CP** violation has been observed in K_L decays. Secondly, our phenomenological and theoretical descriptions have been refined to the point that we can predict with confidence that the known forces of nature will generate huge **CP** asymmetries, which could even be close to 100%, in the decays of so-called beauty mesons. Dedicated experiments are being set up to start taking data that would reveal such effects before the turn of the millennium. What they observe – or do not observe – will shape our knowledge of nature's fundamental forces.

We consider it thus an opportune time to take stock, to represent **CP** invariance and its limitations in its full multi-layered complexity. In our presentation we pursue three goals.

- We want to provide a detailed frame of reference for properly evaluating the role of **CP** violation in fundamental physics and to prepare us for digesting the upcoming observations and discoveries.

- We will show that an in-depth treatment of **CP** violation draws on most concepts and tools of particle physics. It thus serves as an unorthodox introduction to quantum field theories (and beyond).
- We want to communicate to the reader that the quest for understanding **CP** violation is more than just an important scholarly task. It represents a most exciting intellectual adventure of which we do not know the outcome. For this very purpose we provide historical perspectives from the last half century.

Accordingly our intended readership is manifold: we want

- to give (theoretical) guidance to the workers in the field;
- to provide an introduction for people who would like to become researchers in this field or at least educated observers;
- to present material which could serve as a supplementary text for courses on quantum field theory; and
- to allow people interested in the history and development of fundamental science to glean maybe some new insights.

We are not pretending our book makes easy reading. We hope, however, that the committed reader will find gratifying the way we start from the basics, give numerous homework problems as an integral part of the learning process and enrich – we think – the narrative with historical remarks. We actually believe that more than one reading will be necessary for a full understanding. To facilitate such an approach we designate sections which can be left out in a first reading by placing their title between the symbols ♠.

As theorists we cannot do full justice to experimental endeavours. Yet we try to communicate our conviction that physics is so wonderfully exciting exactly because it is an empirical science where theory and experiment play an interactive role.

We have benefitted greatly from interacting with many of our colleagues. In particular we would like to acknowledge Dr N. Uraltsev and Dr Z-Z. Xing for their advice and collaboration, Dr A. Garcia and Dr U. Sarid for their suggestions concerning the text. We also express our gratitude to Bernie and Theresa Vonderschmidt for their hospitality during the period in which part of the book was written.

Part I
Basics of CP violation

1
Prologue

All animals are equal.
But some animals are
more equal than others!

G. Orwell, Animal Farm

The sciences in general and physics in particular are full of fascinating phenomena; this is why they have attracted intense human interest early on and have kept it ever since. Yet even so we feel that the question to which degree nature is invariant under time reversal and **CP** transformations is so fundamental that it richly deserves its own comprehensive monograph. Two lines of reasoning – different, though not unrelated to each other – lead us to this conclusion. The first relies on multi-layered considerations, the second is based on a property inferred for the whole universe.

- The first line of reasoning centres on the important role symmetries have always played in physics. It has been recognized only last century, though, how central and crucial this role actually is, and this insight forms one of the lasting legacies of modern physics to human perception of nature and thus to human culture. The connection between continuous symmetries – like translational and rotational invariance – and conserved quantities – momentum and angular momentum for these examples – has been formulated through Noether's theorems. The pioneering work of Wigner and others revealed how atomic and nuclear spectra that appeared at first sight to be quite complicated could be understood through an analysis of underlying symmetry groups, even when they hold only in an approximate sense. This line of reasoning was successfully applied to nuclear and elementary particle physics through the introduction of isospin

3

symmetry $SU(2)$, which was later generalized to $SU(3)$ symmetry in particle physics.

A completely new chapter has been opened through the introduction of *local* gauge theories. In particular Yang–Mills theories were introduced as a formal extrapolation of abelian gauge theories like QED. It was realized only considerably later that the gauge principle plays an essential role in constructing fully relativistic quantum theories that are both non-trivial and renormalizable. Furthermore, it was understood that symmetry breaking can be realized in two different modes, namely *manifestly* and *spontaneously*.

Similarly, *discrete* symmetries have formed an important part of our understanding of the physical world around us – as in crystallography and chemistry. The weight of such considerations is emphasized further with the imposition of permutation symmetries in quantum theory as expressed through Bose–Einstein and Fermi–Dirac statistics. Embedding discrete symmetries into local gauge theories has led to intriguing consequences; among them is the emergence of anomalies, which will be discussed in later chapters.

The primary subject of this book will be discrete symmetries which are of general and fundamental relevance for physics:

– parity **P** – reflecting the space coordinate $\vec{x}$ into $-\vec{x}$;
– charge conjugation **C** – transforming a particle into its antiparticle;
– the combined transformation of **CP**;
– time reversal **T** – changing the time coordinate t into $-t$; it thus amounts to reversal of motion.

We have learnt that nature is largely, but not completely, invariant under these transformations. Although these insights were at first less than eagerly accepted by our community, they form an essential element of what is called the Standard Model (SM) of high energy physics. Yet the story is far from over. For the SM contains 20-odd parameters; those actually exhibit a rather intriguing pattern that cannot be accidental. They must be shaped by some unknown New Physics, and we consider it very likely that a comprehensive analysis of how these discrete symmetries are implemented in nature will reveal the intervention of New Physics.

Furthermore we believe that time reversal, **T**, and the combined transformation of **CP** occupy a very unique place in the pantheon of symmetries. The fact that their violation has been observed in nature has consequences the importance of which cannot be overestimated. Once it was realized that **P** and **C** are violated – and actually violated maximally – it was noted with considerable relief that **CP** was apparently still conserved. For it had been suggested that microscopic **T** invariance follows from Mach's principle; because of **CPT** invariance that holds so naturally in quantum field theories, **CP** violation could not occur without **T** violation.

There is a subtle point about time reversal that is to be understood. **T** can be viewed as reversal of motion. The notion that the laws of nature might be invariant under such a transformation seems absurd at first sight. When watching a filmed sequence of events we can usually tell with *great confidence* whether the film is being played forward or backward. For example, a house of cards will collapse into a disordered pile rather than rise out from such a pile by itself. However this disparity can be understood by realizing that while both sequences are in principle equally possible – as demanded by microscopic **T** invariance – the second one is so unlikely to occur for a macroscopic system as to make it practically impossible. That is why the expression 'with great confidence' was used above. We will address this point in more detail later on.

It came as a great shock that microscopic **T** invariance is violated in nature, that 'nature makes a difference between past and future' even on the most fundamental level. We might feel that such a statement is sensationalist rather than scientific; yet there is indeed something very special about a violation of the invariance under **T** or **CP**. We offer the following observations in support of our view. Elucidating them will be one of the central themes of our book. They might carry little meaning for the reader at this point; yet we expect this to have changed after she or he has finished the book.

– **CP** violation is more fundamental than **C** violation in the following sense: **C** violation as it was discovered can be described by saying that only left-handed neutrinos and right-handed anti-neutrinos interact. This, however, does not allow a genuine distinction of matter and anti-matter:[1] for their difference is expressed in terms of 'left' and 'right handed' which represents a convention – as long as **CP** is conserved! However once **CP** is violated – even if ever so slightly – then matter and antimatter can be distinguished in an absolute, convention-independent way. The practical realization of this general observation goes as follows: while the K_L meson can decay into a positron or an electron together with a pion of the opposite charge and a neutrino or anti-neutrino, it exhibits a slight preference for the mode $K_L \rightarrow e^+ \nu_e \pi^-$. The positron is then called an antilepton; matter and antimatter are thus distinguished by *nature* rather than by convention.

– Time reversal is described by an anti-unitary rather than a unitary operator, which introduces many intriguing subtleties. Among

---

[1] In our world the electron is defined as matter and the positron is defined as antimatter.

them is Kramer's degeneracy: from $\mathbf{T}^2 = \pm 1$ we deduce that two distinct classes of states can exist; those with $\mathbf{T}^2 = -1$ are interpreted as fermions, in contrast to bosons for which $\mathbf{T}^2 = 1$ holds.

– **CP** violation represents the most delicately broken symmetry observed so far in nature and provides us with a powerful phenomenological probe. Consider the historical precedent: the observation of $K_L \to \pi\pi$ in 1964 led to the prediction (in 1972) that a *third* family of quarks and leptons had to exist – before the existence of the final member of the *second* family, the charm quark, had been accepted. It took until 1995 before the top quark, the last member of the third family, had been discovered – with a mass about 400 times as much as the K_L meson!

– While **P** violation has not been understood in a profound way, it can unequivocably be embedded into the gauge sector through chiral couplings to the gauge bosons. Among other things, we can give a natural and meaningful definition of a 'maximal' violation of **P** or **C** invariance, namely that all interacting neutrinos [antineutrinos] are left-handed [right-handed]. Furthermore, the 'see-saw' mechanism provides us with an intriguing dynamical scenario invoking the restauration of **P** invariance at high energies to explain the smallness of neutrino masses. The situation is completely different for **CP** and **T** violation. In general it can enter through gauge or Yukawa interactions. In the Kobayashi–Maskawa (KM) ansatz it is implemented through complex Yukawa couplings; thus it is connected to the least understood part of the SM. The best that can be said is that the SM with three quark families allows for **CP** violation; yet the latter appears as a mere 'add-on'. We are not even able to give real meaning to the notion of 'maximal' **CP** violation.

• The second argument focuses on the observation that while the universe is almost empty, it is not completely empty and actually in a decidedly biased way so! To use a more traditional scientific language: for every trillion or so photons there is just a single baryon in the universe – apparently without any sight of an antibaryon that cannot be explained as the product of a primary collision between matter particles:

$$\text{today}: \quad N(\text{antibaryons}) \ll N(\text{baryons}) \ll N(\text{photons}) \qquad (1.1)$$

Of course we have no reason to complain about this state of affairs. Life could not have developed, we could not exist if nature had been more even-handed in its matter–antimatter distribution.

As first pointed out in a seminal paper by A. Sakharov, there are three essential elements in any attempt to understand the excess of baryons over antibaryons in the universe as a dynamical quantity that can be *calculated* rather than merely *attributed to the initial conditions*:

(1) reactions that change baryon number have to occur;

(2) they cannot be constrained by **CP** invariance;

(3) they must proceed outside thermal equilibrium.

CP violation is thus one essential element in any attempt to achieve such an ambitious goal, and as it turns out, it is the one area that we can best subject to further experimental scrutiny.

In summary: understanding **CP** and **T** violation will bring with it both practical benefits and profound insights since it represents an essential and unalienable element in the fabric of nature's grand design, as sketched in Fig. 1.1. A second glance at this sketch shows that we are dealing with a highly interwoven as well as dense fabric, as is true for all high quality fabrics. To understand its structure, to exploit the inter-relationships among its elements and to interpret data, we obviously need a guiding principle (or two); the concept of symmetry in all its implementations can serve as such. We feel very strongly that progress towards understanding this tapestry can be made only through a feed-back between further dedicated and comprehensive experimental studies and theoretical analysis.

Synopsis

The aim of this book is to show that **T** and **CP** invariance and their violations are much more than exotic phenomena existing in their own little reservation. As indicated above (and discussed in more detail later on) these subjects are, despite their subtle appearance, intimately connected with nature's fundamental structure. Their proper treatment therefore requires a full understanding and usage of our most advanced theoretical tools, namely quantum mechanics and quantum field theory.[2] At the same time we will insist that close contact with experiment has to be maintained.

To pursue this goal our presentation will proceed as follows: first we will describe in considerable detail how **P**, **C** and **T** transformations are implemented in classical physics, and in theories with first and second quantization. Then we will briefly recapitulate how the study of strange particles initiated the observation of **P** non-invariance and led to the

---

[2] Even superstring theories might be called upon in the future to provide the substratum for the relevant field theory, as briefly mentioned in our discussion of **CPT** invariance and of New Physics scenarios.

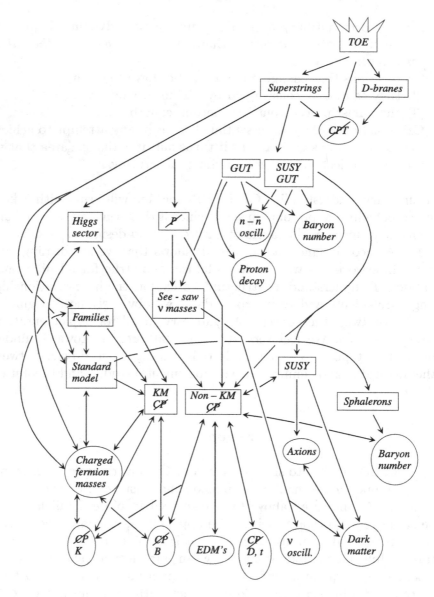

Figure 1.1 Nature's grand tapestry.

discovery of **CP** violation, before describing the phenomenology of the neutral kaon system in detail. After addressing other searches for **T** non-invariance in K decays and through electric dipole moments of neutrons and atoms, we introduce Kobayashi–Maskawa (KM) theory as the minimal implementation of **CP** violation in the SM of high energy physics, and apply it to the description of strange decays and electric dipole moments. We will emphasize how essential it is that a dedicated pursuit of searches for **CP** violation in light quark systems continue in the foreseeable future.

On the other hand, KM theory leads unequivocally to a 'paradigm of large **CP** asymmetries' in B decays. The rich phenomenology for beauty decays can be characterized by six points:

- some predictions enjoy high *parametric* reliability, in particular $B_d \rightarrow J/\psi K_S$ and $B_s \rightarrow J/\psi \eta$, $J/\psi \phi$;
- for others – as in $B_d \rightarrow \pi^+\pi^-$ – such reliability can be achieved through measuring related transitions;
- *parametric* reliability can be turned into *numerical* precision;
- there are many promising ways to search for indirect manifestations of 'New Physics', the most obvious ones being the analysis of $B_s \rightarrow J/\psi \eta$, $J/\psi \phi$, $D_s \bar{D}_s$, $\phi\phi$ and $B_d \rightarrow \phi K_S$, $\eta' K_S$;
- completion of such a program requires a long-term commitment;
- most importantly: *This KM paradigm has been verified experimentally even on the quantitative level!*

Even *before* this validation in B decays KM theory had exhibited attractive, or at least intriguing, aspects.

(A): It provides a natural gateway for **CP** violation to enter; no new degrees of freedom have to be postulated. Three complete quark families have been found experimentally. The SM then does *not automatically* conserve **CP**: it has enough structure to support the existence of a physical weak phase. It could still have turned out that this phase vanishes; yet within the SM context this would appear to be 'unnatural' – it would have to have a dynamical origin beyond the SM.

(B): It had accommodated the observed phenomenology – quite meagre in its positive signals *at that time* – within the experimental and theoretical uncertainties. It had made predictions borne out by the data, namely the elusiveness of direct **CP** violation and the tiny size of electric dipole moments. It had achieved this with some of the fundamental parameters – V_{td}, V_{ts} and m_t – observed to be of a numerical size that before (when the KM ansatz was conceived and for many years thereafter) would have appeared to be quite unreasonable.

Despite these intriguing features and the impressive validation in B decays we consider it quite unlikely that the KM ansatz could remain the final word on **CP** violation – far from it! We are willing to stake our reputation[3] on the prediction that dedicated and comprehensive studies of **CP** violation will reveal the presence of New Physics.

---

[3] Of course it is merely a theorists' reputation.

- The KM ansatz constitutes merely a *parametrization* of a profound phenomenon. The KM matrix actually reflects the mismatch in the alignment of the up-type and down-type quark mass matrices; its elements and thus also the origin of **CP** violation are, therefore, related to two of the central mysteries of the SM: why are there quark (and lepton) families and how do their masses get generated? Because of this connection it is not surprising that we cannot claim a true understanding of **CP** violation. On the other hand, without detailed knowledge of the physical elements of the KM matrix, we do not make full use of the information that nature is allowing us to acquire on the dynamics underlying the generation of fermion masses.
- What we already know about these matrix elements – mostly concerning their moduli – strongly suggests that some very specific dynamics was generating them. For the matrix, rather than being merely unitary, exhibits a very peculiar structure, as outlined before, that is quite unlikely to have come about by accident. The matrix thus contains information on New Physics – albeit in a highly coded form.
- Many extensions to the SM have been suggested to cure perceived, i.e. 'theoretical' ills of the SM. Now we have uncovered *experimental* evidence for the SM's incompleteness:
  (1) neutrinos do oscillate;
  (2) the SM has no candidate for Dark Matter;
  (3) the SM cannot drive baryogenesis.
  We do not even list the bizarre phenomenon referred to as Dark Energy.

CP studies represent highly sensitive probes for manifestations of such New Physics, and we must make the best use of it.

Beyond the pragmatic motivation to discuss other theories for **CP** breaking dynamics sketched above, there is an intellectual one as well. Those theories provide us with an opportunity to address general aspects of the way in which **CP** violation can be realized in nature.

(i) **CP** symmetry can be broken in a 'hard' way, i.e. through dimension-four operators in the Lagrangian, namely gauge couplings to fermions or Yukawa couplings. The KM ansatz is an implementation of the latter variant of such a scenario. For quark mass matrices are derived from the Yukawa couplings through the Higgs mechanism; since in the SM we introduce a single Higgs doublet field, we need the Yukawa couplings to exhibit an irreducible complex phase. This phase is then a free parameter and cannot be calculated.

(ii) **CP** invariance can be broken explicitly in a 'soft' manner, i.e. through operators of dimension below four. We will see that SUSY extensions provide for such scenarios which hold out the promise – or at least the hope – that the basic **CP** violating parameters could be understood dynamically.

(iii) **CP** symmetry is realized in a spontaneous fashion; this is also referred to – sloppily – as spontaneous **CP** violation: while the Lagrangian conserves **CP** (the gauge and Yukawa couplings can be made real), the ground state does not; the vacuum expectation value of neutral Higgs fields develop complex phases. Again, we entertain the hope that the relevant quantities can be derived from the dynamics – in principle, and some day. Models with an extended Higgs or gauge sector allow us to realize such scenarios.

We sketch various extensions of the KM implementation of **CP** violation – among them SUSY scenarios and extra (space) dimension models – and describe processes where realistically only the intervention of New Physics can produce observable **CP** asymmetries, namely in:

- the decays of charm hadrons and τ leptons;
- production and decay of the top quark;
- ν oscillations.

Finally, we address the most ambitious problem, namely baryogenesis in the Big Bang universe.

2

Prelude: C, P and T in classical dynamics

Time reversal – more
than meets the eye

In this chapter we study **C**, **P** and **T** symmetry in classical mechanics and electrodynamics. First we restate the definitions of these transformations together with some comments.

- Parity transformations **P** change a space coordinate $\vec{x}$ into $-\vec{x}$. This is equivalent to a mirror reflection followed by a rotation. For momentum and angular momentum we have, respectively:

$$\vec{p} \xrightarrow{\mathbf{P}} -\vec{p}, \quad \vec{l} \equiv \vec{x} \times \vec{p} \xrightarrow{\mathbf{P}} \vec{l}. \tag{2.1}$$

  This is an example for a general classification: among vectors, which are defined by their transformations under rotations, we can distinguish between polar vectors that change their sign under parity ($\vec{V} \xrightarrow{\mathbf{P}} -\vec{V}$) and axial vectors that do not ($\vec{A} \xrightarrow{\mathbf{P}} \vec{A}$). Likewise we have scalars S – like $\vec{p}_1 \cdot \vec{p}_2$ – and pseudoscalars P – like $\vec{p} \cdot \vec{l}$ – with $S \xrightarrow{\mathbf{P}} S$ whereas $P \xrightarrow{\mathbf{P}} -P$.

- Time reversal **T** reflects t into $-t$ while leaving $\vec{x}$ unchanged and therefore

$$\vec{p} \xrightarrow{\mathbf{T}} -\vec{p}, \quad \vec{l} \xrightarrow{\mathbf{T}} -\vec{l}. \tag{2.2}$$

  T thus represents *reversal of motion*.

- Charge conjugation **C** transforms a particle into its antiparticle of equal mass, momentum and spin, but opposite quantum numbers like electric charge. The notion of an antiparticle is actually ad hoc and even foreign in the realm of *non*-relativistic dynamics. It arose first in the context of the Dirac equation before it was realized that

12

the existence of such antiparticles is a general necessity in quantum field theory.

2.1 Classical mechanics

The motion of an object with mass m is controlled by Newton's equation

$$\vec{F} = m\frac{d^2\vec{x}}{dt^2} \tag{2.3}$$

for given initial position $x(0)$ and velocity $\dot{x}(0)$; $\vec{F}$ denotes the force acting on the object, and $\vec{x}$ is the coordinate of the object at time t.

2.1.1 Parity

Equation (2.3) is clearly invariant under parity $\vec{x}(t) \xrightarrow{\mathbf{P}} -\vec{x}(t)$ because the force is described by a polar vector

$$\vec{F} \xrightarrow{\mathbf{P}} -\vec{F}. \tag{2.4}$$

The impact of parity can be visualized in the following way. Consider the motion of a billiard ball. The ball is hit with a cue and given an initial velocity $\vec{v}_I$. It moves around the table bouncing off its side walls and ends up with a final velocity $\vec{v}_F$ as shown in Fig. 2.1(a). Call this sequence the *genuine* motion. If we take a movie of this motion we can (re)view the genuine motion. If we had made the movie using an old-fashioned film

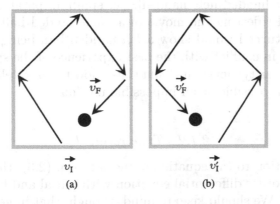

Figure 2.1 Take a movie of the motion of a billiard ball shown in (a). We get (b) by turning the film around and projecting the light from the reverse side. It is the mirror image of (a).

(rather than a video tape) we could turn the film and project the light through it *backward*; the screen would then show a *fake* motion as shown in Fig. 2.1(b), namely a mirror image of the genuine one. However, just watching the screen, we would *not* be able to tell the difference as long as parity is conserved. The fake motion constitutes a physically possible sequence: parity constitutes a symmetry! If we painted the side walls of the billiard table in different colours and communicated this information to the spectator, the fake motion could be distinguished from the genuine one. Yet in that case we would only have exchanged a *convention* for what is left and right!

At this point the curious reader might wonder how parity *non*-invariance could actually reveal itself in this scenario. Consider a gedanken world permeated by a time-independent and homogeneous electric field and billiard balls that are electrically charged. With this electric field pulling charged billiard balls in its direction, their trajectories will be curved rather than straight lines. More importantly, they and their mirror images are quite distinct in their time evolution, even in the points where the balls hit the walls of the table; based on their past experience in such a world the spectators will without a doubt be able to tell whether the movie shows the real motion or the film has been turned before the projection. This situation is described by saying that parity is broken explicitly by a background field.

You might view this example as contrived since the background field first of all breaks rotational invariance. Let us consider a more refined example *without* a background field. The billiard balls are now riding on an air cushion and spinning around their direction of motion. (The air cushion is introduced simply to separate the spinning from an overall forward motion.) Assume the interaction to be such that when a billiard ball hits a wall it is reflected at a smaller [larger] angle than its incoming angle when it is right handed [left handed], i.e. spinning parallel [antiparallel] to its motion. Projecting the movie of a *right*-handed ball *after* the film tape has been turned would show a *left*-handed ball being reflected at a *smaller* angle – in conflict with the past experience of the spectators! The intervention of parity violation would thus allow them to tell whether they were watching a possible or an impossible motion.

2.1.2   Time reversal

If $\vec{x}(t)$ is a solution to the equation of motion, Eq. (2.3), then so is $\vec{x}(-t)$ for this second order differential equation with initial and final conditions switching roles. We should keep in mind, though, that in general $\vec{x}(-t) \neq \pm\vec{x}(t)$; we will give an example in Section 2.2.3.

It turns out to be particularly instructive to discuss the example with billiard balls and how their motion would appear in a movie.

Consider first the case of a single billiard ball. In the genuine motion the ball starts out with initial velocity $\vec{v}_I$, bounces off the side walls of the billiard table and ends up with final velocity $\vec{v}_F$. However, if we watched the sequence projected onto the screen from a film running backwards we could not decide if we watched a fake motion or a genuine one where the ball starts out with $-\vec{v}_F$ and ends up with $-\vec{v}_I$, see Fig. 2.2 or Fig. 2.1(b).

Next we consider the situation in which a stationary ball is hit by another ball, as shown in Fig. 2.3. Watching the filmed sequence we would – with *considerable confidence* – single out the genuine motion as the following one: a ball comes in, hits a stationary ball and both balls move with, in general, different velocities and under different angles, as shown

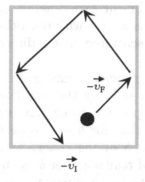

Figure 2.2 Motion of the billiard ball when our film is run backwards. Again, this motion looks genuine to us. Over the years we have learned to take the time reversal symmetry of classical mechanics for granted.

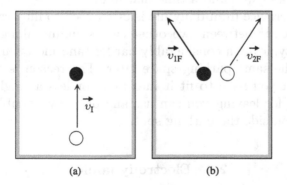

Figure 2.3 Now imagine two balls on a billiard table. In (a) a white ball is hit towards the black ball. When they collide, they fly apart (b). Now reverse this motion and observe it running backwards in time. All of us can tell when the film is run backwards. This is because we know from experience that adjusting the motion of two balls so that one of them stops just after the collision is very very difficult and requires lots of practice.

in Fig. 2.3. For the reverse motion where two balls come together under different angles and with different velocities and hit each other causing one of the balls to come to a complete stop is, as we know from experience, a very improbable one.

We have performed this or a similar kind of experiment before and have learnt that it takes lots of practice and many tries to realize the reversed motion, since it requires a very careful tuning of the initial conditions. To summarize our discussion of this example: both the motion and its reversed version represent sequences *allowed* by the dynamics, i.e. *microscopic* time reversal invariance holds; however, the reversed version is much *less likely* to occur, leading to an apparent macroscopic asymmetry! Thus we can identify the fake motion with considerable confidence, as stated above. If we go one step further in our experiments, namely hit a collection of densely packed billiard balls at rest with one other billiard ball and watch the billiard balls getting scattered in all directions, confidence quickly turns into certainty!

The central message in general terms is the following: the motion of an object is controlled not only by the equation of motion, but also by *initial conditions.* Microscopic time reversal invariance means that the rate for reaction $a \to b$ equals that for $b \to a$ *once the initial configurations,* namely a in one case and b in the other, *have been precisely realized!* However, the probability of realizing a or b as an initial state is in general different, and in fact vastly so for complex processes. The likelihood of the time-reversed version of a *complex* reaction to happen is very low indeed. This observation is one element – though not necessarily the only one – in resolving the following puzzle: if physical laws are invariant under time reversal, why can't we build a time machine?

Another experience from daily life is often used to illustrate this point: manoeuvring a car between two other cars standing along the kerbside for parking is typically a considerably harder (and more frustrating) task than leaving the same parking space later. The reason is the following: to park the car you have to fit it into more or less a single cell of final configurations; for leaving you can use any of many possible trajectories to final states outside the parking spot.

2.2   Electrodynamics

Electrodynamics is governed by Maxwell's equations:

$$\vec{\nabla} \cdot \vec{E} = 4\pi\rho$$

$$\vec{\nabla} \times \vec{B} - \frac{1}{c}\frac{\partial \vec{E}}{\partial t} = \frac{4\pi}{c}\vec{J}$$

$$\vec{\nabla} \cdot \vec{B} = 0$$

$$\vec{\nabla} \times \vec{E} + \frac{1}{c} \frac{\partial \vec{B}}{\partial t} = 0. \tag{2.5}$$

2.2.1 Charge conjugation

The equations remain manifestly invariant under sign reversal of charge density, current, electric field and magnetic field:

$$\rho \xrightarrow{\text{C}} -\rho, \quad \vec{J} \xrightarrow{\text{C}} -\vec{J}$$

$$\vec{E} \xrightarrow{\text{C}} -\vec{E}, \quad \vec{B} \xrightarrow{\text{C}} -\vec{B}. \tag{2.6}$$

These symmetry transformations define charge conjugation **C** for classical electrodynamics.

2.2.2 Parity

The electric field between two oppositely charged particles changes sign when the positions of these two particles are reversed. Similarly, we expect the current density $\vec{J}$ to change sign. The following transformations thus represent parity reflection **P**:

$$\rho \xrightarrow{\text{P}} \rho, \quad \vec{J} \xrightarrow{\text{P}} -\vec{J}$$

$$\vec{E} \xrightarrow{\text{P}} -\vec{E}, \quad \vec{B} \xrightarrow{\text{P}} \vec{B}. \tag{2.7}$$

Note that $\vec{B}$ does not change sign. This is consistent with the fact that $\vec{B} = \vec{\nabla} \times \vec{A}$, where $\vec{A}$ is the vector potential. $\vec{B}$ is said to be an axial vector. Obviously Maxwell's equations are invariant under the combined transformations of Eq. (2.7).

2.2.3 Time reversal

Under time reversal, we expect the current and thus the $\vec{B}$ field to reverse direction, whereas charge density and the electric field remain invariant. Maxwell's equations indeed possess a symmetry

$$\rho \xrightarrow{\text{T}} \rho, \quad \vec{J} \xrightarrow{\text{T}} -\vec{J}$$

$$\vec{E} \xrightarrow{\text{T}} \vec{E}, \quad \vec{B} \xrightarrow{\text{T}} -\vec{B}, \tag{2.8}$$

consistent with our expectations. A particle carrying charge q and moving with velocity $\vec{v}$ experiences a **T** invariant Lorentz force:

$$\vec{F} = q \left(\vec{E} + \vec{v} \times \vec{B} \right) \xrightarrow{\text{T}} \vec{F}. \tag{2.9}$$

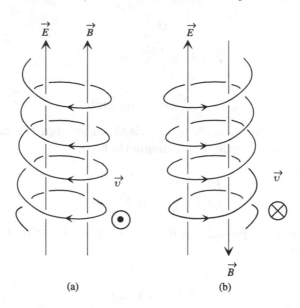

Figure 2.4   (a) Motion of a particle, with $q > 0$, in static electric and magnetic fields, where the initial velocity $\vec{v}$ is out of the plane of the paper as shown. (b) Under time reversal, $\vec{E} \to \vec{E}$, $\vec{B} \to -\vec{B}$, $\vec{v} \to -\vec{v}$. Note that the time-reversed motion cannot be obtained by taking a movie of motion (a) and running the film backwards.

As shown in Fig. 2.4 (a), its motion under the influence of homogenous and time-independent electric and magnetic fields *parallel* to each other is described by a *clockwise* (for $q > 0$) screw around the direction of $\vec{E}$ if the initial momentum of the particle was perpendicular to the two fields.

As shown in Fig. 2.4 (b), under **T**, $\vec{B}$ flips its direction, thus becoming *anti*parallel to $\vec{E}$: the particle now describes a *counter-clockwise* screw around $\vec{E}$.

Running the movie of an *upward clockwise* screw motion backwards will show a *downward clockwise* screw, which is intrinsically distinct from the truly time reversed motion which is a *counter-clockwise* screw! So what went wrong?

Without the $\vec{E}$ field, the particle describes a clockwise closed orbit around the $\vec{B}$ field. After time reversal, which flips both $\vec{B}$ and the initial momentum it still describes a clockwise orbit around the new direction of $\vec{B}$, which amounts to a counter-clockwise orbit around the original direction; this sequence is reproduced when the film is played backwards. The problem is the motion of a particle under a static $\vec{E}$ field (Problem 2.6).

2.3 Résumé

The discrete transformations **C**, **P** and **T** can be defined in classical dynamics in a rather straightforward manner (although the concept of charge conjugation is ad hoc at this level) and simple criteria for them to represent symmetries can be stated. The question of time-reversal symmetry requires a clear distinction between, on the one hand, the equation of motion and, on the other, the likelihood for certain initial conditions to be established. *Microscopic* **T** invariance means first of all that if $a \rightarrow b$ can happen, so can $b \rightarrow a$! Both processes have to occur with equal likelihood only if a and b can equally be realized as *initial* states. The apparent **T** asymmetries encountered in daily life are ascribed to asymmetries in the realizability of the corresponding initial conditions. For further reading we suggest Ref. [1].

Problems

2.1. In elementary courses on electricity and magnetism we learn that the direction of the magnetic field $\vec{B}$ is given by the *right-hand rule*: The right-hand thumb points to the direction of the current and the other fingers wrapping around the current vector give the direction of the $\vec{B}$ field. Under parity $\vec{J} \rightarrow -\vec{J}$. The right-hand rule would seem to imply that $\vec{B} \rightarrow -\vec{B}$, contradicting Eq. (2.7). Resolve the apparent contradiction.

2.2. Discuss the transformation property of the vector potential $\vec{A}(x,t)$ under **C**, **P** and **T**.

2.3. Discuss the transformation properties of a magnetic monopole under **C**, **P** and **T**.

2.4. Ohm's law,

$$\vec{j} = \sigma \vec{E}, \tag{2.10}$$

stating that the current density is proportional to the electric field strength, appears to violate time reversal invariance since $\vec{j} \rightarrow -\vec{j}$, yet $\vec{E} \rightarrow \vec{E}$ under **T**. Resolve the apparent paradox. *Hint*: note that Ohm's law does not represent a microscopic identity; it is based on a time average. Is the underlying equation of motion purely second order in time?

2.5. Since the forces driving our metabolism are electromagnetic in nature, they conserve **T**. This suggests that we can, in principle, get younger. Why has this not been observed?

2.6. Consider the motion of a particle in parallel $\vec{E}$ and $\vec{B}$ fields, as shown in Fig. 2.4(a) where the initial velocity is coming out of the plane of the paper. Convince yourselves that without the $\vec{E}$ field there is no problem, as discussed in the text. Now consider a motion of a particle in the background $\vec{E}$ field. By integrating the equation of motion, we get

$$m\vec{\ddot{x}} = q\vec{E}t. \tag{2.11}$$

How can we interpret this equation in connection with the motion obtained by running the film backwards?

3

C, P and T in non-relativistic quantum mechanics

Subtle is the Lord –
but malicious she is not (we hope)!

Einstein

One of the basic concepts of quantum mechanics is the superposition principle: If $|a\rangle$ and $|b\rangle$ are vectors in a Hilbert space, so are $|\psi\rangle = \alpha|a\rangle + \beta|b\rangle$, $|\psi'\rangle = \alpha'|a\rangle + \beta'|b\rangle$. For an operator $\mathcal{O}$ to represent a symmetry, we must have

$$|\langle\psi|\mathcal{O}^\dagger\mathcal{O}|\psi'\rangle|^2 = |\langle\psi|\psi'\rangle|^2 \qquad (3.1)$$

for no measurement outcome to be affected. This can be satisfied by[1]

$$\langle\psi|\mathcal{O}^\dagger\mathcal{O}|\psi'\rangle = \langle\psi|\psi'\rangle \quad \text{or} \quad \langle\psi|\mathcal{O}^\dagger\mathcal{O}|\psi'\rangle = \langle\psi|\psi'\rangle^*. \qquad (3.2)$$

Assume $|a\rangle$ and $|b\rangle$ to be orthogonal; then $\mathcal{O}|a\rangle$ and $\mathcal{O}|b\rangle$ have to be orthogonal as well. While the first relation of Eq. (3.2) implies that $\mathcal{O}$ is unitary, one infers from the second that $\mathcal{O}$ is *anti*-unitary:

$$\mathcal{O}|\psi\rangle = \alpha^*\mathcal{O}|a\rangle + \beta^*\mathcal{O}|b\rangle. \qquad (3.3)$$

An anti-unitary operator can be defined by the product

$$\mathcal{O} = UK, \qquad (3.4)$$

where U is unitary and K is the complex conjugation operator defined by

$$K[\alpha|a\rangle + \beta|b\rangle] = \alpha^*K|a\rangle + \beta^*K|b\rangle, \qquad (3.5)$$

---

[1] An alert reader may say that an additional arbitrary phase can be introduced on the left-hand side of Eq. (3.2). This is indeed true and is always present. Our argument goes through in spite of this freedom.

where α and β are complex constants. The Hermitian conjugate of K always acts to the left:

$$[\alpha\langle a| + \beta\langle b|]K^\dagger = \alpha^*\langle a|K^\dagger + \beta^*\langle b|K^\dagger. \tag{3.6}$$

The inverse operator is defined by

$$K^2 = 1. \tag{3.7}$$

We can also define K by its action on the wave function:

$$\langle x|K|\psi\rangle = \psi(x)^*. \tag{3.8}$$

We will see below that parity and charge conjugation are described by unitary operators, whereas time reversal is implemented by an anti-unitary one.

Concerning symmetry transformations, two questions have to be addressed.

- Are the admissible initial states invariant?
- Do the dynamics obey the symmetry?

If both conditions are satisfied, the possible final states will remain invariant.

3.1 Parity

Consider the Schrödinger equation for the state vectors $|\psi; t\rangle$,

$$i\hbar\frac{\partial}{\partial t}|\psi; t\rangle = H|\psi; t\rangle$$

$$H = \frac{|\vec{P}|^2}{2m} + V(\vec{X}), \tag{3.9}$$

where $\vec{X}$ is the position operator, and assume parity transformations to be implemented by a *unitary* operator **P** commuting with H:

$$[\mathbf{P}, H] = 0. \tag{3.10}$$

Parity is then conserved in Eq. (3.9): if $|\psi; t\rangle$ represents a solution, so does $\mathbf{P}|\psi; t\rangle$. For this conclusion to be correct

$$\mathbf{P}^{-1}i\mathbf{P} = i \tag{3.11}$$

has to hold, i.e. **P** has to be linear rather than antilinear.

There are simple rules telling us how observables behave under $\mathbf{P}$. Based on the correspondence principle, we require the expectation value of the position operator $\vec{X}$ to change sign under parity,[2]

$$\langle\psi;t|\mathbf{P}^\dagger\vec{X}\mathbf{P}|\psi;t\rangle = -\langle\psi;t|\vec{X}|\psi;t\rangle. \tag{3.12}$$

This is guaranteed to happen if

$$\mathbf{P}^\dagger\vec{X}\mathbf{P} = -\vec{X} \quad \text{or} \quad \{\vec{X},\mathbf{P}\} = 0, \tag{3.13}$$

where $\{a,b\}$ denotes the anticommutator, holds as an operator relation. Hence

$$\mathbf{P}|\vec{x}\rangle = e^{i\delta}|-\vec{x}\rangle \tag{3.14}$$

for eigenstates of the position operator with $e^{i\delta}$ being an arbitrary phase. From $\vec{X}|\vec{x}\rangle = \vec{x}|\vec{x}\rangle$ in conjunction with Eq. (3.13) we obtain

$$\vec{X}\mathbf{P}|\vec{x}\rangle = -\mathbf{P}\vec{X}|\vec{x}\rangle = (-\vec{x})\mathbf{P}|\vec{x}\rangle; \tag{3.15}$$

i.e. $\mathbf{P}|\vec{x}\rangle$ is an eigenvector of $\vec{X}$ with eigenvalue $-\vec{x}$. Conveniently, we adopt the phase convention $e^{i\delta} = 1$. Thus

$$\mathbf{P}^2 = 1, \tag{3.16}$$

meaning that $\mathbf{P}$ is Hermitian as well as unitary,

$$\mathbf{P}^\dagger = \mathbf{P}^{-1} = \mathbf{P} \tag{3.17}$$

and its eigenvalues are ± 1.

The momentum operator is best introduced as the generator of (infinitesimal) translations $d\vec{x}$ in space:

$$\mathcal{T}(d\vec{x}) \simeq 1 + \frac{i}{\hbar}\vec{P}\cdot d\vec{x} \tag{3.18}$$

A translation followed by a parity transformation is equivalent to a parity reflection followed by a translation in the opposite direction, $\mathcal{T}(-d\vec{x})$:

$$\mathbf{P}\mathcal{T}(d\vec{x}) = \mathcal{T}(-d\vec{x})\mathbf{P} \tag{3.19}$$

$$\mathbf{P}\left(1 + \frac{i}{\hbar}\vec{P}\cdot d\vec{x}\right) \simeq \left(1 - \frac{i}{\hbar}\vec{P}\cdot d\vec{x}\right)\mathbf{P}; \tag{3.20}$$

thus we have

$$\mathbf{P}^\dagger\vec{P}\mathbf{P} = -\vec{P} \quad \text{or} \quad \{\vec{P},\mathbf{P}\} = 0. \tag{3.21}$$

---

[2] The unitarity of $\mathbf{P}$ stated above – $\mathbf{P}^\dagger = \mathbf{P}^{-1}$ – is used here and below.

Since **P** anticommutes with both $\vec{X}$ and $\vec{P}$, it leaves the quantization condition

$$[X_i, P_j] = i\hbar\delta_{ij} \tag{3.22}$$

invariant, provided that **P** is linear and thus commutes with i.

Angular momentum $\vec{J}$ is introduced as the generator of rotations. Since rotations and parity commute, so do parity and angular momentum:

$$\mathbf{P}^\dagger \vec{J} \mathbf{P} = \vec{J} \quad \text{or} \quad [\vec{J}, \mathbf{P}] = 0. \tag{3.23}$$

This holds in particular for the orbital angular momentum operator $\vec{L} = \vec{X} \times \vec{P}$:

$$[\vec{L}, \mathbf{P}] = 0 \tag{3.24}$$

which can be deduced also directly using Eq. (3.13) and Eq. (3.21).

Rewriting Eq. (3.9) in a configuration space representation, we arrive at

$$i\hbar\frac{\partial}{\partial t}\psi(\vec{x}, t) = H_x\psi(\vec{x}, t), \tag{3.25}$$

where

$$\psi(\vec{x}, t) \equiv \langle\vec{x}|\psi; t\rangle \tag{3.26}$$

$$H_x = -\frac{\hbar^2}{2m}\nabla^2 + V(\vec{x}). \tag{3.27}$$

If $\psi(t, \vec{x})$ is a solution of this Schrödinger equation, then $\psi(t, -\vec{x})$ solves

$$i\hbar\frac{\partial}{\partial t}\psi(t, -\vec{x}) = \left[-\frac{\hbar^2}{2m}\nabla^2 + V(-\vec{x})\right]\psi(t, -\vec{x}). \tag{3.28}$$

Let us now assume that we have a parity even potential

$$V(\vec{X}) = V(-\vec{X}). \tag{3.29}$$

Then $\psi(t, \vec{x})$ and $\psi(t, -\vec{x})$ solve the same equation – as do the combinations $\psi_\pm(t, \vec{x}) = \psi(t, \vec{x}) \pm \psi(t, -\vec{x})$; i.e. for a parity even potential we can express all solutions as eigenstates of parity. If $|\psi\rangle$ is a parity eigenstate with eigenvalue $+1$ or -1, then its wavefunction is parity even or odd, respectively:

$$\psi(-\vec{x}) \equiv \langle-\vec{x}|\psi\rangle = \langle\vec{x}|\mathbf{P}|\psi\rangle = \pm\langle\vec{x}|\psi\rangle \equiv \pm\psi(\vec{x}). \tag{3.30}$$

Parity considerations provide many powerful constraints. A few typical ones are listed under Problems 3.1–3.4.

For elementary particles or fields we can define also an *intrinsic* parity. This can be done in a strict and unambiguous manner as long as all forces conserve parity. Yet we can go one step further: we can assign an intrinsic parity as determined by the *strong* forces that produce the particle in question – even if the subsequent *weak* decay does not obey this symmetry. A case in point is pions and kaons. As explained in a later chapter, they are associated with the spontaneous realization of chiral invariance and have to act as pseudoscalar states, i.e. they carry odd intrinsic parity. This is confirmed by direct observation; on the other hand we will see that parity is violated even maximally in weak decays of pions and kaons.

With an S-wave $\pi\pi$ state necessarily carrying even parity – see Problem 3.2 – the decay $K \to \pi\pi$ could *not* occur – if the weak decays conserved parity!

3.2 Charge conjugation

To discuss a non-trivial consequence of **C** invariance we introduce the minimal electromagnetic coupling:

$$H = \frac{|\vec{P}|^2}{2m} - \frac{e}{2mc}[\vec{A} \cdot \vec{P} + \vec{P} \cdot \vec{A}] + \frac{e^2}{2mc^2}\vec{A}^2 + e\phi; \qquad (3.31)$$

e is the charge of the particle, $\vec{A}$ the vector potential, and ϕ the electric potential. Note that the potential $V = H - |\vec{P}|^2/2m$ is velocity or momentum dependent here. The Hamilton operator is invariant under the following transformation:

$$e \xrightarrow{\mathbf{C}} -e, \quad \vec{A} \xrightarrow{\mathbf{C}} -\vec{A}, \quad \phi \xrightarrow{\mathbf{C}} -\phi, \qquad (3.32)$$

with $\vec{P}$ and $\vec{X}$ remaining unchanged. That is the motion of a particle remains unchanged if the sign of the external field $\vec{A}$, and ϕ, as well as the charge of the particle, are reversed. It can be seen that:

$$\mathbf{C}H\mathbf{C}^{-1} = H, \qquad (3.33)$$

and if $\psi(x, t)$ is a solution to a Schrödinger equation, so is $\mathbf{C}\psi(x, t)$ if $\mathbf{C}$ is a linear operator. Note also that the operation defined in Eq. (3.32) is such that operating it twice gets back to the original configuration to within a phase factor: $\mathbf{C}^2 = e^{2i\delta}\mathbf{1}$. As in the case of parity, we can adopt the phase convention $\delta = 0$. Accordingly

$$\mathbf{C}^2 = 1. \qquad (3.34)$$

Then the operator $\mathbf{C}$ has eigenvalues ± 1. We will extend the discussion of charge conjugation later when discussing relativistic quantum theory.

3.3   Time reversal

In our discussion of time reversal in classical mechanics we have emphasized one point in particular. *Microscopic* **T** invariance means that the rates for the two processes $a \to b$ and $b \to a$ are the same if b, the final state in the first process, is arranged identically – *in all its aspects* – as the initial state for the second process. We pointed out that such an arrangement requires fine-tuning the initial conditions that is typically very difficult to achieve. This difficulty in corpuscular mechanics turns into a real impossibility in wave mechanics.

The best example we know of is given by Lee in Ref. [2], see Fig. 3.1. Consider the decay

$$\mu^-(\Uparrow) \to e_L^- + \bar{\nu}_{e,R} + \nu_{\mu,L}, \tag{3.35}$$

where the muon is polarized 'up', i.e. perpendicular to the line of flight of the decay products. Now consider a time-reversed collision of three particles forming a muon. It leads to a muon polarized in the line-of-flight direction of the initial beams and *not* a muon polarized 'down':

$$e_L^- + \bar{\nu}_{e,R} + \nu_{\mu,L} \to \mu^-(\Leftarrow) \neq \mu^-(\Downarrow). \tag{3.36}$$

The fact that we cannot realize the fully time reversed reaction goes well beyond the *practical* impossibility of creating such neutrino beams and colliding them with an electron beam, which we encounter already in classical physics. The underlying reason is a quantum mechanical effect, as can be seen in two ways. (i) Spin 'up' in the, say, x direction is *not* orthogonal to spin 'up' in the z direction. (ii) The truly time reversed version of muon decay has an initial state $e_L^- + \bar{\nu}_{eR} + \nu_{\mu,L}$ that consists of incoming *spherical* waves that furthermore have to be coherent;

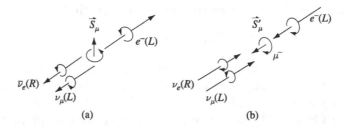

Figure 3.1   (a) A collinear decay of a vertically polarized muon. (b) The time-reversed reaction results in a muon at rest with its polarization vector along the electron momentum. So, in quantum mechanics, we cannot realize the fully time-reversed reaction. Reproduced from Ref. [2] by permission of Harwood Academic Publishers.

i.e. we had to reverse the momenta and spins of the three leptons in *all possible directions* while maintaining the required phase relationships among their amplitudes! This is obviously impossible rather than 'merely' unfeasible.

Let us now prepare the necessary formalism. The task at hand is to find a unitary or anti-unitary operator that under $t \to \bar{t} = -t$ generates reversal of motion at the operator level; i.e.

$$\begin{aligned}
\mathbf{T}^{-1}\vec{X}\mathbf{T} &= \vec{X} &&\text{or} && [\vec{X}, \mathbf{T}] = 0 \\
\mathbf{T}^{-1}\vec{P}\mathbf{T} &= -\vec{P} &&\text{or} && \{\vec{P}, \mathbf{T}\} = 0 \\
\mathbf{T}^{-1}\vec{J}\mathbf{T} &= -\vec{J} &&\text{or} && \{\vec{J}, \mathbf{T}\} = 0,
\end{aligned} \tag{3.37}$$

and that through its action upon a solution of the Schrödinger equation provides us with another solution, to be interpreted as the time-reversed one:

$$|\bar{\psi}; -t\rangle = \mathbf{T} |\psi; t\rangle. \tag{3.38}$$

A glance at the most fundamental equation of quantum mechanics

$$[X_i, P_j] = i\hbar\delta_{ij} \tag{3.39}$$

leads us to conclude that something unusual happens. According to Eq. (3.37), the left-hand side changes sign. What happens with the right-hand side which is a c-number? If $\mathbf{T}$ is anti-unitary we have $\mathbf{T}^{-1}i\mathbf{T} = -i$ and the commutation relation is invariant under $\mathbf{T}$ [3]. If we had overlooked that Eq. (3.1) allows an anti-unitary operator to implement a symmetry transformation, we would have concluded that time reversal symmetry is inconsistent with quantum mechanics. For a moment consider $\mathbf{T} = K$, the complex conjugation operator. We then have

$$\mathbf{T}^{-1}\vec{X}\mathbf{T} = \vec{X} \tag{3.40}$$

$$\mathbf{T}^{-1}\left(-i\hbar\frac{\partial}{\partial\vec{X}}\right)\mathbf{T} = i\hbar\frac{\partial}{\partial\vec{X}} \quad \text{i.e.} \quad \mathbf{T}^{\dagger}\vec{P}\mathbf{T} = -\vec{P}, \tag{3.41}$$

as required by Eq. (3.37). From the defining algebraic relation for angular momentum,

$$[J_i, J_j] = i\hbar\epsilon_{ijk}J_k, \tag{3.42}$$

we read off that

$$\mathbf{T}^{-1}\vec{J}\mathbf{T} = -\vec{J} \tag{3.43}$$

indeed follows with the antiunitary nature of $\mathbf{T}$ again being essential. For the orbital angular momentum operator this property can be derived also directly from the definition $\vec{L} \equiv \vec{X} \times \vec{P}$.

What remains to be seen is the following: how is the solution of the time reversed Schrödinger equation related to the solution of the original one if the Hamilton operator commutes with **T**? Under time reversal the Schrödinger equation Eq. (3.25) changes into

$$\mathbf{T}H|\psi;t\rangle = \mathbf{T}i\hbar\frac{\partial}{\partial t}|\psi;t\rangle = i\hbar\frac{\partial}{\partial \bar{t}}\mathbf{T}|\psi;t\rangle = H\mathbf{T}|\psi;t\rangle, \qquad (3.44)$$

since $\mathbf{T}i\hbar(\partial/\partial t) = -i\hbar(\partial/\partial t)\mathbf{T}$ and $[H, \mathbf{T}] = 0$. With the definition

$$|\bar{\psi};\bar{t} = -t\rangle \equiv \mathbf{T}|\psi;t\rangle, \qquad (3.45)$$

this turns into

$$i\hbar\frac{\partial}{\partial \bar{t}}|\bar{\psi};\bar{t}\rangle = H|\bar{\psi};\bar{t}\rangle. \qquad (3.46)$$

To see the relationship between the solutions $|\bar{\psi}\rangle$ and $|\psi\rangle$ explicitly we expand the state vectors $|\psi\rangle$ and $|\bar{\psi}\rangle$ in terms of the position eigenstates,

$$|\psi\rangle = \int d^3x'|x'\rangle\langle x'|\psi\rangle \qquad (3.47)$$

$$\mathbf{T}|\psi\rangle = \int d^3x'\mathbf{T}(|x'\rangle\langle x'|\psi\rangle) = \int d^3x'|x'\rangle\langle x'|\psi\rangle^*. \qquad (3.48)$$

If $\psi(\vec{x}, t) \equiv \langle\vec{x}|\psi;t\rangle$ is a solution of the Schrödinger equation Eq. (3.25), then its complex conjugate $\psi^*(\vec{x}, t) = \langle\vec{x}|\psi;t\rangle^*$ satisfies

$$i\hbar\frac{\partial}{\partial(-t)}\psi^*(\vec{x}, t) = H\psi^*(\vec{x}, t). \qquad (3.49)$$

Equation (3.49) can of course be obtained directly from Eq. (3.25) through complex conjugation, which effectively achieves $t \to -t$ through $i\partial/\partial t \to (i\partial/\partial t)^* = i\partial/\partial(-t)$. What we have shown here is how it follows from the formal properties of the antiunitary operator **T**.

In many cases the ansatz $\mathbf{T} = K$ provides a satisfactory representation of time reversal. However, this simple ansatz is not sufficient when spin degrees of freedom are involved. Consider, for example, a system with spin 1. The spin operator $\vec{S}$ can then be represented as

$$S_x = i\hbar\begin{pmatrix} 0 & 0 & 0 \\ 0 & 0 & -1 \\ 0 & 1 & 0 \end{pmatrix}, \quad S_y = i\hbar\begin{pmatrix} 0 & 0 & 1 \\ 0 & 0 & 0 \\ -1 & 0 & 0 \end{pmatrix}, \quad S_z = i\hbar\begin{pmatrix} 0 & -1 & 0 \\ 1 & 0 & 0 \\ 0 & 0 & 0 \end{pmatrix}. \qquad (3.50)$$

Since in this representation

$$\vec{S}^* = -\vec{S}, \qquad (3.51)$$

$\mathbf{T} = K$ transforms $\vec{S}$ in the required manner. Let $\psi(m)$ be the wavefunction of an eigenstate of, for example, S_z:

$$S_z \psi(m) = m \psi(m). \tag{3.52}$$

Taking the complex conjugate, we obtain

$$- S_z \psi^*(m) = m \psi^*(m), \tag{3.53}$$

i.e. $\psi^*(m)$ is the wavefunction of an eigenstate of S_z with eigenvalue $-m$, and

$$\mathbf{T}\psi(m) = \psi^*(m) = \psi(-m) \tag{3.54}$$

is indeed satisfied for $\mathbf{T} = K$ – if S_z is defined by Eq. (3.50).

Yet the representation of Eq. (3.50) is not very illuminating. In quantum mechanics, we are allowed to have at most one component of $\vec{J}$ commuting with H. To reveal the underlying physics we choose a maximal set of commuting operators, like

$$[H, J^2] = [H, J_z] = [J^2, J_z] = 0. \tag{3.55}$$

It is then more natural to choose a representation in which one of the components of, in our example, the spin operator $\vec{S}$ is diagonal:

$$S_x = \frac{\hbar}{\sqrt{2}} \begin{pmatrix} 0 & 1 & 0 \\ 1 & 0 & 1 \\ 0 & 1 & 0 \end{pmatrix}, \quad S_y = \frac{\hbar}{\sqrt{2}} \begin{pmatrix} 0 & -i & 0 \\ i & 0 & -i \\ 0 & i & 0 \end{pmatrix}, \quad S_z = \hbar \begin{pmatrix} 1 & 0 & 0 \\ 0 & 0 & 0 \\ 0 & 0 & -1 \end{pmatrix}. \tag{3.56}$$

But now Eq. (3.51) is no longer satisfied, and we must modify the operator $\mathbf{T}$:

$$\mathbf{T} = \exp\left[-i\frac{\pi}{\hbar}S_y\right] K, \tag{3.57}$$

where $\exp[-i\frac{\pi}{\hbar}S_y]$ rotates a vector by an angle π around the y axis so that $S_{x,z} \to -S_{x,z}$, but $S_y \to S_y$. The choice of Eq. (3.57) satisfies $\mathbf{T}\vec{S}\mathbf{T}^{-1} = -\vec{S}$ for the representation of Eq. (3.56).

For future reference, we list useful properties of $\mathbf{T}$.

(1)

$$\mathbf{T}^\dagger\mathbf{T} = K^\dagger U^\dagger U K = K^\dagger K. \tag{3.58}$$

(2)

$$\langle A|\mathbf{T}^\dagger\mathbf{T}|B\rangle = \langle A|K^\dagger K|B\rangle = \langle A|B\rangle^* = \langle B|A\rangle. \tag{3.59}$$

3.4 Kramers' degeneracy

The apparently simple fact that **T** is anti-unitary has a very important consequence we might not have anticipated: it tells us that all possible physical states can be divided into two distinct classes. These two classes turn out to be the bosonic and the fermionic degrees of freedom, yet the intriguing aspect is that the reasoning proceeds without reference to spin or the quantization condition; [1, 4, 5].

The argument goes as follows: if we perform the time reversal operation twice, a state can change at most by a phase factor:

$$\mathbf{T}^2 = UKUK = UU^*KK = UU^* = U(U^T)^{-1} = \eta, \qquad (3.60)$$

where η is an $n \times n$ diagonal matrix with diagonal elements $(e^{i\eta_1}, \ldots, e^{i\eta_n})$. The last equality in Eq. (3.60) implies

$$U = \eta U^T \quad \text{or} \quad U^T = U\eta. \qquad (3.61)$$

Substituting the second relation into the first one, we obtain

$$U = \eta U \eta. \qquad (3.62)$$

Then η_i must satisfy

$$1 = e^{i(\eta_i + \eta_j)} \quad \text{for } i, \, j = 1, \ldots, n. \qquad (3.63)$$

Since this equation is valid for all i, j, we set $i = j$ and find that

$$e^{i\eta_j} = \pm 1 \qquad (3.64)$$

must hold. Note that there is no such restriction for unitary operators. $\mathbf{P}^2 = e^{i\alpha}$, $\mathbf{C}^2 = e^{i\beta}$, where α and β are arbitrary phases.

There is a clear physical distinction between $\mathbf{T}^2 = +1$ and $\mathbf{T}^2 = -1$. Suppose a system is invariant under time reversal and let $|E\rangle$ and $|E^{(T)}\rangle = \mathbf{T}|E\rangle$ be eigenstates with the same eigenvalue. We now show that if $\mathbf{T}^2 = -1$, $|E\rangle$ and $|E^{(T)}\rangle$ are orthogonal. The proof is very simple:

$$\langle E|E^{(T)}\rangle = \langle E^{(T)}|\mathbf{T}^\dagger \mathbf{T}|E\rangle = \langle E|(\mathbf{T}^\dagger)^2 \mathbf{T}|E\rangle = -\langle E|\mathbf{T}|E\rangle, \qquad (3.65)$$

i.e.

$$\langle E|\mathbf{T}|E\rangle = 0, \qquad (3.66)$$

meaning that the two energy-degenerate vectors $|E\rangle$ and $\mathbf{T}|E\rangle$ do *not* describe the same physical state. They are actually orthogonal to each other. *This is referred to as Kramers' degeneracy.*

This implies that $|E\rangle$ carries an internal degree of freedom that is changed by $\mathbf{T}$, if $\mathbf{T}^2 = -1$. If $\mathbf{T}^2 = +1$ holds, such a conclusion cannot be drawn, since the argument given above then yields a tautology. This was all derived *without* reference to spin, half-integer or otherwise. Of course it is completely consistent with results explictly derived for spin degrees of freedom, see Eq. (3.54).

Consider a state with n spin-1/2 particles with z component of the polarization s_i. Noting that $(e^{-i\frac{\pi}{2}\sigma_y})^2|\pm\rangle = -|\pm\rangle$, we obtain

$$\mathbf{T}^2|x_1, s_1; \ldots; x_n, s_n\rangle = (-1)^n|x_1, s_1; \ldots; x_n, s_n\rangle; \qquad (3.67)$$

i.e. $\mathbf{T}^2 = -1$ applies for a system with an odd number of fermions.

Kramers' degeneracy has two consequences. One is conceptual and we have discussed it above. It can be rephrased in the following way. Physical systems can be eigenstates of $\mathbf{T}^2$ with eigenvalues of either $+1$ or -1. A world where only the eigenvalue $+1$ were realized would be conceivable. However, it turns out that both classes of states are realized in nature. Similarly, it would be conceivable that only integer-spin states occurred in nature, or only systems obeying Bose–Einstein statistics. Yet the general experience is that dynamical structures – unless explicitly forbidden – are implemented, and actually in a very efficient way: odd-integer states are dubbed as fermions[3] and now are also seen as carrying the eigenvalue -1 of $\mathbf{T}^2$.

The second consequence of Kramers' degeneracy is of a more practical nature. Often it merely expresses the degeneracy between odd-integer spin up and down. But it can be more intriguing and emerge as a useful tool for treating problems in solid state physics with few manifest symmetries. Consider, for example, a system of electrons in an external electrostatic field $\vec{E}$. Such a field breaks rotational invariance; therefore neither spin nor orbital angular momentum are conserved in this situation. Yet time reversal invariance is still preserved. No matter how complicated this field is, (at least) a *two-fold degeneracy* is necessarily maintained for an *odd* number of electrons due to their spin-orbit couplings, but not for an *even* one; odd-number and even-number electron systems therefore exhibit very different behaviour in such external fields. This degeneracy is lifted once the electrons interact through their magnetic moments with an *external magnetic* field. For a magnetic field is not invariant under $\mathbf{T}$. This observation allows us to use Kramers' degeneracy as a sensitive probe of the electronic states of paramagnetic crystals [6]. As just stated, odd-number systems of electrons exhibit at least a two-fold degeneracy that remains immune to the *electric* fields inside the crystals; it is, however, lifted by

---

[3] In quantum field theory this is understood through the spin-statistics theorem.

magnetic fields. Measuring the energy splittings then allows us to deduce the local magnetic fields inside the crystal.

3.5 Detailed balance

When we are thinking of how to test time-reversal symmetry, the first thing we might come up with is to see if a reaction

$$a + b \to c + d, \tag{3.68}$$

and its reverse

$$c + d \to a + b, \tag{3.69}$$

occur with equal probability in nature. This symmetry is called *detailed balance*. Microscopically, the time-reversal symmetry implies detailed balance, i.e. the equality of S matrix elements for the two processes (see Problems 3.6 and 3.9). The term detailed balance, strictly speaking, is actually used only when there are no spin degrees of freedom; when they are present, one averages over the initial spins and sums over the final ones. For clarity Eq. (3.101) is often referred to as the *reciprocity relation*. We have to keep in mind, however, that time-reversal invariance is a sufficient, but not necessary, condition for detailed balance.

- If the **T** violating interaction H_- is so weak that it is relevant in first order only, then the hermiticity of H_- is sufficient to impose detailed balance:

$$|\langle ab|H_-|cd\rangle|^2 = |\langle cd|H_-|ab\rangle|^2. \tag{3.70}$$

- Another symmetry such as parity might ensure it in the specific case under study. Consider elastic pion–nucleon scattering $\pi_i + N_i \to \pi_f + N_f$ where the phase space factors are obviously the same for a given c.m. energy. With two different spin configurations in the initial and final state the transition operator T is described by a 2×2 matrix; thus it can be expanded in terms of the Pauli matrices.

$$T(\vec{p}_{N_f}, p_{\pi_f}; \vec{p}_{N_i}, \vec{p}_{\pi_i}) = g_1(s,t) + g_2(\vec{p}_{N_i} + \vec{p}_{\pi_i}) \cdot \vec{\sigma}$$
$$+ g_3(s,t)(\vec{p}_{N_i} - \vec{p}_{N_f}) \cdot \vec{\sigma} + g_4(s,t)(\vec{p}_{N_i} \times \vec{p}_{N_f}) \cdot \vec{\sigma}, \tag{3.71}$$

where $s = (p_{N_i} + p_{\pi_i})^2$ and $t = (p_{N_i} - p_{N_f})^2$.
With

$$\vec{\sigma} \overset{\mathbf{T}}{\leftrightarrow} -\vec{\sigma}, \ \vec{p}_{\pi_i} \overset{\mathbf{T}}{\leftrightarrow} -\vec{p}_{\pi_f}, \ \vec{p}_{N_i} \overset{\mathbf{T}}{\leftrightarrow} -\vec{p}_{N_f}, \tag{3.72}$$

time-reversal invariance implies

$$g_3(s,t) = 0. \tag{3.73}$$

Yet this term also violates parity. If the scattering interactions conserve parity, detailed balance therefore holds even if time-reversal symmetry is violated. For further reading we suggest Refs. [1, 7, 8].

3.6 Electric dipole moments

The most direct tests of **T** invariance arise for single particle transitions. Later we will discuss neutrino oscillations in this context. Here we address a *static* quantity describing a particle or state. For further reading we suggest Ref. [8].

Consider a system, such as an elementary particle, an atom or a molecule in a weak electric field $\vec{E}$. The energy shift of the system due to the external electric field can be expanded in a power series in $\vec{E}$,

$$\Delta \mathcal{E} = d_i E_i + d_{ij} E_i E_j + \cdots . \tag{3.74}$$

The coefficient of the linear term in $\vec{E}$ is called the electric dipole moment (EDM), and that of the quadratic term is called the induced dipole moment.

We will show that if an elementary particle possesses an EDM, it implies both **P** and **T** violations. We will further show that detection of an EDM for an atom can imply **P** and **T** violation. This may sound surprising, for we can imagine an atom to behave like a dumb-bells with opposite charges on the ends. We can easily construct such dumb-bells without violating any fundamental symmetry of nature, and it possesses an EDM by definition. We will explain that the *non*-degenerate groundstate of an atom cannot possess an EDM that comes from such a dumb-bell like structure.

3.6.1 The neutron EDM

An EDM is a measure of charge polarization within a particle. It is an expectation value of an operator

$$\vec{d} = \sum_i \vec{r}_i e_i. \tag{3.75}$$

For **P** conserved we have:

$$\langle N, s | \vec{d} | N, s \rangle = \langle N, s | \mathbf{P}^\dagger \mathbf{P} \vec{d} \mathbf{P}^\dagger \mathbf{P} | N, s \rangle = -\langle N, s | \vec{d} | N, s \rangle. \tag{3.76}$$

So, it must vanish. A non-vanishing expectation value of this operator then implies that the dynamics acting on the particle violates parity symmetry.

In the early 1950s, Purcell and Ramsey used this argument to test parity conservation in nuclear forces [9]. At the time the existing limit on the neutron EDM, obtained by Rabi, Havens and Rainwater, and by Fermi and Hughes, was 10^{-17} e cm. With an experiment at the Oak Ridge reactor, they had improved the limit by a factor of 10^3.

After the discovery of parity violation in 1957 this experiment seemed ideally suited for an independent search for parity violation. Landau, Lee and Schwinger, however, pointed out that a non-vanishing EDM required violation of yet another symmetry: **T**. Since $\vec{d}$ is a vector,

$$\langle N, s | \vec{d} | N, s \rangle = C \langle N, s | \vec{J} | N, s \rangle \tag{3.77}$$

must hold, where $\vec{J}$ is the angular momentum operator. The proportionality follows from the fact that $\vec{J}$ is the only three-component object which characterizes a neutron; it is *not* based on treating the neutron as an elementary rather than a bound state. Let us be more precise and specific: if in addition to $\vec{J}$ we could specify a different direction, say $\vec{U}$, for the system under study, to which the EDM could be proportional, then the system would be *degenerate* by its dependance on $\vec{U}$. The issue thus is *non-degeneracy* rather than *elementarity*! Under time reversal invariance, $\vec{d}$ transforms as

$$\mathbf{T}\vec{d}\mathbf{T}^{-1} = \vec{d}, \tag{3.78}$$

compared to

$$\mathbf{T}\vec{J}\mathbf{T}^{-1} = -\vec{J}; \tag{3.79}$$

i.e. the EDM has to vanish: $C = 0$. For an EDM to emerge, both parity and time reversal must be violated. This did not discourage Ramsey. He states [9],

> ... just as parity had been an assumed symmetry which should be based on experiment, time reversal symmetry should be similarly based.

The energy shift of a neutron electric dipole in an electric field $\vec{E}(\vec{r})$ is given by

$$\Delta \mathcal{E} = C \vec{J} \cdot \vec{E}(\vec{r}). \tag{3.80}$$

As $\vec{E}(\vec{r})$ is even under **T** and $\vec{J}$ is odd, such a term in a Hamiltonian violates time reversal invariance.

State-of-the-art measurements of the neutron EDM are done with ultra-cold neutrons: neutrons from a reactor are cooled down to an energy of the order of 10^{-7} eV and then stored in a magnetic bottle. This energy is comparable to the kinetic energy gained by a neutron when it drops 1 m in the earth's gravitational field. Ultra-cold neutrons undergo total reflection from the surfaces of most materials. Neutron spin is also preserved during these collisions. When these neutrons are placed in parallel static electric and magnetic fields, the energy shift given by Eq. (3.80) changes the neutron Larmor frequency. This shift in frequency is measured by the magnetic resonance technique [10]. The 2006 PDG [11] average is:

$$d_n < 2.9 \times 10^{-26} \, \text{e cm}. \qquad (3.81)$$

Ramsey ends his talk [9] by saying:

> After 43 years of searching for, but not finding, a neutron electric dipole moment, I suppose I should be discouraged and believe that no particle dipole moment will ever be discovered and the search should be abandoned. On the contrary, I am now quite optimistic.... For the most of the past 43 years, the searches have been lonely ones. Now there are promising experiments with atoms, electrons, and protons as well. I sincerely hope that someone will hit the jackpot soon... at age 76 I cannot wait another 43 years unless there is a way to achieve real time reversal for biological clocks.

3.6.2   Water molecules and atoms

In the previous section, we pointed out that the EDM $\vec{d}$ must be proportional to the total angular momentum $\vec{J}$ of a *non*-degenerate system. Only then does the non-vanishing of $\vec{d}$ imply **T** violation. Yet for a complex system like an atom or a molecule, there surely must be other vectors characterizing the system.

Two hydrogen atoms bonded to an oxygen atom form a water molecule. These hydrogen atoms can move like coupled harmonic oscillators. In a symmetric oscillation the two hydrogen atoms oscillate in the same direction and in an antisymmetric one they move toward each other or away from each other. The symmetric [antisymmetric] state is in a parity + [−] state.

To discuss the EDM of a water molecule, consider it placed in an external uniform electric field. Let us denote the two parity eigenstates discussed above as $|\pm\rangle$ with energies $\mathcal{E}_\pm$. These states are not degenerate. The symmetric energy state has lower energy: $\mathcal{E}_+ < \mathcal{E}_-$. In a constant external electric field along the z direction, these opposite parity states mix. Denote the mixing matrix element of the electric potential associated with E_z by $\langle +|zE_z|-\rangle = \Delta$. The Hamilton operator can be written in matrix notation as

$$H = \begin{pmatrix} \mathcal{E}_+ & \Delta \\ \Delta & \mathcal{E}_- \end{pmatrix} \qquad (3.82)$$

with eigenvalues

$$\mathcal{E}_{1,2} = \frac{1}{2}(\mathcal{E}_+ + \mathcal{E}_-) \pm \left[\frac{1}{4}(\mathcal{E}_+ - \mathcal{E}_-)^2 + \Delta^2\right]^{\frac{1}{2}}. \qquad (3.83)$$

With a *sufficiently weak* electric field $\vec{E}$ we can arrange for $\langle +|zE_z|-\rangle \equiv \Delta \ll |\mathcal{E}_- - \mathcal{E}_+|$, which leads to

$$\mathcal{E}_1 \simeq \mathcal{E}_- + \frac{\Delta^2}{\mathcal{E}_- - \mathcal{E}_+}, \quad \mathcal{E}_2 \simeq \mathcal{E}_+ - \frac{\Delta^2}{\mathcal{E}_- - \mathcal{E}_+}; \qquad (3.84)$$

and thus

$$\mathcal{E}_2 - \mathcal{E}_1 = \mathcal{E}_- - \mathcal{E}_+ + \frac{2\Delta^2}{\mathcal{E}_- - \mathcal{E}_+}; \qquad (3.85)$$

i.e. the induced energy shift is *quadratic* in $\vec{E}$ rather than linear. In the terminology of Eq. (3.74), it is an *induced* electric dipole moment that does *not* imply **T** violation. Accordingly water molecules do not have an EDM.

The ground state of an atom or a molecule is non-degenerate. The above argument thus applies, and if an EDM is detected, it implies that its elementary constituents – electrons or nucleons – must possess an EDM.

3.6.3 Dumb-bells

Dumb-bells with a positive charge on one end and a negative one on the other possess an electric dipole by definition. Yet that cannot imply **P** (or **T**) violation, since electrodynamics – actually electrostatics in this case – conserve both. This apparent paradox is resolved by the system being *degenerate* – another qualification for the definition in Eq. (3.74).

Consider exact energy degeneracy for the two parity eigenstates $|+\rangle$ and $|-\rangle$. Placing this system inside a constant external electric field as before leads to a Hamilton operator given by Eq. (3.82) with $\mathcal{E}_+ = \mathcal{E}_-$. Then, instead of Eq. (3.85), we have

$$\Delta\mathcal{E} = \mathcal{E}_2 - \mathcal{E}_1 = 2\Delta. \qquad (3.86)$$

The new eigenstates are the linear combinations $|+\rangle \pm |-\rangle$ of the parity eigenstates due to

$$d_z = \langle +|z|-\rangle \neq 0; \qquad (3.87)$$

i.e. the energy shift is linear in the electric field. This mixing of the parity eigenstates due to the external force is fully consistent with parity symmetry. Schiff calls this a permanent EDM [12] as well. Degenerate states can possess a permanent EDM that does *not* imply **T** violation. Atomic states such as $2s_{\frac{1}{2}}$, $2p_{\frac{1}{2}}$ states of the hydrogen atom can possess an EDM without violating **T**, as these are degenerate states.

Dumb-bells realize this abstract scenario in the following way. Center the dumb-bell on the origin and align it with the z axis. If the *positive* charge is located at $z > 0$ or $z < 0$ we denote the state by $| \Uparrow \, \rangle$ or $| \Downarrow \, \rangle$, respectively. *Without* the external electric field, $| \Uparrow \, \rangle$ and $| \Downarrow \, \rangle$ are degenerate – as is any linear combination of them. This applies in particular for the parity eigenstates $| \pm \rangle = \frac{1}{\sqrt{2}}[| \Uparrow \, \rangle \pm | \Downarrow \, \rangle]$. Yet inside the external field the degeneracy is lifted, and $| \Uparrow \, \rangle$ and $| \Downarrow \, \rangle$ become the unique energy eigenstates. None of this creates a conflict with **P** and **T** invariance. As seen from Eq. (3.82), the degeneracy of the ground states opens the way for generating an electric dipole moment due to its structure. We will refer to this effect as a structural EDM.

3.6.4 Schiff's theorem

As the ground state of an atom is non-degenerate, it cannot possess a structural EDM. The observation of an EDM therefore establishes **T** violation in the dynamics of its constituents. Yet effects due to an electron EDM can get screened in an atom: for the charge carriers inside an atom placed into an external electric field get shifted; this distortion creates a polarization and thus an induced electric field that shields the EDM.

Schiff's theorem states that this shielding is *complete* and lists the conditions for its validity [12, 13]: an atomic EDM vanishes under the following conditions.

(1) Atoms consist of *non*-relativistic particles, which interact only electro*statically*.
(2) The EDM distribution of each atomic constituent is identical to its charge distribution.

The electric field at r_i – the coordinate of electron i – is given by

$$\vec{E}_i = -\vec{\nabla}_{r_i} V(r_1, \ldots, r_n), \qquad (3.88)$$

and the term in the Hamilton operator associated with the EDM reads

$$H_{\text{edm}} = -d \sum_i \vec{\sigma} \cdot \vec{\nabla}_{r_i} V(r_1, \ldots, r_n) = -id[Q, H_0], \qquad (3.89)$$

where $H_0 = (p^2/2mc) + V$ and $Q = -i\sum_i \vec{\sigma} \cdot \vec{\nabla}_{r_i}$. Then we can write the Hamiltonian for the entire atom as:

$$H = H_0 - id[Q, H_0] = e^{-idQ} H_0 e^{idQ} + \mathcal{O}(Q^2). \qquad (3.90)$$

If ψ_n is an eigenstate of H_0 with eigenvalue E_n, then obviously so is $e^{-idQ}\psi_n$ for H – with the *same* eigenvalue E_n! Thus there is no shift in the atomic levels of non-relativistic systems due to a possible EDM of the atomic constituents.

In heavy and thus high Z atoms, on the other hand, valence electrons feel strong Coulomb fields when they come close to the nucleus. Their motion is then highly relativistic and one of the conditions of Schiff's theorem breaks down. As the EDM interaction violates parity, it generates strong mixing between opposite parity states at short-distance. We might have expected that the complete shielding of the electron EDM expressed in Schiff's theorem gets translated into a partial shielding when relativistic corrections are included. Yet it turns out that for heavy atoms such as Cs, Tl, Rb, and Fr, there actually emerge huge *enhancement* factors that are as large as two orders of magnitude or even more! For the alkali metal Cesium with its single valence electron one finds $d_{\text{Cs}} \simeq 100 \cdot d_e$ [14]. The PDG 2006 value is obtained from thallium:

$$d_e = (0.07 \pm 0.07) \times 10^{-26}\,\text{e cm}. \qquad (3.91)$$

We can realistically expect further improvements in sensitivity by orders of magnitude in the foreseeable future, due to advances in experimental techniques and the identification of atoms or molecules with even larger enhancement factors relating the EDM of the atom to that of the electron.

3.7 Résumé

Defining charge conjugation, parity and time reversal transformations in quantum mechanics and stating which criteria have to be satisfied for **C, P** and **T** to represent symmetries of the Hilbert space requires more technical ado than for classical dynamics. A special feature arises in that time reversal is described by an *antilinear* operator. Of special interest is Kramer's degeneracy, which anticipates the existence of fermions.

Problems

3.1. Show that a necessary condition for

$$\langle \phi | \vec{X} | \psi \rangle \neq 0 \qquad (3.92)$$

is that $|\psi\rangle$ and $|\phi\rangle$ are eigenstates of *opposite* parity. This is referred to as *Laporte's rule*.

3.2. Consider an eigenstate of $\vec{L}^2$ and L_z denoted by $|l, m\rangle$. Prove that

$$\mathbf{P}|l, m\rangle = (-1)^l|l, m\rangle \qquad (3.93)$$

holds.

3.3. Show that if the dynamics conserve parity – $[H, \mathbf{P}] = 0$ – and if $|\psi\rangle$ denotes a *non-degenerate* eigenstate of H, then it is also a parity eigenstate.

3.4. Demonstrate that for $[H, \mathbf{P}] = 0$ no transition can occur between states of opposite parity.

3.5. For an *orbital* angular momentum eigenstate, establish the identity

$$\mathbf{T}\,|l, m\rangle = (-1)^m|l, -m\rangle \qquad (3.94)$$

from the properties of spherical harmonics.

3.6. Prove that

$$\langle\psi_1|H|\psi_2\rangle = \langle\psi_2|\mathbf{T}^\dagger H\mathbf{T}|\psi_1\rangle \qquad (3.95)$$

if $\mathbf{T}$ commutes with H.

3.7. Show that the charge conjugation operator $\mathbf{C}$ must be linear.

3.8. Suppose

$$\sum_n |n; \text{out}\rangle\langle n; \text{out}| = \sum_n |n; \text{in}\rangle\langle n; \text{in}| = 1, \qquad (3.96)$$

– i.e. both 'in' and 'out' states form a complete set of states – and

$$\sum_n \mathcal{O}|n; \text{out}\rangle\langle n; \text{out}|\mathcal{O}^\dagger = 1, \qquad (3.97)$$

where $\mathcal{O}$ stands for $\mathbf{C}$, $\mathbf{P}$ or $\mathbf{T}$. Finally, define

$$\Gamma_{12} = 2\pi \sum_F \delta(M_P - M_F)\langle P^0|H_{\text{weak}}|n; \text{out}\rangle\langle n; \text{out}|H_{\text{weak}}|\overline{P}^0\rangle.$$

$$(3.98)$$

Show that $\mathbf{CP}$, $\mathbf{T}$ and $\mathbf{CPT}$ symmetries imply $\Gamma_{12} = \Gamma_{21}$, $\Gamma_{12} = \Gamma_{12}^*$ and $\Gamma_{12} = \Gamma_{21}^*$, respectively.

3.9. The time reversal operator transforms

$$\mathbf{T}|\vec{p}_a, \vec{s}_a; \vec{p}_b, \vec{s}_b; \text{in}\rangle = |-\vec{p}_a, -\vec{s}_a; -\vec{p}_b, -\vec{s}_b; \text{out}\rangle, \qquad (3.99)$$

and the scattering $a + b \to c + d$ is controlled by the S matrix

$$S(a + b \to c + d) = \langle\vec{p}_c, \vec{s}_c; \vec{p}_d, \vec{s}_d; \text{out}|\vec{p}_a, \vec{s}_a; \vec{p}_b, \vec{s}_b; \text{in}\rangle. \quad (3.100)$$

Show that **T** invariance implies:

$$S(a + b \rightarrow c + d) = S(c + d \rightarrow a + b), \qquad (3.101)$$

where

$$S(c + d \rightarrow a + b) = \langle -\vec{p}_a, -\vec{s}_a; -\vec{p}_b, -\vec{s}_b; \text{out} | -\vec{p}_c, -\vec{s}_c; -\vec{p}_d, -\vec{s}_d; \text{in} \rangle.$$
$$\qquad (3.102)$$

3.10. Consider an electron at rest and write an operator $R(2\pi)$ which rotates the spin of the electron by 2π. Discuss the relationship between $R(2\pi)$ and $\mathbf{T}^2$.

4

C, P and T in relativistic
quantum theories

In describing symmetries in relativistic quantum theories we usually start
out by showing how relativistic extensions of the Schrödinger equation
like the Klein–Gordon and Dirac equations transform under the symmetry; subsequently we discuss quantum field theory. We do not feel that
following the historical development is always the most illuminating way
to introduce new concepts. Instead we will begin with quantum field theories, i.e. second quantized theory, and subsequently come back to the first
quantized version.

To introduce symmetries in quantum field theory, we assume that there
are operators which transform states in a way that leave all physical
observables unchanged.

In particular, we formulate the theory such that

(1) The groundstate or 'vacuum' remains invariant:

$$\mathbf{P}\,|0\rangle = |0\rangle, \quad \mathbf{C}\,|0\rangle = |0\rangle, \quad \mathbf{T}\,|0\rangle = |0\rangle. \tag{4.1}$$

(2) Likewise for the action:

$$S = \int d^4x \mathcal{L}(t, \vec{x}) \to S \tag{4.2}$$

or equivalently,

$$[\mathbf{P}, H] = [\mathbf{C}, H] = [\mathbf{T}, H] = 0. \tag{4.3}$$

(3) The quantization conditions must remain invariant.

As can be inferred from the title of the book, the symmetries we are
discussing are not exact. We must eventually consider how to break them.

41

When symmetries are broken, the above conditions are relaxed. Nature has chosen two ways to do this. We say that a symmetry is broken:

- *explicitly* if condition (2) is no longer valid;
- *spontaneously* if condition (1) is no longer valid. It is actually more appropriate to call it a spontaneous realization of the symmetry, since the dynamics still obey the latter. We will use both terms interchangeably.

The gauge symmetry of the SM is broken spontaneously. QCD with massless quarks possesses an $SU(3)_L \times SU(3)_R$, chiral symmetry. This symmetry is broken spontaneously to $SU(3)_{L+R}$, giving rise to massless Goldstone bosons that double as pion and kaon pseudoscalar mesons; those acquire masses when chiral symmetry is broken explicitly through quark masses.

How **C, P** and **T** are broken is an interesting question to which we do not know the answer. Let us hope that one of the readers will eventually settle this question. With this comment, we shall postpone the discussion of symmetry breaking to later chapters.

4.1   Notation

Before we go on, let us define some notation [15]. Throughout the remainder of this book we will adopt the unit system that is most convenient for high energy physics, where the speed of light provides the natural scale and quantum effects are not necessarily small; i.e. we set

$$c = 1 = \hbar. \tag{4.4}$$

We write four-vectors with upper and lower indices, where the metric tensor $g_{\mu\nu}$, defined by

$$g_{\mu\nu} = \mathrm{diag}[1, -1, -1, -1] \tag{4.5}$$

is used to raise or lower indices:

$$x^\mu = (t, \vec{x}), \qquad x_\mu = g_{\mu\nu}x^\nu = (t, -\vec{x}), \tag{4.6}$$

with the usual convention of summing over repeated indices. Likewise

$$p^\mu = (E, \vec{p}), \quad p_\mu = (E, -\vec{p})$$
$$\partial^\mu = \frac{\partial}{\partial x_\mu}, \quad \partial_\mu = \frac{\partial}{\partial x^\mu}, \tag{4.7}$$

and for the identification of the four-momentum operator with the derivative operator in configuration space:

$$p^\mu = i\left(\frac{\partial}{\partial t}, -\frac{\partial}{\partial x^1}, -\frac{\partial}{\partial x^2}, -\frac{\partial}{\partial x^3}\right) = i\partial^\mu. \tag{4.8}$$

For the Dirac matrices we shall choose the representation

$$\gamma^0 = \begin{pmatrix} 1 & 0 \\ 0 & -1 \end{pmatrix}, \quad \gamma^i = \begin{pmatrix} 0 & \sigma^i \\ -\sigma^i & 0 \end{pmatrix}, \tag{4.9}$$

where

$$\mathbf{1} = \begin{pmatrix} 1 & 0 \\ 0 & 1 \end{pmatrix}, \sigma^1 = \begin{pmatrix} 0 & 1 \\ 1 & 0 \end{pmatrix}, \sigma^2 = \begin{pmatrix} 0 & -i \\ i & 0 \end{pmatrix}, \sigma^3 = \begin{pmatrix} 1 & 0 \\ 0 & -1 \end{pmatrix} \tag{4.10}$$

are the 2×2 unit matrix and Pauli matrices, respectively. Also,

$$\gamma^5 = i\gamma^0\gamma^1\gamma^2\gamma^3 = \begin{pmatrix} 0 & 1 \\ 1 & 0 \end{pmatrix}. \tag{4.11}$$

Finally, all operators which are products of field operators are understood to be normal ordered.

4.2 Spin-1 fields

Maxwell's equations can be expressed in a compact and manifestly Lorentz covariant fashion:

$$\partial_\mu F^{\mu\nu} = eJ^\nu, \tag{4.12}$$

with $F^{\mu\nu} \equiv \partial^\mu A^\nu - \partial^\nu A^\mu$ being the field strength tensor and J^ν the current. The equation of motion, Eq. (4.12), can be obtained from the Lagrangian:

$$\mathcal{L} = -\frac{1}{4}F_{\mu\nu}F^{\mu\nu} - eJ^\mu A_\mu. \tag{4.13}$$

Electrodynamics are quantized by first introducing a gauge condition; for example, the Coulomb gauge condition reads:

$$\vec{\nabla} \cdot \vec{A} = 0. \tag{4.14}$$

The conjugate momentum associated with $\vec{A}$ is given by

$$\vec{\Pi}(t, \vec{x}) = -\frac{\partial}{\partial t}\vec{A}(t, \vec{x}) - \vec{\nabla}A_0 = \vec{E}, \tag{4.15}$$

and we postulate the quantization condition

$$[A_i(t,\vec{x}), \Pi_j(t,\vec{y})] = -i\left(\delta_{ij}\delta^{(3)}(\vec{x}-\vec{y}) + \frac{\partial^2}{\partial x^i \partial x^j}\frac{1}{4\pi|\vec{x}-\vec{y}|}\right), \quad (4.16)$$

$$[A_i(t,\vec{x}), A_j(t,\vec{y})] = 0 = [\Pi_i(t,\vec{x}), \Pi_j(t,\vec{y})], \quad i,j = 1,2,3. \quad (4.17)$$

The last term on the right-hand side of Eq. (4.16) is required for the quantization condition to be consistent with Gauss's Law: $\vec{\nabla}\cdot\vec{E} = 0$.

To show that quantum electrodynamics is invariant under **P**, **C** and **T**, we have to investigate the symmetry of the Lagrangian as well as the quantization conditions.

From our experience with classical fields, we can postulate the transformation properties for the quantized fields:

$$\begin{array}{ll} \textbf{P } A_\mu(t,\vec{x})\textbf{P}^\dagger = A^\mu(t,-\vec{x}), & \textbf{P } J^\mu(t,\vec{x})\textbf{P}^\dagger = J_\mu(t,-\vec{x}), \\ \textbf{C } A_\mu(t,\vec{x})\textbf{C}^\dagger = -A_\mu(t,\vec{x}), & \textbf{C } J^\mu(t,\vec{x})\textbf{C}^\dagger = -J^\mu(t,\vec{x}), \\ \textbf{T } A_\mu(t,\vec{x})\textbf{T}^{-1} = A^\mu(-t,\vec{x}), & \textbf{T } J^\mu(t,\vec{x})\textbf{T}^{-1} = J_\mu(-t,\vec{x}). \end{array} \quad (4.18)$$

Now let us see if the action remains invariant under these transformations. First observe that

$$\textbf{P}\partial_\mu\textbf{P}^\dagger = \partial^\mu, \quad (4.19)$$

where the raising of the Lorentz indices should be noted. Thus

$$\textbf{P}\mathcal{L}(t,\vec{x})\textbf{P}^\dagger = \mathcal{L}(t,-\vec{x}) \quad (4.20)$$

$$\textbf{C}\mathcal{L}(t,\vec{x})\textbf{C}^\dagger = \mathcal{L}(t,\vec{x}) \quad (4.21)$$

$$\textbf{T}\mathcal{L}(t,\vec{x})\textbf{T}^{-1} = \mathcal{L}(-t,\vec{x}). \quad (4.22)$$

Since the sign of integration variables can be changed, as we integrate over all space–time, the action Eq. (4.2) is invariant under these symmetries.

P, **C** and **T** leave the quantization conditions unchanged. Let us illustrate this by showing that **T** leaves Eq. (4.16) invariant:

$$\textbf{T}[A_i(t,\vec{x}), \Pi_j(t,\vec{y})]\textbf{T}^{-1} = [-A_i(-t,\vec{x}), \Pi_j(-t,\vec{y})]$$

$$= i\left(\delta_{ij}\delta^{(3)}(\vec{x}-\vec{y}) + \frac{\partial^2}{\partial x^i \partial x^j}\frac{1}{4\pi|\vec{x}-\vec{y}|}\right). \quad (4.23)$$

Note that the right-hand side of Eq. (4.23) acquires a minus sign since **T** is anti-unitary. On the left-hand side we have used the fact that

$$\textbf{T}\vec{\Pi}(t,\vec{x})\textbf{T}^{-1} = \frac{\partial}{\partial t}\vec{A}(-t,\vec{x}) - \vec{\nabla}A_0(-t,\vec{x})$$

$$= \vec{\Pi}(-t,\vec{x}). \quad (4.24)$$

The invariance of the quantization condition under **C** and **P** is left as an exercise (see Problem 4.1).

A photon field operator can be expanded in terms of creation and annihilation operators:

$$A^\mu(t, \vec{x}) = \int \frac{d^3\vec{p}}{\sqrt{(2\pi)^3 2E_p}} \sum_{s=\pm} [\mathbf{a}(\vec{p}, s)\epsilon_s^\mu e^{-ip\cdot x} + \mathbf{a}^\dagger(\vec{p}, s)\epsilon_s^{\mu*} e^{ip\cdot x}], \quad (4.25)$$

where $\epsilon_s^\mu(\vec{p})$ and $\mathbf{a}_s^\dagger$ are a polarization vector, and a creation operator for a photon with momentum $\vec{p}$ and polarization s, respectively.

For an on-shell photon with four-momentum $p^\mu = (p, 0, 0, p)$, the polarization vector is given by [16]:

$$\epsilon_\pm^\mu = \frac{1}{\sqrt{2}}(0, 1, \pm i, 0), \quad (4.26)$$

where we have chosen $\epsilon_\pm^i$ to be an eigenstate of S_z as given in Eq. (3.50). The photon polarization must transform as spin:

$$\mathbf{T}\vec{\epsilon}_\pm\mathbf{T}^{-1} = (\vec{\epsilon}_\pm)^* = \vec{\epsilon}_\mp$$

$$\mathbf{P}\vec{\epsilon}_\pm\mathbf{P}^{-1} = \vec{\epsilon}_\pm$$

$$\mathbf{C}\vec{\epsilon}_\pm\mathbf{C}^{-1} = \vec{\epsilon}_\pm. \quad (4.27)$$

Now transforming $A^\mu(t, \vec{x})$ under **C**, **P** and **T** and changing variables, we can deduce

$$\mathbf{CP}\mathbf{a}(\vec{p}, \pm)\mathbf{CP}^\dagger = -\mathbf{a}(-\vec{p}, \pm),$$

$$\mathbf{C}\mathbf{a}(\vec{p}, \pm)\mathbf{C}^\dagger = -\mathbf{a}(\vec{p}, \pm),$$

$$\mathbf{T}\mathbf{a}(\vec{p}, \pm)\mathbf{T}^{-1} = -\mathbf{a}(-\vec{p}, \mp). \quad (4.28)$$

Thus we can say that a one-photon state carries odd *intrinsic* parity and **C** parity:

$$\mathbf{P}|\gamma; \vec{p}\rangle = -|\gamma; -\vec{p}\rangle \quad (4.29)$$

$$\mathbf{C}|\gamma; \vec{p}\rangle = -|\gamma; \vec{p}\rangle. \quad (4.30)$$

Other spin-1 fields can carry internal quantum numbers – as is the case for non-abelian gauge bosons. For those we cannot define intrinsic **C** parity. However, they can still carry intrinsic parity; the latter is odd [even] for vector [axial vector] fields.

4.3 Spin-0 fields

A charged spinless particle can be described by a complex field $\phi(t, \vec{x})$. Consider a Lagrangian for such a spin-0 field in an electromagnetic field,

$$\mathcal{L} = \frac{1}{2}(\partial_\mu + ieA_\mu)\phi^\dagger(t, \vec{x})(\partial^\mu - ieA^\mu)\phi(t, \vec{x}) - \frac{1}{2}m^2\phi^\dagger(t, \vec{x})\phi(t, \vec{x}) \quad (4.31)$$

with quantization conditions

$$[\phi(t, \vec{x}), \phi(t, \vec{y})] = [\phi^\dagger(t, \vec{x}), \phi^\dagger(t, \vec{y})] = 0, \quad (4.32)$$

$$[\phi(t, \vec{x}), \partial_t\phi^\dagger(t, \vec{y})] = [\phi^\dagger(t, \vec{x}), \partial_t\phi(t, \vec{y})] = i\delta^{(3)}(\vec{x} - \vec{y}). \quad (4.33)$$

As in the previous example, the field can be expanded in terms of creation and annihilation operators:

$$\phi(t, \vec{x}) = \int \frac{d^3\vec{p}}{\sqrt{(2\pi)^3 2E_p}}[\mathbf{b}(\vec{p})e^{-ipx} + \mathbf{d}^\dagger(\vec{p})e^{ipx}], \quad (4.34)$$

where $\mathbf{b}(\vec{p})$, and $\mathbf{d}(\vec{p})$ are annihilation operators for particle and antiparticle, respectively.

We shall find that the quantized theory exhibits **C, P** and **T** symmetry.

The transformation properties of $\phi(t, \vec{x})$ can be guessed by examining the current:

$$J^\mu(t, \vec{x}) = i[\phi^\dagger(t, \vec{x})\partial^\mu\phi(t, \vec{x}) - \phi(t, \vec{x})\partial^\mu\phi^\dagger(t, \vec{x})]. \quad (4.35)$$

4.3.1 Parity

Let us postulate

$$\phi^P(t, \vec{x}) \equiv \mathbf{P}\phi(t, \vec{x})\mathbf{P}^\dagger = \phi(t, -\vec{x}). \quad (4.36)$$

With this definition, using Eq. (4.18) we can check that the action remains invariant. The same goes for the quantization conditions Eq. (4.32) and Eq. (4.33).

The transformation property, Eq. (4.36), leads to

$$\mathbf{P}\mathbf{b}(\vec{p})\mathbf{P}^\dagger = \mathbf{b}(-\vec{p}), \quad \mathbf{P}\mathbf{d}(\vec{p})\mathbf{P}^\dagger = \mathbf{d}(-\vec{p}). \quad (4.37)$$

So, an n-particle state transforms as:

$$\mathbf{P}|\vec{p}_1, \ldots, \vec{p}_n\rangle = |-\vec{p}_1, \ldots, -\vec{p}_n\rangle. \quad (4.38)$$

4.3.2 Charge conjugation

The charge conjugation transformation for the current is given by Eq. (4.18), and this can be realized by

$$\phi^C(t, \vec{x}) \equiv \mathbf{C}\phi(t, \vec{x})\mathbf{C}^\dagger = \phi^\dagger(t, \vec{x}). \tag{4.39}$$

By taking the Hermitian conjugate of Eq. (4.32) and Eq. (4.33), we can easily show that $\phi^C(t, \vec{x})$ satisfies the quantization condition. Also, by taking the Hermitian conjugate of $\mathcal{L}(t, \vec{x})$ given in Eq. (4.31), we see that the Lagrangian written in terms of $\phi^C(t, \vec{x})$ remains invariant except for the sign of the charge.

Under $\mathbf{C}$, the creation and annihilation operators transform into each other:

$$\mathbf{C}\mathbf{b}(\vec{p})\mathbf{C}^\dagger = \mathbf{d}(\vec{p}). \tag{4.40}$$

Thus charge conjugation $\mathbf{C}$ changes a particle into its antiparticle.

4.3.3 Time reversal

Let us show that

$$\phi^{\mathbf{T}}(t, \vec{x}) \equiv \mathbf{T}\phi(t, \vec{x})\mathbf{T}^{-1} = \phi(-t, \vec{x}), \quad \mathbf{T}\phi^\dagger(t, \vec{x})\mathbf{T}^{-1} = \phi^\dagger(-t, \vec{x}), \tag{4.41}$$

gives the correct transformation property of the current, Eq. (4.18). The time component of the current transforms as

$$\begin{aligned} \mathbf{T}J^0(t, \vec{x})\mathbf{T}^{-1} &= -i[\phi^\dagger(-t, \vec{x})\partial^0\phi(-t, \vec{x}) - \phi(-t, \vec{x})\partial^0\phi^\dagger(-t, \vec{x})] \\ &= J^0(-t, \vec{x}). \end{aligned} \tag{4.42}$$

The space component can be analysed in a similar manner, and indeed Eq. (4.18) is satisfied. The transformation properties of the creation and annihilation operators can be deduced from Eq. (4.41),

$$\mathbf{T}\mathbf{b}(\vec{p})\mathbf{T}^{-1} = \mathbf{b}(-\vec{p}), \quad \mathbf{T}\mathbf{d}(\vec{p})\mathbf{T}^{-1} = \mathbf{d}(-\vec{p}). \tag{4.43}$$

At this point the reader should notice a curious difference between how a wave function and a field operator transform under $\mathbf{T}$. In the previous chapter, we pointed out that a wave function transforms as

$$\Psi(t, \vec{x}, \vec{p}) \xrightarrow{\mathbf{T}} \Psi^*(-t, \vec{x}, -\vec{p}) \tag{4.44}$$

in contrast to Eq. (4.41). This will be addressed later in Section 4.7.

The quantization conditions Eq. (4.32) and Eq. (4.33) remain invariant under time reversal, whereas the Lagrangian transforms as

$$
\begin{aligned}
\mathbf{T}\mathcal{L}(t,\vec{x})\mathbf{T}^{-1} &= \frac{1}{2}(\partial_\mu - ieA^\mu(-t,\vec{x}))\phi^\dagger(-t,\vec{x})(\partial^\mu + ieA_\mu(-t,\vec{x}))\phi(-t,\vec{x}) \\
&\quad - \frac{1}{2}m^2\phi^\dagger(-t,\vec{x})\phi(-t,\vec{x}) \\
&= \frac{1}{2}(\overline{\partial}^\mu + ieA^\mu(-t,\vec{x}))\phi^\dagger(-t,\vec{x})(\overline{\partial}_\mu - ieA_\mu(-t,\vec{x}))\phi(-t,\vec{x}) \\
&\quad - \frac{1}{2}m^2\phi^\dagger(-t,\vec{x})\phi(-t,\vec{x}) \\
&= \mathcal{L}(-t,\vec{x}), \quad\quad\quad\quad\quad\quad\quad\quad\quad\quad\quad\quad (4.45)
\end{aligned}
$$

where $\overline{\partial}$ denotes ∂ with the sign of time reversed. Thus the action remains invariant.

4.4 Spin-1/2 fields

The Lagrangian for a free spin-1/2 field reads as follows:

$$
\mathcal{L} = \overline{\psi}(t,\vec{x})\left(i\gamma_\mu\partial^\mu - m\right)\psi(t,\vec{x}), \quad\quad\quad (4.46)
$$

and the current is given by

$$
J^\mu(t,\vec{x}) = \overline{\psi}(t,\vec{x})\gamma^\mu\psi(t,\vec{x}). \quad\quad\quad (4.47)
$$

We postulate the canonical anticommutation relations for the fields are given by

$$
\{\psi_\alpha(t,\vec{x}), \psi_\beta^\dagger(t,\vec{y})\} = \delta^3(\vec{x}-\vec{y})\delta_{\alpha\beta}
$$
$$
\{\psi_\alpha(t,\vec{x}), \psi_\beta(t,\vec{y})\} = \{\psi_\alpha^\dagger(t,\vec{x}), \psi_\beta^\dagger(t,\vec{y})\} = 0. \quad (4.48)
$$

The spinor field can conveniently be expressed through its Fourier components:

$$
\psi_\alpha(t,\vec{x}) = \int \frac{d^3\vec{p}}{(2\pi)^{\frac{3}{2}}}\sqrt{\frac{m}{E_p}}\sum_{s=\pm}[\mathbf{b}(\vec{p},s)u_\alpha(\vec{p},s)e^{-ip\cdot x} + \mathbf{d}^\dagger(\vec{p},s)v_\alpha(\vec{p},s)e^{ip\cdot x}],
$$
$$
(4.49)
$$

with $\mathbf{b}[\mathbf{b}^\dagger]$ and $\mathbf{d}[\mathbf{d}^\dagger]$ denoting annihilation [creation] operators for particles and antiparticles, respectively.

The Dirac spinors $u(\vec{p},s)$ and $v(\vec{p},s)$ represent solutions to the Dirac equation in momentum space:

$$
(\not{p} - m)u(\vec{p},s) = 0, \quad\quad\quad\quad\quad\quad (4.50)
$$

$$(\not p + m)v(\vec{p}, s) = 0. \qquad (4.51)$$

They are given by

$$u(\vec{p}, +) = \sqrt{\frac{E+m}{2m}} \begin{pmatrix} 1 \\ 0 \\ \frac{p_z}{E+m} \\ \frac{p_x+ip_y}{E+m} \end{pmatrix} ; \qquad u(\vec{p}, -) = \sqrt{\frac{E+m}{2m}} \begin{pmatrix} 0 \\ 1 \\ \frac{p_x-ip_y}{E+m} \\ \frac{-p_z}{E+m} \end{pmatrix} ,$$

$$v(\vec{p}, +) = \sqrt{\frac{E+m}{2m}} \begin{pmatrix} \frac{p_z}{E+m} \\ \frac{p_x+ip_y}{E+m} \\ 1 \\ 0 \end{pmatrix} ; \qquad v(\vec{p}, -) = \sqrt{\frac{E+m}{2m}} \begin{pmatrix} \frac{p_x-ip_y}{E+m} \\ \frac{-p_z}{E+m} \\ 0 \\ 1 \end{pmatrix} ,$$

$$(4.52)$$

with the spin quantized along the z axis.

The following relations can easily be read off from Eq. (4.52):

$$u(-\vec{p}, s) = \gamma_0 u(\vec{p}, s)$$
$$v(-\vec{p}, s) = -\gamma_0 v(\vec{p}, s) \qquad (4.53)$$
$$i\gamma^2 u(\vec{p}, s)^* = sv(\vec{p}, -s)$$
$$i\gamma^2 v(\vec{p}, s)^* = -su(\vec{p}, -s) \qquad (4.54)$$
$$\gamma_1 \gamma_3 u(\vec{p}, s)^* = -su(-\vec{p}, -s)$$
$$\gamma_1 \gamma_3 v(\vec{p}, s)^* = -sv(-\vec{p}, -s). \qquad (4.55)$$

Equations (4.52) to (4.55), which were derived for free Dirac spinors, can be employed also in discussing *asymptotic* fields.

4.4.1 Parity

We want to find a transformation law for $\psi(t, \vec{x})$ which guarantees the transformation property of the current given in Eq. (4.18). Noting that

$$\gamma_0 \gamma^\mu \gamma_0 = \gamma_\mu, \qquad (4.56)$$

we make the ansatz:

$$\psi^{\mathbf{P}}(t, \vec{x}) \equiv \mathbf{P}\psi(t, \vec{x})\mathbf{P}^\dagger = \gamma_0 \psi(t, -\vec{x}). \qquad (4.57)$$

Then we have

$$\overline{\psi^{\mathbf{P}}} \equiv \mathbf{P}\overline{\psi}(t, \vec{x})\mathbf{P}^\dagger = \overline{\psi}(t, -\vec{x})\gamma_0, \qquad (4.58)$$

and the current transforms as

$$\mathbf{P}J^\mu(t, \vec{x})\mathbf{P}^\dagger = \overline{\psi}(t, -\vec{x})\gamma_0\gamma^\mu\gamma_0\psi(t, -\vec{x}) = J_\mu(t, -\vec{x}), \qquad (4.59)$$

as desired. Since

$$\mathbf{P}\mathcal{L}(t, \vec{x})\mathbf{P}^\dagger = \mathcal{L}(t, -\vec{x}), \tag{4.60}$$

the action remains invariant. Likewise for the anticommutation relations:

$$\begin{aligned}
\{\psi_\alpha^\mathbf{P}(t, \vec{x}), (\psi^\mathbf{P})_\beta^\dagger(t, \vec{y})\} &= (\gamma_0)_{\alpha\rho}(\gamma_0)_{\sigma\beta}\{\psi_\rho(t, -\vec{x}), \psi_\sigma^\dagger(t, -\vec{y})\} \\
&= (\gamma_0)_{\alpha\rho}(\gamma_0)_{\rho\beta}\delta^{(3)}(\vec{x} - \vec{y}) \\
&= \delta^{(3)}(\vec{x} - \vec{y})\delta_{\alpha\beta}. \tag{4.61}
\end{aligned}$$

Likewise,

$$\{\psi_\alpha^\mathbf{P}(t, \vec{x}), \psi_\beta^\mathbf{P}(t, \vec{y})\} = 0 = \{(\psi^\mathbf{P})_\alpha^\dagger(t, \vec{x}), (\psi^\mathbf{P})_\beta^\dagger(t, \vec{y})\}. \tag{4.62}$$

The transformation properties of the fields can also – and maybe more transparently – be expressed in terms of the creation and annihilation operators. Applying the parity operator to the Dirac field, we have

$$\begin{aligned}
\mathbf{P}\,\psi(t, \vec{x})\mathbf{P}^\dagger &= \int \frac{d^3\vec{p}}{(2\pi)^{\frac{3}{2}}} \sqrt{\frac{m}{E_p}} \sum_{s=\pm} \Big[\mathbf{P}\,\mathbf{b}(\vec{p}, s)\mathbf{P}^\dagger u(\vec{p}, s)e^{-ip\cdot x} \\
&\quad + \mathbf{P}\,\mathbf{d}^\dagger(\vec{p}, s)\mathbf{P}^\dagger v(\vec{p}, s)e^{ip\cdot x}\Big] \\
&= \int \frac{d^3\vec{p}}{(2\pi)^{\frac{3}{2}}} \sqrt{\frac{m}{E_p}} \sum_{s=\pm} \Big[\mathbf{P}\,\mathbf{b}(-\vec{p}, s)\mathbf{P}^\dagger u(-\vec{p}, s)e^{-i(E_p t+\vec{p}\cdot\vec{x})} \\
&\quad + \mathbf{P}\,\mathbf{d}^\dagger(-\vec{p}, s)\mathbf{P}^\dagger v(-\vec{p}, s)e^{i(E_p t+\vec{p}\cdot\vec{x})}\Big] \\
&= \int \frac{d^3\vec{p}}{(2\pi)^{\frac{3}{2}}} \sqrt{\frac{m}{E_p}} \sum_{s=\pm} \Big[\mathbf{P}\,\mathbf{b}(-\vec{p}, s)\mathbf{P}^\dagger \gamma_0 u(\vec{p}, s)e^{-i(E_p t+\vec{p}\cdot\vec{x})} \\
&\quad - \mathbf{P}\,\mathbf{d}^\dagger(-\vec{p}, s)\mathbf{P}^\dagger \gamma_0 v(\vec{p}, s)e^{i(E_p t+\vec{p}\cdot\vec{x})}\Big]. \tag{4.63}
\end{aligned}$$

To satisfy Eq. (4.57) with

$$\begin{aligned}
\gamma_0\psi(t, -\vec{x}) = \gamma_0 \int \frac{d^3\vec{p}}{(2\pi)^{\frac{3}{2}}} \sqrt{\frac{m}{E_p}} \sum_{s=\pm} \Big[&\mathbf{b}(\vec{p}, s)u(\vec{p}, s)e^{-i(E_p t+\vec{p}\cdot\vec{x})} \\
&+ \mathbf{d}^\dagger(\vec{p}, s)v(\vec{p}, s)e^{i(E_p t+\vec{p}\cdot\vec{x})}\Big], \tag{4.64}
\end{aligned}$$

we have to require

$$\mathbf{P}\,\mathbf{b}(\vec{p}, s)\mathbf{P}^\dagger = \mathbf{b}(-\vec{p}, s), \quad \mathbf{P}\,\mathbf{d}(\vec{p}, s)\mathbf{P}^\dagger = -\mathbf{d}(-\vec{p}, s). \tag{4.65}$$

The second equality in Eq. (4.65) is particularly intriguing: we have found that fermions and antifermions carry *opposite* intrinsic parity.

4.4.2 Charge conjugation

In analogy with the spin-0 case, where $\mathbf{C}\phi\mathbf{C}^\dagger = \phi^\dagger$, we seek a transformation

$$\psi^{\mathbf{C}}(t,\vec{x}) = \mathbf{C}\psi(t,\vec{x})\mathbf{C}^\dagger = C\overline{\psi}^{\mathrm{tr}}(t,\vec{x}), \qquad (4.66)$$

where the superscript tr denotes transposition, whereas T refers to time reversal. Noting that

$$\overline{\psi^{\mathbf{C}}}(t,\vec{x}) = \psi^{\mathrm{tr}}\gamma^0 C^\dagger \gamma^0, \qquad (4.67)$$

we find

$$\begin{aligned}
\mathbf{C}J^\mu(t,\vec{x})\mathbf{C}^\dagger &= \overline{\psi^{\mathbf{C}}}(t,\vec{x})\gamma^\mu\psi^{\mathbf{C}}(t,\vec{x}) \\
&= \psi_\alpha(t,\vec{x})[\gamma^0 C^\dagger \gamma^0 \gamma^\mu C]_{\alpha\beta}\overline{\psi}_\beta(t,\vec{x}) \\
&= -\overline{\psi}_\beta(t,\vec{x})[\gamma^0 C^\dagger \gamma^0 \gamma^\mu C]_{\alpha\beta}\psi_\alpha(t,\vec{x}) \\
&= -J^\mu(t,\vec{x}),
\end{aligned} \qquad (4.68)$$

provided that

$$\gamma^0 C^\dagger \gamma^0 \gamma_\mu C = \gamma_\mu^{\mathrm{tr}}. \qquad (4.69)$$

So, if

$$C = i\gamma^2\gamma^0, \qquad (4.70)$$

Equation (4.69) is satisfied, and we recover the transformation property of the current under $\mathbf{C}$ given in Eq. (4.18).

With our choice of phase, $C^2 = -1$ and

$$C = -C^\dagger = -C^{-1} = -C^{\mathrm{tr}}. \qquad (4.71)$$

The free Lagrangian Eq. (4.46) is invariant under $\mathbf{C}$.

$$\begin{aligned}
\mathbf{C}\mathcal{L}\mathbf{C}^\dagger &= -\psi(t,\vec{x})_\alpha[C^\dagger\,(i\gamma_\mu\partial^\mu - m)\,C]_{\alpha\beta}\overline{\psi}_\beta(t,\vec{x}) \\
&= \overline{\psi}(t,\vec{x})_\beta[C^\dagger\left(-i\gamma_\mu\overleftarrow{\partial}^\mu - m\right)C]_{\alpha\beta}\psi_\alpha(t,\vec{x}) \\
&= \mathcal{L}.
\end{aligned} \qquad (4.72)$$

Through an integration by parts in evaluating the action, we see that it is invariant under $\mathbf{C}$.

We find again that the anticommutators stay the same:

$$\begin{aligned}
\{\psi^{\mathbf{C}}_\alpha(t,\vec{x}), (\psi^{\mathbf{C}})^\dagger_\beta(t,\vec{y})\} &= (C\gamma_0)_{\alpha\rho}(\gamma_0 C^\dagger)_{\sigma\beta}\{\psi^*_\rho(t,\vec{x}), \psi^{\mathrm{tr}}_\sigma(t,\vec{y})\} \\
&= (C\gamma_0)_{\alpha\rho}(\gamma_0 C^\dagger)_{\rho\beta}\delta^{(3)}(\vec{x}-\vec{y}) \\
&= \delta^{(3)}(\vec{x}-\vec{y})\delta_{\alpha\beta},
\end{aligned} \qquad (4.73)$$

and likewise

$$\{\psi_\alpha^C(t,\vec{x}), \psi_\beta^C(t,\vec{y})\} = 0 = \{(\psi^C)_\alpha^\dagger(t,\vec{x}), (\psi^C)_\beta^\dagger(t,\vec{y})\}. \tag{4.74}$$

The transformation properties of the creation and annihilation operators are deduced by considering the expression in momentum space:

$$\begin{aligned}
\mathbf{C}\psi(t,\vec{x})\mathbf{C}^\dagger &= \sum_s \int \frac{d^3p}{(2\pi)^{\frac{3}{2}}} \sqrt{\frac{m}{E_p}} \left[\mathbf{C}\mathbf{b}(\vec{p},s)\mathbf{C}^\dagger u(\vec{p},s)e^{-ip\cdot x} \right. \\
&\qquad \left. + \mathbf{C}\mathbf{d}^\dagger(\vec{p},s)\mathbf{C}^\dagger v(\vec{p},s)e^{ip\cdot x} \right] \\
&= \sum_s \int \frac{d^3p}{(2\pi)^{\frac{3}{2}}} \sqrt{\frac{m}{E_p}} \left[\mathbf{b}^\dagger(\vec{p},s)i\gamma^2 u^*(\vec{p},s)e^{ip\cdot x} \right. \\
&\qquad \left. + \mathbf{d}(\vec{p},s)i\gamma^2 v^*(\vec{p},s)e^{-ip\cdot x} \right] \\
&= \sum_s \int \frac{d^3p}{(2\pi)^{\frac{3}{2}}} \sqrt{\frac{m}{E_p}} \left[\mathbf{b}^\dagger(\vec{p},s)(s)v(\vec{p},-s)e^{ip\cdot x} \right. \\
&\qquad \left. + \mathbf{d}(\vec{p},s)(-s)u(\vec{p},-s)e^{-ip\cdot x} \right]. \tag{4.75}
\end{aligned}$$

This leads to the transformation property:

$$\mathbf{C}\mathbf{b}(\vec{p},s)\mathbf{C}^\dagger = s\mathbf{d}(\vec{p},-s). \tag{4.76}$$

4.4.3 Time reversal

Let us consider the time reversal operator

$$\psi^\mathbf{T}(t,\vec{x}) = \mathbf{T}\psi(t,\vec{x})\mathbf{T}^{-1} = U\psi(-t,\vec{x}). \tag{4.77}$$

The electromagnetic current transforms as

$$\begin{aligned}
\mathbf{T}J^\mu(t,\vec{x})\mathbf{T}^{-1} &= \mathbf{T}\overline{\psi}(t,\vec{x})\mathbf{T}^{-1}\gamma^{\mu*}\mathbf{T}\psi(t,\vec{x})\mathbf{T}^{-1} \\
&= \overline{\psi}(-t,\vec{x})U^\dagger\gamma^{\mu*}U\psi(-t,\vec{x}). \tag{4.78}
\end{aligned}$$

So, we need to find a matrix $\mathbf{T} = UK$ such that

$$U\gamma^{\mu*}U^\dagger = \gamma_\mu. \tag{4.79}$$

A simple computation shows that

$$U = \gamma^1\gamma^3 \tag{4.80}$$

satisfies Eq. (4.79).

Since

$$\mathbf{T}\mathcal{L}\mathbf{T}^{-1} = \overline{\psi}(-t, \vec{x})U^{\dagger}\left(-i\gamma_{\mu}^{*}\partial^{\mu} - m\right)U\psi(-t, \vec{x})$$
$$= \overline{\psi}(-t, \vec{x})\left(-i\gamma^{\mu}\partial^{\mu} - m\right)\psi(-t, \vec{x})$$
$$= \mathcal{L}(-t, \vec{x}), \tag{4.81}$$

the action remains invariant under $\mathbf{T}$. Likewise for the quantization conditions; Eq. (4.48) can be easily verified. We obtain:

$$\{\psi_{\alpha}^{\mathbf{T}}(t, \vec{x}), (\psi^{\mathbf{T}})_{\beta}^{\dagger}(t, \vec{y})\} = (\gamma^{1}\gamma^{3})_{\alpha\sigma}(\gamma^{3}\gamma^{1})_{\kappa\beta}\{\psi_{\sigma}(-t, \vec{x}), \psi_{\kappa}^{\dagger}(-t, \vec{y})\}$$
$$= (\gamma^{1}\gamma^{3})_{\alpha\sigma}(\gamma^{3}\gamma^{1})_{\sigma\beta}\delta^{(3)}(\vec{x} - \vec{y}) = \delta^{(3)}(\vec{x} - \vec{y})\delta_{\alpha\beta}$$
$$\{\psi_{\alpha}^{\mathbf{T}}(t, \vec{x}), \psi_{\beta}^{\mathbf{T}}(t, \vec{y})\} = 0 = \{(\psi^{\mathbf{T}})_{\alpha}^{\dagger}(t, \vec{x}), (\psi^{\mathbf{T}})_{\beta}^{\dagger}(t, \vec{y})\}. \tag{4.82}$$

In momentum space, we expect

$$\mathbf{T}\psi(t, \vec{x})\mathbf{T}^{-1} = \sum_{s}\int \frac{d^{3}p}{(2\pi)^{\frac{3}{2}}}\sqrt{\frac{m}{E_{p}}}[\mathbf{T}\mathbf{b}(\vec{p}, s)\mathbf{T}^{-1}u^{*}(\vec{p}, s)e^{ipx}$$
$$+ \mathbf{T}\mathbf{d}^{\dagger}(\vec{p}, s)\mathbf{T}^{-1}v^{*}(\vec{p}, s)e^{-ipx}]$$
$$= \sum_{s}\gamma^{1}\gamma^{3}s\int \frac{d^{3}p}{(2\pi)^{\frac{3}{2}}}\sqrt{\frac{m}{E_{p}}}[\mathbf{T}\mathbf{b}(\vec{p}, s)\mathbf{T}^{-1}u(-\vec{p}, -s)e^{ipx}$$
$$+ \mathbf{T}\mathbf{d}^{\dagger}(\vec{p}, s)\mathbf{T}^{-1}v(-\vec{p}, -s)e^{-ipx}]. \tag{4.83}$$

Using a relation given in Eq. (4.77),

$$\psi^{\mathbf{T}}(t', \vec{x}) = \sum_{s}\int \frac{d^{3}p}{(2\pi)^{\frac{3}{2}}}\sqrt{\frac{m}{E_{p}}}\gamma^{1}\gamma^{3}[\mathbf{b}(\vec{p}, s)u(\vec{p}, s)e^{-i(-E_{p}t - \vec{p}\cdot\vec{x})}$$
$$+ \mathbf{d}(\vec{p}, s)^{\dagger}v(\vec{p}, s)e^{i(-E_{p}t - \vec{p}\cdot\vec{x})}]$$
$$= \gamma^{1}\gamma^{3}\sum_{s}\int \frac{d^{3}p}{(2\pi)^{\frac{3}{2}}}\sqrt{\frac{m}{E_{p}}}[\mathbf{b}(-\vec{p}, -s)u(-\vec{p}, -s)e^{-i(-E_{p}t + \vec{p}\cdot\vec{x})}$$
$$+ \mathbf{d}^{\dagger}(-\vec{p}, -s)v(-\vec{p}, -s)e^{i(-E_{p}t + \vec{p}\cdot\vec{x})}]. \tag{4.84}$$

Comparing these two equations, we obtain

$$\mathbf{T}\mathbf{b}(\vec{p}, s)\mathbf{T}^{-1} = s\mathbf{b}(-\vec{p}, -s), \qquad \mathbf{T}\mathbf{d}(\vec{p}, s)\mathbf{T}^{-1} = s\mathbf{d}(-\vec{p}, -s). \tag{4.85}$$

4.5 CP and CPT transformations

The transformation properties of scalar, pseudoscalar, vector and axial-vector fields – S, P, V_{μ} and A_{μ}, respectively – are summarized in Table 4.1.

Table 4.1   _Summary of how charged fields transform under_ **C, P** _and_ **T**

F	$PFP^\dagger$	$CFC^\dagger$	TFT^{-1}	$CPFCP^\dagger$	$CPTFCPT^{-1}$
$S^+(t,\vec{x})$	$S^+(t,-\vec{x})$	$S^-(t,\vec{x})$	$S^+(-t,\vec{x})$	$S^-(t,-\vec{x})$	$S^-(-t,-\vec{x})$
$P^+(t,\vec{x})$	$-P^+(t,-\vec{x})$	$P^-(t,\vec{x})$	$P^+(-t,\vec{x})$	$-P^-(t,-\vec{x})$	$-P^-(-t,-\vec{x})$
$V_\mu^+(t,\vec{x})$	$V^{+\mu}(t,-\vec{x})$	$-V_\mu^-(t,\vec{x})$	$V^{+\mu}(-t,\vec{x})$	$-V^{-\mu}(t,-\vec{x})$	$-V_\mu^-(-t,-\vec{x})$
$A_\mu^+(t,\vec{x})$	$-A^{+\mu}(t,-\vec{x})$	$A_\mu^-(t,\vec{x})$	$A^{+\mu}(-t,\vec{x})$	$-A^{-\mu}(t,-\vec{x})$	$-A_\mu^-(-t,-\vec{x})$

Next we consider how fermion bilinears transform under these discrete symmetries. First define

$$\overline{\psi}_a^{\mathbf{P}}(t,\vec{x})\Gamma_i\psi_b^{\mathbf{P}}(t,\vec{x}) = \overline{\psi}_a(t,-\vec{x})\Gamma_i^{\mathbf{P}}\psi_b(t,-\vec{x});$$

$$\overline{\psi}_a^{\mathbf{C}}(t,\vec{x})\Gamma_i\psi_b^{\mathbf{C}}(t,\vec{x}) = \overline{\psi}_b(t,\vec{x})\Gamma_i^{\mathbf{C}}\psi_a(t,\vec{x});$$

$$\overline{\psi}_a^{\mathbf{T}}(t,\vec{x})\Gamma_i\psi_b^{\mathbf{T}}(t,\vec{x}) = \overline{\psi}_a(-t,\vec{x})\Gamma_i^{\mathbf{T}}\psi_b(-t,\vec{x});$$

$$\overline{\psi}_a^{\mathbf{CP}}(t,\vec{x})\Gamma_i\psi_b^{\mathbf{CP}}(t,\vec{x}) = \overline{\psi}_b(t,-\vec{x})\Gamma_i^{\mathbf{CP}}\psi_a(t,-\vec{x});$$

$$\overline{\psi}_a^{\mathbf{CPT}}(t,\vec{x})\Gamma_i\psi_b^{\mathbf{CPT}}(t,\vec{x}) = \overline{\psi}_b(-t,-\vec{x})\Gamma_i^{\mathbf{CPT}}\psi_a(-t,-\vec{x}), \quad (4.86)$$

where Γ stands for all possible combinations of γ matrices. Table 4.2 summarizes these transformation properties.

Note that we can identify

$$\begin{aligned} \overline{\psi}\psi = S, &\qquad \overline{\psi}\gamma_\mu\psi = V_\mu, \\ \overline{\psi}\gamma_5\psi = P, &\qquad \overline{\psi}\gamma_\mu\gamma_5\psi = A_\mu. \end{aligned} \qquad (4.87)$$

The transformation properties of a Lagrangian written in terms of bosons is thus precisely the same as that for a Lagrangian involving fermion bilinears. We shall, therefore, confine our discussion to the transformation

Table 4.2   _Summary of how fermion bilinears transform under_ **C, P** _and_ **T**

Γ	$\Gamma^{\mathbf{P}}$ $\gamma_0\Gamma\gamma_0$	$\Gamma^{\mathbf{C}}$ $(\gamma^2\gamma^0\Gamma\gamma^2\gamma^0)^{tr}$	$\Gamma^{\mathbf{T}}$ $-\gamma^1\gamma^3\Gamma^*\gamma^1\gamma^3$	$\Gamma^{\mathbf{CP}}$ $-(\gamma^2\Gamma\gamma^2)^{tr}$	$\Gamma^{\mathbf{CPT}}$ $(\gamma_5\gamma_0\Gamma^\dagger\gamma_0\gamma_5)^{tr}$
1	1	1	1	1	1
γ_5	$-\gamma_5$	γ_5	γ_5	$-\gamma_5$	$-\gamma_5$
γ_μ	γ^μ	$-\gamma_\mu$	γ^μ	$-\gamma^\mu$	$-\gamma_\mu$
$\gamma_\mu\gamma_5$	$-\gamma^\mu\gamma_5$	$\gamma_\mu\gamma_5$	$\gamma^\mu\gamma_5$	$-\gamma^\mu\gamma_5$	$-\gamma_\mu\gamma_5$
$\sigma_{\mu\nu}$	$\sigma^{\mu\nu}$	$-\sigma_{\mu\nu}$	$-\sigma^{\mu\nu}$	$-\sigma^{\mu\nu}$	$\sigma_{\mu\nu}$
$\sigma_{\mu\nu}\gamma_5$	$-\sigma^{\mu\nu}\gamma_5$	$-\sigma_{\mu\nu}\gamma_5$	$-\sigma^{\mu\nu}\gamma_5$	$\sigma^{\mu\nu}\gamma_5$	$-\sigma_{\mu\nu}\gamma_5$

properties of a Lagrangian expressed through boson fields. The notation is simpler that way and the essence of the argument becomes more transparent.

Let us consider the simple interaction Lagrangian

$$\mathcal{L}_T = aV_\mu^+(t, \vec{x})V^{\mu,-}(t, \vec{x}) + bA_\mu^+(t, \vec{x})A^{\mu,-}(t, \vec{x})$$
$$+ cV_\mu^+(t, \vec{x})A^{\mu,-}(t, \vec{x}) + c^*A_\mu^+(t, \vec{x})V^{\mu,-}(t, \vec{x}), \qquad (4.88)$$

which under **CP** transforms as follows:

$$\mathbf{CP}\mathcal{L}_T\mathbf{CP}^\dagger = aV^{\mu,-}(t, -\vec{x})V_\mu^+(t, -\vec{x}) + bA^{\mu,-}(t, -\vec{x})A_\mu^+(t, -\vec{x})$$
$$+ cV^{\mu,-}(t, -\vec{x})A_\mu^+(t, -\vec{x}) + c^*A^{\mu,-}(t, -\vec{x})V_\mu^+(t, -\vec{x}), \qquad (4.89)$$

i.e. **CP** *is conserved if the coupling parameter c is real!* Similar statements can be made for Lagrangians written in terms of scalar and pseudoscalar boson fields. (See Problem 4.12).

Up to now we have framed our discussion in terms of Lagrangians, since they have the nice feature of being Lorentz invariant. Now we will switch over to using mostly Hamiltonians. They are not Lorentz scalars, yet the discussion of oscillation phenomena is naturally conducted in the Hamiltonian formalism, since the mass of a particle is its energy in its rest frame.[1] One can also point out that as far as the interaction terms are concerned the two quantities are identical apart from a sign: $H_I = -\mathcal{L}_I$.

Our previous discussion can be generalized by considering the following Hamiltonian expressed in terms of local operators H_i:

$$H = \sum_i a_iH_i + \text{h.c.} \qquad (4.90)$$

where $\mathbf{CP}H_i\mathbf{CP}^\dagger = H_i^\dagger$ and **CP** is conserved for a_i real.

Reconciling the demands of quantum mechanics with those of special relativity within a *local* description requires the existence of antiparticles. A considerably stronger statement can actually be made concerning the relationship between particles and antiparticles: the combined transformation **CPT** can always be defined – as an anti-unitary operator – in such a way for a local quantum field theory that it represents a symmetry [17], i.e.

$$\mathbf{CPT}\,\mathcal{L}(t, \vec{x})\,(\mathbf{CPT})^{-1} = \mathcal{L}(-t, -\vec{x}). \qquad (4.91)$$

---

[1] To the charge we are not consistent we reply with Oscar Wilde: 'Consistency is the last refuge of the unimaginative'.

This important theorem can be proven rigorously in axiomatic field theory based on the assumptions of:

- Lorentz invariance;
- the existence of a unique vacuum state;
- weak local commutativity obeying the 'right' statistics.

The Lagrangian of Eq. (4.88) transforms as

$$\mathbf{CPT}\mathcal{L}_T\mathbf{CPT}^{-1}$$
$$= aV^{\mu-}(-t,-\vec{x})V_\mu^+(-t,-\vec{x}) + bA^{\mu-}(-t,-\vec{x})A_\mu^+(-t,-\vec{x})$$
$$+ c^*V^{\mu-}(-t,-\vec{x})A_\mu^+(-t,-\vec{x}) + cA^{\mu-}(-t,-\vec{x})V_\mu^+(-t,-\vec{x}) \quad (4.92)$$

i.e. **CPT** is indeed conserved, no matter what the coupling parameters a, b and c are.

The argument is easily repeated for fermions by noting again that each bosonic field can be written in terms of fermionic bilinears which transform in exactly in the same way under **C**, **P** and **T**; likewise for more realistic Lagrangians or Hamiltonians.

4.6 Some consequences of the CPT theorem

Although the proof of this theorem, at least for the basic cases, appears rather simple, its consequences are far-reaching. The most celebrated ones are presented here.

- The equality of masses and total widths or lifetimes for particles P and antiparticles $\overline{P}$:

$$M(P) = M(\overline{P}), \quad \Gamma(P) = \Gamma(\overline{P}), \quad (4.93)$$

which are easily proved.

–
$$M(P) = \langle P|H|P\rangle$$
$$= \langle P|(\mathbf{CPT})^\dagger\mathbf{CPT}\, H(\mathbf{CPT})^{-1}\mathbf{CPT}\,|P\rangle^*$$
$$= \langle \overline{P}|\mathbf{CPT}H\mathbf{CPT}^{-1}|\overline{P}\rangle^*$$
$$= \langle \overline{P}|H|\overline{P}\rangle^* = M(\overline{P}). \quad (4.94)$$

– Under time reversal, an incoming spherical wave transforms to an outgoing spherical wave. So, under time reversal multi-particle states transform as

$$\mathbf{T}|p_1, p_2, \ldots; \text{out}\rangle = |-p_1, -p_2, \ldots; \text{in}\rangle. \quad (4.95)$$

$$\Gamma(P) = 2\pi \sum_f \delta(M_P - E_f) |\langle f; \text{out}|H_{\text{decay}}|P\rangle|^2$$

$$= 2\pi \sum_f \delta(M_P - E_f) |\langle f; \text{out}|\mathbf{CPT}^\dagger \mathbf{CPT} H_{\text{decay}} \mathbf{CPT}^{-1}\mathbf{CPT}|P\rangle^*|^2$$

$$= 2\pi \sum_f \delta(M_P - E_f) |\langle \overline{f}; \text{in}|H_{\text{decay}}|\overline{P}\rangle|^2$$

$$= 2\pi \sum_f \delta(M_P - E_f) |\langle \overline{f}; \text{out}|H_{\text{decay}}|\overline{P}\rangle|^2$$

$$= \Gamma(\overline{P}). \tag{4.96}$$

We have used the fact that both 'in' and 'out' states form complete sets of states:

$$\sum_f |f; \text{in}\rangle\langle f; \text{in}| = \sum_f |f; \text{out}\rangle\langle f; \text{out}| = 1. \tag{4.97}$$

Proofs for these relations are quite elementary. Nevertheless their contents are far from trivial, namely that these equalities are guaranteed already by **CPT** symmetry irrespective of whether **CP** is conserved or not.

Noting that the completeness condition (see Problem 4.8) was used in the proof given above, we can deduce the following: all decay channels can be divided into classes F_i containing final states $f_\alpha^{(i)}$ that are mutually distinct under the strong interactions; i.e. for $f_\alpha^{(i)} \in F_i$ and $f_\rho^{(j)} \in F_j$ with $i \neq j$ neither $f_\alpha^{(i)} \to f_\rho^{(j)}$ nor $f_\rho^{(j)} \to f_\alpha^{(i)}$ can be driven by the strong interactions, irrespective of α and ρ. **CPT** invariance then implies

$$\sum_{f_\alpha^{(i)} \in F_i} \Gamma(P \to f_\alpha^{(i)}) = \sum_{\overline{f}_\alpha^{(i)} \in \overline{F}_i} \Gamma(\overline{P} \to \overline{f}_\alpha^{(i)}), \tag{4.98}$$

where $\overline{f}_\alpha^{(i)}$ and $\overline{F}_i$ are **CP** conjugate to $f_\alpha^{(i)}$ and F_i, respectively. This feature will be proved in Section 4.10.2.

– As an example, let us neglect *weak* and *electromagnetic* final state interactions. Then, 'in' and 'out' states cannot be distinguished, and we have a weaker form of **CPT** invariance. Under this assumption **CPT** implies not only that the *total* sum of all partial decay rates be the same for particles and antiparticles (Problem 4.10); it actually tells us that the sums over certain *subsets* of all decay rates have to coincide for particles and antiparticles:

$$\Gamma(\mu^- \to e^- \overline{\nu}_e \nu_\mu) = \Gamma(\mu^+ \to e^+ \nu_e \overline{\nu}_\mu) \tag{4.99}$$

beyond

$$\Gamma_{\text{tot}}(\mu^-) = \Gamma_{\text{tot}}(\mu^+). \tag{4.100}$$

Likewise,

$$\Gamma(K^- \to e^- \bar{\nu} \pi^0) = \Gamma(K^+ \to e^+ \nu \pi^0), \tag{4.101}$$

$$\Gamma(K^- \to \pi^- \pi^0) = \Gamma(K^+ \to \pi^+ \pi^0), \tag{4.102}$$

$$\Gamma(K^0 \to \pi^+ \pi^- + \pi^0 \pi^0) = \Gamma(\overline{K}^0 \to \pi^+ \pi^- + \pi^0 \pi^0). \tag{4.103}$$

On the other hand, we *cannot* ignore *strong* final state interaction and the equality of $\Gamma(K^0 \to \pi^+ \pi^-)$ and $\Gamma(\overline{K}^0 \to \pi^+ \pi^-)$ is *not* guaranteed by **CPT** invariance.

- Magnetic dipole moments are equal in magnitude, but opposite in sign for particles and antiparticles (Problem 4.10)

$$\mu_{\text{mag}}(P) = -\mu_{\text{mag}}(\overline{P}). \tag{4.104}$$

4.7 ♠ Back to first quantization ♠

With the discrete transformations implemented in a quantum field theory, we can translate them easily into the language of quantum mechanics, i.e. a theory with first quantization only, as expressed through the Dirac or Klein–Gordon equations etc. for wave functions. The ground state is assumed to be invariant under **C**, **P** and **T**:

$$\mathbf{C}|0\rangle = |0\rangle, \quad \mathbf{P}|0\rangle = |0\rangle, \quad \mathbf{T}|0\rangle = |0\rangle. \tag{4.105}$$

The transformation properties for an n-particle state are determined by the transformation properties of the appropriate creation operators; i.e. for a single electron [positron] state with momentum p and spin s:

$$|e^-(\vec{p}, s)\rangle = \mathbf{b}^\dagger(\vec{p}, s)|0\rangle, \qquad \left[|e^+(\vec{p}, s)\rangle = \mathbf{d}^\dagger(\vec{p}, s)|0\rangle \right].$$

The wavefunction Ψ describing such a one-particle state produced by the field operator ψ is then given by

$$\Psi(t, \vec{x}) \equiv \langle 0|\psi(t, \vec{x})|\vec{p}, s\rangle, \tag{4.106}$$

which is easily generalized to the case of an n-particle state

$$|\vec{p}_1, s_1; \dots; \vec{p}_n, s_n\rangle.$$

A transformation is described by an operator $\mathcal{O}$ acting in the following way:

$$\Psi(t, \vec{x}) = \langle 0|\psi(t, \vec{x})|\vec{p}_1, s_1; \ldots; \vec{p}_n, s_n\rangle$$

$$\xrightarrow{\mathcal{O}} \Psi^{\mathcal{O}}(t, \vec{x}) = \langle 0|\mathcal{O}\psi(t, \vec{x})\mathcal{O}^{-1}|\vec{p}_1, s_1; \ldots; \vec{p}_n, s_n\rangle. \quad (4.107)$$

In quantum mechanics we have found

$$\Psi(t, \vec{x}) \xrightarrow{\mathbf{T}} \Psi^*(-t, \vec{x}). \quad (4.108)$$

For time reversal symmetry, using its anti-unitary property, we obtain

$$\begin{aligned}
\Psi(t, \vec{x}) &= \langle 0|\mathbf{T}^\dagger\mathbf{T}\psi(t, \vec{x})\mathbf{T}^{-1}\mathbf{T}|\vec{p}_1, s_1; \ldots; \vec{p}_n, s_n\rangle^* \\
&\xrightarrow{\mathbf{T}} \langle 0|\psi(\vec{x}, -t)| - \vec{p}_1, -s_1; \ldots; -\vec{p}_n, -s_n\rangle^* \\
&= \Psi^*(-t, \vec{x}),
\end{aligned} \quad (4.109)$$

which is consistent with Eq. (4.108)!

A similar relationhip between the operator equation Eq. (4.57) and Eq. (3.30) can easily be obtained.

4.8 ♠ Phase conventions for C and P ♠

From the discussion of Kramer's degeneracy we have found that $\mathbf{T}^2$ must be ± 1, so we are not free to change the phase of $\mathbf{T}$. But no such constraint exists for **C** and **P**.

We have been somewhat cavalier about the phases for **C** and **P** up to now. For example, Eq. (4.57), and Eq. (4.66) can be replaced by

$$\mathbf{P}\psi(t, \vec{x})\mathbf{P}^\dagger = \eta^P\gamma^0\psi(t, -\vec{x})$$

$$\mathbf{C}\psi(t, \vec{x})\mathbf{C}^\dagger = \eta^C C\overline{\psi}^{tr}(t, \vec{x}), \quad (4.110)$$

where η^P and η^C are arbitrary phases; similarly for fields with spin 0 or 1. It can easily be seen that these phases do not change any of the conclusions reached above in an essential way.

There is a reason for not worrying too much about the phases other than innate sloppiness. For there is no unique definition of, say, parity: it can be redefined by combining it with any conserved internal quantum number like baryon number B, lepton number L or electric charge Q:

$$\mathbf{P}' = \mathbf{P}\, e^{i(bB + lL + qQ)}, \quad (4.111)$$

with b, l and q being arbitrary real numbers (as long as interactions do not change B, L, or Q). (See Problem 4.4.) **P** and **P**′ can lay equivalent claims to be *the* parity operator.

For the neutral pion with $B = L = Q = 0$ the two definitions coincide; its parity is thus well defined *ab initio* (and turns out to be negative). The situation is intrinsically different for the proton p, neutron n, electron e or the charged pion: **P** and **P'** are not equivalent for those states. This apparent vice can however be turned into a virtue. By judiciously adjusting the parameters b, q and l we can assign an intrinsic parity of $+1$ to p, n and e. With this definition, the intrinsic parity of the charged pion can be determined from [7]

$$\pi^- + d \to n + n, \tag{4.112}$$

where π^- is captured in an S wave state. Since π^- is spin-less and deuterium has $J = 1$, the parity of the initial state is equal to the intrinsic parity of π^-. The final two-neutron state forms a $^3P_1$ configuration which carries odd parity; thus

$$\mathbf{P}|\pi^-\rangle = -|\pi^-\rangle. \tag{4.113}$$

The parity of π^0 can be detemined by analysing correlations between the polarizations of the two gamma rays in $\pi^0 \to \gamma\gamma$ decay [7]. Experiments give

$$\mathbf{P}|\pi^0\rangle = -|\pi^0\rangle. \tag{4.114}$$

This is gratifying since it is consistent with combining π^+, π^0 and π^- into an iso-triplet.

For charge conjugation transformation, we recall that $\mathbf{C}|\gamma\rangle = -|\gamma\rangle$. Since the decay

$$\pi^0 \to \gamma\gamma \tag{4.115}$$

is the major decay mode of π^0, we have

$$\mathbf{C}|\pi^0\rangle = +|\pi^0\rangle \tag{4.116}$$

and the decay $\pi^0 \to 3\gamma$ is forbidden by charge conjugation symmetry.

4.9 ♠ Internal symmetries ♠

Continuous internal symmetries like isospin and $SU(3)_{\text{Fl}}$ introduce other charge–like quantum numbers that change under charge conjugation. That is described in terms of an octet of complex scalar or pseudoscalar fields, Φ_i, where 'i' labels elements of the group representation under study: the charge conjugation operation can be defined as

$$\mathbf{C}\Phi_i(t, \vec{x})\mathbf{C}^\dagger = \eta_i^C \Phi_i^\dagger(t, \vec{x}) \tag{4.117}$$

i.e. we can choose a different phase η_i^C for each field. This freedom is important, for example, when we write Φ_i in terms of the real $SU(3)_{Fl}$ octet fields:

$$\pi^\pm = \frac{1}{\sqrt{2}}(\phi_1 \mp i\phi_2), \quad K^\pm = \frac{1}{\sqrt{2}}(\phi_4 \mp i\phi_5),$$

$$K^0 = \frac{1}{\sqrt{2}}(\phi_6 - i\phi_7), \quad \overline{K} = \frac{1}{\sqrt{2}}(\phi_6 + i\phi_7),$$

$$\pi^0 = \phi_3, \qquad\qquad \eta = \phi_8. \tag{4.118}$$

Note that (K^+, K^0) is an $SU(2)$ doublet. Under an $SU(2)$ rotation around an axis $\vec{n}$ by an angle θ, a doublet representation transforms as follows:

$$\begin{pmatrix} \Phi_+ \\ \Phi_- \end{pmatrix} \to e^{-i\frac{\theta}{2}\vec{\sigma}\cdot\vec{n}} \begin{pmatrix} \Phi_+ \\ \Phi_- \end{pmatrix}$$

$$i\sigma^2 \begin{pmatrix} \Phi_+^* \\ \Phi_-^* \end{pmatrix} \to e^{-i\frac{\theta}{2}\vec{\sigma}\cdot\vec{n}} i\sigma^2 \begin{pmatrix} \Phi_+^* \\ \Phi_-^* \end{pmatrix}. \tag{4.119}$$

It turns out to be more convenient to introduce an extra minus sign into the definition of the charge-conjugated doublet:

$$\begin{pmatrix} K^+ \\ K^0 \end{pmatrix}^C = \begin{pmatrix} -\overline{K}^0 \\ K^- \end{pmatrix}, \tag{4.120}$$

i.e. making use of the freedom defined in Eq. (4.117), we can adopt a different phase convention for the charge conjugate of (K^0, K^+) since they form a distinct iso-doublet.

$$\mathbf{C}|K^+\rangle = +|K^-\rangle$$
$$\mathbf{C}|K^0\rangle = -|\overline{K}^0\rangle. \tag{4.121}$$

For our later discussion the concept of G-parity will turn out to be very powerful. Consider the operator $\exp[-i\pi T_i]$, where T_i is a generator of isospin rotations around the axis i. Under this rotation,

$$e^{-i\pi T_2} \begin{pmatrix} \phi_1 \\ \phi_2 \\ \phi_3 \end{pmatrix} = \begin{pmatrix} -\phi_1 \\ \phi_2 \\ -\phi_3 \end{pmatrix}. \tag{4.122}$$

This has the effect of

$$e^{-i\pi T_2}|\pi^\pm\rangle = -|\pi^\mp\rangle, \quad e^{-i\pi T_2}|\pi^0\rangle = -|\pi^0\rangle. \tag{4.123}$$

So, defining a G-parity as $G = \mathbf{C}\exp[-i\pi T_2]$, we find

$$G|\pi^{\pm}\rangle = -|\pi^{\pm}\rangle, \quad G|\pi^0\rangle = -|\pi^0\rangle. \tag{4.124}$$

It follows immediately that a state with n pions is an eigenstate of G-parity:

$$G|\pi_1 ... \pi_n\rangle = (-1)^n |\pi_1 ... \pi_n\rangle. \tag{4.125}$$

With the strong interaction obeying isospin and charge invariance, G-parity is also conserved.

4.10   The role of final state interactions

4.10.1   **T** *invariance and Watson's theorem*

A strangeness changing transition has to be *initiated* by weak forces which can be treated perturbatively. On the other hand, the final state is shaped largely by strong dynamics which is mostly beyond the reach of a perturbative description. Yet, even so, we can make some reliable theoretical statements based on symmetry considerations.

On the one hand, the G-parity introduced above tells us that a state of an *even* number of pions cannot evolve *strongly* into a state with an *odd* number. The two-step process where a K meson decays *weakly* into two pions which subsequently interact *strongly* to evolve into a three pion final state therefore cannot happen:

$$K \xrightarrow{H_{\text{weak}}} 2\pi \xrightarrow{H_{\text{str}}} 3\pi. \tag{4.126}$$

On the other hand, the two pions emerging from the weak decay $K \to \pi\pi$ are not asymptotic states yet; due to the strong interactions they will undergo rescattering before they lose sight of each other. Deriving the properties of these strong final state interactions from first principles is beyond our present computational capabilities. However, we can relate some of their properties reliably to other observables.

We have learnt from the discussion preceding Eq. (4.90) that **T** violation enters through phases in the coefficients a_i:

$$H_W = \sum_i a_i H_W^i + h.c, \tag{4.127}$$

where $\mathbf{T}H_W^i\mathbf{T}^{-1} = H_W^i$. In the following discussion of Watson's theorem, we can consider **T** invariant H_W^i without loss of generality:

$$\mathbf{T}H_W^i\mathbf{T}^{-1} = H_W^i.$$

The amplitude for $K^0 \to 2\pi$ caused by H_W^i can be written as

$$\langle (2\pi)_I^{\text{out}} | H_W^i | K^0 \rangle = |A_I^i| e^{i\phi_I},\qquad(4.128)$$

where I denotes the isospin of the 2π state. We will show now that the phase ϕ_I generated by the strong interactions actually coincides with the S wave $\pi\pi$ phase shift δ_I taken at energy M_K [18]. That is, the amplitude is real except for the fact that the two pions interact before becoming asymptotic states.

With $\mathbf{T}$ being an *anti*-unitary operator and using Eq. (3.58) we can write down:

$$\langle (\pi\pi)_I; \text{out} | H_W^i | K \rangle = \langle (\pi\pi)_I; \text{out} | \mathbf{T}^\dagger \mathbf{T} \, H_W^i \mathbf{T}^{-1} \mathbf{T} \, | K \rangle^*$$
$$= \langle (\pi\pi)_I; \text{in} | H_W^i | K \rangle^*,\qquad(4.129)$$

since for a single particle state – K in this case – there is no distinction between an in and an out state. Next we insert a complete set of out states:

$$\langle (\pi\pi)_I; \text{out} | H_W^i | K \rangle = \sum_n (\langle (\pi\pi)_I; \text{in} | n; \text{out} \rangle \langle n; \text{out} | H_W^i | K \rangle)^*. \qquad(4.130)$$

$\langle n; \text{out} | (\pi\pi)_I; \text{in} \rangle$ is an S matrix element which contains the energy momentum conserving delta function.

We can now analyse the possible final states. Kinematically, the only allowed hadronic states in the sum over n are 2π and 3π combinations. G-parity, which is conserved by the strong interactions, enforces

$$\langle (\pi\pi)_I; \text{in} | (3\pi); \text{out} \rangle = 0. \qquad(4.131)$$

Therefore only the 2π out state can contribute in the sum:

$$\langle (\pi\pi)_I; \text{in} | n; \text{out} \rangle = 0 \quad \text{for} \quad n \neq (2\pi)_I. \qquad(4.132)$$

This is usually referred to as the condition of *elastic* unitarity, and is shown in Fig. 4.1. With the S matrix for $(\pi\pi)_I \to (\pi\pi)_I$ given by

$$S_{\text{elastic}} \equiv \langle (\pi\pi)_I; \text{out} | (\pi\pi)_I; \text{in} \rangle = e^{2i\delta_I}, \qquad(4.133)$$

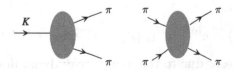

Figure 4.1 Only 2π states contribute to the sum of intermediate states in the amplitude for $K \to \pi\pi$. $K \to 3\pi$ is forbidden by G-parity and $K \to 4\pi$ is forbidden by energy conservation.

where δ_I denotes the $\pi\pi$ phase shift, we find from Eq. (4.130)

$$\langle(\pi\pi)_I; \text{out}|H_W^i|K\rangle = e^{2i\delta_I}\langle(\pi\pi)_I; \text{out}|H_W^i|K\rangle^*, \tag{4.134}$$

or equivalently

$$\langle(\pi\pi)_I; \text{out}|H_W^i|K^0\rangle = \langle(\pi\pi)_I; \text{out}|H_W^i|K^0\rangle e^{i\delta_I} \equiv \mathcal{A}_I^i e^{i\delta_I}. \tag{4.135}$$

Since this reasoning holds for all H_W^i, we can write:

$$\langle(\pi\pi)_I; \text{out}|H_W|K^0\rangle = \mathcal{A}_I e^{i\delta_I}. \tag{4.136}$$

where $\mathcal{A}_I e^{i\delta_I} = \sum_i a_i \langle(\pi\pi)_I; \text{out}|H_W^i|K^0\rangle$ with the complex coefficients a_i providing the gateways for **CP** violation to enter.

These arguments can be retraced for $\overline{K}^0 \to \pi\pi$, yielding

$$\langle(\pi\pi)_I; \text{out}|H_W|\overline{K}^0\rangle = \overline{\mathcal{A}}_I e^{i\delta_I}. \tag{4.137}$$

As long as H_W conserves **T**, the two amplitudes $\mathcal{A}_I$ and $\overline{\mathcal{A}}_I$ remain real after having the strong phase shift factored out. Once, however, **T** is violated in H_W, additional phases appear in $\mathcal{A}_I$ and $\overline{\mathcal{A}}_I$ as discussed later.

Strong final state interactions effect also the decays of heavy flavour hadrons, yet we *cannot* apply Watson's theorem there blindly. In particular there is no reason why *elastic* unitarity should apply in two-body or even quasi-two-body *beauty* decays: strong final state interactions are actually quite likely to generate additional hadrons in the final state. The decays of *charm* hadrons represent a borderline case: while the final state interactions can change the identity of the emerging particles and can produce additional hadrons, their impact is somewhat moderated since the available phase space is less than abundant. As discussed in more detail in the chapter on charm decays, introducing the concept of *absorption* might provide a useful approximation here.

4.10.2 Final state interactions and partial widths

Final state interactions become indispensible in making **CP** violation observable in partial widths. Consider a weak decay channel $P \to f$ receiving contributions from two coherent processes. The transition amplitudes then reads as follows:

$$A(P \to f) = e^{i\phi_1^{\text{weak}}} e^{i\delta_1^{\text{FSI}}} |\mathcal{A}_1| + e^{i\phi_2^{\text{weak}}} e^{i\delta_2^{\text{FSI}}} |\mathcal{A}_2|, \tag{4.138}$$

where $\phi_{1,2}^{\text{weak}}$ are phases due to the weak decay dynamics and $\delta_{1,2}^{\text{FSI}}$ are due to strong (or electromagnetic) final state interactions. For **CP** conjugate amplitude, we have

$$A(\overline{P} \to \overline{f}) = e^{-i\phi_1^{\text{weak}}} e^{i\delta_1^{\text{FSI}}} |\mathcal{A}_1| + e^{-i\phi_2^{\text{weak}}} e^{i\delta_2^{\text{FSI}}} |\mathcal{A}_2|, \tag{4.139}$$

and therefore, for the partial widths,

$$\frac{\Gamma(P \to f) - \Gamma(\overline{P} \to \overline{f})}{\Gamma(P \to f) + \Gamma(\overline{P} \to \overline{f})}$$

$$= -\frac{2\sin(\Delta\phi_W)\sin(\Delta\delta^{\text{FSI}})|\mathcal{A}_2/\mathcal{A}_1|}{1 + |\mathcal{A}_2/\mathcal{A}_1|^2 + 2|\mathcal{A}_2/\mathcal{A}_1|\cos(\Delta\phi_W)\cos(\Delta\delta^{\text{FSI}})} \quad (4.140)$$

where $\Delta\phi_W = \phi_1^{\text{weak}} - \phi_2^{\text{weak}}$ and $\Delta\delta^{\text{FSI}} = \delta_1^{\text{FSI}} - \delta_2^{\text{FSI}}$.

Thus there are two requirements for a decay to reveal **CP** violating effects.

(1) **CP** violation enters through weak dynamics: $\Delta\phi_W = \phi_1^{\text{weak}} - \phi_2^{\text{weak}}$.
(2) Final state interactions (FSI) induce a non-trivial phase shift: $\Delta\delta^{\text{FSI}} = \delta_1^{\text{FSI}} - \delta_2^{\text{FSI}}$. This will happen in particular when the two transition amplitudes differ in their isospin content.

The asymmetry gets larger if the two interfering amplitudes are of comparable size: $\mathcal{A}_2/\mathcal{A}_1 \sim \mathcal{O}(1)$.

♠ *A subtlety concerning the* **CPT** *constraint* ♠

There is a subtlety hiding in the expressions of Eq. (4.138) and Eq. (4.139): the constraint from **CPT** invariance is not apparent. Assume f_α and f_β are the only decay channels producing the total width; then, as seen in Eq. (4.98), **CPT** invariance already enforces $\Gamma(P \to f_\alpha) + \Gamma(P \to f_\beta) = \Gamma(\overline{P} \to \overline{f}_\alpha) + \Gamma(\overline{P} \to \overline{f}_\beta)$. How do we see that the asymmetry given in Eq. (4.140) vanishes when summed over f_α and f_β?

To see the physics, let us illustrate it with a simple model of only two final states f_α and f_β. Let us further assume that strong (and electromagnetic) forces drive transitions among f_α and f_β – and likewise for $\overline{f}_\alpha$ and $\overline{f}_\beta$ – as described by an S matrix:

$$\mathcal{S} = \begin{pmatrix} e^{2i\delta_\alpha} & i\mathcal{T}e^{i(\delta_\alpha + \delta_\beta)} \\ i\mathcal{T}e^{i(\delta_\alpha + \delta_\beta)} & e^{2i\delta_\beta} \end{pmatrix}. \quad (4.141)$$

The symmetric nature of $\mathcal{S}$ can easily be derived by applying **T** symmetry to $\langle f_\alpha; out | f_\beta; in \rangle$. We also assume that $\mathcal{T}$ is small and work with only the first order in $\mathcal{T}$. $\mathcal{S}$ is unitary to that order.

Using the fact that f_α and f_β form a complete set of states that couple to P, we write

$$\langle f_\alpha; out | H | P \rangle = \langle f_\alpha; out | f_\alpha; in \rangle \langle f_\alpha; in | H | P \rangle + \langle f_\alpha; out | f_\beta; in \rangle \langle f_\beta; in | H | P \rangle \quad (4.142)$$

Writing $\langle f_\alpha; {}^{out}_{in}|H|P\rangle = e^{\pm i\delta_\alpha}\mathcal{A}_\alpha$, which follows from the Watson's theorem for small $\mathcal{T}$, we can write

$$\langle f_\alpha : out|H|P\rangle = e^{i\delta_\alpha}\left[\mathcal{A}_\alpha + i\mathcal{T}\mathcal{A}_\beta\right]$$
$$\langle \overline{f}_\alpha : out|H|\overline{P}\rangle = e^{i\delta_\alpha}\left[\mathcal{A}_\alpha^* + i\mathcal{T}\mathcal{A}_\beta^*\right] \qquad (4.143)$$

where the second line follows from **CPT** invariance. Then we obtain

$$\Gamma(\overline{P} \to \overline{f}_\alpha) - \Gamma(P \to f_\alpha) = 2\mathcal{T}\text{Im}T_\alpha^*T_\beta;$$
$$\Gamma(\overline{P} \to \overline{f}_\beta) - \Gamma(P \to f_\beta) = 2\mathcal{T}\text{Im}T_\beta^*T_\alpha, \qquad (4.144)$$

and $\Gamma(P \to f_\alpha) + \Gamma(P \to f_\beta) = \Gamma(\overline{P} \to \overline{f}_\alpha) + \Gamma(\overline{P} \to \overline{f}_\beta)$ indeed holds as it has to.

This simple model can be extended into a more realistic scenario: consider two sets A and B of final states such that all states a in set A have the same weak coupling and therefore can rescatter into each other; likewise for the states b in set B. Yet no (significant) rescattering can occur between states in sets A and B. Then we can write down for final states a and $\tilde{a}$ from A:

$$T(P \to a) = e^{i\delta_a}\left[T_a + \sum_{a \neq \tilde{a}} T_{\tilde{a}} i\mathcal{T}^{\text{resc}}_{\tilde{a}a}\right] \qquad (4.145)$$

$$T(\bar{P} \to \bar{a}) = e^{i\delta_a}\left[T_a^* + \sum_{a \neq \tilde{a}} T_{\tilde{a}}^* i\mathcal{T}^{\text{resc}}_{\tilde{a}a}\right] \qquad (4.146)$$

$$\Delta\gamma(a) \equiv |T(\bar{P} \to \bar{a})|^2 - |T(P \to a)|^2 = 4\sum_{a \neq \tilde{a}} \mathcal{T}^{\text{resc}}_{\tilde{a}a}\text{Im}T_a^*T_{\tilde{a}}; \qquad (4.147)$$

This **CP** asymmetry has to vanish upon summing over all such states a:

$$\sum_a \Delta\gamma(a) = 4\sum_a \sum_{a \neq \tilde{a}} \mathcal{T}^{\text{resc}}_{\tilde{a}a}\text{Im}T_a^*T_{\tilde{a}} = 0, \qquad (4.148)$$

since $\mathcal{T}^{\text{resc}}_{\tilde{a}a}$ and $\text{Im}T_a^*T_{\tilde{a}}$ are symmetric and antisymmetric, respectively, in the indices a and $\tilde{a}$. This illustrates that **CPT** symmetry imposes equality not only between total widths of particles and antiparticles, but also between subclasses of partial widths. Those subclasses are defined by their quantum numbers, yet can still be very heterogeneous namely consisting of two- and quasi-two-body modes, three-body channels and other multi-body decays.

The salient points of this discussion are summarized in the following list.

- While we do not know (yet) how to calculate strong FSI and thus $\mathcal{S}_{ab}$ reliably and therefore cannot predict the size of direct **CP** asymmetries, even when the weak phases are known, **CPT** symmetry implies some relations between **CP** asymmetries in different channels. Writing down arbitrary strong phases for the amplitudes in Eq. (4.138) and Eq. (4.139) does *not automatically* satisfy **CPT** constraints.
- If the two channels that rescatter have comparable widths – $\Gamma(P \to a) \sim \Gamma(P \to b)$ – one would like the rescattering $b \leftrightarrow a$ to proceed via the usual strong forces; for otherwise the asymmetry $\Delta\Gamma$ is suppressed relative to these widths by the electromagnetic coupling.
- If on the other hand the channels command very different widths – say $\Gamma(P \to a) \gg \Gamma(P \to b)$ – then a large *relative* asymmetry in $P \to b$ is accompanied by a tiny one in $P \to a$.
- If one finds a direct **CP** asymmetry in one channel, one can infer – based on rather general grounds – which other channels have to exhibit the compensating asymmetry as required by **CPT** invariance. Observing them would enhance the significance of the measurements very considerably.
- Typically there can be several classes of rescattering channels. The SM weak dynamics select a subset of those where the compensating asymmetries have to emerge. QCD frameworks can be invoked to estimate the relative weight of the asymmetries in the different classes. Analysing them can teach us important lessons about the inner workings of QCD.

4.10.3 ♠ T *symmetry and final state interactions* ♠

In decay processes it is impossible to realize the time-reversed sequence, as discussed in Section 3.3; tests of detailed balance thus cannot be performed. Instead we search for **T**-odd correlations. Consider, for example, $\Lambda \to P + \pi^-$ and $\overline{\Lambda} \to \overline{P} + \pi^+$ decays. The final state $\pi^- P$ is a linear combination of an isospin $\frac{1}{2}$ and $\frac{3}{2}$ state. The $\Delta I = \frac{1}{2}$ rule to be discussed later states that in $\Lambda \to \pi^- P$, only the $\Delta I = \frac{1}{2}$ channel contributes to a very good approximation. Then we can say that the $\pi^- P$ state is nearly an eigenstate of the strong interactions with $I = \frac{1}{2}$, and the decay amplitudes are given by

$$\langle P\pi^- | \mathcal{H} | \Lambda \rangle = i\overline{u}(p_P, s_P)(\mathcal{A}_S e^{i\delta_S} + \mathcal{A}_P e^{i\delta_P}\gamma_5)u(p_\Lambda, s_\Lambda)$$
$$\langle \overline{P}\pi^+ | \mathcal{H} | \overline{\Lambda} \rangle = i\overline{v}(p_{\overline{\Lambda}}, s_{\overline{\Lambda}})(\overline{\mathcal{A}}_S e^{i\delta_S} + \overline{\mathcal{A}}_P e^{i\delta_P}\gamma_5)v(p_{\overline{P}}, s_{\overline{P}}). \quad (4.149)$$

We have used Watson's theorem to extract out the strong phase. $\mathcal{A}_S$, $\mathcal{A}_P$, $\overline{\mathcal{A}}_S$ and $\overline{\mathcal{A}}_P$ are scalar functions with the weight of the S- and P-wave decay amplitudes. These amplitudes contain the weak dynamics. Under **P** symmetry, $\mathcal{A}_S = 0$; **CP** or **T** symmetry, $\mathcal{A}_{S,P} = \mathcal{A}_{S,P}^*$; **CPT** symmetry, $\overline{\mathcal{A}}_S = -\mathcal{A}_S^*$ and $\overline{\mathcal{A}}_P = \mathcal{A}_P^*$ (See Problem 4.13).

Since we are interested in probabilities, we square these amplitudes by inserting the spin projection operators for both s_P and s_Λ and perform the sum over these spins. We then see that

$$|\langle P(p_P, s_P)\pi^-|H|\Lambda(p_\Lambda, s_\lambda)\rangle|^2 \sim |i\overline{u}(p_P, s_P)(\mathcal{A}_S e^{i\delta_S} + \mathcal{A}_P e^{i\delta_P}\gamma_5)u(p_\Lambda, s_\Lambda)|^2$$
(4.150)

contains terms giving rise to spin and angular correlations:

$$\sim \text{Im}\,(\mathcal{A}_P^* \mathcal{A}_S e^{i(\delta_S - \delta_P)})\epsilon_{\alpha\beta\gamma\delta}p_\Lambda^\alpha p_P^\beta s_\Lambda^\gamma s_P^\delta. \qquad (4.151)$$

The kinematic term $\epsilon_{\alpha\beta\gamma\delta}p_\Lambda^\alpha p_P^\beta s_\Lambda^\gamma s_P^\delta$ reduces to $-M_\Lambda \vec{s}_\Lambda \cdot (\vec{s}_P \times \vec{p}_P)$ in the rest frame of the Λ. This quantity changes sign under **T** (as well as **P**).

This is an example where a **T**-odd correlation can exist with*out* **T** violation. Of course, an analogous term arises in the antiparticle decay. A triple correlation that is *odd* under **T** can arise in two fundamentally different ways (and any combination thereof).

- The intervention of **T** violating forces can create complex phases in $\mathcal{A}_P$ relative to $\mathcal{A}_S$.
- On the other hand, even **T** conserving dynamics (like the strong and electromagnetic forces) can induce $\text{Im}[e^{i(\delta_S - \delta_P)}] \neq 0$.

Measurements show indeed that $\langle \vec{p}_P \cdot (\sigma_P \times \sigma_\Lambda)\rangle \neq 0$, and they yield

$$\arg\frac{A_P}{A_S} + \delta_P - \delta_S = -6.5 \pm 3.5^\circ, \qquad (4.152)$$

which is quite consistent with what we independently know about πN phase shifts, $\delta_P - \delta_S$; i.e. Eq. (4.152) is fully compatible with $\arg(A_P/A_S) = 0$.

The reader is likely to raise the following question: how come the observation of a **P**-odd correlation establishes **P** violation, whereas that of a **T**-odd correlation does not carry the same unequivocal message about **T** invariance?

The answer is built on several elements of a conceptual as well as technical nature.

- As seen from Eq. (4.95), the substitution $(\vec{p}, \vec{s}) \longrightarrow (-\vec{p}, -\vec{s})$ does not fully implement time reversal for decay and scattering experiments – an incoming spherical wave is transformed to an outgoing spherical wave. See Problem 4.13.
- If the interactions under study are so weak that only first-order effects have to be considered, then a **T**-odd correlation does actually signal **T** violation.
- The situation changes, however, qualitatively, if higher order contributions are – or have to be – included. Those effects are derived through iteration factors $\exp[-i \int dt\, H]$. These evolution operators do *not* commute with an *anti*linear operator like **T**, even if H does.

To disentangle a **CP** violating correlation from one due to final state interaction, we can compare **CP** conjugate observables. If **CP** is conserved, we get

$$\text{Im}\,[e^{i(\delta_S - \delta_P)}(\mathcal{A}_S \mathcal{A}_P^*)] = -\text{Im}\,[e^{i(\delta_S - \delta_P)}(\overline{\mathcal{A}}_S \overline{\mathcal{A}}_P^*)], \qquad (4.153)$$

i.e. equal **T**-odd correlations for particle and antiparticle decays.

4.11 Résumé and outlook

Quantum field theories provide us with all the necessary tools for implementing discrete transformations in a natural way. Charge conjugation is no longer ad hoc; the existence of antiparticles is required by the demands of Lorentz invariance. The combined transformation **CPT** constitutes an almost unavoidable symmetry of any quantum field theory. **C** and **P** invariance, on the other hand, can easily be broken. **CP** symmetry can be violated, but requires attention to subtle details.

Due to **CPT** constraints, **CP** violation manifests itself through complex phases in the underlying dynamics. These phases enter the amplitudes of a reaction and its **CP** conjugate version with *opposite* signs and are customarily referred to as *weak* phases. In our preceding discussion of Watson's theorem we have learnt that a *second* class of phases arises through final state interaction, i.e. wherever we treat an interaction beyond the lowest order. This applies in particular to the strong force, but sometimes also to electromagnetic forces. They generate phase shifts that are the same for a given process and its **CP** conjugate.

This second class of phases can induce a **T**-odd correlation even with vanishing weak phases, i.e. if the underlying dynamics is **T** invariant. This effect can be disentangled by comparing such correlation in **CP** conjugate transitions. On the other hand, in partial widths asymmetries, these phase shifts act as a necessary evil – necessary since, in their absence, an existing

CP violation will remain unobservable, and evil since their values cannot be computed.

Problems

4.1. Show the quantization condition Eq. (4.16) is invariant under **C** and **P**.

4.2. Going back to the definition $\mathbf{T} = UK$, show that the two equations given in Eq. (4.41) are consistent with each other.

4.3. Verify $[\psi(t, \vec{x})^{\mathbf{C}}]^{\mathbf{C}} = \psi(t, \vec{x})$ and $[\psi(t, \vec{x})^{\mathbf{T}}]^{\mathbf{T}} = \psi(t, \vec{x})$ for a spinor field.

4.4. Assume that a theory is invariant under parity and also under a global gauge transformation. Then show that the parity operation is defined only up to a phase. For example

$$\mathbf{P}\phi(t, \vec{x})\mathbf{P}^{\dagger} = e^{i\alpha}\phi(t, -\vec{x}).$$

4.5. The field equation for a complex scalar field reads in scalar QED

$$[(i\partial_{\mu} - eA_{\mu})(i\partial^{\mu} - eA^{\mu}) - m^2]\phi(t, \vec{x}) = 0. \qquad (4.154)$$

Show that if ϕ is a solution to Eq. (4.154), then so is $\phi^{\dagger}$ – with the same mass, but opposite charge $-e$.

4.6. Apply **T** to Eq. (4.154) and derive

$$\left[-\partial^2 + ie\left(\bar{\partial}_{\mu}A^{\mu}(\bar{t}, \vec{x}) + A_{\mu}(\bar{t}, \vec{x})\bar{\partial}^{\mu}\right) + e^2 A_{\mu}(\bar{t}, \vec{x})A^{\mu}(\bar{t}, \vec{x}) - m^2\right]$$
$$\times \mathbf{T}\phi(t, \vec{x})\mathbf{T}^{-1} = 0, \qquad (4.155)$$

where $\bar{\partial} = (\partial_{\bar{t}}, \vec{\partial})$. Thus we can conclude the following: if $\phi(t, \vec{x})$ – as a solution to the Klein–Gordon equation – describes the motion of a scalar particle with charge e and mass m in an electromagnetic field, then $\mathbf{T}\phi(-t, \vec{x})\mathbf{T}^{-1}$ traces the original motion of the particle backwards in time.

4.7. The behaviour of a spin-1/2 field coupled to an electromagnetic field is described by

$$[i\gamma_{\mu}(\partial^{\mu} + ieA^{\mu}) - m]\psi(t, \vec{x}) = 0; \qquad (4.156)$$

show that

$$\left[i\gamma_{\mu}\left(\partial^{\mu} + ie\mathbf{P}A^{\mu}\mathbf{P}^{\dagger}\right) - m\right]\mathbf{P}\psi(t, \vec{x})\mathbf{P}^{\dagger} = 0; \qquad (4.157)$$

where $\mathbf{P}\psi(t, \vec{x})\mathbf{P}^{\dagger} = \gamma_0\psi(t, -\vec{x})$.

4.8. Take the complex conjugate of Eq. (4.156). Show that $\psi^*(x)$ satisfies the following equation:

$$[i(\partial\!\!\!/ - ie\,A\!\!\!/ - m]\gamma_2\psi^*(x) = 0. \qquad (4.158)$$

Discuss the physical interpretation of $\gamma^2\psi^*(x)$.

4.9. Applying the anti-unitary operator $\mathbf{T}$ to the Dirac equation, show that

$$\left[i\gamma_0\left(\partial_{-t} + ieA_0(-t,\vec{x})\right) - i\vec{\gamma}\cdot\left(\vec{\partial} + ie\vec{A}(-t,\vec{x})\right) - m\right]$$
$$\times \mathbf{T}\psi(t,\vec{x})\mathbf{T}^{-1} = 0. \qquad (4.159)$$

Thus we have

$$\mathbf{T}\psi(t,\vec{x})\mathbf{T}^{-1} = \psi(-t,\vec{x}). \qquad (4.160)$$

4.10. Investigate how the magnetic moment $\vec{\mu}_{\mathrm{mag}}(P)$, defined by

$$\vec{\mu}_{\mathrm{mag}}(P)\cdot\vec{B} = \langle P,\vec{s}|\vec{\sigma}|P,\vec{s}\rangle\cdot\vec{B},$$

transforms under $\mathbf{CPT}$.

4.11. By using the completeness of the 'in' and 'out' states, show that $\mathbf{CPT}$ symmetry implies Eq. (4.102).

4.12. Consider the P, S, V_μ and A_μ fields defined in Eq. (4.87). Discuss how they transform under $\mathbf{C}$, $\mathbf{P}$ and $\mathbf{T}$. Consider a toy Hamiltonian

$$H = (aS + bP)(aS + bP)^\dagger + (cV_\mu + dA_\mu)\cdot(cV^\mu + dA^\mu)^\dagger. \quad (4.161)$$

Under what condition is H invariant under $\mathbf{C}$, $\mathbf{P}$, $\mathbf{T}$, $\mathbf{CP}$ and $\mathbf{CPT}$?

4.13. Consider the Hamiltonian for the nonleptonic decay $\Lambda \to \pi^- P$:

$$H_W = i\overline{\psi}_P(\mathcal{A}_S + \mathcal{A}_P\gamma_5)\psi_\Lambda\phi_\pi^+ + i\overline{\psi}_\Lambda(\overline{\mathcal{A}}_S + \overline{\mathcal{A}}_P\gamma_5)\psi_P\phi_\pi^- \quad (4.162)$$

(a) Since the Hamiltonian must be Hermitian, require that the second term is the Hermitian conjugate of the first term and derive: $\overline{\mathcal{A}}_S = -\mathcal{A}_S^*$ and $\overline{\mathcal{A}}_P = \mathcal{A}_P^*$.

(b) Using the transformation properties of $\phi_\pi(t,\vec{x})$:

$$\mathbf{P}\phi_\pi^\pm(t,\vec{x})\mathbf{P}^\dagger = -\phi_\pi^\pm(t,-\vec{x})$$
$$\mathbf{C}\phi_\pi^\pm(t,\vec{x})\mathbf{C}^\dagger = +\phi_\pi^\mp(t,\vec{x})$$
$$\mathbf{T}\phi_\pi^\pm(t,\vec{x})\mathbf{T}^{-1} = -\phi_\pi^\pm(-t,\vec{x}) \qquad (4.163)$$

Show that

$$\mathbf{P}H_W\mathbf{P}^\dagger = i\overline{\psi}_P(\mathcal{A}_S - \mathcal{A}_P\gamma_5)\psi_\Lambda(-\phi_\pi^+)$$
$$+ i\overline{\psi}_\Lambda(\overline{\mathcal{A}}_S - \overline{\mathcal{A}}_P\gamma_5)\psi_P(-\phi_\pi^-) \qquad (4.164)$$

and $\mathbf{P}$ invariance implies $\mathcal{A}_S = 0$;

(c) **CP** invariance implies

$$\mathbf{CP} H_W \mathbf{CP}^\dagger = i\overline{\psi}_\Lambda (\mathcal{A}_S - \mathcal{A}_P \gamma_5)\psi_P(-\phi_\pi^-)$$
$$+ i\overline{\psi}_P (\overline{\mathcal{A}}_S - \overline{\mathcal{A}}_P \gamma_5)\psi_\Lambda(-\phi_\pi^+) \quad (4.165)$$

with $\mathcal{A}_S = -\overline{\mathcal{A}}_S$, and $\mathcal{A}_P = \overline{\mathcal{A}}_P$.

(d) Show that **T** symmetry implies

$$\mathbf{T} H_W \mathbf{T}^{-1} = - i\overline{\psi}_P (\mathcal{A}_S^* + \mathcal{A}_P^* \gamma_5)\psi_\Lambda(-\phi_\pi^+)$$
$$- i\overline{\psi}_\Lambda (\overline{\mathcal{A}}_S + \overline{\mathcal{A}}_P \gamma_5)\psi_P(-\phi_\pi^-). \quad (4.166)$$

with $\mathcal{A}_S$ and $\mathcal{A}_P$ being real.

5

The arrival of strange particles

The discovery of hadrons with the internal quantum number 'strangeness' marks the beginning of a most exciting epoch in particle physics that even now, 50 years later, has not yet reached its conclusion. As we will describe in detail, the discoveries made in this area of research have proved essential in formulating the SM of high energy physics. This is *not* to say, however, that it has been a particularly glorious chapter for theorists always making successful predictions. On the contrary, we have to admit that by and large experiments have driven the development, and that major discoveries came unexpectedly or even against expectations expressed by theorists. Theorists on the whole have fared better in making *post*dictions than *pre*dictions.[1] Some of the relevant questions remain unanswered today and/or have led to even more profound puzzles.

These discoveries were not spread out evenly over time; some periods were clearly more fertile than others. In this chapter we will give a brief overview of important developments. We do not aim at giving a complete history of this period. Instead we try to give a flavour of the excitement accompanying these discoveries, thus illustrating the thrill and the unexpected turns that can occur in the basic sciences in general and in high energy physics in particular.[2]

5.1 The discovery of strange particles

Tracking chambers, being based on ionization, cannot record the passage of a neutral particle. Yet when it decays into two charged particles, the tracks of the decay products can be recorded, and a so-called V pattern

---

[1] Such an achievement, which is not so uncommon in the basic sciences, should not be belittled, though!

[2] For an exciting historical account of the progress of our field during this period, see Ref. [19].

arises. Such a pattern was observed by Rochester and Butler in October 1946 when they exposed cloud chambers (the state-of-the-art tracking chambers of that period) to cosmic rays. They concluded that they had observed the decay of a primary particle with a mass of (435 ± 100) MeV into two secondary particles with a mass around 100 MeV; or in today's language:

$$K^0 \rightarrow \pi^+\pi^-. \tag{5.1}$$

The discovery of charged pions – in a different experiment – was reported later, in May 1947, and Rochester and Butler reported their observation in December 1947 after having found one event with a kink in its charged track; the latter event today would be referred to as

$$K^+ \rightarrow \pi^+\pi^0. \tag{5.2}$$

It should be noted that the neutral pion was officially not discovered until 1950.

These objects were produced and studied in experiments at the new accelerators at Brookhaven National Laboratory (hereafter referred to as BNL) starting in 1953, and at the Berkeley Laboratory from 1955 onward. Another neutral particle was seen to decay into two charged particles, this time with a mass somewhat larger than the proton mass. Today we refer to it as the Λ hyperon, and thus

$$\Lambda \rightarrow p\pi^-. \tag{5.3}$$

This led immediately to a puzzle: it was observed that the production rate for these new particles greatly exceeded their decay rate. To be more specific: hyperons are produced copiously in πp collisions; i.e.

$$\pi p \rightarrow \Lambda + X \tag{5.4}$$

is driven by the *strong* interactions, whereas the decay of Eq. (5.3) exhibits a lifetime of about 10^{-10} s, which represents an eternity relative to strong transition times of order 10^{-23} s. This apparent paradox, which gave rise to the name 'strange' particles, was resolved by Pais in 1952 through the concept of *associated production* [20]. A new quantum number – not surprisingly called 'strangeness' – is introduced, which is conserved by the strong, although *not* the weak, interaction. Particles carrying this quantum number are then produced pairwise by the strong interactions from a non-strange initial state, as in

$$pp \rightarrow K\overline{K} + X$$
$$\pi p \rightarrow \Lambda K + X. \tag{5.5}$$

The decays $K \to \pi\pi$ or $\Lambda \to p\pi$ changing this quantum number have to proceed *weakly*. There are actually two ways in which strangeness can be introduced: we could assign a 'strange parity' +1 to all non-strange particles like nucleons and pions and −1 to all strange particles like Λ and K. Hadronic collisions would yield an even number of strange particles, whereas $\Lambda \to p\pi$ and $K \to \pi\pi$ have to proceed weakly. Yet the discovery of the so-called cascade particles through their weak decay

$$\Xi^- \to \Lambda\pi^- \qquad (5.6)$$

closed this option of strangeness being a *multiplicative* quantum number. With the weak forces flipping the sign of the strange parity, we had to assign strange parity +1 to Ξ^- due to Eq. (5.6). Yet then the transition

$$\Xi^- \to n\pi^- \qquad (5.7)$$

would not only be allowed, it would proceed *strongly* and thus dominate weak transitions. However, Eq. (5.7) has never been observed! Thus – as pointed out independently by Gell-Mann [21] and Nakano and Nishijima [22] in 1953 – we have to introduce strangeness as an *additive* quantum number: we assign strangeness −1 to Λ and K^-, +1 to K^+, −2 to Ξ^-, and 0 to all non-strange hadrons. This satisfies the observed pattern of production and decay.

5.2 The $\theta - \tau$ puzzle

The second period is characterized by the $\theta - \tau$ puzzle. Two decay reactions had been found for charged strange mesons, namely

$$\theta^+ \to \pi^+\pi^0,$$
$$\tau^+ \to \pi^+\pi^+\pi^-. \qquad (5.8)$$

With θ^+ being spinless, the π^+ and π^0 have to form an S-wave; the 2π final state thus necessarily carries positive parity. The angular distributions of the three pions from τ^+ decay reveal the final state to carry zero total angular momentum as well, but with negative parity! It was assumed that parity, like angular momentum, was conserved by the relevant forces. The parity of the initial state then coincides with that of the final state. With θ and τ exhibiting different parity, they had to be distinct objects and thus indeed deserved different names. The problem arose when ever more precise measurements failed to find any significant difference in either the mass or the lifetime of the θ and τ mesons. This constituted the $\theta - \tau$ puzzle: how could nature – short of being malicious – assign the same mass to two distinct particles? Or even more baffling: how could nature contrive to generate the same lifetime to two distinct particles, the major decay channels of which possess totally different phase spaces?

This Gordian knot could be cut in one stroke if parity were not absolutely conserved. For then θ and τ could represent merely two decay modes of the same particle. Parity conservation had been tested extensively in strong and electromagnetic transitions. Yet the breakthrough came when Lee and Yang pointed out in 1956 [23] that this symmetry had *not* been probed yet in weak transitions; they proceeded to suggest a list of relevant tests. Soon after this theoretical analysis, Wu and collaborators [24] found in nuclear β decays that parity and charge conjugation invariance were indeed broken by the weak forces. This epochal discovery was almost instantly confirmed by other groups in different processes, and this allowed the identification of both θ^+ and τ^+ with the K^+ meson.

5.3 The $\Delta I = \frac{1}{2}$ rule

Two decays which are seemingly no different have quite different decay rates:

$$\frac{\Gamma(K_S \to \pi^+\pi^-)}{\Gamma(K^+ \to \pi^+\pi^0)} \sim 450. \tag{5.9}$$

Why is one so much bigger than the other? A close examination tells us that there is a difference. Bose statistics constrain the 2π system to carry isospin 0 or 2 with the charged combination $\pi^+\pi^0$ forming a pure I$=2$ state.[3] $K^+ \to \pi^+\pi^0$ is thus a pure $\Delta I = 3/2$ transition, whereas for $K_S \to \pi\pi$, a $\Delta I = 1/2$ amplitude can contribute. Equation (5.9) thus reveals a huge enhancement of $\Delta I = 1/2$ transitions over $\Delta I = 3/2$ ones. Gell-Mann and Pais [25] coined the name $\Delta I = 1/2$ rule for this effect over 50 years ago. The mysterious number given in Eq. (5.9) implies that the ratio of two amplitudes A_0, and A_2, labelled by the isospin of the final state pions is 0, and 2, respectively is:

$$\left|\frac{A_2}{A_0}\right| = \omega \sim \frac{1}{20}. \tag{5.10}$$

This number will be important in Chapter 7.

Such a big difference usually accompanies a discovery of some selection rule. Many years went by without full understanding of this number. We shall come back to this in Chapter 9.

---

[3] We ignore here isospin breaking effects produced by, e.g. QED corrections.

5.4 The existence of two different neutral kaons

At about the same time another important lesson was learnt, again in response to a challenge. As mentioned in the beginning of this chapter, neutral strange mesons had been found to exist:

$$K^{\text{neut}} \to \pi^+ \pi^-. \tag{5.11}$$

Since these mesons carry non-zero strangeness they cannot be – unlike the π^0 – their own antiparticle. Therefore two neutral kaons had to exist, K^0 and $\overline{K}^0$, differing by two units of strangeness. The obvious question then arises: how can you establish their separate existence? This challenge was successfully taken up by Gell-Mann and Pais [26]. Their analysis – through careful quantum mechanical reasoning – has yielded some of the more glorious pages in the theoretical development.

It appeared natural to assume **CP** invariance for the analysis[4] – partly for simplicity and partly because illustrious theorists had made strong-worded pronouncements why nature had better obey that symmetry. We will come back to that aspect in the next chapter.

Manifest **CP** symmetry can be realized in one of two scenarios: the asymptotic states, i.e. the states with definite mass and lifetimes, either are

- **CP** *eigenstates* or
- pairs of *mass degenerate* particles and antiparticles with equal mass.

This will be illustrated by the following explicit discussion. Let us, for now, turn off weak forces. K^0 and $\overline{K}^0$ can then neither decay nor transform into each other. Denote a time-dependent $K^0 - \overline{K}^0$ state by

$$\Psi(t) = a(t)|K^0\rangle + b(t)|\overline{K}^0\rangle \equiv \begin{pmatrix} a(t) \\ b(t) \end{pmatrix}. \tag{5.12}$$

We consider the free Schrödinger equation for the $\Psi(t)$:

$$i\hbar \frac{\partial}{\partial t} \Psi = H\Psi, \tag{5.13}$$

where H denotes the Hamilton operator. For the single particle system under discussion, H is a generalized mass matrix, which is diagonal and real in the absence of weak forces:

$$H = \begin{pmatrix} M_K & 0 \\ 0 & M_K \end{pmatrix}, \tag{5.14}$$

---

[4] Gell-Mann and Pais actually assumed **C** conservation at first.

where **CPT** symmetry enforces the equality of the two diagonal elements; K^0 and $\overline{K}^0$ are thus two *degenerate* mass eigenstates.

The situation changes *qualitatively* once weak forces are included that can change strangeness: decay processes with $|\Delta S| = 1$ take place. Consider, for example, decays:

$$K^0 \to \pi\pi, \qquad \overline{K}^0 \to \pi\pi. \tag{5.15}$$

The dynamics then mixes K^0 and $\overline{K}^0$ through the chain $K^0 \to \pi\pi \to \overline{K}^0$. We denote the mixing term of the Hamiltonian as Δ. For now, let us forget the fact that we have to enlarge our Hilbert space to include (multi)pion states, and discuss the mixing effect – we shall come back to this point in Chapter 6. Then

$$H = \begin{pmatrix} M_K & \Delta \\ \Delta & M_K \end{pmatrix}. \tag{5.16}$$

Since Δ is second order in the weak interactions, it is truly infinitesimal compared to M_K. Nevertheless it dictates the eigenstate to be

$$|K_1\rangle \equiv \frac{1}{\sqrt{2}} \left(|K^0\rangle + |\overline{K}^0\rangle \right),$$

$$|K_2\rangle \equiv \frac{1}{\sqrt{2}} \left(|K^0\rangle - |\overline{K}^0\rangle \right). \tag{5.17}$$

With

$$\mathbf{CP}|K^0\rangle = |\overline{K}^0\rangle, \tag{5.18}$$

we have $\mathbf{CP}|K_{\frac{1}{2}}\rangle = \pm|K_{\frac{1}{2}}\rangle$. It can be shown that (Problem 5.1)

$$\mathbf{CP}|\pi\pi\rangle = +|\pi\pi\rangle. \tag{5.19}$$

Therefore the 2π final state is fed by K_1 decays only,

$$K_1 \to 2\pi,$$
$$K_2 \not\to 2\pi. \tag{5.20}$$

The leading non-leptonic channel for K_2 is then

$$K_2 \to 3\pi. \tag{5.21}$$

The phase space for Eq. (5.21) is very restricted – $3 \cdot M_\pi \simeq 420\,\text{MeV}$ versus $M(K_2) \simeq 500\,\text{MeV}$. Thus we expect the lifetime for the **CP** odd state K_2 to be much longer than for the **CP** even K_1. The long-lived meson thus predicted was discovered by Lederman and his collaborators in 1956

[27]. Since K_1 and K_2 possess quite different lifetimes, it is customary to refer to them as K_S and K_L, respectively, with S [L] referring to short-lived [long-lived]. Likewise we use $M(K_1) = M(K_S) = M_S$ and $M(K_2) = M(K_L) = M_L$. State-of-the-art measurements yield for their lifetimes [11]:

$$\tau_S \equiv \tau(K_S) = (0.8958 \pm 0.0005) \times 10^{-10}\,\text{s}$$
$$\tau_L \equiv \tau(K_L) = (5.099 \pm 0.021) \times 10^{-8}\,\text{s}. \tag{5.22}$$

It should be kept in mind, though, that this huge difference in lifetimes, namely $\tau(K_L)/\tau(K_S) \sim 600$, reflects a dynamical accident. If pions were massless, $\tau_L \sim \tau_S$ and we might not know about **CP** violation even today!

5.5 CP invariant $K^0 - \overline{K}^0$ oscillations

There are many more intriguing facets than the existence of two (vastly) different lifetimes for neutral kaons. Their behaviour can be described in terms of two equivalent bases, namely the strong interaction or strangeness eigenstates K^0 and $\overline{K}^0$ or the mass eigenstates K_L and K_S. The latter have a simple exponential evolution in (proper) time t:

$$|K_L(t)\rangle = e^{-i(M_L - \frac{i}{2}\Gamma_L)t}|K_L(0)\rangle$$
$$|K_S(t)\rangle = e^{-i(M_S - \frac{i}{2}\Gamma_S)t}|K_S(0)\rangle. \tag{5.23}$$

Thus

$$I(K_S \to K_S; t) = I_0 \cdot |\langle K_S|K_S(t)\rangle|^2 = I_0 e^{-\Gamma_S t}$$
$$I(K_S \to K_L; t) = I_0 \cdot |\langle K_L|K_S(t)\rangle|^2 = 0\,, \tag{5.24}$$

i.e. an initially pure K_S beam of intensity I_0 will subsequently contain only K_S mesons plus their decay products; likewise for initial K_L beams.

Yet for an initially pure K^0 beam the pattern is much more complex. The time evolution for a $|K^0(t)\rangle$ is obtained by inverting Eq. (5.17), i.e.

$$|K^0(t)\rangle = \frac{1}{\sqrt{2}} \left(|K_S(t)\rangle + |K_L(t)\rangle \right)$$
$$= \frac{1}{\sqrt{2}} \left(e^{-i(M_S - \frac{i}{2}\Gamma_S)t}|K_S(0)\rangle + e^{-i(M_L - \frac{i}{2}\Gamma_L)t}|K_L(0)\rangle \right)$$
$$= f_+(t)|K^0\rangle + f_-(t)|\overline{K}^0\rangle, \tag{5.25}$$

and likewise for the antiparticle

$$|\overline{K}^0(t)\rangle = \frac{1}{\sqrt{2}} \left(|K_S(t)\rangle - |K_L(t)\rangle \right)$$

$$= \frac{1}{\sqrt{2}} \left(e^{-i(M_S - \frac{i}{2}\Gamma_S)t}|K_S(0)\rangle - e^{-i(M_L - \frac{i}{2}\Gamma_L)t}|K_L(0)\rangle \right)$$

$$= f_-(t)|K^0\rangle + f_+(t)|\overline{K}^0\rangle, \tag{5.26}$$

where

$$f_\pm(t) = \frac{1}{2}\left[e^{-i(M_S - \frac{i}{2}\Gamma_S)t} \pm e^{-i(M_L - \frac{i}{2}\Gamma_L)t} \right]. \tag{5.27}$$

The probability of finding a K_S in an initially pure K^0 beam, $I(K^0 \to K_S; t)$, or $\overline{K}^0$ beam, $I(\overline{K}^0 \to K_S; t)$, still follows an exponential decay law:

$$I(K^0 \to K_S; t) = I_0 \cdot |\langle K_S|K^0(t)\rangle|^2 = \frac{1}{2}I_0 e^{-\Gamma_S t} = I(\overline{K}^0 \to K_S; t), \tag{5.28}$$

with the factor $1/2$ reflecting the presence of the K_L component in the beam:

$$I(K^0 \to K_L; t) = \frac{1}{2}I_0 e^{-\Gamma_L t} = I(\overline{K}^0 \to K_L; t). \tag{5.29}$$

A non-trivial pattern arises when we ask for the probability of finding a K^0 in an initially pure K^0 beam:

$$I(K^0 \to K^0; t) = I_0 \cdot |\langle K^0|K^0(t)\rangle|^2 = I_0 \cdot |f_+(t)|^2$$

$$= \frac{1}{4}I_0 \left[e^{-\Gamma_L t} + e^{-\Gamma_S t} + 2e^{-\frac{1}{2}(\Gamma_S + \Gamma_L)t} \cos \Delta M_k t \right], \tag{5.30}$$

where

$$\Delta M_K = M_L - M_S. \tag{5.31}$$

The last term in the square bracket of Eq. (5.30) reflects the fact that it cannot be decided as a matter of principle whether the observed K^0 came from the K_S or K_L component. Therefore $I(K^0 \to K^0; t)$ is *not* given by the average of $I(K^0 \to K_S; t)$ and $I(K^0 \to K_L; t)$ – an interference term is also present. The law of probabilities is not violated of course, since we have

$$I(K^0 \to \overline{K}^0; t) = \frac{1}{4}I_0 \left[e^{-\Gamma_L t} + e^{-\Gamma_S t} - 2e^{-\frac{1}{2}(\Gamma_S + \Gamma_L)t} \cos \Delta M_K t \right] \tag{5.32}$$

and therefore

$$I(K^0 \to \overline{K}^0; t) + I(K^0 \to K^0; t) = \frac{1}{2}I_0 \left[e^{-\Gamma_L t} + e^{-\Gamma_S t} \right]. \tag{5.33}$$

These functions are shown in Fig. 5.1. From Eq. (5.32) we read off that $I(K^0 \to \overline{K}^0; t = 0) = 0$, as it has to be since the initial beam consists of K^0 mesons only. Yet at later times $t > 0$ we have

$$I(K^0 \to \overline{K}^0; t > 0) > 0 ; \tag{5.34}$$

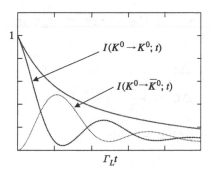

Figure 5.1   The probability of finding K^0 in an initial K^0 beam as a function of time, and the probability of finding $\overline{K}^0$ in the same beam.

i.e. a state that is orthogonal to the initial state gets spontaneously 'generated' at later times!

From the preceding discussion it is clear how the K_S and K_L are identified, namely through the decay modes $K_S \to 2\pi$ and $K_L \to 3\pi$ exhibiting the life-times τ_S and τ_L, respectively. The states of definite strangeness, namely K^0 and $\overline{K}^0$, can be identified through the flavour-specific semileptonic channels

$$K^0 \to l^+\nu\pi^-, \quad \overline{K}^0 \to l^-\overline{\nu}\pi^+. \tag{5.35}$$

It is an experimental fact [11] that semileptonic decays of $K^0[\overline{K}^0]$ always contain $l^+[l^-]$:

$$x = \frac{\langle l^-\overline{\nu}\pi^+|\mathcal{H}|K^0\rangle}{\langle l^+\nu\pi^-|\mathcal{H}|K^0\rangle} \tag{5.36}$$

$$\text{Re}\,x = -0.002 \pm 0.006$$
$$\text{Im}\,x = 0.0012 \pm 0.0021. \tag{5.37}$$

For the hyperon decay,

$$\frac{\Gamma(\Sigma^+ \to nl^+\nu)}{\Gamma(\Sigma^+ \to nl^-\overline{\nu})} < 0.043. \tag{5.38}$$

Within the quark model it is automatic since semileptonic strange decays are driven by

$$s \to ul^-\overline{\nu}\,. \tag{5.39}$$

Changes in strangeness and the quark charge in this decay are then correlated:

$$\Delta S \quad = 0 - (-1) \quad = +1$$
$$\Delta Q = +\frac{2}{3} - \left(-\frac{1}{3}\right) = +1 \; ; \tag{5.40}$$

i.e. it obeys the $\Delta S = \Delta Q = +1$ rule.

Thus we find

$$\Gamma(K^0(t) \to l^- \bar{\nu}\pi^+) = |f_-(t)|^2 \Gamma(\overline{K}^0 \to l^- \bar{\nu}\pi^+)$$
$$\Gamma(\overline{K}^0(t) \to l^- \bar{\nu}\pi^+) = |f_+(t)|^2 \Gamma(\overline{K}^0 \to l^- \bar{\nu}\pi^+). \tag{5.41}$$

Analysing the decay rate evolution of these modes in time, we can extract ΔM_K in addition to Γ_S and Γ_L. The present value is [11]

$$\Delta M_K = M_L - M_S = (3.449 \pm 0.001) \times 10^{-12} \, \text{MeV}$$
$$\Delta \Gamma_K = \Gamma_S - \Gamma_L = (7.335 \pm .004) \times 10^{-12} \, \text{MeV}. \tag{5.42}$$

It should be seen as utterly amazing that such a tiny mass difference ΔM_K can be measured, in particular if we quote it relative to the average of the K_S and K_L mass

$$\frac{\Delta M_K}{M_K} \simeq 7 \times 10^{-15}. \tag{5.43}$$

What this shows is the power of quantum mechanical interference effects tracked over 'macroscopic' distances. To understand it, first note that

$$\cos(\Delta M_K t) = \cos\left(\frac{\Delta M_K}{\Gamma_S} \frac{t}{\tau_S}\right). \tag{5.44}$$

If the beam energy is chosen high enough then the neutral kaon – after its production and before its decay – travels a distance that can be resolved and measured. Knowing its momentum we can determine t, its proper time of decay. Its scale is set by $\tau_S = \Gamma_S^{-1}$ and the scale for ΔM_K is then given by Γ_S. The width Γ_S is obviously of second order in the weak interactions. ΔM_K represents a $\Delta S = 2$ transition; as explained later in detail, it is generated by iterating two $\Delta S = 1$ reactions and thus is also of second order in the weak interactions. Therefore ΔM_K and Γ_S can be expected to be of roughly comparable order of magnitude. Since $\Gamma_S \simeq 7.3 \cdot 10^{-12} \, \text{MeV}$ such experiments are sensitive to mass differences $\sim \mathcal{O}\left(10^{-12} \, \text{MeV}\right)$. Indeed we find

$$\frac{\Delta M_K}{\Gamma_S} \simeq 0.49. \tag{5.45}$$

Finally, we should note that in the 1960s tracking technologies were not sufficiently refined for resolving reliably the differences in the production and decay vertices for particles with $c\tau_S \simeq 2.7\,\mathrm{cm}$. Another technique was used then, to be discussed next.

5.6 Regeneration – which is heavier: K_L or K_S?

So far we have implicitly assumed that the neutral kaons travel through a trivial medium, namely empty space, which does not affect the particle–antiparticle balance. However, if they travel through nuclear matter, the situation will change dramatically since K^0 and $\overline{K}^0$ interact quite differently with a nucleus. A K^0 with the quark content $\bar{s}d$ can, at low energies, interact only (quasi-)elastically with nucleons N:

$$K^0 + N \rightarrow K^0 + N^{(*)}, \tag{5.46}$$

where the notation $N^{(*)}$ allows for the nucleon to be excited into a non-strange resonance N^*. A $\overline{K}^0 = (s\bar{d})$ on the other hand can – in addition to its elastic reaction $\overline{K}^0 + N \rightarrow \overline{K}^0 + N^{(*)}$ – also excite a nucleon into a Λ or Σ hyperon or a Y^* resonance:

$$\overline{K}^0[s\bar{d}] + N[uud/udd] \rightarrow \pi^0[d\bar{d}] + \Lambda/\Sigma/Y^*[suu/sud]\,. \tag{5.47}$$

K^0 and $\overline{K}^0$ will therefore propagate quite differently through nuclear matter. To obtain an intuitive picture, let us assume that $\overline{K}^0$ is totally absorbed by a nucleus whereas K^0 is not. The nucleus then acts like a Stern–Gerlach filter letting K^0 pass but not $\overline{K}^0$. Pais and Piccioni [28] presented a beautiful analysis of this basic quantum mechanical problem briefly sketched here.

Consider a π^- beam hitting a target and producing a mixture of K^0 and $\overline{K}^0$ or, equivalently, of K_S and K_L. Travelling through a vacuum, the mass eigenstates will decay through $K_S \rightarrow 2\pi$ and $K_L \rightarrow 3\pi$. Since $\tau_S \ll \tau_L$, the K_S component will quickly decay away leaving a practically pure K_L beam behind. If the latter hits nuclear matter, only its K^0 component will not be absorbed; the emerging state is thus the linear combination $|K_S\rangle + |K_L\rangle$ – i.e. the previously extinct K_S component has been *regenerated* through rescattering with nuclear matter! This is depicted in Fig. 5.2.

To be more specific, we can write down the time evolution for a K^0 or a $\overline{K}^0$ travelling through a medium by modifying Eq. (5.25) and Eq. (5.26) as follows:

$$|K^0(t_{\mathrm{med}})\rangle = f_+(t_{\mathrm{med}})g|K^0\rangle + f_-(t_{\mathrm{med}})\bar{g}|\overline{K}^0\rangle$$
$$|\overline{K}^0(t_{\mathrm{med}})\rangle = f_-(t_{\mathrm{med}})g|K^0\rangle + f_+(t_{\mathrm{med}})\bar{g}|\overline{K}^0\rangle. \tag{5.48}$$

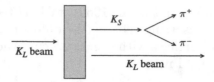

Figure 5.2 Regeneration of K_S from a K_L beam going through matter. K_L is a coherent linear combination of the K^0 and $\overline{K}^0$ states. When it travels through matter, it interacts with nuclei through strong interaction. The strong interaction eigenstates K^0 and $\overline{K}^0$ scatter differently and the coherence is lost. The resulting beam is a mixture of K_L and K_S – K_S is regenerated.

The quantities g and $\overline{g}$ are the forward-scattering amplitudes for K^0 and $\overline{K}^0$ in the medium, respectively. We then find

$$\Gamma(K^0(t_{\mathrm{med}})) \sim e^{-\Gamma_L t_{\mathrm{med}}}|g_-|^2 + e^{-\Gamma_S t_{\mathrm{med}}}|g_+|^2$$
$$+ 2e^{-\frac{1}{2}(\Gamma_L+\Gamma_S)t_{\mathrm{med}}}\mathrm{Re}\left(e^{-i\Delta M_K t_{\mathrm{med}}}g_- g_+^*\right), \qquad (5.49)$$

where $g_- = \frac{1}{2}(g - \overline{g})$ and $g_+ = \frac{1}{2}(g + \overline{g})$. By introducing plates of varying thickness, measurements of $-\Delta M_K \tau + \arg(g_+) - \arg(g_-)$ can be made. Finally, the phases of g_+ and g_- are determined from known phase shifts of $K^0 N$ and $\overline{K}^0 N$ scattering. This allows us to determine not only $|\Delta M_K|$, but even the sign of ΔM_K. It turns out that

$$M_L > M_S, \qquad (5.50)$$

leading to the nice mnemonic that in K_L and K_S, L means both *larger* (mass) and *longer* (lifetime) whereas S denotes *smaller* (mass) and *shorter* (lifetime).

5.7 The quiet before the storm

We might forgive theorists of that time for feeling quite smug about the situation. Many essential insights had been gained from analysing the dynamics of strange particles, or were at least prompted by it.

- A new quantum number – strangeness – conserved by the strong and electromagnetic forces had been discovered. In modern parlance we would say that the first hadrons of the second family had been found.
- It had been realized that parity as well as charge conjugation were violated by the weak forces and actually in a maximal way (as far as the charged currents are concerned).

- It appeared, however, that we did not have to fall back all the way to **CPT** symmetry: the violations observed for **C** and **P** were such that they compensated each other, i.e. **CP** invariance apparently was satisfied.
- General arguments had been advanced as to why nature had to be **CP** symmetric.
- The predictions concerning the existence of two kinds of neutral kaons with vastly different lifetimes had been confirmed.
- The mass difference for those two states had been determined with spectacular sensitivity through a judicious application of ingenious quantum mechanical reasoning and beautiful experimentation.

Yet this harmonious picture was about to receive a shattering blow![5]

5.8 The discovery of CP violation

At a conference held in 1989 celebrating the XXVth anniversary of CP violation, Pais stated in his retrospective lecture [19]: 'It was a very good year, 1964, both in theory and in experimental high energy physics'. For the following discoveries and breakthroughs were made that year.

(1) The Higgs mechanism for the *spontaneous* realization of a symmetry was first developed.
(2) The quark model and the first elements of current algebra were put forward.
(3) The charm quark was first introduced to establish quark–lepton symmetry.
(4) $SU(6)$ symmetry was proposed.
(5) The first storage ring for e^+e^- collisions was built in Frascati.
(6) The Ω^- baryon was found at Brookhaven National Laboratory.
(7) CP violation was discovered at the same laboratory.

Finding the Ω^- was seen as a breakthrough confirmation of the 'eight-fold way' of $SU(3)$ symmetry. Being composed of three strange quarks, it also exhibits rather directly the need for colour as a new internal degree of freedom. The theoretical concepts listed under items (1)–(3) turn out to be very relevant for our attempts to deal with **CP** invariance and its limitations.

---

[5] It should be noted that not everybody subscribed to this orthodox view. Okun, in his 1963 text book [29], explicitly listed the search for $K_L \to \pi\pi$ as a priority for the future then. In response to his suggestion, an experiment was performed; however, it did not succeed in accumulating statistics sufficient to see a signal.

As mentioned in the preceding chapter, **CP** violation had not been expected. Its observation thus came as a shock to the community, as illustrated in Pais' personal account [19]:

> I had been aware of that result since one morning in early June when I was having breakfast in the old, rather dilapidated, Brookhaven Cafeteria which some of us remember with slight nostalgia. That morning Jim and Val had walked up to me and had asked could they talk to me. Naturally, I said. They proceeded to tell me that they had found K_2-decays into $\pi^+\pi^-$. How could that be I asked; it violates CP-invariance. They knew that, they said, but there it was. Why was the effect not due to regeneration of short-lived K_S, I wanted to know. Because, they said, that effect was far too small in the helium bag where the 2π had been found. I asked many more questions, why were they not seeing $2\pi\gamma$ decays with a soft photon, or $\pi\mu\nu$ decays with a soft neutrino and perhaps some confusion about mass. They had thought long and hard about these and other alternatives (the actual experiment had been concluded the previous July) and had ruled them out one by one. After they left I had another coffee. I was shaken by the news. I knew quite well that a small amount of CP-violation would not drastically alter the earlier discussions, based on CP-invariance, of the neutral K-complex. Also, the experience of seeing a symmetry fall by the wayside was not new to anyone who had lived through the 1956–7 period. At that time, however, there had at once been the consolation that P- and C-violation could be embraced by a new pretty concept, CP-invariance. What shook all concerned was that with CP gone there was nothing elegant to replace it with. The very smallness of the 2π rate, CP-invariance as a near miss, made the news even harder to digest.

While this discovery was completely unexpected, it did not occur in a vacuum, but marked the culmination of a long series of experiments, the highlights of which were the following events [30]:

- at the BNL Cosmotron, the BNL-Columbia group discovered what is today known as K_L [27];
- Piccioni and his collaborators [31] demonstrated the regeneration phenomenon at the Berkeley Betatron;
- Adair and his collaborators [32] claimed to have seen an excess of K_S regeneration in the forward direction.

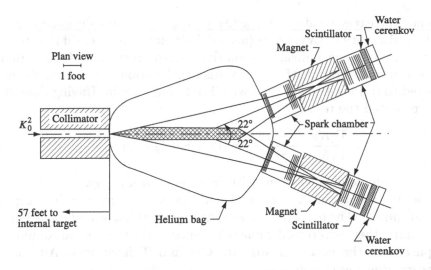

Figure 5.3   A schematic drawing of an apparatus used by Cronin *et al.* K^0 and $\overline{K}^0$ are created on the internal target. They travel through the collimator so that only the ones travelling parallel to the axis of the apparatus are accepted. Furthermore, most of the K_S component has decayed away and we have almost a pure K_L beam. Two detectors which consist of magnets and spark chambers measure three-momenta of the two-body decay product $K_L \to \pi^+\pi^-$. The invariant mass of the $\pi^+\pi^-$ pair as well as the direction of K_L have been checked.

Cronin, Fitch, Christenson and Turlay [33] set out to investigate this anomalous K_S regeneration. Their apparatus and its location with respect to the AGS proton beam is shown in Fig. 5.3.[6]

K^0 and $\overline{K}^0$ mesons are produced in a target 57 feet in front of the spectrometer. Over that distance the K_S component has practically completely faded away and a pure K_L beam remains. The latter passes through a lead collimator into a bag filled with helium gas. Knowing the parameters controlling regeneration we compute the K_S regeneration rate due to K_L scattering off helium nuclei: it is found to be entirely negligible.

By inserting a regenerator they could study the excess K_S production in the forward direction reported by Adair *et al.* [32]. With much better statistics, they were able to rule out any anomalous regeneration of K_S mesons. As a secondary objective they set out to put a better upper limit on the CP-violating $K_L \to 2\pi$ channel. They had claimed in their proposal that it '... can set a limit of about one in a thousand for the partial decay of $K_2 \to 2\pi$ in one hour of operation' [34]. Instead they saw, to their astonishment, the decay $K_L \to 2\pi$.

---

[6] Reminiscing in March 1997 about how the experiment was operated, Val Fitch revealed a certain nonchalance concerning safety: 'Sometimes we had to duck in and out of the beam line. But we were of course protected by our film badges'.

Since the experimenters were acutely aware of the revolutionary nature of their discovery if it were true (and of their embarrassment if it were not) they explored even unlikely alternative interpretations of their signal – like exotic regeneration of K_S or a misidentification of the final state, as referred to in Pais' conversation with Fitch and Cronin. Having done that they reported the result

$$\frac{\Gamma(K_L \to \pi^+\pi^-)}{\Gamma(K_S \to \pi^+\pi^-)} = [(2.0 \pm 0.4) \times 10^{-3}]^2, \tag{5.51}$$

which is completely consistent with what is known today.

It appears that the *experimental findings* per se were readily accepted. As Pais put it: 'The perpetrators were widely regarded as real pros.' On the other hand there was considerable reluctance in the theoretical community to part with the notion of absolute **CP** and **T** invariance. Alternative *interpretations* were suggested.

- Observing $K_L \to \pi\pi$ to occur *by itself* does *not* establish **CP** violation. What is needed for that conclusion to take hold is to have two mass *non*-degenerate states K_L and K_S[7] with $K_S \to \pi\pi$ and $K_L \to 2\pi, 3\pi$ – within the framework of *linear* quantum mechanics with its superposition principle! We could attempt to reconcile the observations with **CP** symmetry by going beyond the usual quantum mechanics and introduce *non-linear* terms into the Schrödinger equation. This would violate the superposition principle – yet that had not been tested to that sensitivity before. This escape route was actually tried [35] – which actually shows once more how unprepared the theoretical community was for accepting CP violation.

- It was suggested that a new particle U emerged unobserved from the decay:

$$K_L \to K_S + U \to (\pi\pi)_{K_S} + U. \tag{5.52}$$

Observation of an interference between $K_L \to \pi\pi$ and $K_S \to \pi\pi$ with K_S being coherently regenerated from the K_L beam also ruled out this scenario [35]. It is curious to note – from a historical perspective – that here it has turned out to be correct to abandon an apparently sacrosanct symmetry principle rather than introduce an exotic new particle. In an earlier situation when faced with the paradox of *continuous* electron spectra in β decay one – actually Pauli – had been vindicated in retaining the symmetry leading to momentum conservation even at the price of postulating an exotic particle, the neutrino, with apparently little prospect of ever establishing its existence.

---

[7] It does not matter at this point how tiny ΔM_K is.

- It was pointed out that the observed **CP** asymmetry might be an *environmental* effect due to the preponderance of matter over anti-matter in our corner of the universe. More specifically, the existence of a new long-range force of cosmological origin was postulated.

While not all conceivable alternatives were ruled out experimentally in a completely rigorous fashion, the community came quickly and decisively around to accepting **CP** violation as a fact.

The observations can be summarized by saying that the quantum mechanical K_L state contains a small admixture of a **CP** even component in addition to its dominant **CP** odd part:

$$|K_L\rangle = \frac{1}{\sqrt{1 + |\bar{\epsilon}|^2}} \left(|K_2\rangle + \bar{\epsilon}|K_1\rangle \right). \tag{5.53}$$

CPT invariance then tells us that the K_S state in turn contains a **CP** odd component controlled by the same (complex) impurity parameter $\bar{\epsilon}$:

$$|K_S\rangle = \frac{1}{\sqrt{1 + |\bar{\epsilon}|^2}} \left(|K_1\rangle - \bar{\epsilon}|K_2\rangle \right). \tag{5.54}$$

It should be noted that because $\mathrm{Br}(K_L \to \pi\pi)$ is so tiny, the argument used to explain the large lifetime difference still applies.

Problems

5.1. Show that the final state in $K \to \pi^+\pi^-, \pi^0\pi^0$ is **CP** even. Next consider $K \to 3\pi^0, \pi^+\pi^-\pi^0$; what can you now say about the **CP** parity of the final state?

5.2. Imagine a world in which the pion is massless. Do you think **CP** violation would have been discovered?

5.3. Imagine instead a world with $M_\pi = 200\,\mathrm{MeV}$. How would this affect the $\tau - \theta$ puzzle and the discovery of **CP** violation?

5.4. By diagonalizing the Hamiltonian shown in Eq. (5.16), show that $M_{K_1} - M_{K_2} = 2\Delta$. Is the sign of Δ observable? Can you distinguish the states K_1 and K_2 physically? If so, how? In discussing Eq. (3.74) we had argued that for $E_+ = E_-$, the sign of Δ there had no meaning. Compare the two situations.

6

Quantum mechanics of
neutral particles

In the preceding chapter we described the discovery of **CP** violation from
a historical point of view; we depended largely on intuition rather than
on the formalism. The latter will be discussed in detail now.

6.1 The effective Hamiltonian

For the sake of simplicity, we have been intentionally sloppy in describing
$K - \overline{K}$ mixing in Section 5.4. We simply stated the time dependences,
Eq. (5.23), of $|K_L(t)\rangle$ and $|K_S(t)\rangle$ states assuming **CP** was conserved.

Here we shall present the quantum mechanical formalism for particle–
antiparticle oscillations, and show how **CP** violation can manifest itself
there. The formalism of particle–antiparticle oscillations presented in this
section is very general. It describes a situation where particles and antiparticles – unlike the case with π^0 mesons – are distinguished by an internal
quantum number like strangeness, beauty, charm, etc., or lepton number,
baryon number, etc.; weak or superweak interactions not conserving these
quantum numbers can drive a particle–antiparticle transition where the
internal quantum number changes by two units. We will not contemplate
violation of electric charge conservation and therefore study only electri-
cally neutral states P^0 and $\overline{P}^0$. Yet P^0 may be K^0, D^0, B^0, a neutron or
a neutrino;[1] for the original discussion, see Ref. [36, 37]. P^0 and $\overline{P}^0$ are
distinguished merely by an internal quantum number F with the prop-
erty that $\Delta F = 0$ for H_{strong} and H_{QED} while $\Delta F \neq 0$ for H_{weak}. Let
us suppose that H_{weak} couples P^0 and $\overline{P}^0$ to a common state I: Then

---

[1] The formalism described here applies directly to neutron–antineutron and $\nu - \overline{\nu}$ oscillations; as
shown later it can easily be generalized to the more 'conventional' case of $\nu_\alpha - \nu_\beta$ oscillations.

$H_{\text{weak}} \otimes H_{\text{weak}}$ will induce a transition $P^0 \leftrightarrow \overline{P}^0$ as a one-step process or as an iteration of two $\Delta F = 1$ reactions through an intermediate state I:

$$P^0 \xrightarrow{\Delta F=1} I \xrightarrow{\Delta F=1} \overline{P}^0 \quad \text{or} \quad \overline{P}^0 \xrightarrow{\Delta F=-1} I \xrightarrow{\Delta F=-1} P^0. \qquad (6.1)$$

I can represent a real on-shell state common to P^0 and $\overline{P}^0$ decays or it can denote a virtual off-shell state. All this can be taken into account by considering the Hamiltonian for a state carrying the quantum number F:

$$H = H_{\Delta F=0} + H_{\Delta F=1} + H_{\Delta F=2}. \qquad (6.2)$$

$H_{\Delta F=0}$ contains the strong and electromagnetic forces conserving F;[2] $H_{\Delta F=1}$ denotes the weak forces changing F by one unit; $H_{\Delta F=2}$ allows for the presence of some so-called superweak forces changing F by two units, thus producing $P^0 \to \overline{P}^0$ as a one-step process.

The time evolution of the $P^0 \leftrightarrow \overline{P}^0$ system, including its decays, is given by a vector in the Hilbert space

$$|\tilde{\Psi}(t)\rangle = a(t)|P^0\rangle + b(t)|\overline{P}^0\rangle + c(t)|\pi\pi\rangle + d(t)|3\pi\rangle + e(t)|\pi l \,\overline{\nu}_l\rangle + \cdots, \quad (6.3)$$

where l stands for an electron or muon, and the Schrödinger equation reads

$$i\hbar\frac{\partial}{\partial t}\tilde{\Psi} = H\tilde{\Psi}, \qquad (6.4)$$

with H being an *infinite*-dimensional Hermitian matrix in the Hilbert space. The full time dependence of $\tilde{\Psi}(t)$ cannot be obtained rigorously, since that would require subjugating strong dynamics completely to theoretical control, which at present is beyond our capabilities.

The situation simplifies dramatically and becomes tractable if we reduce our demands. Let us consider the following scenario [36, 37].[3]

- The initial state is a linear combination of P^0 and $\overline{P}^0$ alone: $|\Psi(0)\rangle = a(0)|P^0\rangle + b(0)|\overline{P}^0\rangle$.
- We are interested only in $a(t)$ and $b(t)$ and not in any other coefficients.
- We restrict ourselves to times that are much larger than a typical strong interaction scale. This is called the Weisskopf–Wigner approximation [39].

---

[2] There are also weak corrections with $\Delta F = 0$, but they can safely be ignored here.
[3] Another formulation proposed in Ref. [38] relies on the analysis of the neutral kaon propagator in which the proper self-energy diagram $\Pi^*(k^2)$ must be approximated by $\Pi^*(m_{p^o}^2 + \text{io}^+)$ near poles of the propagator.

Then we can write

$$i\hbar \frac{\partial}{\partial t}\Psi(t) = \mathcal{H}\Psi(t) \tag{6.5}$$

where $\Psi(t)$ is restricted to the subspace of P^0 and $\overline{P}^0$:

$$\Psi(t) = \begin{pmatrix} a(t) \\ b(t) \end{pmatrix}. \tag{6.6}$$

The matrix $\mathcal{H}$ is given by

$$\mathcal{H} = \mathbf{M} - \frac{i}{2}\mathbf{\Gamma} = \begin{pmatrix} M_{11} - \frac{i}{2}\Gamma_{11} & M_{12} - \frac{i}{2}\Gamma_{12} \\ M_{21} - \frac{i}{2}\Gamma_{21} & M_{22} - \frac{i}{2}\Gamma_{22} \end{pmatrix}, \tag{6.7}$$

where

$$M_{11} = M_P + \sum_n \mathcal{P}\left[\frac{|\langle n; \text{out}|H_{\Delta F=1}|P^0\rangle|^2}{M_P - M_n}\right]$$

$$M_{22} = M_{\overline{P}} + \sum_n \mathcal{P}\left[\frac{|\langle n; \text{out}|H_{\Delta F=1}|\overline{P}^0\rangle|^2}{M_{\overline{P}} - M_n}\right]$$

$$M_{12} = M_{21}^* = \langle P^0|H_{SW}|\overline{P}^0\rangle$$

$$\qquad + \sum_n \mathcal{P}\left[\frac{\langle P^0|H_{\Delta F=1}|n; \text{out}\rangle\langle n; \text{out}|H_{\Delta F=1}|\overline{P}^0\rangle}{M_P - M_n}\right]$$

$$\Gamma_{11} = 2\pi\sum_n \delta(M_P - M_n)|\langle n; \text{out}|H_{\Delta F=1}|P^0\rangle|^2$$

$$\Gamma_{22} = 2\pi\sum_n \delta(M_P - M_n)|\langle n; \text{out}|H_{\Delta F=1}|\overline{P}^0\rangle|^2$$

$$\Gamma_{12} = \Gamma_{21}^* = 2\pi\sum_n \delta(M_P - M_n)\langle P^0|H_{\Delta F=1}|n; \text{out}\rangle\langle n; \text{out}|H_{\Delta F=1}|\overline{P}^0\rangle. \tag{6.8}$$

$\mathcal{P}$ stands for the *principal part* prescription.[4] A detailed derivation can be found in [40], while [38] gives an alternative approach.

----

[4] A comment on the normalization of states is in order. By $\langle \vec{k}_1 \ldots \vec{k}_n|H_{\Delta F=1}|P^0\rangle$, we mean

$$\frac{(2\pi)^3\delta^3(\vec{p}-\vec{k}_1-\ldots\vec{k}_n)\langle\vec{k}_1\ldots\vec{k}_n|H(0)|P\rangle}{\sqrt{2MV2E_1V\ldots2E_nV}},$$

where $\langle \vec{k}_1 \ldots \vec{k}_n|H(0)|P\rangle$ is a Lorentz scalar, and $H_{\Delta F=1} = \int d^3x H(x)$.

6.2   Constraints from CPT, CP and T

Let us assume that H commutes with **C**, **P** and **T**. We shall define these operators such that

$$\mathbf{C}|P^0\rangle = -|\overline{P}^0\rangle, \quad \mathbf{P}|P^0\rangle = -|P^0\rangle, \quad \mathbf{T}|P^0\rangle = |P^0\rangle, \qquad (6.9)$$

with

$$\mathbf{CP}|n; \text{out(in)}\rangle = |\overline{n}; \text{out(in)}\rangle$$
$$\mathbf{T}|n; \text{in(out)}\rangle = |n; \text{out(in)}\rangle. \qquad (6.10)$$

Assuming as usual that the 'in' and 'out' states form equivalent complete bases, we can easily show (see Problem 6.1) that $M_{11} = M_{22}$ and $\Gamma_{11} = \Gamma_{22}$ from **CP** and **CPT** symmetry. Also, $M_{12} = M_{21}$ and $\Gamma_{12} = \Gamma_{21}$ from **CP** and **T** symmetry. Thus we have (see Problems 6.8 and 6.1)

$$\mathbf{CPT} \text{ or } \mathbf{CP} \text{ invariance} \implies M_{11} = M_{22}, \ \Gamma_{11} = \Gamma_{22},$$
$$\mathbf{CP} \text{ or } \mathbf{T} \text{ invariance} \implies \operatorname{Im} M_{12} = 0 = \operatorname{Im} \Gamma_{12}. \qquad (6.11)$$

6.3   Spherical coordinates

The Schrödinger equation is best solved by diagonalizing the matrix $\mathcal{H}$; the solutions to the then decoupled equations represent the two mass eigenstates. Since **M** and $\boldsymbol{\Gamma}$ are Hermitian they can be diagonalized by a unitary transformation. It is however not immediately obvious that the same transformation can diagonalize both **M** and $\boldsymbol{\Gamma}$. The formulism for tackling this problem was developed in Refs. [41] and [42].

Any 2×2 matrix can be expanded in terms of the three Pauli matrices σ_i and the unit matrix using complex coefficients:

$$\mathbf{M} - \frac{i}{2}\boldsymbol{\Gamma} = E_1\sigma_1 + E_2\sigma_2 + E_3\sigma_3 - iD\mathbf{1}, \qquad (6.12)$$

where the Pauli matricies are as defined in Eq. (4.10). Comparing both sides of Eq. (6.12) we arrive at

$$E_1 = \operatorname{Re} M_{12} - \frac{i}{2}\operatorname{Re}\Gamma_{12}$$

$$E_2 = -\operatorname{Im} M_{12} + \frac{i}{2}\operatorname{Im}\Gamma_{12}$$

$$E_3 = \frac{1}{2}(M_{11} - M_{22}) - \frac{i}{4}(\Gamma_{11} - \Gamma_{22})$$

$$D = \frac{i}{2}(M_{11} + M_{22}) + \frac{1}{4}(\Gamma_{11} + \Gamma_{22}). \qquad (6.13)$$

We can define *complex* numbers E, θ, and ϕ such that

$$E = \sqrt{E_1^2 + E_2^2 + E_3^2}, \tag{6.14}$$

$$E_1 = E \sin\theta \cos\phi, \quad E_2 = E \sin\theta \sin\phi, \quad E_3 = E \cos\theta. \tag{6.15}$$

Cosine and sine of a complex number z are defined in the usual way:

$$\cos z = \frac{1}{2}\left(e^{iz} + e^{-iz}\right), \quad \sin z = \frac{1}{2i}\left(e^{iz} - e^{-iz}\right). \tag{6.16}$$

The condition for $P^0 - \overline{P}^0$ oscillations to occur is then expressed by

$$E \neq 0, \quad \sin\theta \neq 0. \tag{6.17}$$

The constraints on $\mathbf{M}$ and $\mathbf{\Gamma}$ imposed by the discrete symmetries are expressed as follows:

$$\begin{aligned} \mathbf{CPT} \text{ or } \mathbf{CP} \text{ invariance} &\implies \cos\theta = 0 \\ \mathbf{CP} \text{ or } \mathbf{T} \text{ invariance} &\implies \phi = 0. \end{aligned} \tag{6.18}$$

It is important to note that these arrows apply in one direction only; for example we can adopt a phase convention in which $\cos\theta = 0$ or $\phi = 0$ even when $\mathbf{CP}$ is violated.

This representation in (complex) spherical coordinates allows us to write down the mass eigenstates in a rather compact fashion:

$$\begin{aligned} |P_1\rangle &= p_1 |P^0\rangle + q_1 |\overline{P}^0\rangle, \\ |P_2\rangle &= p_2 |P^0\rangle - q_2 |\overline{P}^0\rangle, \end{aligned} \tag{6.19}$$

where

$$\begin{aligned} p_1 &= N_1 \cos\frac{\theta}{2}, \quad q_1 = N_1 e^{i\phi} \sin\frac{\theta}{2}, \\ p_2 &= N_2 \sin\frac{\theta}{2}, \quad q_2 = N_2 e^{i\phi} \cos\frac{\theta}{2}, \end{aligned} \tag{6.20}$$

with the normalization factors (θ and ϕ are complex in general!):

$$\begin{aligned} N_1 &= \frac{1}{\sqrt{|\cos\frac{\theta}{2}|^2 + |e^{i\phi}\sin\frac{\theta}{2}|^2}}, \\ N_2 &= \frac{1}{\sqrt{|\sin\frac{\theta}{2}|^2 + |e^{i\phi}\cos\frac{\theta}{2}|^2}}. \end{aligned} \tag{6.21}$$

With **CPT** invariance imposing $\cos \theta = 0$, considerable simplifications arise: $p_1 = p_2$ and $q_1 = q_2$, and we can drop the subscripts. The states

$$|P_1\rangle = p|P^0\rangle + q|\overline{P}^0\rangle,$$
$$|P_2\rangle = p|P^0\rangle - q|\overline{P}^0\rangle, \qquad (6.22)$$

are mass eigenstates with eigenvalues

$$M_1 - \frac{i}{2}\Gamma_1 = -iD + E = M_{11} - \frac{i}{2}\Gamma_{11} + \frac{q}{p}\left(M_{12} - \frac{i}{2}\Gamma_{12}\right),$$

$$M_2 - \frac{i}{2}\Gamma_2 = -iD - E = M_{11} - \frac{i}{2}\Gamma_{11} - \frac{q}{p}\left(M_{12} - \frac{i}{2}\Gamma_{12}\right). \quad (6.23)$$

with

$$\left(\frac{q}{p}\right)^2 = \frac{M_{12}^* - \frac{i}{2}\Gamma_{12}^*}{M_{12} - \frac{i}{2}\Gamma_{12}}. \qquad (6.24)$$

Obviously there exist *two* solutions:

$$\frac{q}{p} = \pm\sqrt{\frac{M_{12}^* - \frac{i}{2}\Gamma_{12}^*}{M_{12} - \frac{i}{2}\Gamma_{12}}}. \qquad (6.25)$$

Choosing the negative rather than the positive sign in Eq. (6.25) is equivalent to interchanging the labels $1 \leftrightarrow 2$ of the mass eigenstates, see Eq. (6.22) and Eq. (6.23).

6.4 ♠ On phase conventions ♠

The binary ambiguity just mentioned is a special case of a more general one. For antiparticles are defined up to a phase only; adopting a different phase convention – e.g. going from $|\overline{P}^0\rangle$ to $e^{i\xi}|\overline{P}^0\rangle$ – will modify the off-diagonal elements of **M** and **Γ**:

$$M_{12} - \frac{i}{2}\Gamma_{12} \longrightarrow e^{i\xi}\left(M_{12} - \frac{i}{2}\Gamma_{12}\right) \qquad (6.26)$$

and thus

$$\frac{q}{p} \longrightarrow e^{-i\xi}\frac{q}{p} \qquad (6.27)$$

yet leave their product invariant:

$$\frac{q}{p}\left(M_{12} - \frac{i}{2}\Gamma_{12}\right) \longrightarrow \frac{q}{p}\left(M_{12} - \frac{i}{2}\Gamma_{12}\right). \qquad (6.28)$$

This is as it should be since the differences in mass and width, see Eq. (6.23),

$$M_2 - M_1 = -2\text{Re}\left(\frac{q}{p}(M_{12} - \frac{i}{2}\Gamma_{12})\right)$$

$$\Gamma_2 - \Gamma_1 = 4\text{Im}\left(\frac{q}{p}(M_{12} - \frac{i}{2}\Gamma_{12})\right), \tag{6.29}$$

being observables, have to be insensitive to the *arbitrary* phase of $\overline{P}^0$!

The following comments will turn out to be important for the subsequent discussion on **CP** asymmetries.

- The mass eigenstates P_1 and P_2 will in general not be orthogonal to each other without **CP** invariance:

$$\langle P_1 | P_2 \rangle = |p|^2 - |q|^2 \neq 0 \tag{6.30}$$

  for $|q/p| \neq 1$ or $\text{Im}\phi \neq 0$.
- At this point, the physical meaning of labels '1' and '2' are not clear. Once we know the sign of $M_1 - M_2$, it becomes an empirical question whether

$$\Gamma_2 > \Gamma_1 \quad \text{or} \quad \Gamma_2 < \Gamma_1 \tag{6.31}$$

  holds, i.e. whether the heavier state is shorter or longer lived.
- In the limit of **CP** invariance the two mass eigenstates are **CP** eigenstates as well, and we can raise another meaningful question: is the heavier state **CP** even or odd? With **CP** symmetry implying $\arg(M_{12}/\Gamma_{12}) = 0$, q/p becomes a pure phase: $|q/p| = 1$. It is then convenient to adopt a phase *convention* s.t. M_{12} is real; $q/p = \pm 1$ and $\mathbf{CP}|P^0\rangle = \pm|\overline{P}^0\rangle$ are then the remaining choices.
  - With $q/p = +1$ we have

$$|P_1\rangle = \frac{1}{\sqrt{2}}\left(|P^0\rangle + |\overline{P}^0\rangle\right)$$

$$|P_2\rangle = \frac{1}{\sqrt{2}}\left(|P^0\rangle - |\overline{P}^0\rangle\right). \tag{6.32}$$

For $\mathbf{CP}|P^0\rangle = |\overline{P}^0\rangle$, P_1 and P_2 are **CP** even and odd, respectively; therefore

$$M_- - M_+ = M_2 - M_1 = -2\text{Re}\left(\frac{q}{p}\left(M_{12} - \frac{i}{2}\Gamma_{12}\right)\right) = -2M_{12} \tag{6.33}$$

For $\mathbf{CP}|P^0\rangle = -|\overline{P}^0\rangle$, P_1 and P_2 switch roles; i.e. P_1 and P_2 are now $\mathbf{CP}$ odd and even. Thus

$$M_- - M_+ = M_1 - M_2 = 2M_{12}. \qquad (6.34)$$

– Alternatively we can set $q/p = -1$:

$$|P_1\rangle = \frac{1}{\sqrt{2}}\left(|P^0\rangle - |\overline{P}^0\rangle\right)$$

$$|P_2\rangle = \frac{1}{\sqrt{2}}\left(|P^0\rangle + |\overline{P}^0\rangle\right) \qquad (6.35)$$

while maintaining $\mathbf{CP}|P^0\rangle = |\overline{P}^0\rangle$; P_1 and P_2 are then $\mathbf{CP}$ odd and even, respectively. Accordingly

$$M_- - M_+ = M_1 - M_2 = 2\mathrm{Re}\left(\frac{q}{p}\left(M_{12} - \frac{i}{2}\Gamma_{12}\right)\right) = -2M_{12}. \qquad (6.36)$$

– We leave the fourth possibility as an exercise.
– Equation (6.33) and Eq. (6.36) on one side and Eq. (6.34) on the other apparently do not coincide; yet below we will see that the theoretical prediction for M_{12} changes sign depending on the choice of $\mathbf{CP}|P^0\rangle = \pm|\overline{P}^0\rangle$. These expressions therefore all agree, of course. If the lifetime difference is too small to be observed or $\mathbf{CP}$ is too badly broken to be of use in classifying states, then we are limited to stating that one mass eigenstate is heavier than the other.
• Later we will discuss how to evaluate M_{12} within a given theory for the $P - \overline{P}$ complex. The examples above have illustrated that some care has to be applied in interpreting such a result.
– Expressing the mass eigenstates explicitly in terms of flavour eigenstates involves some conventions. Once we adopt a certain definition, we have to stick with it; yet our choice cannot influence observables.

6.5   ♠ ΔM and ΔΓ ♠

The *relative* phase between M_{12} and Γ_{12} represents an observable quantity describing indirect $\mathbf{CP}$ violation. The following notation turns out to be convenient:

$$M_{12} = \overline{M}_{12}e^{i\xi}, \quad \Gamma_{12} = \overline{\Gamma}_{12}e^{i\xi}e^{i\zeta}, \quad \frac{\Gamma_{12}}{M_{12}} = \frac{\overline{\Gamma}_{12}}{\overline{M}_{12}}e^{i\zeta} = re^{i\zeta}. \qquad (6.37)$$

The signs of $\overline{M}_{12}$ and $\overline{\Gamma}_{12}$ are fixed such that both ξ and $\xi + \zeta$ are restricted to lie between $-\pi/2$ and $\pi/2$; i.e. the real quantities $\overline{M}_{12}$ and $\overline{\Gamma}_{12}$ are a priori allowed to be negative as well as positive. A relative minus sign between M_{12} and Γ_{12} is of course physically significant, while the absolute sign is not. Yet we will see that the absolute sign provides us with a useful bookkeeping device.

Throughout this book, we will set

$$\Delta M = M_2 - M_1, \quad \text{and} \quad \Delta\Gamma = \Gamma_1 - \Gamma_2 \tag{6.38}$$

since then both ΔM and $\Delta\Gamma$ are positive for kaons. Let us sketch three complementary scenarios.

- For the K meson system, we have (see Problem 6.10)

$$0 < \Delta M_K = M_L - M_S = -2\overline{M}^K_{12}$$
$$0 < \Delta\Gamma_K = \Gamma_S - \Gamma_L = 2\overline{\Gamma}^K_{12}. \tag{6.39}$$

The data, Eq. (5.45), and the fact that $\Delta\Gamma \simeq \Gamma_S$ due to $\Gamma_S \gg \Gamma_L$ lead to $r \sim 2$, where r is defined in Eq. (6.37). In this approximation,

$$\frac{q}{p} \simeq \sqrt{\frac{M^*_{12}}{M_{12}}} \left(1 - \frac{(1+i)\zeta_K}{2} \right) \tag{6.40}$$

and

$$\left| \frac{q}{p} \right| \simeq 1 - \frac{\zeta_K}{2}. \tag{6.41}$$

- We can have $\Delta\Gamma \ll \Delta M$ together with $\Delta\Gamma \ll \Gamma$; this is the situation predicted for the B_d complex, as explained later. We then have

$$\Delta M_B = M_2 - M_1 = -2\,\overline{M}^B_{12}$$
$$\Delta\Gamma = \Gamma_1 - \Gamma_2 = 2\,\overline{\Gamma}^B_{12} \cos\zeta_B \tag{6.42}$$

$$\frac{q}{p} \simeq \sqrt{\frac{M^*_{12}}{M_{12}}} \left(1 - \frac{r}{2} \sin\zeta_B \right), \quad \left| \frac{q}{p} \right| \simeq 1 - \frac{r}{2}\sin\zeta. \tag{6.43}$$

For later purposes, we record here

$$\frac{q}{p} \simeq sign\left(\overline{M}^B_{12} \right) e^{-i\xi} \left(1 - \frac{r}{2} \sin\zeta_B \right), \tag{6.44}$$

We find $|q/p| \simeq 1$ similar to the K^0 case, yet for a different reason, namely $r \ll 1$ rather than $\zeta \ll 1$. With $\Delta\Gamma/\Gamma$ too small to be observable and/or ζ sizeable, the mass eigenstates can be characterized by their masses only without any other empirical label.

- A slight variation arises if $\Delta\Gamma \ll \Gamma$ holds, yet $\Delta\Gamma/\Gamma$ can still be measured. While the expressions listed above apply, their consequences are different. $\Delta\Gamma$ is an observable now in *practice* rather than merely in principle. This situation might be realized for B_s mesons, to be discussed later.

The authors of Ref. [43] define

$$\tilde{\Delta}M \equiv M_H - M_L, \tag{6.45}$$

where the subscript H [L] stands for heavy [light]; $\tilde{\Delta}M$ is thus positive by definition. They also define

$$|P_L\rangle = p|P^0\rangle + q|\overline{P}^0\rangle$$
$$|P_H\rangle = p|P^0\rangle - q|\overline{P}^0\rangle \tag{6.46}$$

with

$$\frac{q}{p} = \pm\sqrt{\frac{M_{12}^* - \frac{i}{2}\Gamma_{12}^*}{M_{12} - \frac{i}{2}\Gamma_{12}}}. \tag{6.47}$$

If **CP** symmetry (approximately) holds we can sensibly ask if the **CP** even state is heavier or lighter than the **CP** odd state. If the calculation yields a negative value for the difference $M_2 - M_1$ – yet we want to keep the convention $\mathbf{CP}|P^0\rangle = |\overline{P}^0\rangle$ – we must choose the minus sign in Eq. (6.47) to remain consistent with Eq. (6.45). To say it differently: if we require $\tilde{\Delta}M > 0$ we have to compute the sign of $M_2 - M_1$ to decide on the sign for q/p in Eq. (6.47). Alternatively we can keep the plus sign if we adopt $\mathbf{CP}|P^0\rangle = -|\overline{P}^0\rangle$. In either case we have to calculate the sign of $M_2 - M_1$ *before* we can fix the convention for q/p or the phase of $|\overline{P}^0\rangle$. We feel it is more natural to define a convention independent of prior theoretical calculations.

6.6 Master equations of time evolution

A procedure similar to the one used in deriving Eq. (5.25) and Eq. (5.26) leads to expressions for the time evolution of states starting out as P^0 or $\overline{P}^0$:

$$|P^0(t)\rangle = f_+(t)|P^0\rangle + \frac{q}{p}f_-(t)|\overline{P}^0\rangle$$
$$|\overline{P}^0(t)\rangle = f_+(t)|\overline{P}^0\rangle + \frac{p}{q}f_-(t)|P^0\rangle, \tag{6.48}$$

with

$$f_\pm(t) = \frac{1}{2} e^{-iM_1 t} e^{-\frac{1}{2}\Gamma_1 t} \left[1 \pm e^{-i\Delta Mt} e^{\frac{1}{2}\Delta\Gamma t} \right]. \tag{6.49}$$

Denoting by $A(f)$ and $\overline{A}(f)$ the amplitude for the decay of P^0 and $\overline{P}^0$, respectively, into a final state f, and by $\overline{\rho}(f)$ and $\rho(f)$ their ratios, i.e.

$$A(f) = \langle f|H_{\Delta F=1}|P^0\rangle, \quad \overline{A}(f) = \langle f|H_{\Delta F=1}|\overline{P}^0\rangle,$$

$$\overline{\rho}(f) = \frac{\overline{A}(f)}{A(f)} = \frac{1}{\rho(f)}, \tag{6.50}$$

we write down

$$\Gamma(P^0(t) \to f) \propto e^{-\Gamma_1 t}|A(f)|^2 \left[K_+(t) + K_-(t) \left|\frac{q}{p}\right|^2 |\overline{\rho}(f)|^2 \right.$$

$$\left. + 2\mathrm{Re}\left[L^*(t) \left(\frac{q}{p}\right) \overline{\rho}(f) \right] \right] \tag{6.51}$$

$$\Gamma(\overline{P}^0(t) \to f) \propto e^{-\Gamma_1 t}|\overline{A}(f)|^2 \left[K_+(t) + K_-(t) \left|\frac{p}{q}\right|^2 |\rho(f)|^2 \right.$$

$$\left. + 2\mathrm{Re}\left[L^*(t) \left(\frac{p}{q}\right) \rho(f) \right] \right], \tag{6.52}$$

where

$$|f_\pm(t)|^2 = \frac{1}{4} e^{-\Gamma_1 t} K_\pm(t),$$

$$f_-(t)f_+^*(t) = \frac{1}{4} e^{-\Gamma_1 t} L^*(t),$$

$$K_\pm(t) = 1 + e^{\Delta\Gamma t} \pm 2e^{\frac{1}{2}\Delta\Gamma t}\cos\Delta Mt,$$

$$L^*(t) = 1 - e^{\Delta\Gamma t} + 2ie^{\frac{1}{2}\Delta\Gamma t}\sin\Delta Mt. \tag{6.53}$$

Particle–antiparticle oscillations thus provide for the presence of two different amplitudes contributing coherently, with their relative weight varying with the time of decay. This is illustrated in Fig. 6.1.

It is convenient to separate out the main exponential time dependence of the decay rate [44]:

$$\Gamma(B^0(t) \to f) \propto \frac{1}{2} e^{-\Gamma_1 t} \cdot G_f(t),$$

$$G_f(t) = a + be^{\Delta\Gamma_B t} + ce^{\Delta\Gamma_B t/2} \cos \Delta M_B t + de^{\Delta\Gamma_B t/2} \sin \Delta M_B t, \tag{6.54}$$

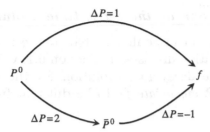

Figure 6.1 With $P - \overline{P}$ oscillation and non-vanishing amplitudes for both $P^0 \to f$ and $\overline{P}^0 \to f$ decays there are two decay chains $P^0 \to f$ and $P^0 \to \overline{P}^0 \to f$. This is another version of Fig. 6.2(c). Just as in the two-slit interference pattern observed in Young's experiment in optics, these two decay chains interfere according to the principle of quantum mechanics.

$$a = |A(f)|^2 \left[\frac{1}{2} \left(1 + \left| \frac{q}{p}\overline{\rho}(f) \right|^2 \right) + \mathrm{Re} \left(\frac{q}{p}\overline{\rho}(f) \right) \right], \qquad (6.55)$$

$$b = |A(f)|^2 \left[\frac{1}{2} \left(1 + \left| \frac{q}{p}\overline{\rho}(f) \right|^2 \right) - \mathrm{Re} \left(\frac{q}{p}\overline{\rho}(f) \right) \right], \qquad (6.56)$$

$$c = |A(f)|^2 \left[1 - \left| \frac{q}{p}\overline{\rho}(f) \right|^2 \right], \quad d = -2|A(f)|^2 \mathrm{Im} \left(\frac{q}{p}\overline{\rho}(f) \right); \qquad (6.57)$$

$$\Gamma(\overline{B}^0(t) \to \overline{f}) \propto \frac{1}{2} e^{-\Gamma_1 t} \cdot \overline{G}_{\overline{f}}(t),$$

$$\overline{G}_{\overline{f}}(t) = \overline{a} + \overline{b}e^{\Delta \Gamma_B t} + \overline{c}e^{\Delta \Gamma_B t/2} \cos \Delta M_B t + \overline{d}e^{\Delta \Gamma_B t/2} \sin \Delta M_B t, \quad (6.58)$$

$$\overline{a} = |\overline{A}(\overline{f})|^2 \left[\frac{1}{2} \left(1 + \left| \frac{p}{q}\rho(\overline{f}) \right|^2 \right) + \mathrm{Re} \left(\frac{p}{q}\rho(\overline{f}) \right) \right], \qquad (6.59)$$

$$\overline{b} = |\overline{A}(\overline{f})|^2 \left[\frac{1}{2} \left(1 + \left| \frac{p}{q}\rho(\overline{f}) \right|^2 \right) - \mathrm{Re} \left(\frac{p}{q}\rho(\overline{f}) \right) \right], \qquad (6.60)$$

$$\overline{c} = |\overline{A}(\overline{f})|^2 \left[1 - \left| \frac{p}{q}\rho(\overline{f}) \right|^2 \right], \quad \overline{d} = -2|\overline{A}(\overline{f})|^2 \mathrm{Im} \left(\frac{p}{q}\rho(\overline{f}) \right). \qquad (6.61)$$

Obviously **CP** invariance is violated if

$$G_f(t) \neq \overline{G}_{\overline{f}}(t). \qquad (6.62)$$

Theorem on the decay time evolution

The underlying physics can be illuminated by the following observation: within the approximation discussed in Section 6.1, **CP** violation is established by finding the decay rate evolution for the decay of a neutral meson P^0 into a **CP** *eigenstate* $f_\pm$ to be different from any *single pure* exponential; i.e.

$$\frac{d}{dt} e^{\Gamma t} \text{rate}(P^0 \to f_\pm; t) \neq 0 \quad \forall \text{ real } \Gamma \quad \Longrightarrow \quad \textbf{CP} \text{ violation.} \quad (6.63)$$

The proof is completely elementary. Assume **CP** to be conserved. Then the mass eigenstates of P^0 are **CP** eigenstates as well: $\textbf{CP}\left|P_{\frac{0}{1}}^0\right\rangle = \pm\left|P_{\frac{0}{1}}^0\right\rangle$; furthermore P_1^0 can decay into f_+, but P_2^0 cannot. Therefore

$$\text{rate}(P^0 \to f_+; t) = \frac{N_1}{N_1 + N_2} \cdot \text{rate}(P^0 \to f_+; t) = e^{-\Gamma_1 t} \cdot \text{const.}, \quad (6.64)$$

where $N_1 [N_2]$ denotes the original number of $P_1^0 [P_2^0]$ mesons in the beam; q.e.d. Such a deviation from a pure exponential time evolution in $K^0 \to \pi\pi$ has been observed, as shown in Fig. 7.2; the departure from the exponential at large t which demonstrates **CP** violation is caused by the $K_L \to \pi\pi$ amplitude.

This can be seen also by realizing that if **CP** is not violated, it implies $q/p = 1$, and $\overline{A}(f) = A(t)$ and all but a and $\bar{a}$ vanish in the master equations.

6.7 CP violation: classes (A), (B) and (C)

The decay rate evolution is in general rather complex, where the time dependence is described in terms of factors $\cos \Delta M t$, $\sin \Delta M t$, $e^{-\Gamma t}$, $e^{-\Gamma t} \cos \Delta M t$ and $e^{-\Gamma t} \sin \Delta M t$.

Staring at the most general expression is not always very illuminating. The physics can become clearer through examining simpler special cases; we will discuss three complementary categories.

(A) No (or no appreciable) oscillations occur, i.e.

$$\Delta M = \Delta \Gamma = 0. \quad (6.65)$$

Then we have

$$K_+(t) \equiv 4, \ K_-(t) \equiv L(t) \equiv 0. \quad (6.66)$$

This situation is depicted in Fig. 6.2(A). The time evolutions are then purely exponential in t; a **CP** asymmetry can still arise, if

$$|A(f)| \neq |\overline{A}(\bar{f})| \quad (6.67)$$

were to hold.

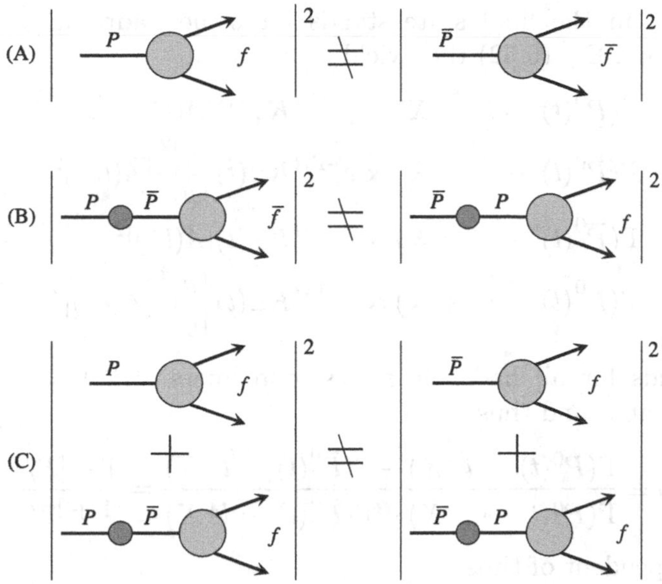

Figure 6.2 Three ways in which **CP** violation can show its face. (A) Without $P - \overline{P}$ mixing, **CP** asymmetry may arise if there is **CP** violation in the $P \to f$ and $\overline{P} \to \overline{f}$ decay amplitudes. This is called direct **CP** violation. (B) In the presence of $P - \overline{P}$ oscillations, flavour specific final states can reveal **CP** violation in M_{12}. (C) If we choose a **CP** eigenstate f, both $P \to f$ and $\overline{P} \to f$ decays occur. Then there are two possible decay chains for $P \to f$ and $\overline{P} \to f$ decays. These two decay chains interfere and this may result in a **CP** asymmetry.

(B) *Flavour-specific* decays are those that can come from either P^0 or $\overline{P}^0$, but not both:

$$P^0 \to f \not\leftarrow \overline{P}^0 \quad \text{or} \quad P^0 \not\to f \leftarrow \overline{P}^0. \tag{6.68}$$

This situation is depicted in Fig. 6.2(B). Prominent flavour-specific channels for neutral mesons like K^{neut}, D^{neut} or B^{neut} are provided by semileptonic decays. This is condensed into the following notation:

$$P^0 \to l^+ + X \not\leftarrow \overline{P}^0, \quad P^0 \not\to l^- + X \leftarrow \overline{P}^0, \tag{6.69}$$

i.e.

$$A(l^-) = \overline{A}(l^+) = 0. \tag{6.70}$$

Here the X in the final state stands for some hadronic state. Equation (6.51) and Eq. (6.52) then yield

$$\Gamma(P^0(t) \to l^+ + X) \propto e^{-\Gamma_1 t} K_+(t) |A(l^+)|^2$$

$$\Gamma(P^0(t) \to l^- + X) \propto e^{-\Gamma_1 t} K_-(t) \left|\frac{q}{p}\right|^2 |\overline{A}(l^-)|^2$$

$$\Gamma(\overline{P}^0(t) \to l^- + X) \propto e^{-\Gamma_1 t} K_+(t) |\overline{A}(l^-)|^2$$

$$\Gamma(\overline{P}^0(t) \to l^+ + X) \propto e^{-\Gamma_1 t} K_-(t) \left|\frac{p}{q}\right|^2 |A(l^+)|^2. \tag{6.71}$$

If X stands for all hadronic states, one infers $A(l^+) = \overline{A}(l^-)$ from **CPT** invariance and thus

$$A_{SL}(P^0) \equiv \frac{\Gamma(P^0(t) \to l^- X) - \Gamma(\overline{P}^0(t) \to l^+ X)}{\Gamma(P^0(t) \to l^- X) + \Gamma(\overline{P}^0(t) \to l^+ X)} = \frac{1 - |P/q|^4}{1 + |p/q|^4} \tag{6.72}$$

and is independent of time.

(C) *Flavour-non-specific* final states are those that are fed by both P^0 and $\overline{P}^0$ decays, although not necessarily with the same rate:

$$P^0 \to f \leftarrow \overline{P}^0. \tag{6.73}$$

This situation is depicted in Fig. 6.2(C). Examples are

$$K^0 \to \pi\pi \leftarrow \overline{K}^0$$
$$D^0 \to K\overline{K}, \, \pi\pi, \, K\pi, \, \overline{K}\pi \leftarrow \overline{D}^0$$
$$B^0 \to \psi K_S, \, D\overline{D}, \, \pi\pi \leftarrow \overline{B}^0. \tag{6.74}$$

CP eigenstates – **CP**$|f_\pm\rangle = \pm|f_\pm\rangle$ – fall into this category, but such final states are not necessarily **CP** eigenstates: for example, doubly Cabibbo suppressed transitions allow the channels $K^-\pi^+$ and $K^+\pi^-$ to be fed from D^0 as well as $\overline{D}^0$ decays; this subject will be addressed in more detail later on.

- **Case (C1)** $|A(f)| = |\overline{A}(f)|$
  When the final state is a **CP** eigenstate there is a possibility that this holds. Consider a case in which r defined in Eq. (6.43) is negligible. Then we have $|q/p| = 1$ and $|\overline{\rho}(f)| = 1$. The master equations Eq. (6.51) and Eq. (6.52) yield:

$$\Gamma(P^0(t) \to f) \propto 2e^{-\Gamma_1 t} |A(f)|^2$$
$$\times \left(1 + e^{\Delta\Gamma t} + \mathrm{Re}\left(\frac{q}{p}\overline{\rho}(f)\right)[1 - e^{\Delta\Gamma t}] - 2\mathrm{Im}\left(\frac{q}{p}\overline{\rho}(f)\right)e^{\frac{1}{2}\Delta\Gamma t}\sin\Delta M t\right) \tag{6.75}$$

$$\Gamma(\overline{P}^0(t) \to f) \propto 2e^{-\Gamma_1 t} |A(f)|^2$$
$$\times \left(1 + e^{\Delta\Gamma t} + \mathrm{Re}\left(\frac{q}{p}\overline{\rho}(f)\right)[1 - e^{\Delta\Gamma t}] + 2\mathrm{Im}\left(\frac{q}{p}\overline{\rho}(f)\right)e^{\frac{1}{2}\Delta\Gamma t}\sin\Delta M t \right)$$

(6.76)

and

$$\frac{\Gamma(P^0(t) \to f) - \Gamma(\overline{P}^0(t) \to f)}{\Gamma(P^0(t) \to f) + \Gamma(\overline{P}^0(t) \to f)} = \frac{-2\sin(\phi_{\Delta F=2} + \phi_{\Delta F=1})e^{\frac{1}{2}\Delta\Gamma t}\sin\Delta M t}{1 + e^{\Delta\Gamma t} + \cos(\phi_{\Delta F=2} + \phi_{\Delta F=1})[1 - e^{\Delta\Gamma t}]}$$

(6.77)

where we have written $q/p = e^{i\phi_{\Delta F=2}}$ and $\overline{\rho} = e^{i\phi_{\Delta F=1}}$, and used the fact that $\frac{q}{p}\overline{\rho}(f) = (\frac{p}{q}\rho(f))^*$. Thus we see that even if $|q/p| = 1$ and $|A(f)| = |\overline{A}(f)|$ hold, a **CP** asymmetry arises if two conditions are satisfied:
- $P^0 - \overline{P}^0$ oscillations occur generating $\Delta M \neq 0$;
- $\arg(q/p) + \arg(\overline{\rho}(f)) \neq 0$.
- **Case (C2)** $|A(f)| \neq |\overline{A}(f)|$

$$G_f(t) = |A(f)|^2 \left[1 + |\overline{\rho}(f)|^2 + \left(1 - |\overline{\rho}(f)|^2\right)\cos\Delta M t \right.$$
$$\left. - 2\mathrm{Im}\left(\frac{q}{p}\overline{\rho}(f)\right)\sin\Delta M t \right],$$

(6.78)

$$\overline{G}_f(t) = |\overline{A}(f)|^2 \left[1 + |\rho(f)|^2 + \left(1 - |\rho(f)|^2\right)\cos\Delta M t \right.$$
$$\left. - 2\mathrm{Im}\left(\frac{p}{q}\rho(f)\right)\sin\Delta M t \right].$$

(6.79)

The **CP** asymmetry can then be expressed as follows:

$$\frac{G_f(t) - \overline{G}_f(t)}{G_f(t) + \overline{G}_f(t)} = C_f \cos\Delta M t - S_f \sin\Delta M t$$

(6.80)

$$C_f = \frac{1 - |\overline{\rho}(f)|^2}{1 + |\overline{\rho}(f)|^2}, \quad S_f = \frac{2\mathrm{Im}((q/p)\overline{\rho}(f))}{1 + |\overline{\rho}(f)|^2}.$$

(6.81)

Equation (6.81) shows explicitly that the observable **CP** asymmetry in this case can have two sources, which can be clearly separated by the time dependence, namely the coefficients of the cos and sin terms. The former clearly requires $|\bar{\rho}(f)|^2 \neq 1$, i.e. *direct* **CP** violation residing in $\mathcal{H}(\Delta B = 1)$.

A useful consistency check for the data is provided by

$$C_f^2 + S_f^2 \leq 1 \tag{6.82}$$

as shown in Problem 6.8.

6.8 ♠ On the sign of the CP asymmetry ♠

As discussed in Section 6.5 the sign of ΔM is convention dependent and thus not an observable, unless $\Delta\Gamma$ can be measured. One might then think that outside the kaon complex the observable asymmetry $\sin\Delta Mt \, \mathrm{Im}\left(\frac{q}{p}\bar{\rho}(f)\right)$ contains an ambiguity concerning its sign. For the term $\sin\Delta Mt = \sin(M_2 - M_1)t$ flips its sign under exchanging $1 \leftrightarrow 2$, and the subscripts 1 and 2 are mere labels as long as we *cannot* distinguish P_1 and P_2 through their lifetimes. However, further reflection shows that the exchange $1 \leftrightarrow 2$ flips the sign of ΔM as well as that of $\frac{q}{p}\bar{\rho}(f)$. The observable $\sin\Delta Mt \, \mathrm{Im}\frac{q}{p}\bar{\rho}(f)$ thus remains *unaffected* by such a switch in labels. This can be seen in the following manner.

(1) Changing $q/p \to -q/p$ maintains the defining property $(q/p)^2 = (M_{12}^* - \frac{i}{2}\Gamma_{12}^*)/(M_{12} - \frac{i}{2}\Gamma_{12})$, see Eq. (6.24).
(2) Yet the two mass eigenstates labelled by subscripts 1 and 2 exchange places, see Eq. (6.22).
(3) The mass difference $\Delta M \equiv M_2 - M_1 = -2\mathrm{Re}[\frac{q}{p}(M_{12} - \frac{i}{2}\Gamma_{12})]$ then flips its sign – yet $\sin\Delta Mt\mathrm{Im}\frac{q}{p}\bar{\rho}(f)$ remains invariant!
(4) Alternatively, $|\overline{P}^0\rangle \to -|\overline{P}^0\rangle$ leads to the switch $1 \leftrightarrow 2$, see Eq. (6.19). Thus $\Delta M \to -\Delta M$, $\bar{\rho}(f) \to -\bar{\rho}(f)$ again without affecting $\sin\Delta Mt\mathrm{Im}\frac{q}{p}\bar{\rho}(f)$!

Hence we infer that *within a given theory of $\Delta F = 1, 2$ dynamics* we can calculate the product $\sin\Delta Mt \, \mathrm{Im}\left(\frac{q}{p}\bar{\rho}(f)\right)$ *including its sign*. Later this will be demonstrated explicitly. Here we want to make only two additional statements.

- **CP**$|f_\pm\rangle = \pm|f_\pm\rangle$. We then have

$$\langle f_\pm|H|\overline{P}^0\rangle = \langle f_\pm|(\mathbf{CP})^\dagger\mathbf{CP}H(\mathbf{CP})^\dagger\mathbf{CP}|\overline{P}^0\rangle$$
$$= \pm\langle f_\pm|H^{CP}|P^0\rangle \tag{6.83}$$

and therefore

$$\bar{\rho}(f_\pm) = \pm \frac{\langle f_\pm | H^{CP} | P^0 \rangle}{\langle f_\pm | H | P^0 \rangle} \tag{6.84}$$

i.e. the sign of the asymmetry depends on the **CP** quantum number of the final state.

- Consider a decay $B \rightarrow f_\pm$. It is important to distinguish this decay from $B \rightarrow f_\pm \pi^0$, which is often experimentally non-trivial. Because $\mathbf{CP} |\pi^0\rangle = -|\pi^0\rangle$, certain kinematic configurations will lead to $\bar{\rho}(f_\pm \pi^0) = -\bar{\rho}(f_\pm)$. This will have an effect of reducing the asymmetry.

6.9 What happens if you don't observe the decay time?

For the K meson system it is relatively easy to observe the time dependence of the instantaneous decay rate. Yet the tracks of short-lived particles such as B mesons, which live for only about 1.5 ps, are much more difficult to measure since $c\tau_B \sim 0.5$ mm. As explained later, so-called micro-vertex detectors have been developed that allow us to track such decays; yet typical detectors effectively integrate over all times of decay. Therefore we give here time-integrated versions of our predictions. We can express time-integrated versions of Eq. (6.51)–(6.52) and Eq. (6.75)–(6.76) in terms of

$$\langle K_\pm \rangle \equiv \int_0^\infty dt \, e^{-\Gamma_1 t} K_\pm(t)$$
$$= \frac{2}{\Gamma} \left[\frac{1}{1-y^2} \pm \frac{1}{1+x^2} \right],$$
$$\langle L^* \rangle \equiv \int_0^\infty dt \, e^{-\Gamma_1 t} L^*(t)$$
$$= \frac{2}{\Gamma} \left[-\frac{y}{1-y^2} + i \frac{x}{1+x^2} \right], \tag{6.85}$$

where

$$x = \frac{\Delta M}{\Gamma}, \quad y = \frac{\Delta \Gamma}{2\Gamma}, \quad \text{and} \quad \Gamma = \frac{1}{2}(\Gamma_1 + \Gamma_2) \tag{6.86}$$

denote the oscillation rate calibrated by the decay rate. We record here the time-integrated version of Eq. (6.71), Eq. (6.75) and Eq. (6.76), the latter two for $y = 0$:

$$\int_0^\infty \Gamma(P^0(t) \rightarrow l^+ + X) dt \propto \frac{2 + x^2 - y^2}{(1-y^2)(1+x^2)} |A(l^+)|^2,$$

$$\int_0^\infty \Gamma(P^0(t) \to l^- + X)dt \propto \frac{x^2 + y^2}{(1 - y^2)(1 + x^2)} \left|\frac{q}{p}\right|^2 |\overline{A}(l^-)|^2,$$

$$\int_0^\infty \Gamma(\overline{P}^0(t) \to l^- + X)dt \propto \frac{2 + x^2 - y^2}{(1 - y^2)(1 + x^2)} |\overline{A}(l^-)|^2,$$

$$\int_0^\infty \Gamma(\overline{P}^0(t) \to l^+ + X)dt \propto \frac{x^2 + y^2}{(1 - y^2)(1 + x^2)} \left|\frac{p}{q}\right|^2 |A(l^+)|^2. \quad (6.87)$$

If $|\overline{\rho}| = |\frac{q}{p}| = 1$ and $y = 0$,

$$\int_0^\infty \Gamma(P^0(t) \to f)dt \propto 4|A(f)|^2 \cdot \left(1 - \frac{x}{1 + x^2}\mathrm{Im}\frac{q}{p}\overline{\rho}(f)\right),$$

$$\int_0^\infty \Gamma(\overline{P}^0(t) \to f)dt \propto 4|A(f)|^2 \cdot \left(1 + \frac{x}{1 + x^2}\mathrm{Im}\frac{q}{p}\rho(f)\right). \quad (6.88)$$

The observable **CP** asymmetry thus depends primarily on the ratio x rather than on ΔM. This is as expected: for with the **CP** asymmetry being due to the interference between the decay and the oscillation amplitude, it is the *relative* weight of the latter that counts. On the other hand, it remains highly desirable to resolve the decay vertices and thus track the time evolution of these decays. Later we will explain why this is even absolutely essential in certain cases like $\Upsilon(4S) \to B^0\overline{B}^0$.

6.10 Regeneration

Let us consider how a beam made up of P states evolves in time when passing through nuclear matter along the z axis; z and t are related by $z = \gamma\beta t$, $\gamma = 1/\sqrt{1 - \beta^2}$ and β denotes the velocity of P. We start with defining $\Psi(t)$ similarly to Eq. (5.12) with K replaced by P. The behaviour of P^0 and $\overline{P}^0$ passing through a medium can be described in terms of indices of refraction n and $\overline{n}$ defined by Ref. [42],

$$\frac{d|\Psi(t)\rangle}{dz} = i\,k\,\begin{pmatrix} n - 1 & 0 \\ 0 & \overline{n} - 1 \end{pmatrix} |\Psi(t)\rangle, \quad (6.89)$$

where k is the momentum of the particle. The evolution equations then read as follows:

$$\frac{d|\Psi(t)\rangle}{dt} = -\left[(\tfrac{1}{2}\Gamma + i\,\mathbf{M}) - ik\beta\gamma\begin{pmatrix} n - 1 & 0 \\ 0 & \overline{n} - 1 \end{pmatrix}\right] |\Psi(t)\rangle. \quad (6.90)$$

Refraction effectively adds a term $-k\beta\gamma(n-1)$ to M_{11} and $-k\beta\gamma(\overline{n}-1)$ to M_{22}, while keeping the off-diagonal matrix element unchanged:

$$\frac{d|\Psi(t)\rangle}{dt} = -\left[(\tfrac{1}{2}\Gamma + i\,\tilde{\mathbf{M}})\right] |\Psi(t)\rangle \quad (6.91)$$

$$\tilde{\mathbf{M}} = \begin{pmatrix} M_{11} - k\beta\gamma(n-1) & M_{12} \\ M_{12}^* & M_{22} - k\beta\gamma(\bar{n}-1) \end{pmatrix}. \qquad (6.92)$$

We can employ the same procedure as developed in the preceding chapter to decouple the differential equations. The eigenvectors and eigenvalues of $\tilde{\mathbf{M}}$ will differ from those of $\mathbf{M}$. Yet there is one important *qualitative* distinction: since

$$\tilde{M}_{11} \neq \tilde{M}_{22} \qquad (6.93)$$

we are dealing with a situation where **CPT** invariance is *effectively* broken: the fact that P^0 and $\overline{P}^0$ propagate through matter rather than antimatter mimics **CPT** violation. This observation recalls a theme we have sounded before in analysing time reversal invariance (and to which we will return): *an observed difference in two conjugate transition rates can be caused by an asymmetry in the prevailing boundary conditions rather than a difference in the fundamental dynamics.*

Using spherical coordinates as introduced in Section 6.3, we have

$$\tilde{E}_1 = E_1, \quad \tilde{E}_2 = E_2,$$
$$\tilde{E}_3 = -\frac{1}{2}k\beta\gamma(n-\bar{n}), \quad \tilde{E} = \sqrt{E_1^2 + E_2^2 + \tilde{E}_3^2}, \qquad (6.94)$$

and therefore

$$\cos\tilde{\theta} = \frac{\tilde{E}_3}{\tilde{E}} \neq 0, \qquad (6.95)$$

representing the effective **CPT** asymmetry alluded to above. We then find for a state that started out as pure $|P_1\rangle$ at time $t = 0$, at later times

$$|P_1(t)\rangle = (\tilde{f}_+(t) + \sin\tilde{\theta}\tilde{f}_-(t))|P_1\rangle + \cos\tilde{\theta}\tilde{f}_-(t)|P_2\rangle, \qquad (6.96)$$

where the tilde over $\tilde{f}_\pm(t)$ reminds us that these functions are determined by the eigenvalues of the mass matrix $\tilde{\mathbf{M}}$ given in Eq. (6.92). Likewise when the initial state is pure $|P_2\rangle$:

$$|P_2(t)\rangle = (\tilde{f}_+(t) - \sin\tilde{\theta}\tilde{f}_-(t))|P_2\rangle + \cos\tilde{\theta}\tilde{f}_-(t)|P_1\rangle. \qquad (6.97)$$

This means that due to $\cos\tilde{\theta} \neq 0$ a P_2 [P_1] is regenerated from an initially pure P_1 [P_2] beam when traversing matter.

This is also of practical interest for actual tests of **CPT** invariance. Real experiments are undertaken not in a perfect vacuum, but in an environment that is dominated by matter; in particular the detector is *not* **CPT** invariant! This places some inherent limitations on **CPT** tests.

6.11 The Bell–Steinberger inequality

With P_1, P_2 denoting the two mass eigenstates of the $P^0 - \overline{P}^0$ complex we have

$$\langle P_1 | \tfrac{1}{2}\mathbf{\Gamma} + i\mathbf{M} | P_2 \rangle = (\tfrac{1}{2}\Gamma_2 + iM_2)\langle P_1 | P_2 \rangle,$$
$$\langle P_2 | \tfrac{1}{2}\mathbf{\Gamma} + i\mathbf{M} | P_1 \rangle = (\tfrac{1}{2}\Gamma_1 + iM_1)\langle P_2 | P_1 \rangle. \tag{6.98}$$

Taking the complex conjugate of the first equation and adding it to the second one yields

$$\langle P_2 | \mathbf{\Gamma} | P_1 \rangle = \left[\frac{1}{2}(\Gamma_1 + \Gamma_2) + i(M_1 - M_2) \right] \langle P_2 | P_1 \rangle. \tag{6.99}$$

Recalling the definition Eq. (6.8), we have

$$\langle P_2 | \Gamma | P_1 \rangle = 2\pi \sum_f \delta(M_P - M_f)\langle f | H | P_2 \rangle^* \langle f | H | P_1 \rangle, \tag{6.100}$$

and using the Schwartz inequality,

$$|\langle P_2 | \Gamma | P_1 \rangle|^2 \leq \sum_f \Gamma_1^f \Gamma_2^f, \tag{6.101}$$

we finally arrive at [45]

$$|\langle P_2 | P_1 \rangle| \leq \sqrt{\frac{\sum_f 4\Gamma_1^f \Gamma_2^f}{(\Gamma_1 + \Gamma_2)^2 + 4(M_1 - M_2)^2}}. \tag{6.102}$$

Equation (6.102) is called the Bell–Steinberger relation or inequality, which was first derived for neutral kaons. In the preceding section we have learnt that the two mass eigenstates are no longer orthogonal to each other if **CP** (or **CPT**) invariance is violated. The inequality in Eq. (6.102) tells us that the amount of that non-orthogonality is constrained by a sum over exclusive P_1 and P_2 widths. We see that this inequality is numerically quite relevant for kaons.

$$|\langle K_L | K_S \rangle| \leq \sqrt{\frac{\sum_f \Gamma_L^f \Gamma_S^f}{(\Gamma_L + \Gamma_S)^2 + 4(\Delta M_K)^2}} \simeq \sqrt{\frac{\sum_f 2\Gamma_L^f \Gamma_S^f}{\Gamma_S^2}}, \tag{6.103}$$

where again we have used $\Gamma_L \ll \Gamma_S \simeq 2\Delta M_K$. Since $\Gamma_L \Gamma_S = \sum_{f_L, f_S} \Gamma_L^{f_L} \Gamma_S^{f_S} \geq \sum_f \Gamma_L^f \Gamma_S^f$, we obtain, using the experimental result Eq. (5.22), a very conservative bound

$$|\langle K_L | K_S \rangle| \leq \sqrt{\frac{2\Gamma_L}{\Gamma_S}} \simeq 0.06. \tag{6.104}$$

Since **CP** invariance requires $\langle K_L | K_S \rangle = 0$, the following message is contained in Eq. (6.104): unitarity as expressed through the Bell–Steinberger relation together with the experimental findings $\Gamma_L \ll \Gamma_S \sim 2\Delta M_K$ already imposes a near-orthogonality of K_L and K_S, irrespective of **CP** violation!

6.12 Résumé on $P^0 - \overline{P}^0$ oscillations

Now we are in a position to summarize our discussion on $P^0 - \overline{P}^0$ oscillations. There are four classes of **CP** asymmetries.

(1)

$$|A(f)| \neq |\overline{A}(\overline{f})|, \tag{6.105}$$

or for f being **CP** self-conjugate: $|\overline{\rho}_f| \neq 1$. This condition is obviously independent of the phase convention for $\overline{P}^0$. Such an asymmetry unambiguously reflects $\Delta F = 1$ dynamics and can occur also in the decays of charged mesons, baryons and leptons. It is referred to as *direct* **CP** *violation*.

(2)

$$\left| \frac{q}{p} \right| \neq 1 . \tag{6.106}$$

This can be detected experimentally by measuring the lepton asymmetry, see Eq. (6.72). This is unambiguously driven by the dynamics in the $\Delta F = 2$ sector. It thus represents **CP** violation in $P^0 - \overline{P}^0$ oscillations and is often referred to as *superweak* **CP** *violation*, as explained later.

(3)

$$\text{Im} \left[\frac{q}{p} \overline{\rho}_f \right] = \left| \frac{q}{p} \overline{\rho}_f \right| \sin(\arg(q/p) + \arg\overline{\rho}(f)) \neq 0 . \tag{6.107}$$

It reflects the combined effect of $\Delta F = 2$ and $\Delta F = 1$ dynamics and can be referred to as *CP violation involving* $P^0 - \overline{P}^0$ *oscillations*. As long as such an asymmetry has been studied for a *single* final-state f only (or in a single pair of **CP** conjugate states f and $\overline{f}$), it is meaningless to differentiate between the effects of $\Delta F = 2$ and $\Delta F = 1$ forces, i.e. between *superweak* and *direct* **CP** violation. For a change in the phase convention will shift the weight between $\arg(q/p)$ and $\arg\overline{\rho}(f)$. Changing

$$\text{CP}|P^0\rangle \equiv |\overline{P}^0\rangle \tag{6.108}$$

to the equivalent definition

$$\mathbf{CP}|P^0\rangle \equiv e^{i\xi}|\overline{P}^0\rangle, \quad \xi \text{ real}, \tag{6.109}$$

will obviously have no effect on $|\overline{\rho}_f|$ or $|q/p|$. Yet q/p and $\overline{\rho}_f$ are affected by it: on one hand we have

$$\overline{\rho}(f) = \frac{\overline{A}(f)}{A(f)} = \frac{\langle f|H_{\Delta F=1}|\overline{P}^0\rangle}{\langle f|H_{\Delta F=1}|P^0\rangle} \to e^{2i\xi}\overline{\rho}(f), \tag{6.110}$$

whereas on the other hand we find

$$(M_{12}, \Gamma_{12}) \to e^{2i\xi}(M_{12}, \Gamma_{12}), \tag{6.111}$$

leading to

$$\frac{q}{p} = \sqrt{\frac{M_{12}^* - \frac{i}{2}\Gamma_{12}^*}{M_{12} - \frac{i}{2}\Gamma_{12}}} \to e^{-i2\xi}\sqrt{\frac{M_{12}^* - \frac{i}{2}\Gamma_{12}^*}{M_{12} - \frac{i}{2}\Gamma_{12}}} = e^{-i2\xi}\frac{q}{p}. \tag{6.112}$$

Yet their product remains invariant:

$$\frac{q}{p}\overline{\rho}(f) \to \frac{q}{p}\overline{\rho}(f), \tag{6.113}$$

i.e. the sum $\arg(q/p) + \arg\overline{\rho}(f)$ – in contrast to its individual terms – is *not* sensitive to changes in the phase convention for the antiparticle and thus qualifies as an observable – as is clear also from the explicit calculation leading to Eq. (6.77).

(4) Once we have found a **CP** asymmetry in two different final states f_1 and f_2 (or in two different conjugate pairs), the issue of superweak vs direct **CP** violation can be addressed in a meaningful way. For if we observe

$$\sin(\arg(q/p) + \arg\overline{\rho}(f_1)) \neq \sin(\arg(q/p) + \arg\overline{\rho}(f_2)), \tag{6.114}$$

then we know unambiguously that direct **CP** violation is present.

These observations suggest the following more meaningful characterization: *if we can choose the phase for the antiparticle in such a way that all* **CP** *asymmetries for this particle–antiparticle complex can be assigned to* $\phi_{\Delta F=2}$, *then we are dealing with a superweak scenario.*

Problems

6.1. Derive $M_{12} = M_{21}$ from **CP** and **T** symmetry.

6.2. Considering the normalization of the state, show that

$$\Gamma_{11} = \frac{1}{2M} \int \prod_{i=1}^{n} \frac{d^3 k_i}{(2\pi)^3 2E_i} (2\pi)^4 \delta^4 (P - k_1 - \cdots - k_n)$$
$$|\langle \vec{k}_1 \ldots k_n | \mathcal{H}(0) | P \rangle|^2 \qquad (6.115)$$

as expected for the width of P decay into n particles.

6.3. In quantum mechanics we have learnt that two eigenstates which correspond to different eigenvalues are orthogonal. Explain Eq. (6.30).

6.4. With $|P_{1,2}\rangle \sim |P^0\rangle \pm e^{i\phi} |\overline{P}^0\rangle$, derive Eq. (6.96) and Eq. (6.97).

6.5. Explain how the experimental fact $3M_\pi \simeq M_K$ makes the Bell–Steinberger bound very powerful.

6.6. For $\Delta\Gamma \ll \Delta M$, show that

$$f_\pm(t) = \begin{pmatrix} \cos \frac{\Delta Mt}{2} \\ i \sin \frac{\Delta Mt}{2} \end{pmatrix} e^{-\frac{1}{2}\Gamma_1 t} e^{-i\frac{1}{2}(M_1 + M_2)t}. \qquad (6.116)$$

$$\Gamma(P^0(t) \to l^+ + X) \propto e^{-\Gamma_1 t} \cos^2 \frac{\Delta Mt}{2} |A(l^+)|^2$$

$$\Gamma(P^0(t) \to l^- + X) \propto e^{-\Gamma_1 t} \sin^2 \frac{\Delta Mt}{2} \left|\frac{q}{p}\right|^2 |\overline{A}(l^-)|^2$$

$$\Gamma(\overline{P}^0(t) \to l^- + X) \propto e^{-\Gamma_1 t} \cos^2 \frac{\Delta Mt}{2} |\overline{A}(l^-)|^2$$

$$\Gamma(\overline{P}^0(t) \to l^+ + X) \propto e^{-\Gamma_1 t} \sin^2 \frac{\Delta Mt}{2} \left|\frac{p}{q}\right|^2 |A(l^+)|^2. \qquad (6.117)$$

6.7. Show that for $\Delta\Gamma \ll \Delta M$, Eq. (6.52), Eq. (6.51) and Eq. (6.77) simplify to

$$\Gamma(P^0 \to f) \sim e^{-\Gamma t} |A(f)|^2 \left[1 - \text{Im}\left(\frac{q}{p}\rho(f)\right) \sin \Delta Mt\right]$$

$$\Gamma(\overline{P}^0 \to f)| \sim e^{-\Gamma t} |A(f)|^2 \left[1 + \text{Im}\left(\frac{q}{p}\rho(f)\right) \sin \Delta Mt\right] \qquad (6.118)$$

$$\frac{\Gamma(P^0(t) \to f) - \Gamma(\overline{P}^0(t) \to f)}{\Gamma(P^0(t) \to f) + \Gamma(\overline{P}^0(t) \to f)} = -\sin(\phi_{\Delta F=2} + \phi_{\Delta F=1}) \sin\Delta Mt. \qquad (6.119)$$

6.8. Show that

$$C_f^2 + S_f^2 \leq 1 \tag{6.120}$$

6.9. There is a special case of considerable interest for neutral kaons: waiting long enough for the K_S component to decay away and thus preparing a K_L beam, we track its transition rate into a $\pi\pi$ final state. Show that

$$\begin{pmatrix} \Gamma(K^0 \to \pi\pi) \\ \Gamma(\overline{K}^0 \to \pi\pi) \end{pmatrix} \propto e^{-\Gamma_S t} + e^{-\Gamma_L t}|\eta|^2 \pm 2e^{-\frac{1}{2}(\Gamma_L + \Gamma_S)t}\mathrm{Re}\left[e^{-i\Delta M_K t}\eta\right], \tag{6.121}$$

where the third term describes the $K_L - K_S$ interference.

6.10. Diagonalize the Hamiltonian

$$\begin{pmatrix} M & \Delta e^{i\phi} \\ \Delta e^{-i\phi} & M \end{pmatrix} \tag{6.122}$$

where Δ is a positive or negative real number and ϕ is the phase of the off-diagonal element, which is defined to be between 0 and π. Show that the mass eigenstate is $(1, \ \pm e^{-i\phi})$ with eigenvalue $M \pm \Delta$.

Part II
Theory and experiments

7

The quest for CP violation in K decays – a marathon

The discovery of **CP** violation raised many more questions than it gave answers. Some of the experimental questions were: Is **CP** violation in $K_L \rightarrow \pi^0 \pi^0$ decay different from that in $K_L \rightarrow \pi^+ \pi^-$ decay? Is there a **CP** asymmetry in semileptonic K_L decays? How do we detect pure **T** violation? What about other **CP** and **T** violating observables like spin correlations $\langle (\vec{p}_1 \times \vec{p}_2) \cdot \vec{s} \rangle$?

This field of physics has been a vibrant one producing fascinating results for the past 40 years – like in an Olympic marathon: there was much at stake; truly major discoveries have been made; some have been rewarded; others were left by the way side. In the preceding chapter we have given a general presentation of various **CP** and **T** asymmetries. Here we will sketch the experimental landscape and present the relevant phenomenology in non-leptonic as well as semileptonic modes. The truly theoretical treatment will be given in Chapter 9.

7.1 The landscape

Where has CP violation emerged?

The burning question after the discovery of **CP** violation was: how does it enter? Does it do so exclusively via the $\Delta S = 2$ amplitude M_{12} driving the $K - \overline{K}$ transition? Or is there a phase in the $\Delta S = 1$ decay amplitude $A(K \rightarrow \pi\pi)$? If it is just in M_{12}, it should not distinguish between the two observables:

$$\eta_{+-} \equiv \frac{\langle \pi^+ \pi^- |H_W|K_L \rangle}{\langle \pi^+ \pi^- |H_W|K_S \rangle}, \quad \eta_{00} \equiv \frac{\langle \pi^0 \pi^0 |H_W|K_L \rangle}{\langle \pi^0 \pi^0 |H_W|K_S \rangle}. \tag{7.1}$$

Direct **CP** violation on the other hand can manifest itself through a difference in the normalized decay amplitudes for $K_L \to \pi^+\pi^-$ vs $\pi^0\pi^0$: $|\eta_{+-}| \neq |\eta_{00}|$. The conventional notation

$$\eta_{+-} \equiv \epsilon + \epsilon', \quad \eta_{00} \equiv \epsilon - 2\epsilon' \tag{7.2}$$

makes this explicit: ϵ' describes the *channel dependent* **CP** violation whereas ϵ characterizes **CP** violating phase in $K - \overline{K}$ oscillations. Direct **CP** violation is expressed through the ratio:

$$\mathrm{Re}\frac{\epsilon'}{\epsilon} = \frac{1}{6}\frac{|\eta_{+-}|^2 - |\eta_{00}|^2}{|\eta_{+-}|^2}. \tag{7.3}$$

The 2006 PDG average [11] reads:

$$|\epsilon| = (2.232 \pm 0.007) \times 10^{-3}. \tag{7.4}$$

As we now know, the technology did not exist to make a meaningful measurement of ϵ'/ϵ in the 1960s and for many years after. Yet many of our friends tried. In those times these searches were referred to as 'Nothing in and nothing out experiments'. For in the $K_L \to \pi^0\pi^0$ decay an experimenter cannot see the K_L going into the detector. The π^0 can only be inferred by detecting photons. Some of our friends, unfortunately, quit physics in frustration.

Even 29 years after the discovery of $\epsilon \neq 0$, the experimental verdict for $\epsilon' \neq 0$ and thus for direct **CP** violation was ambiguous:

$$\mathrm{Re}\frac{\epsilon'}{\epsilon} = \begin{cases} (23 \pm 6.5) \times 10^{-4} & \text{NA31 [46]} \\ (7.4 \pm 5.9) \times 10^{-4} & \text{E731 [47]} \end{cases}; \tag{7.5}$$

It took another 6 years and a new generation of experiments to establish the result on both sides of the Atlantic:

$$\mathrm{Re}\frac{\epsilon'}{\epsilon} = \begin{cases} (18.5 \pm 4.5 \pm 5.8) \times 10^{-4} & \text{NA48 [48]} \\ (28.0 \pm 3.0 \pm 2.8) \times 10^{-4} & \text{KTeV [49]} \end{cases}. \tag{7.6}$$

The tale shows the subtlety of statistics and the power of bias. Had E731's central number been three times larger, which is statistically possible, the community might have embraced the numbers from NA31 and E731 as strong evidence for the existence of direct **CP** violation. On the other hand most theoretical predictions on the 'market' suggested numbers not exceeding 10^{-3}, which created a bias against NA31's findings. The 2006 world average yields [11][1]

$$\mathrm{Re}\frac{\epsilon'}{\epsilon} = (16.6 \pm 2.6) \times 10^{-4}; \tag{7.7}$$

---

[1] KTeV's final number has been presented in early 2008: $\mathrm{Re}\,\epsilon'/\epsilon = (19.2 \pm 2.1) \times 10^{-4}$.

i.e. direct **CP** violation has been finally established! A new paradigm had thus emerged after more than 30 years of dedicated experimentation: the **CP** phenomenology of K_L decays has to be described in terms of *two* numbers rather than one: ϵ and ϵ'!

Quoting the result as in Eq. (7.7) does not do justice to the experimental achievement, since ϵ_K is a very small number itself. The sensitivity achieved becomes more transparent, when quoted in terms of the underlying partial widths:

$$\frac{\Gamma(K^0 \to \pi^+\pi^-) - \Gamma(\bar{K}^0 \to \pi^+\pi^-)}{\Gamma(K^0 \to \pi^+\pi^-) - \Gamma(\bar{K}^0 \to \pi^+\pi^-)} = (5.16 \pm 0.71) \cdot 10^{-6}. \tag{7.8}$$

Establishing $\epsilon' \neq 0$ is a discovery of the very first rank. Its significance does not depend on whether the SM can reproduce it or not – the most concise confirmation of how important it is. The HEP community can take pride in this achievement. The two experimental collaborations and their predecessors deserve the community's respect; they have certainly earned our admiration.

Chasing after K^0 *and* $\overline{K}^0$

When analysing K_L decays, 'patience pays off'; i.e. because of $\Gamma(K_S) \gg \Gamma(K_L)$ we can wait till practically all K_S mesons have decayed away. The task we face in comparing $K^0 \to \overline{K}^0$ with $\overline{K}^0 \to K^0$ is quite different: the decay products reveal the flavour identity of the *final* state kaon:

$$K^0 \Rightarrow \overline{K}^0 \to l^-\nu\pi^+ \quad \text{vs} \quad \overline{K}^0 \Rightarrow K^0 \to l^+\nu\pi^-. \tag{7.9}$$

We then need independent information on the flavour identity of the *initial* kaon. This is achieved through correlations imposed by associated production. The CPLEAR collaboration studied low energy proton–antiproton annihilation:

$$P\overline{P} \to K^+\overline{K}^0\pi^- \quad \text{vs} \quad P\overline{P} \to K^-K^0\pi^+. \tag{7.10}$$

It is the sign of the charged kaon that reveals the flavour identity of the neutral kaon produced in association with it. The ϕ factory DAΦNE allows the study of coherent $K^0 \overline{K}^0$ pairs

$$e^+e^- \to \phi(1020) \to K^0\overline{K}^0. \tag{7.11}$$

The first observation of quantum interference in the process $\phi \to K_S K_L \to \pi^+\pi^-\pi^+\pi^-$ has been announced by the KLOE collaboration at DAΦNE [50].

The time evolution of the decay rate

The CPLEAR Collaboration [51] measured the difference in the decay rate evolution of $K^0 \to \pi^+\pi^-$ and $\overline{K}^0 \to \pi^+\pi^-$ as a function of (proper) time of decay:

$$A_{+-}(t) = \frac{\Gamma(K^0(t) \to \pi^+\pi^-) - \Gamma(\overline{K}^0(t) \to \pi^+\pi^-)}{\Gamma(K^0(t) \to \pi^+\pi^-) + \Gamma(\overline{K}^0(t) \to \pi^+\pi^-)}, \qquad (7.12)$$

shown in Fig. 7.1. We show this figure because it is quite analogous to what we search for and find in B^0 decays. The experimental challenge for measuring it in B^0 decays is provided by the much smaller branching ratio $Br(B \to \Psi K_S) \sim 10^{-4}$ and the much shorter lifetime of $\tau_B \sim 1.5\,\mathrm{ps}$ – meaning there is no significant statistics beyond $\tau \sim 4\tau_B \sim 6\,\mathrm{ps}$. Yet there is a redeeming feature as well: the asymmetry in the B system can be larger by a factor of 100 (or even more) than the 0.2% for the K meson system.

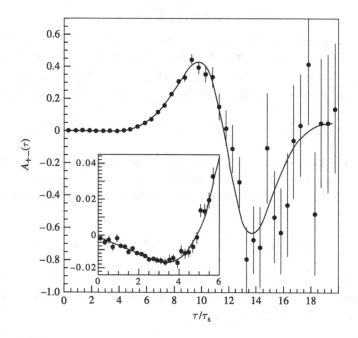

Figure 7.1   Time dependence of $A_{+-}(t)$ in units of K_S lifetime. Later we shall stress that the time-dependent asymmetry between $B \to \Psi K_S$ and $\overline{B} \to \Psi K_S$ is the gold-plated mode for **CP** violation searches in B decays. This figure represents the equivalent asymmetry for K meson decays. This figure was reproduced from *Physics Letters* by permission of Elsevier Science.

Direct test of T invariance

Can we also infer that **T** is violated – without assuming **CPT** symmetry? Yes, we can! The Kabir test [52] has revealed a time asymmetry for the first time, summarized in the final results from the CPLEAR collaboration [53]:

$$A_T = \frac{\text{rate}(\overline{K}^0 \to K^0) - \text{rate}(K^0 \to \overline{K}^0)}{\text{rate}(\overline{K}^0 \to K^0) + \text{rate}(K^0 \to \overline{K}^0)} = (6.6 \pm 1.3 \pm 1.0) \times 10^{-3}. \quad (7.13)$$

A sceptic might point out that in tagging the initial flavour of the neutral kaon, one has assumed that the strong interactions producing the kaons conserve strangeness exactly. Whether that feature or **CPT** invariance is a more conservative assumption, is open for debate.

Difference in integrated rates

CP *asymmetries* were subsequently found in semileptonic K_L decays [11]:

$$A_l \equiv \frac{\Gamma(K_L \to l^+ \nu_l \pi^-) - \Gamma(K_L \to l^- \overline{\nu}_l \pi^+)}{\Gamma(K_L \to l^+ \nu_l \pi^-) + \Gamma(K_L \to l^- \overline{\nu}_l \pi^+)} = (3.32 \pm 0.06) \times 10^{-3}. \quad (7.14)$$

Later we will explain why this asymmetry has to be half of that in Eq. (7.13).

7.2 $K_L \to \pi\pi$ decays

After this survey of the experimental landscape we turn to the phenomenological treatment.

7.2.1 Decay amplitudes

The two pion states in K decays can be classified in terms of their isospin I. Due to Bose statistics two S wave pions in $K_{L,S} \to \pi\pi$ can form an $I = 0$ or 2, yet not a $I = 1$ configuration. As explained in Chapter 4, Watson's theorem tells us that the amplitude for a K^0 decaying into two pions with total isospin I can be expressed as follows:

$$\langle (2\pi)_I | H_W | K^0 \rangle = \mathcal{A}_I e^{i\delta_I}, \quad (7.15)$$

where δ_I denotes the S wave $(\pi\pi)_I$ phase shift at energy M_K and $\mathcal{A}_I$ is real if H_W conserves **CP**; i.e. in that case the phase of the decay amplitude is determined completely by the strong final state interactions. Likewise we find for the **CP** conjugate reactions

$$\langle (2\pi)_I | H_W | \overline{K}^0 \rangle = \overline{\mathcal{A}}_I e^{i\delta_I}. \quad (7.16)$$

The same phase δ_I appears in both Eq. (7.15) and Eq. (7.16) since the strong forces obey **CP** and **T** invariance.

With **CP** violation arising in the weak dynamics an additional phase will emerge in $\mathcal{A}_I$ and $\overline{\mathcal{A}}_I$; it will be referred to as *weak* phase.

Throughout this section we will assume **CPT** invariance until explicitly stated otherwise. Thus we have

$$\langle (2\pi)_I | H_W | K_L \rangle = e^{i\delta_I} \left[p\mathcal{A}_I - q\overline{\mathcal{A}}_I \right]$$

$$\langle (2\pi)_I | H_W | K_S \rangle = e^{i\delta_I} \left[p\mathcal{A}_I + q\overline{\mathcal{A}}_I \right]. \tag{7.17}$$

We now want to write the decay amplitudes in terms of isospin amplitudes:

$$A(K^0 \rightarrow \pi^+\pi^-) = \frac{1}{\sqrt{3}} (\mathcal{A}_2 e^{i\delta_2} + \sqrt{2}\mathcal{A}_0 e^{i\delta_0})$$

$$A(K^0 \rightarrow \pi^0\pi^0) = \sqrt{\frac{2}{3}} (\sqrt{2}\mathcal{A}_2 e^{i\delta_2} - \mathcal{A}_0 e^{i\delta_0})$$

$$A(K^+ \rightarrow \pi^+\pi^0) = \sqrt{\frac{3}{2}} \mathcal{A}_2 e^{i\delta_2}, \tag{7.18}$$

where the phases are obtained using Watson's theorem.

Expressing the state $\pi^+\pi^-$ in terms of $(2\pi)_{I=0,2}$, we have

$$\langle \pi^+\pi^- | H_W | K_L \rangle = \sqrt{\frac{2}{3}} \langle (2\pi)_0 | H_W | K_L \rangle + \frac{1}{\sqrt{3}} \langle (2\pi)_2 | H_W | K_L \rangle$$

$$= \sqrt{\frac{2}{3}} e^{i\delta_0} p\mathcal{A}_0 \left[\left(1 - \frac{q}{p}\frac{\overline{\mathcal{A}}_0}{\mathcal{A}_0} \right) + \frac{1}{\sqrt{2}} e^{i(\delta_2 - \delta_0)} \frac{p\mathcal{A}_2 - q\overline{\mathcal{A}}_2}{p\mathcal{A}_0} \right]$$

$$= \sqrt{\frac{2}{3}} e^{i\delta_0} p\mathcal{A}_0 \left[\Delta_0 + \frac{1}{\sqrt{2}} e^{i(\delta_2 - \delta_0)} \omega \Delta_2 \right], \tag{7.19}$$

where

$$\Delta_I = 1 - \frac{q}{p}\frac{\overline{\mathcal{A}}_I}{\mathcal{A}_I}, \quad \omega = \frac{\mathcal{A}_2}{\mathcal{A}_0}.$$

Similarly

$$\langle \pi^0\pi^0 | H_W | K_L \rangle = -\frac{1}{\sqrt{3}} \langle (2\pi)_0 | H_W | K_L \rangle + \sqrt{\frac{2}{3}} \langle (2\pi)_2 | H_W | K_L \rangle$$

$$= \sqrt{\frac{2}{3}} e^{i\delta_0} p\mathcal{A}_0 \left[-\left(1 - \frac{q}{p}\frac{\overline{\mathcal{A}}_0}{\mathcal{A}_0} \right) + \sqrt{2} e^{i(\delta_2 - \delta_0)} \frac{p\mathcal{A}_2 - q\overline{\mathcal{A}}_2}{p\mathcal{A}_0} \right]$$

$$= -\sqrt{\frac{2}{3}} e^{i\delta_0} p\mathcal{A}_0 \left[\Delta_0 - \sqrt{2} e^{i(\delta_2 - \delta_0)} \omega \Delta_2 \right]. \tag{7.20}$$

Since we know already from Eq. (7.4) that $|\eta_{+-}|$ and $|\eta_{00}|$ are tiny quantities, it suffices to treat these amplitude ratios merely to first order in **CP** violation. Therefore we need to retain the **CP** conserving parts in the $K_S \rightarrow \pi\pi$ amplitude only. Remembering that $p = q = 1/\sqrt{2}$ if we ignore **CP** violation,

$$
\begin{aligned}
\langle \pi^+\pi^- | H_W | K_S \rangle &= \sqrt{\frac{2}{3}} \langle (2\pi)_0 | H_W | K_S \rangle + \frac{1}{\sqrt{3}} \langle (2\pi)_2 | H_W | K_S \rangle \\
&= \frac{2}{\sqrt{3}} e^{i\delta_0} \mathcal{A}_0 \left(1 + \frac{1}{\sqrt{2}} \omega e^{i(\delta_2 - \delta_0)} \right) \\
\langle \pi^0\pi^0 | H_W | K_S \rangle &= -\sqrt{\frac{2}{3}} \langle (2\pi)_0 | H_W | K_S \rangle + \frac{2}{\sqrt{3}} \langle (2\pi)_2 | H_W | K_S \rangle \\
&= -\frac{2}{\sqrt{3}} e^{i\delta_0} \mathcal{A}_0 \left(1 - \sqrt{2} \omega e^{i(\delta_2 - \delta_0)} \right).
\end{aligned}
\tag{7.21}
$$

Since $|\omega| \simeq 1/20$, we ignore terms of $\mathcal{O}(\omega^2)$ and find

$$
\begin{aligned}
\eta_{+-} &= \frac{1}{2} \left(\Delta_0 - \frac{1}{\sqrt{2}} \omega e^{i(\delta_2 - \delta_0)} (\Delta_0 - \Delta_2) \right) \\
\eta_{00} &= \frac{1}{2} \left(\Delta_0 + \sqrt{2} \omega e^{i(\delta_2 - \delta_0)} (\Delta_0 - \Delta_2) \right),
\end{aligned}
\tag{7.22}
$$

or

$$
\begin{aligned}
\epsilon &= \frac{1}{2} \Delta_0 = \frac{1}{2} \left(1 - \frac{q}{p} \frac{\overline{\mathcal{A}}_0}{\mathcal{A}_0} \right) \\
\epsilon' &= -\frac{1}{2\sqrt{2}} \omega e^{i(\delta_2 - \delta_0)} (\Delta_0 - \Delta_2) = \frac{1}{2\sqrt{2}} \omega e^{i(\delta_2 - \delta_0)} \frac{q}{p} \left(\frac{\overline{\mathcal{A}}_0}{\mathcal{A}_0} - \frac{\overline{\mathcal{A}}_2}{\mathcal{A}_2} \right).
\end{aligned}
\tag{7.23}
$$

A few comments can illuminate the content of these relations.

- ϵ and ϵ' are written in terms of $\frac{q}{p} \frac{\overline{\mathcal{A}}_I}{\mathcal{A}_I}$, which has been shown to be independent of phase conventions, see Eq. (6.113). Thus also ϵ and ϵ' are phase convention independent – as they should be since they are experimentally measurable quantities.
- In the literature you will often come across the Wu–Yang phase convention, where the phase of K_0 is chosen so that A_0 is real. Note that our formalism allows us to discuss **CP** violation without choosing a phase convention. This is because our ϵ and ϵ' are independent of them.

- The quantity ϵ defined by the ratios of **CP** violating and conserving decay amplitudes, see Eq. (7.2), should *not* be confused with the **CP** impurity parameter $\bar{\epsilon}$ in the K_L state:

$$|K_L\rangle = \frac{1}{\sqrt{2(1+|\bar{\epsilon}|^2)}}[(1+\bar{\epsilon})|K^0\rangle - (1-\bar{\epsilon})|\overline{K}^0\rangle]. \qquad (7.24)$$

Within the Wu–Yang phase convention, and only in this phase convention,

$$\epsilon = \bar{\epsilon}. \qquad (7.25)$$

- The expression for ϵ' makes explicit that ϵ' vanishes unless

$$\frac{\overline{A}_0}{A_0} \neq \frac{\overline{A}_2}{A_2}; \qquad (7.26)$$

i.e. ϵ' represents differences in the **CP** asymmetries for different decay channels thus implying that $H_W(\Delta S = 1)$ violates **CP**.

7.2.2   *Constraints on* A_I *and* $\overline{A}_I$

In the derivation of Watson's theorem we have used that the strong interactions conserve **T** and **CP**. Now we analyse the constraints that discrete symmetries of the weak Hamiltonian H_W place on A_I and $\overline{A}_I$.

Invoking **CPT** invariance, we deduce

$$
\begin{aligned}
A_I e^{i\delta_I} &= \langle (2\pi)_I; \text{out}|H_W|K^0\rangle \\
&= \langle (2\pi)_I; \text{out}|(\mathbf{CPT})^\dagger \mathbf{CPT}\, H_W (\mathbf{CPT})^{-1}\mathbf{CPT}\,|K^0\rangle^* \\
&= \sum_n \langle (2\pi)_I; \text{in}|n; \text{out}\rangle^* \langle n; \text{out}|H_W|\overline{K}^0\rangle^* \\
&= \overline{A}_I^* e^{i\delta_I}. \qquad (7.27)
\end{aligned}
$$

In Eq. (7.27), we have used the fact that $\langle (2\pi)_I; \text{out}|(2\pi)_I; \text{in}\rangle = S((\pi\pi)_I)$, and in the low energy region, $S((\pi\pi)_I) = e^{2i\delta_I}$.

If, on the other hand, H_W obeys **CP** symmetry – $[H_W, \mathbf{CP}] = 0$ – we conclude:

$$
\begin{aligned}
A_I e^{i\delta_I} &= \langle (2\pi)_I; \text{out}|H_W|K^0\rangle \\
&= \langle (2\pi)_I; \text{out}|(\mathbf{CP})^{-1}\mathbf{CP}\, H_W (\mathbf{CP})^{-1}\mathbf{CP}\,|K^0\rangle \\
&= \langle (2\pi)_I; \text{out}|H_W|\overline{K}^0\rangle = \overline{A}_I e^{i\delta_I}, \qquad (7.28)
\end{aligned}
$$

using our usual phase convention $\mathbf{CP}\,|K^0\rangle = |\overline{K}^0\rangle$. Finally, **T** invariance implies

$$A_I e^{i\delta_I} = \langle (2\pi)_I; \text{out}|\mathbf{T}^\dagger \mathbf{T}\, H_W \mathbf{T}^{-1}\mathbf{T}\,|K^0\rangle^*$$

$$= \langle(2\pi)_I; \text{in}|(2\pi)_I; \text{out}\rangle^* \langle(2\pi)_I; \text{out}|H_W|K^0\rangle^*$$
$$= \mathcal{A}_I^* e^{i\delta_I}, \tag{7.29}$$

i.e. $\mathcal{A}_I$ has to be real; likewise for $\overline{\mathcal{A}}_I$.

In summary:[2]

$$[\mathbf{CPT}, H_W] = 0 \implies \mathcal{A}_I = \overline{\mathcal{A}}_I^*$$

$$[\mathbf{CP}, H_W] = 0 \implies \mathcal{A}_I = \overline{\mathcal{A}}_I$$

$$[\mathbf{T}, H_W] = 0 \implies \mathcal{A}_I = \mathcal{A}_I^*. \tag{7.30}$$

7.2.3 Relating ϵ to $M - \frac{i}{2}\Gamma$

With $q/p = e^{i\phi}$, ϕ complex, and **CPT** invariance implying $\overline{\mathcal{A}}_I = \mathcal{A}_I^*$, as just shown, we have

$$\Delta_I \equiv 1 - \frac{q}{p}\frac{\overline{\mathcal{A}}_I}{\mathcal{A}_I} = 1 - \exp[i(\phi - 2\arg\mathcal{A}_I)] \simeq -i\phi + 2i\arg\mathcal{A}_I, \tag{7.31}$$

since **CP** violation is small. Then we find, see Chapter 6 for notation and details:

$$\phi \simeq \frac{E_2}{E_1} = \frac{-\mathrm{Im}M_{12} + \frac{i}{2}\mathrm{Im}\Gamma_{12}}{\mathrm{Re}M_{12} - \frac{i}{2}\mathrm{Re}\Gamma_{12}}. \tag{7.32}$$

Denoting

$$\xi_I = \arg\mathcal{A}_I, \tag{7.33}$$

we obtain

$$\epsilon \simeq i\left[\frac{1}{2}\frac{\mathrm{Im}M_{12} - \frac{i}{2}\mathrm{Im}\Gamma_{12}}{\mathrm{Re}M_{12} - \frac{i}{2}\mathrm{Re}\Gamma_{12}} + \xi_0\right]$$

$$\epsilon' \simeq \frac{-i}{\sqrt{2}}e^{i(\delta_2-\delta_0)}\omega(\xi_0 - \xi_2). \tag{7.34}$$

The expression for ϵ can be simplified further, for the difference in the two eigenvalues of $\mathbf{M} - \frac{i}{2}\mathbf{\Gamma}$ can be well approximated by

$$M_S - M_L - \frac{i}{2}(\Gamma_S - \Gamma_L) = 2E \simeq 2E_1, \tag{7.35}$$

---

[2] We assume here that strong dynamics obey all these symmetries.

since $E_1 = E\cos\phi \simeq E$ for small ϕ. Finally we arrive at

$$\epsilon \simeq -i \left(\frac{\mathrm{Im}M_{12} - \frac{i}{2}\mathrm{Im}\Gamma_{12}}{\Delta M_K + \frac{i}{2}\Delta\Gamma_K} - \xi_0 \right). \tag{7.36}$$

We have to remember that Eq. (7.36) was derived using approximations that are quite specific for the $K - \overline{K}$ system and do *not* hold for a general $P - \overline{P}$ complex.

7.2.4   The phase of ϵ

Again using approximations that apply specifically to K^{neut} decays we can determine the phase of ϵ. For that purpose we consider the states which can contribute to the sum in Γ_{12} defined in Eq. (6.8). These intermediate states are

$$(2\pi)_0, \ (2\pi)_2, \ \pi^+\pi^-\pi^0, \ 3\pi^0, \ \pi l\nu, \dots \tag{7.37}$$

As we will show in the next section, to within 10% accuracy we can ignore all but the $(2\pi)_0$ state

$$\Gamma_{12} \simeq 2\pi\rho_{2\pi}\mathcal{A}_0^*\overline{\mathcal{A}}_0, \tag{7.38}$$

where $\rho_{2\pi}$ is the phase space factor.

With $\overline{\mathcal{A}}_0 = \mathcal{A}_0^* = e^{-i\xi_0}|\mathcal{A}_0|$ imposed by **CPT** invariance, see Eq. (7.30), we then have

$$\Gamma_{12} \simeq 2\pi\rho_{2\pi}e^{-2i\xi_0}|\mathcal{A}_0|^2 \times \text{ phase space.} \tag{7.39}$$

Retaining **CP** violation to first order only, we can set

$$\langle 2\pi|H_W|K^0\rangle = \frac{1}{\sqrt{2}}\langle 2\pi|H_W|K_S\rangle \tag{7.40}$$

and obtain

$$\Gamma_{12} \simeq \frac{1}{2}e^{-2i\xi_0}\Gamma_S \simeq \frac{1}{2}e^{-2i\xi_0}\Delta\Gamma_K. \tag{7.41}$$

Inserting this expression into Eq. (7.36) leads to

$$\epsilon \simeq \frac{1}{\sqrt{1 + (\frac{\Delta\Gamma}{2\Delta M})^2}} e^{i\phi_{sw}} \left(-\frac{\mathrm{Im}M_{12}}{\Delta M_K} + \xi_0 \right), \tag{7.42}$$

where

$$\phi_{SW} = \tan^{-1}\frac{2\Delta M_K}{\Delta\Gamma_K}. \tag{7.43}$$

Using the experimental numbers given in Eq. (5.42) we arrive at [11]

$$\phi_{\rm SW} = \tan^{-1}\frac{2\Delta M_K}{\Delta \Gamma_K} = (43.51 \pm 0.05)^\circ. \tag{7.44}$$

This phase can be measured in a cleverly conceived experiment first performed by Bohm *et al.* at CERN [55]. A proton beam hits a (nuclear) target producing a neutral kaon that subsequently decays into two pions:

$$P + A \rightarrow K^{\rm neut} + X \rightarrow \pi\pi + X. \tag{7.45}$$

The important point to note is that we do not identify $K^{\rm neut}$ as a K^0 or $\overline{K}^0$ through an observation of X. As seen in Eq. (6.121), the decay rate for an initial K^0 or $\overline{K}^0$ state is expressed by

$$N(K^0(t) \rightarrow \pi\pi) = N_0 \left(e^{-\Gamma_S} + e^{-\Gamma_L}|\eta|^2 + 2\cos(\Delta M_K t - \phi_\eta)e^{-\frac{1}{2}(\Gamma_L + \Gamma_S)}|\eta|\right)$$

$$N(\overline{K}^0(t) \rightarrow \pi\pi) = \overline{N}_0 \left(e^{-\Gamma_S} + e^{-\Gamma_L}|\eta|^2 - 2\cos(\Delta M_K t - \phi_\eta)e^{-\frac{1}{2}(\Gamma_L + \Gamma_S)}|\eta|\right), \tag{7.46}$$

where ϕ_η is the phase of η and N_0 and $\overline{N}_0$ denote the initial number of K^0 and $\overline{K}^0$, respectively; we have ignored here terms of order $|\eta|^2$. The number of observed decays is given by (Fig. 7.2)

$$N(p + A \rightarrow [\pi\pi]_{K^{\rm neut}} + X) \sim$$

$$e^{-\Gamma_S} + \frac{2(N^0 - \overline{N}^0)}{N^0 + \overline{N}^0}\cos(\Delta M_K t - \phi_\eta)e^{-\frac{1}{2}(\Gamma_S + \Gamma_L)t}|\eta| + e^{-\Gamma_L t}|\eta|^2. \tag{7.47}$$

From the *time dependence* of these rates we can extract ϕ_η, the phase of η, *irrespective* of N_0 and $\overline{N}_0$, the numbers of K^0 and $\overline{K}^0$ (as long as $N_0 \neq \overline{N}_0$), respectively, produced in the proton collisions. We should note here that the *sign* of $\Delta M t - \phi_\eta$ cannot be measured, and the measurement of the phase is up to $mod(\pi)$ in this early experiment. Recent analysis assuming **CPT** symmetry yields for the phases of η_{+-} and η_{00} [11]:

$$\phi_{+-} = (43.52 \pm 0.05)^\circ$$
$$\phi_{00} = (43.50 \pm 0.06)^\circ. \tag{7.48}$$

7.3 Semileptonic decays

The dynamics are less complex in semileptonic K_L decays. Within the quark model semileptonic kaon decays obey, as discussed in Section 5.5, the $\Delta S = \Delta Q$ rule, which makes semileptonic modes flavour specific;

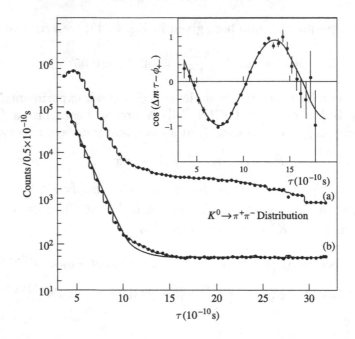

Figure 7.2 The Bohm method was used by Geweniger *et al.* [54]. They measured the $K \to \pi\pi$ rate as a function of time in a reaction $p + A \to K+$ anything. The inset shows the interference term where the exponential factors have been taken out. Note that ϕ_η can be extracted totally independently of the normalization. This figure was reproduced from *Physics Letters* by permission of Elsevier Science.

i.e. transitions yielding 'wrong sign' leptons have to proceed via $K - \overline{K}$ oscillations. This is completely consistent with the observation so far; any future deviation would reveal the intervention of New Physics.

Let us define the following general amplitudes [56]:

$$\langle l^+ \nu \pi^- | \mathcal{H}_W | K^0 \rangle = F_l(1 - y_l)$$
$$\langle l^+ \nu \pi^- | \mathcal{H}_W | \overline{K}^0 \rangle = x_l F_l(1 - y_l)$$
$$\langle l^- \overline{\nu} \pi^+ | \mathcal{H}_W | K^0 \rangle = \overline{x}_l^* F_l^* (1 + y_l^*)$$
$$\langle l^- \overline{\nu} \pi^+ | \mathcal{H}_W | \overline{K}^0 \rangle = F_l^* (1 + y_l^*). \tag{7.49}$$

Discrete symmetries place the following restrictions on these parameters:

$\Delta S = \Delta Q$ rule:	$x_l = \overline{x}_l = 0$
CP invariance:	$x_l = \overline{x}_l^*$; $\quad F_l = F_l^*$; $\quad y_l = -y_l^*$
T invariance:	Im $F = $ Im $y_l = $ Im $x_l = $ Im $\overline{x}_l = 0$
CPT invariance:	$y_l = 0$, $x_l = \overline{x}_l$.

We postpone a discussion of experimental tests of the $\Delta S = \Delta Q$ rule and **CPT** symmetry to Chapter 13. *Till then we assume* **CPT** *symmetry and the* $\Delta S = \Delta Q$ *rule to hold.* We then have

$$\langle l^+ \nu_l \pi^- | \mathcal{H}_W | K_L \rangle = p \langle l^+ \nu_l \pi^- | \mathcal{H}_W | K^0 \rangle$$
$$\langle l^- \bar{\nu}_l \pi^+ | \mathcal{H}_W | K_L \rangle = -q \langle l^- \bar{\nu}_l \pi^+ | \mathcal{H}_W | \overline{K}^0 \rangle, \qquad (7.50)$$

with **CPT** invariance implying

$$\langle l^+ \nu_l \pi^- | \mathcal{H}_W | K^0 \rangle = \langle l^- \nu_l \pi^+ | \mathcal{H}_W | \overline{K}^0 \rangle^*. \qquad (7.51)$$

The asymmetry δ_l defined in Eq. (7.14) is driven by $\Delta S = 2$ dynamics and can be expressed by

$$\delta_l = \frac{|p|^2 - |q|^2}{|p|^2 + |q|^2} \simeq \frac{1}{2} \zeta_K = (3.32 \pm 0.06) \times 10^{-3}, \qquad (7.52)$$

where Eq. (6.41) was used.

As for the Kabir test, the flavour identity of the final state kaon is deduced from its semileptonic decay; its analysis therefore fits in here. Using Eq. (6.48), the probability for $K^0(t)$ to be detected as $\overline{K}^0$, and the probability for $\overline{K}^0(t)$ to be detected as K^0 are given by

$$\text{rate}(K^0 \to \overline{K}^0) = \int_0^\infty \left| \frac{q}{p} f_-(t) \right|^2 dt$$
$$\text{rate}(\overline{K}^0 \to K^0) = \int_0^\infty \left| \frac{p}{q} f_-(t) \right|^2 dt, \qquad (7.53)$$

respectively, so that A_T defined in Eq. (7.13) is given by

$$A_T = \frac{|p/q|^2 - |q/p|^2}{|p/q|^2 + |q/p|^2} = \frac{1 - |q/p|^4}{1 + |q/p|^4} \simeq \zeta_K . \qquad (7.54)$$

Thus we get $A_T = 2\delta_l$, making the results of Eq. (7.13) and Eq. (7.14), quite consistent.

7.4 ♠ $P_\perp$ in K → πμν decays ♠

Another interesting and highly topical example for **CP** violation affecting final state distributions is provided by $K \to \pi\mu\nu$ modes where we measure the polarization of the muon. In the kaon rest frame there are three independent vectors, namely the pion and muon momenta – $\vec{p}_\pi$ and $\vec{p}_\mu$, respectively – and the muon spin vector $\vec{\sigma}_\mu$. We can thus define four non-trivial correlations, namely the scalar $\vec{p}_\pi \cdot \vec{p}_\mu$ and the three pseu-doscalars $\vec{\sigma}_\mu \cdot \vec{p}_\mu$, $\vec{\sigma}_\mu \cdot \vec{p}_\pi$ and $\vec{\sigma}_\mu \cdot (\vec{p}_\mu \times \vec{p}_\pi)$. The correlations $\vec{\sigma}_\mu \cdot \vec{p}_\mu$ and $\vec{\sigma}_\mu \cdot \vec{p}_\pi$ violate parity.

The term $\vec{\sigma}_\mu \cdot (\vec{p}_\mu \times \vec{p}_\pi)$, which counts the net polarization transverse to the decay plane, constitutes also a **T**-odd correlation, i.e. it changes sign under time reversal. Observation of a **P**-odd correlation *unequivocally* establishes parity violation, yet with **T**-odd correlations the situation is not so straightforward, as explained in Section 4.10.3. The physical origin of this difference is that a space-inverted sequence of a decay can readily be implemented, yet a time-inverted one in practice cannot.

Final state interactions mimicking T *violation*

In Section 4.10.3, we discussed how the final state interaction can generate transverse polarization even when the dynamics conserve **T**. The effects of final state interactions have been computed:

$$P_\perp \equiv \left\langle \frac{\vec{s}_\mu \cdot (\vec{p}_\mu \times \vec{p}_\pi)}{|\vec{p}_\pi \times \vec{p}_\mu|} \right\rangle \simeq 10^{-3} \quad \text{for} \quad K^0 \to \pi^- \mu^+ \nu \quad [57]$$

$$\simeq 10^{-6} \quad \text{for} \quad K^+ \to \pi^0 \mu^+ \nu, \quad [58] \quad (7.55)$$

where we have assumed that the underlying dynamics obey time reversal invariance; i.e. the stated non-vanishing values for $P_\perp$ are generated purely by final state interactions. They are of quite different nature in the two transitions: in $K^0 \to \pi^- \mu^+ \nu$ it is mainly Coulomb exchange between the two charged particles in the final state and – not surprisingly – the effect is of order α/π; in $K^+ \to \pi^0 \mu^+ \nu$, on the other hand, the final state interactions are driven by weak forces, which allows for a tiny effect only; for all practical purposes an observation of $P_\perp \neq 0$ in $K^+ \to \pi^0 \mu^+ \nu$ is an unambiguous sign for **T** violation. A recent measurement yields [59]:[3]

$$P_\perp(K^+ \to \pi^0 \mu^+ \nu) = (-1.7 \pm 2.3 \pm 1.1) \times 10^{-3}. \quad (7.56)$$

Phenomenological analysis

The formalism for describing the transverse muon polarization in $K \to \pi \mu \bar{\nu}$ decays has been studied long ago [58, 60, 61] within the context of $V - A$ theory. With the KM ansatz yielding, as discussed later, only an unobservably small value for $P_\perp$, we are interested in looking for effects

---

[3] **CP** invariance unequivocally predicts $P_\perp(K^+ \to \pi^0 \mu^+ \nu) = -P_\perp(K^- \to \pi^0 \mu^- \bar{\nu})$ and $P_\perp(K^0 \to \pi^- \mu^+ \nu) = -P_\perp(\overline{K}^0 \to \pi^+ \mu^- \bar{\nu})$. Yet these relations do not provide practical tests since precise measurements of the polarization of negatively charged muons are not feasible.

beyond the SM. Therefore we start from the general expression for the transition amplitude

$$A(K^- \to \pi^0 \mu^- \bar{\nu}) = \bar{\mu}(p_\mu; \vec{s}_\mu)[F_S + iF_P\gamma_5 + F_V \not{p}_K + F_A \not{p}_K \gamma_5]\nu(p_\nu) \quad (7.57)$$

with $p_{K,\pi,\mu,\nu}$ denoting K, π, μ, ν momenta, respectively, and $\vec{s}_\mu$ the muon spin. We then obtain for the differential decay width [62]

$$d\Gamma = d\Gamma_0 + (s \cdot p_\nu)\, d\Gamma_\nu + (s \cdot p_K)\, d\Gamma_K + \epsilon_{\alpha\beta\gamma\delta} p_K^\alpha s^\beta p_\mu^\gamma p_\nu^\delta \, d\Gamma_\perp, \quad (7.58)$$

where

$$\begin{aligned}
d\Gamma_0 &\propto \frac{1}{2}(p_\mu \cdot p_\nu)[|F_S|^2 + |F_P|^2 - M_K^2(|F_V|^2 + |F_A|^2)] \\
&\quad + (p_\mu \cdot p_K)(p_\nu \cdot p_K)(|F_V|^2 + |F_A|^2) \\
&\quad + m_\mu(p_\nu \cdot p_K)[\mathrm{Re}\,(F_S F_V^*) - \mathrm{Im}\,(F_P F_A^*)] \\
d\Gamma_\nu &\propto -m_\mu[\mathrm{Im}\,(F_S F_P^*) + M_K^2 \mathrm{Re}\,(F_V F_A^*)] \\
&\quad + (p_\mu \cdot p_K)[\mathrm{Im}\,(F_P F_V^*) - \mathrm{Re}\,(F_S F_A^*)]
\end{aligned}$$

$$\begin{aligned}
d\Gamma_K &\propto 2m_\mu(p_\nu \cdot p_K)\mathrm{Re}\,(F_V F_A^*) \\
&\quad + (p_\nu \cdot p_\mu)[\mathrm{Re}\,(F_S F_A^*) - \mathrm{Im}\,(F_P F_V^*)] \\
d\Gamma_\perp &\propto \mathrm{Re}\,(F_P F_A^*) + \mathrm{Im}\,(F_S F_V^*). \quad (7.59)
\end{aligned}$$

These quantities are most conveniently evaluated in the kaon rest frame. The muon polarization s_μ is related to its vector $\vec{\sigma}$ in the muon rest frame as follows:

$$s_0 = \frac{\vec{\sigma} \cdot \vec{p}_\mu}{m_\mu}, \quad \vec{s}_\parallel = \frac{E_\mu}{m_\mu}\vec{n}_\mu(\vec{\sigma} \cdot \vec{n}_\mu) \quad \vec{s}_\perp = \vec{\sigma} - (\vec{\sigma} \cdot \vec{n}_\mu)\vec{n}_\mu \quad (7.60)$$

where $\vec{n}_\mu$ is a unit vector in the direction of $\vec{p}_\mu$. In terms of these vectors,

$$\begin{aligned}
(s \cdot p_K) &= \frac{\vec{\sigma} \cdot \vec{p}_\mu}{m_\mu} M_K \\
(s \cdot p_\nu) &= \vec{\sigma} \cdot \vec{p}_\mu \left(\frac{M_K - E_\pi}{m_\mu} + \frac{\vec{p}_\mu \cdot \vec{p}_\pi}{m_\mu(E + m_\mu)} \right) + \vec{\sigma} \cdot \vec{p}_\pi \\
\epsilon_{\alpha\beta\gamma\delta} p_K^\alpha s^\beta p_\mu^\gamma p_\nu^\delta &= M_K \vec{\sigma} \cdot (\vec{p}_\pi \times \vec{p}_\mu). \quad (7.61)
\end{aligned}$$

It is now trivial to derive the muon polarization in a general direction. In a $V - A$ theory the transition amplitude is given by:

$$A(K^- \to \pi^0 \mu^- \bar{\nu}) = \frac{G_F}{\sqrt{2}} \langle \pi; p_\pi | \bar{u}\gamma_\alpha(1 - \gamma_5)s | K; p_K \rangle \bar{\mu}(p_\mu; \vec{s}_\mu)\gamma^\alpha(1 - \gamma_5)\nu(p_\nu).$$
$$(7.62)$$

Since no axial vector can be constructed from the two available vectors p_π and p_K, only the vector part contributes through two form factors:

$$\langle \pi; p_\pi | \bar{u}\gamma_\alpha(1 - \gamma_5)s | K; p_K \rangle = f_+(q^2)(p_K + p_\pi)_\alpha + f_-(q^2)(p_K - p_\pi)_\alpha \qquad (7.63)$$

Inserting Eq. (7.63) into Eq. (7.62) we obtain

$$A(K^- \to \pi\mu^-(\vec{s})\nu) \simeq \frac{G_F}{\sqrt{2}}[f_+(q^2)\bar{\mu}(p_\mu; \vec{s})(1 + \gamma_5)(2k)\nu(p_\nu)$$
$$+ (f_-(q^2) - f_+(q^2))m_\mu\bar{\mu}(p_\mu; \vec{s})(1 - \gamma_5)\nu(p_\nu)]; \qquad (7.64)$$

i.e. in the general notation of Eq. (7.57):

$$F_S = f_+(\xi - 1)m_\mu, \quad F_P = if_+(\xi - 1)m_\mu, \quad F_V = 2f_+, \quad F_A = -2f_+, \qquad (7.65)$$

where we have defined

$$\xi = \frac{f_-}{f_+}. \qquad (7.66)$$

Therefore [61]

$$d\Gamma \sim a_0 - a_1(\vec{\sigma} \cdot \vec{p}_\mu) + a_2(\vec{\sigma} \cdot p_\nu) - m_\mu M_K \text{Im}\,\xi\, \vec{\sigma} \cdot (\vec{p}_\pi \times \vec{p}_\mu) \qquad (7.67)$$

with

$$a_0 = M_K(E^* - E_\pi)(b^2 m_\mu^2 - M_K^2) + 2M_K^2 E_\nu E_\mu + M_K m_\mu E_\nu(\text{Re } b + \text{Im } b)$$
$$a_1 = 2M_K^2[E_\nu + (E^* - E_\pi)\text{Re } b]$$
$$a_2 = m_\mu[M_K^2 + |b|^2 m_\mu^2 + 2E_\mu M_K \text{Re } b]$$
$$b = \frac{1}{2}(\xi - 1)$$
$$E^* = \frac{M_K^2 + M_\pi^2 - m_\mu}{2M_K}$$
$$P_\perp \simeq \text{Im}\,\xi \frac{m_\mu}{M_K} \left\langle \frac{|\vec{p}_\mu|}{E_\mu + \vec{p}_\mu \cdot \vec{p}_\nu/|\vec{p}_\nu| - m_\mu^2/M_K} \right\rangle. \qquad (7.68)$$

The average transverse muon polarization can thus be expressed through Im ξ:

$$P_\perp \sim (0.2 \sim 0.3)\text{Im}\,\xi. \qquad (7.69)$$

The E246 data then give [11]:

$$\text{Im}\,\xi = -0.006 \pm 0.008. \qquad (7.70)$$

As shown explicitly in Sec. 9.7, observation of a signal here – at the level of $P_\perp \sim \mathcal{O}(10^{-3})$ down to $\mathcal{O}(10^{-5})$ – would be an unequivocal signal for the presence of New Physics. There is thus a strong motivation for a new round of experimentation. A new experiment – E06 (TREK) – has been proposed for the J-PARC facility being constructed in Japan, which aims at a 10^{-4} sensitivity for $\langle P_\perp \rangle$, an improvement by more than an order of magnitude over E246.

7.5 ♠ $K \to 3\pi$ ♠

7.5.1 $K_S \to 3\pi^0$

The transition $K_S \to 3\pi^0$ (like $K_L \to \pi\pi$) requires **CP** violation: with Bose statistics enforcing the three neutral pions to form a symmetric configuration the final state has to be **CP** odd since $\mathbf{CP}|\pi^0\rangle = -|\pi^0\rangle$. A non-zero value for $\eta_{000} \equiv A(K_S \to 3\pi^0)/A(K_L \to 3\pi^0)$ thus constitutes an unambiguous measure for **CP** violation that can be related to $\Gamma(K_L \to 2\pi)$ using **CPT** symmetry:

$$\eta_{000} = \epsilon + i[(\mathrm{Im}A(K_S \to 3\pi^0)/\mathrm{Re}A(K_S \to 3\pi^0)]. \qquad (7.71)$$

The experimental findings are [11]:

$$\mathrm{Im}\,\eta_{000} = (-0.1 \pm 1.6) \times 10^{-2} \qquad (7.72)$$

$$|\eta_{000}| < 0.018 \qquad 90\%\mathrm{C.L.} \qquad (7.73)$$

7.5.2 $K_S \to \pi^+\pi^-\pi^0$

The situation is more complex with $K_S \to \pi^+\pi^-\pi^0$ since the mere existence of this channel does not establish **CP** violation: the **CP** parity of the final state depends, as illustrated in Fig. 7.3, on the orbital angular momentum l of the charged pion pair:

$$\mathbf{CP}|(\pi^+\pi^-)_l\pi^0\rangle = -(-1)^l|(\pi^+\pi^-)_l\pi^0\rangle; \qquad (7.74)$$

for l odd – a configuration allowed for $\pi^+\pi^-$, albeit suppressed by the centrifugal barrier – the mode $K_S \to \pi^+\pi^-\pi^0$ conserves **CP**.[4]

The angular momentum of the charged pion pair is correlated with the isospin of the three pion state, as shown in Table 7.1. The final state is

---

[4] Likewise, $K_L \to \pi^+\pi^-\pi^0$ violates **CP** for l odd; it is also kinematically suppressed relative to the $l = 0$ mode.

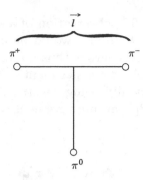

Figure 7.3   Final state configuration for $K^0 \to \pi^+\pi^-\pi^0$. $\vec{l}$ is the angular momentum vector for the $\pi^+\pi^-$ state. Note that $\mathbf{CP}|\pi^+\pi^-\rangle = +|\pi^+\pi^-\rangle$ (see Problem 5.1), $\mathbf{CP}|\pi^0\rangle = -|\pi^0\rangle$. So, $\mathbf{CP}|(\pi^+\pi^-)_l\pi^0\rangle = -(-1)^l|\pi^+\pi^-\pi^0\rangle$.

most conveniently analysed in terms of Dalitz plot variables, namely the pion energies in the kaon rest frame:

$$X = \frac{s_{\pi^+} - s_{\pi^-}}{M_\pi^2}$$

$$Y = \frac{s_{\pi^0} - s_0}{M_\pi^2}, \tag{7.75}$$

where

$$s_i = (p_K - p_i)^2 , \quad i = 1, 2, 3 , \quad s_0 = \frac{1}{3}\sum_i s_i, \tag{7.76}$$

which are odd and even under **CP**, respectively. Defining

$$G(t, X, Y) = \frac{d\Gamma(K^0(t) \to \pi^+\pi^-\pi^0)}{dXdY}$$

$$\overline{G}(X, Y) = \frac{d\Gamma(\overline{K}^0(t) \to \pi^+\pi^-\pi^0)}{dXdY}, \tag{7.77}$$

Table 7.1   **CP**, *l, and isospin properties of the final state* $\pi^+\pi^-\pi^0$

Isospin of $\pi^+\pi^-$	0	1	2
l of $\pi^+\pi^-$	even	odd	even
Isospin of 3π	1	0, 2	1, 3
CP of 3π	odd	even	odd

we can express their difference as follows:

$$
\Delta_{+-0}(t,X,Y) = \frac{G(t,X,Y) - \overline{G}(t,X,Y)}{G(t,X,Y) + \overline{G}(t,X,Y)}
$$
$$
= \frac{Ae^{-\Gamma_S t} + Be^{-\Gamma_L t} + Ce^{-(\Gamma_S+\Gamma_L)t/2}\cos\Delta Mt + De^{-(\Gamma_S+\Gamma_L)t/2}\sin\Delta Mt}{|A_{S+-0}(X,Y)|^2 e^{-\Gamma_S t} + |A_{L+-0}(X,Y)|^2 e^{-\Gamma_L t}}
$$

$$(7.78)$$

with

$$
A(X,Y) = |A_{S+-0}(X,Y)|^2 \left(1 - \left|\frac{q_2}{p_2}\right|^2\right)
$$
$$
B(X,Y) = |A_{L+-0}(X,Y)|^2 \left(1 - \left|\frac{q_1}{p_1}\right|^2\right)
$$
$$
C(X,Y) = -2\mathrm{Re}\left(A_{S+-0}(X,Y)A_{L+-0}(X,Y)^*\right)
$$
$$
D(X,Y) = 2\mathrm{Im}\left(A_{S+-0}(X,Y)A_{L+-0}(X,Y)^*\right). \qquad (7.79)
$$

Here we have used the most general decomposition of the kaon state vectors that does not assume **CPT** symmetry:

$$
|K^0\rangle = \frac{1}{p_1 q_2 + q_1 p_2}[q_1|K_L\rangle + q_2|K_S\rangle]
$$
$$
|\overline{K}^0\rangle = \frac{1}{p_1 q_2 + q_1 p_2}[-p_1|K_L\rangle + p_2|K_S\rangle]; \qquad (7.80)
$$

$A_{S,+-0}(X,Y)$ and $A_{L,+-0}(X,Y)$ denote the decay amplitudes for $K_S \to \pi^+\pi^-\pi^0$ and $K_L \to \pi^+\pi^-\pi^0$, respectively.

We want to emphasize three points about Eq. (7.78).

- Non-vanishing $A(X,Y)$ and $B(X,Y)$ implies *indirect* **CP** violation:

$$
\left|\frac{q_1}{p_1}\right|^2 \neq 1 \neq \left|\frac{q_2}{p_2}\right|^2. \qquad (7.81)
$$

-

$$
\left|\frac{q_1}{p_1}\right|^2 \neq \left|\frac{q_2}{p_2}\right|^2 \qquad (7.82)
$$

signals **CPT** violation as well. Information on $|q_1/p_1|$ is obtained by studying $\Delta_{+-0}(t,X,Y)$ for $t \gg \Gamma_S^{-1}$. Recalling that $|q_2/p_2|$ is extracted from $K_L \to \pi\pi$ or $K_L \to l^\pm\nu\pi^\mp$ it is amusing to note that both $|q_1/p_1|$ and $|q_2/p_2|$ can be determined from K_L transitions.

- A non-vanishing $C(X,Y)$ or $D(X,Y)$ does not necessarily imply **CP** violation, since, as pointed out above, there is a **CP** conserving component to $K_S \to \pi^+\pi^-\pi^0$.

The last point needs to be explained in more detail. With the final state $\pi^+\pi^-\pi^0$ being **CP** even and odd the transition amplitude is even and odd in X, respectively, see Eq. (7.75) [53]:

$$A_{S,+-0}(X,Y) = A_S^+(X,Y) + A_S^-(X,Y)$$
$$A_{L,+-0}(X,Y) = A_L^+(X,Y) + A_L^-(X,Y), \qquad (7.83)$$

where the superscripts $\pm$ denote the **CP** eigenvalue of $\pi^+\pi^-\pi^0$:

$$A_{S[L]}^\pm(X,Y) \equiv \frac{1}{2}\left(A_{S[L],+-0}(X,Y) \pm A_{S[L],+-0}(-X,Y)\right); \qquad (7.84)$$

$A_S^-(X,Y)$ and $A_L^+(X,Y)$ are **CP** violating and the other two are **CP** conserving. The product of amplitudes appearing in $C(X,Y)$ and $D(X,Y)$, see Eq. (7.79), then has four components:

$$
\begin{aligned}
A_{S+-0}(X,Y)A_{L+-0}(X,Y)^* &= A_S^+(X,Y)A_L^-(X,Y)^* &&\text{cons. cons.} \\
&+ A_S^+(X,Y)A_L^+(X,Y)^* &&\text{cons. viol.} \\
&+ A_S^-(X,Y)A_L^-(X,Y)^* &&\text{viol. cons.} \\
&+ A_S^-(X,Y)A_L^+(X,Y)^* &&\text{viol. viol.}
\end{aligned}
$$
$$(7.85)$$

With the first and the last term being odd under the interchange $X \leftrightarrow -X$, they can be eliminated by including events at X and $-X$:

$$\sum_{X,-X} A_{S+-0}(X,Y)A_{L+-0}(X,Y)^* = A_S^+(|X|,Y)A_L^+(|X|,Y)^*$$

$$+ A_S^-(|X|,Y)A_L^-(|X|,Y)^*. \qquad (7.86)$$

The **CP** violating $|A_S^-(X,Y)|$ and $|A_L^+(X,Y)|$ are small relative to the **CP** conserving $|A_L^-(X,Y)|$; likewise for $|A_S^+(X,Y)|$, though for a different reason: with the $\pi^+ - \pi^-$ pair being in an angular momentum $l = 1$ configuration and the π^0 also in an $l = 1$ state in the 3π centre of mass frame, see Table 7.1, this amplitude is reduced by a centrifugal barrier. Therefore we will retain only terms at most linear in $A_{S,+-0}(X,Y)$ in the following. Equation (7.78) then reads as follows:

$$\Delta_{+-0}(t,|X|,Y) \equiv \frac{\sum_{X,-X} G(t,X,Y) - \sum_{X,-X} \overline{G}(t,X,Y)}{\sum_{X,-X} G(t,X,Y) + \sum_{X,-X} \overline{G}(t,X,Y)} \qquad (7.87)$$

$$= \left(1 - \left|\frac{q_1}{p_1}\right|^2\right) + 2e^{-(\Gamma_S + \Gamma_L)t/2}$$

$$\times \left\{ -\cos \Delta Mt \, \mathrm{Re} \left[\frac{A_{\bar{S}}^-(|X|,Y)}{A_L^-(|X|,Y)} + R \frac{A_L^+(|X|,Y)^*}{A_L^-(|X|,Y)^*} \right] \right.$$

$$\left. + \sin \Delta Mt \, \mathrm{Im} \left[\frac{A_{\bar{S}}^-(|X|,Y)}{A_L^-(|X|,Y)} + R \frac{A_L^+(|X|,Y)^*}{A_L^-(|X|,Y)^*} \right] \right\},$$

$$\tag{7.88}$$

where

$$R(|X|,Y) = \frac{A_S^+(|X|,Y)}{A_L^-(|X|,Y)}. \tag{7.89}$$

Note that symmetrization in X must be done before we form the asymmetry.

The second term in the square brackets, being the product of two small ratios, can then be dropped, and we arrive at:

$$\Delta_{+-0}(t,|X|,Y) = \left(1 - \left|\frac{q_1}{p_1}\right|^2\right) + 2e^{-(\Gamma_S + \Gamma_L)t/2}$$

$$\times \left[-\cos \Delta Mt \, \mathrm{Re}\, \eta_{+-0} + \sin \Delta Mt \, \mathrm{Im}\, \eta_{+-0} \right],$$

$$\tag{7.90}$$

where

$$\eta_{+-0} = \frac{A_{\bar{S}}^-(|X|,Y)}{A_L^-(|X|,Y)} \tag{7.91}$$

represents a **CP** violating parameter.

The difference $\Delta_{+-0}(t,|X|,Y)$ between the transition rates $G(t,|X|,Y)$ and $\bar{G}(t,|X|,Y)$ symmetrized in the Dalitz plot variable X thus becomes a **CP** asymmetry and allows the determination of the amplitude ratio η_{+-0}. A main element in this derivation was the observation that $\frac{|A_S^+(|X|,Y)|}{|A_L^-(|X|,Y)|}$ is small due to the centrifugal barrier of the three pion final state and $\frac{|A_L^+(|X|,Y)|}{|A_L^-(|X|,Y)|}$ is tiny due to the approximate validity of **CP** symmetry.

An analysis similar to the derivation of η_{+-} leads to

$$\eta_{+-0} = \frac{1}{2} \left(1 + \frac{q_1}{p_1} \frac{\overline{A}(\pi^+\pi^-\pi^0)_-}{A(\pi^+\pi^-\pi^0)_-} \right), \tag{7.92}$$

where $A(\pi^+\pi^-\pi^0)^-$ and $\overline{A}(\pi^+\pi^-\pi^0)^-$ are $K \to \pi^+\pi^-\pi^0$ and $\overline{K} \to \pi^+\pi^-\pi^0$ amplitudes with the three pions forming a **CP** $= -1$ final state, respectively.

We recover $\eta_{+-0} = 0$ in the limit of **CP** invariance, since in that case $q_1 A_I(\pi^+\pi^-\pi^0) = -p_1 \bar{A}_I(\pi^+\pi^-\pi^0)$ for $I = 1, 3$.

The experimental values are [53]:

$$\text{Re}\,\eta_{+-0} = \left(-2 \pm 7 \begin{smallmatrix} +4 \\ -1 \end{smallmatrix}\right) \times 10^{-3} \tag{7.93}$$

$$\text{Im}\,\eta_{+-0} = \left(-2 \pm 9 \begin{smallmatrix} +2 \\ -1 \end{smallmatrix}\right) \times 10^{-3}. \tag{7.94}$$

7.5.3 $K^\pm \to \pi^\pm \pi^+ \pi^-$

In $K^+ \to \pi^+\pi^+\pi^-$ vs. $K^- \to \pi^-\pi^-\pi^+$ we can search for direct **CP** violation through an analysis of the Dalitz plot population. We can use the following parametrization [63]:

$$|\mathcal{M}|^2 \propto 1 + g\frac{s_3 - s_0}{am_{\pi^+}^2} + h\left(\frac{s_3 - s_0}{am_{\pi^+}^2}\right)^2 + j\frac{s_2 - s_1}{am_{\pi^+}^2} + ak\left(\frac{s_2 - s_1}{am_{\pi^+}^2}\right)^2 + \cdots.$$
$$\tag{7.95}$$

The index 3 is used for the 'odd pion out', i.e. the π^- $[\pi^+]$ in K^+ $[K^-]$ decays. The following constraints are imposed by **CP** invariance on the expansion coefficients:

$$j(K^+) = 0 = j(K^-)\,,\ g(K^+) = g(K^-)$$
$$h(K^+) = h(K^-)\ \ k(K^+) = k(K^-). \tag{7.96}$$

The data have become somewhat 'long in the tooth' [11]:

$$\frac{g(K^+) - g(K^-)}{g(K^+) + g(K^-)} = (1.5 \pm 2.9) \times 10^{-4},$$

and some information, though inconclusive, is also available on h and k. Considering that we are dealing here with direct **CP** violation in non-leptonic decays, which is also probed through searches for ϵ'/ϵ it seems unlikely that an effect will be observed here or a meaningful bound established. Theory will then be spared another embarrassment since it is quite unclear how a reliable description can be developed here after a less than sterling experience in the simpler case of $K_L \to \pi^+\pi^-$ vs $K_L \to \pi^0\pi^0$.

7.6   ♠ Hyperon decays ♠

In Section 4.10.3 we have discussed how final state interactions can generate **T** odd correlations in hyperon decays. In this section, we revisit this problem and discuss how we can separate final state interaction from **T** violating effects. Here we restrict our discussion to $\Lambda \to N\pi$ decays, where N stands for a proton or a neutron. Due to the $\Delta I = 1/2$ rule, both the S and P wave final states $N\pi$ are predominantly in an $I = 1/2$ configuration – an eigenstate of the strong interactions. Everything applies, however, to $\Xi \to \Lambda\pi$ decay as well as other nonleptonic hyperon decays.

The decay amplitude for $\Lambda \to N\pi$ is given by [2, 64]

$$A(\Lambda \to N\pi) = \bar{u}_N(\vec{p}_N, \vec{s}_N)[\mathcal{A}_S e^{i\delta_S} + \mathcal{A}_P e^{i\delta_P}\gamma_5]u_\Lambda(\vec{p}_\Lambda, \vec{s}_\Lambda). \qquad (7.97)$$

$\mathcal{A}_S$ and $\mathcal{A}_P$ stand for S and P wave amplitudes, which are parity violating and conserving, respectively. Watson's theorem relates the strong phase to the πN phase shifts δ_S, and δ_P for a given isospin state.

To derive an expression for the polarization of the final state nucleon, let us go to the Λ rest frame and write Eq. (7.97) in terms of ψ_N and ψ_Λ, which are two component spinors for N and Λ, respectively. Then we can read off ψ_N from Eq. (7.97).

$$\psi_N = [A_S + A_P\vec{\sigma} \cdot \hat{p}_N]\psi_\Lambda, \qquad (7.98)$$

where $\hat{p}$ is a unit vector along the direction of $\vec{p}$. Finally $A_S = \sqrt{\frac{E+M}{2M}}\mathcal{A}_S e^{i\delta_S}$, and $A_P = \sqrt{\frac{E-M}{2M}}\mathcal{A}_P e^{i\delta_P}$. The density matrix [12]

$$\rho_N = \psi_N\psi_N^\dagger = \rho_0(1 + \langle\vec{s}_N\rangle \cdot \vec{\sigma}) \qquad (7.99)$$

gives the nucleon polarization. Writing $\psi_N = \mathcal{M}\psi_\Lambda$, $\mathcal{M} \equiv A_S + A_P\vec{\sigma} \cdot \hat{p}_N$, we have

$$\rho_N = \mathcal{M}\rho_\Lambda\mathcal{M}^\dagger \qquad (7.100)$$

where $\rho_\Lambda = 1 + \langle\vec{s}_\Lambda\rangle \cdot \vec{\sigma}$ denotes the Λ polarization in its rest frame. The polarization of the nucleon is given by

$$\langle\vec{s}_N\rangle = \frac{(\alpha + \langle\vec{s}_\Lambda\rangle \cdot \hat{p})\hat{p} + \beta(\langle\vec{s}_\Lambda\rangle \times \hat{p}) + \gamma\hat{p} \times (\langle\vec{s}_\Lambda\rangle \times \hat{p})}{1 + \alpha\langle\vec{s}_\Lambda\rangle \cdot \hat{p}}, \qquad (7.101)$$

where

$$\alpha = \frac{2\text{Re}(\mathcal{A}_S^*\mathcal{A}_P e^{i(-\delta_S+\delta_P)})}{|\mathcal{A}_S|^2 + |\mathcal{A}_P|^2}, \quad \beta = \frac{2\text{Im}(\mathcal{A}_S^*\mathcal{A}_P e^{i(-\delta_S+\delta_P)})}{|\mathcal{A}_S|^2 + |\mathcal{A}_P|^2}, \quad \gamma = \frac{|\mathcal{A}_S|^2 - |\mathcal{A}_P|^2}{|\mathcal{A}_S|^2 + |\mathcal{A}_P|^2}. \qquad (7.102)$$

For $\bar{\Lambda} \to \bar{N}\pi$ we have the analogous quantities

$$\bar{\alpha} = \frac{2\text{Re}(\bar{\mathcal{A}}_S^*\bar{\mathcal{A}}_P e^{i(-\delta_S+\delta_P)})}{|\bar{\mathcal{A}}_S|^2 + |\bar{\mathcal{A}}_P|^2}, \quad \bar{\beta} = \frac{2\text{Im}(\bar{\mathcal{A}}_S^*\bar{\mathcal{A}}_P e^{i(-\delta_S+\delta_P)})}{|\bar{\mathcal{A}}_S|^2 + |\bar{\mathcal{A}}_P|^2}, \quad \bar{\gamma} = \frac{|\bar{\mathcal{A}}_S|^2 - |\bar{\mathcal{A}}_P|^2}{|\bar{\mathcal{A}}_S|^2 + |\bar{\mathcal{A}}_P|^2}. \qquad (7.103)$$

The strong phases $\delta_{S,P}$ remain unchanged in going from Eq. (7.102) to Eq. (7.103), since the strong forces are **T** invariant. Problem 4.13 tells us:

- **P** invariance $\Longrightarrow \mathcal{A}_S = 0 = \bar{\mathcal{A}}_S$;
- **T** invariance $\Longrightarrow \mathcal{A}_S, \bar{\mathcal{A}}_S, \mathcal{A}_P, \bar{\mathcal{A}}_P$ real;
- **CP** invariance $\Longrightarrow \mathcal{A}_S = -\bar{\mathcal{A}}_S, \mathcal{A}_P = \bar{\mathcal{A}}_P$;
- **CPT** invariance $\Longrightarrow \bar{\mathcal{A}}_S = -(\mathcal{A}_S)^*, \bar{\mathcal{A}}_P = (\mathcal{A}_P)^*$.

Hence we learn:

- Non-vanishing α, $\bar{\alpha}$, $\bar{\beta}$ or β mean **P** is violated; likewise for γ, $\bar{\gamma} \neq -1$;
- A non-vanishing β does not automatically imply **T** violation:

$$\beta = \frac{2\mathcal{A}_S \mathcal{A}_P}{|\mathcal{A}_S|^2 + |\mathcal{A}_P|^2} \cdot \sin(\delta_P - \delta_S); \tag{7.104}$$

  i.e. it can be mimicked by final state interactions as discussed in Section 4.10.3;
- **CP** symmetry implies

$$\alpha = -\bar{\alpha}, \qquad \beta = -\bar{\beta}. \tag{7.105}$$

$$A = \frac{\alpha + \bar{\alpha}}{\alpha - \bar{\alpha}}, \quad B = \frac{\beta + \bar{\beta}}{\beta - \bar{\beta}} \tag{7.106}$$

thus have to vanish – unless **CP** invariance is violated!

Let us discuss the **CP** asymmetry A of Eq. (7.106). The *strong* interaction phases for $P\pi$ final states have been measured [65], those for $\Lambda\pi$ have not. For a theoretical discussion we refer the reader to Ref. [66], which predicts:

$$A(\Lambda \to P\pi^-) \sim (0.05 \div 1.2) \cdot 10^{-4} \tag{7.107}$$

$$A(\Xi \to \Lambda\pi^-) \sim (0.2 \div 3.5) \cdot 10^{-4}. \tag{7.108}$$

The HyperCP experiment at FNAL has searched for **CP** violation in the decay asymmetries combined for Λ and Ξ decays:

$$A_{\Lambda\Xi} \equiv \frac{\alpha_\Lambda \alpha_\Xi - \alpha_{\overline{\Lambda}} \alpha_{\overline{\Xi}}}{\alpha_\Lambda \alpha_\Xi + \alpha_{\overline{\Lambda}} \alpha_{\overline{\Xi}}} \simeq A_\Lambda + A_\Xi, \tag{7.109}$$

where $A_\Lambda = (\alpha_\Lambda + \alpha_{\overline{\Lambda}})/(\alpha_\Lambda - \alpha_{\overline{\Lambda}})$, $A_\Xi = (\alpha_\Xi + \alpha_{\overline{\Xi}})/(\alpha_\Xi - \alpha_{\overline{\Xi}})$. They did not find an effect [67]:

$$A_{\Lambda\Xi} = (0.0 \pm 5.1 \pm 4.4) \cdot 10^{-4}. \tag{7.110}$$

They have searched for a difference in analogous decay asymmetries in $\overline{\Omega}^{+} \to \overline{\Lambda}K^{+} \to \overline{P}\pi^{+}K^{+}$ vs. $\Omega^{-} \to \Lambda K^{-} \to P\pi^{-}K^{-}$ again without seeing a signal [68].

Observing **CP** violation in the decays of strange baryons would be a first rate discovery. Yet due to our inability to predict reliably the strong interaction corrections to the amplitudes it would not be any easier to decide than in the case of ϵ'/ϵ whether it is due to the KM ansatz or due to New Physics.

7.7 The bard's song

We have referred to the search for **CP** violation in strange decays as a marathon, which it certainly was. It might reflect more poetic justice, if instead we called it the heroic period of **CP** studies. For it exhibits many parallels to the hero's journey in mythology and her quest for redemption: it began with a most unusual event that made the continuation of the previous routine impossible – namely the most unexpected discovery of **CP** violation in 1964; it took a long time covering many stages with frequent ups and downs, required overcoming seemingly insurmountable obstacles; finally it reached the 'holy grail', namely establishing direct **CP** violation in 1999.

One can reasonably argue there is still some unfinished business left concerning nonleptonic strange decays, namely finding **CP** asymmetries in $K \to 3\pi$ or in hyperon decays. Yet as theorists we are disinclined to push this view. As we will explain later, the story of the theoretical treatment of ϵ' does not represent one of the glory pages of theoretical physics, to put it mildly: one can say that the SM scored no better than a draw with respect to this observable. We are not eager for another such humbling experience.

Yet there are challenges that still have to be met, namely measuring the ultra-rare decays $K \to \pi\nu\bar{\nu}$ (and the transverse muon polarization in $K^{+} \to \pi^{0}l^{+}\nu$), as we will discuss in later chapters. Those amount to another adventure requiring and deserving a new heroic journey – in analogy to the second Trojan war.

Problems

7.1. Let us take the definition of the **CP** transformation to be:

$$\mathbf{CP}|K^{0}\rangle = i|K^{0}\rangle. \tag{7.111}$$

Assuming **CP** symmetry, show that

$$\bar{\epsilon} = \frac{1+i}{1-i}. \tag{7.112}$$

So, $\bar{\epsilon}$ as defined in Eq. (7.24) depends on phase convention and is not a physical observable. Note also that $\operatorname{Im} M_{12}$ and $\operatorname{Im} \Gamma_{12}$ are non-vanishing. Are they observables? How does $E_1^2 + E_2^2$ get modified? Are ΔM and $\Delta \Gamma$ observables? How does ϵ get modified?

7.2. In Eq. (7.42), we have derived ϵ in a way which is independent of phase conventions. In the literature, we often come across the Wu–Yang phase convention in which A_0 is real. If A_0 had a phase $e^{i\xi_0}$ in some phase convention, show that we must make

$$|K^0\rangle \to e^{-i\xi_0}|K^0\rangle \tag{7.113}$$

to make A_0 real. How does this change M_{12}? Show that Eq. (7.42) remains invariant under this change of phase convention.

7.3. Compute $|A(K \to \pi\mu(\vec{s})\nu)|^2$. The spin dependence can be computed by inserting the spin projection operator $\Lambda_\pm = \frac{1}{2}(1 \pm \gamma_5 \slashed{s})$ and summing over the spin. Show that:

$$
\begin{aligned}
|A(K \to \pi\mu(\vec{s})\nu)|^2 = \frac{G_F^2}{2m_\mu m_\nu} \Big[&|f_+(q^2)|^2 \\
\times \operatorname{Tr}((\slashed{p}_\mu + m_\mu)\gamma_+ (\slashed{p}_K + \slashed{p}_\pi)\slashed{p}_\nu(\slashed{p}_K + \slashed{p}_\pi)\gamma_- \Lambda_-) \\
+ |f_-(q^2)|^2 \operatorname{Tr}((\slashed{p}_\mu + m_\mu)\gamma_- \slashed{p}_\nu \Lambda_-) \\
+ 2\operatorname{Re}\Big(f_+(q^2)f_-^*(q^2)\operatorname{Tr}((\slashed{p}_\mu + m_\mu)\gamma_+ (\slashed{p}_K + \slashed{p}_\pi)\slashed{p}_\nu \Lambda_-) \Big) \Big].
\end{aligned}
\tag{7.114}
$$

Show that the third term contains $\epsilon_{\alpha\beta\gamma\delta}p_K^\alpha s^\beta p_\mu^\gamma p_\pi^\delta$, which is proportional to $\vec{s} \cdot \vec{p}_K \times \vec{p}_\pi$, in the rest frame of the K meson.

7.4. Show that **CP**$|\pi^+\pi^-\pi^0\rangle = (-1)^I|\pi^+\pi^-\pi^0\rangle$ holds for $K_S \to \pi^+\pi^-\pi^0$ decay, where I is the isospin of the 3π state.

7.5. Writing $q/p = 1 + i\phi$, show that

$$\eta_{+-0} = -\frac{i}{2}\phi + \frac{1}{2}\frac{\sum_I \omega_I e^{i(\delta_I - \delta_1)}\operatorname{Im}\frac{\overline{A}_I(\pi^+\pi^-\pi^0)}{A_I(\pi^+\pi^-\pi^0)}}{\sum_I \omega_I e^{i(\delta_I - \delta_1)}}. \tag{7.115}$$

Here ω_I is the Clebsch–Gordan coefficient corresponding to the $(3\pi)_I$ state. If only the final state with $(\pi^+\pi^-)_{I=0}$ contributes:

$$\eta_{+-0} = -\frac{i}{2}\phi + \frac{1}{2}\operatorname{Im}\left(\frac{\overline{A}_1(\pi^+\pi^-\pi^0)}{A_1(\pi^+\pi^-\pi^0)}\right). \tag{7.116}$$

8

The KM implementation of
CP violation

The KM phase is like the Scarlet Pimpernel:
Sometimes here, sometimes there,
Sometimes everywhere![1]

8.1 A bit of history

As described before, the discovery of **CP** violation had not been anticipated – it actually ran counter to rather firm expectations. Nevertheless, the experimental findings as well as their interpretation were soon
accepted. The relevant phenomenology for strange decays was speedily
developed, as described in the two preceding sections, and the community took on the challenge of designing dynamical models for **CP** and
T violation.

A general *classification* of dynamical models was defined using a general
decomposition of the Hamiltonian into **CP** even and odd parts:

$$H = H^+ + H^-, \quad (\mathbf{CP})^\dagger H^\pm \mathbf{CP} = \pm H^\pm. \tag{8.1}$$

- *Millistrong*: **CP** violation is implemented as a small correction of
  order 10^{-3} relative to the strong interactions:

$$H^- = H^-_{MS} \sim 10^{-3} \cdot H_S. \tag{8.2}$$

---

[1] Quote adapted from: 'They seek him here, they seek him there, Those Frenchies seek him
everywhere, Is he in heaven or is he in hell? That damned, elusive Pimpernel!' Sir Percival
Blakeney

H_{MS}^- conserves strangeness and parity like H_S does. $K_L \to \pi\pi$ is understood as a three-step process:

$$K_L \xrightarrow{H_W} I_- \xrightarrow{H_{MS}^-} I_+ \xrightarrow{H_S} \pi\pi, \qquad (8.3)$$

where the states $I_\pm$ are **CP** eigenstates with strangeness zero.

- *Milliweak*: Here we assume

$$H_{MW}^- \sim |\epsilon| \cdot H_W \sim 10^{-3} \cdot H_W \qquad (8.4)$$

and H_{MW}^- can change parity and strangeness. It then follows that H_{MW}^- constitutes a $\Delta S = 1$ interaction like H_W, and $\Delta S = 2$ transitions are produced by iteration:

$$\langle K^0 | H_{MW}^- | \overline{K}^0 \rangle = 0 \qquad (8.5)$$

$$\arg\frac{M_{12}}{\Gamma_{12}} \sim \frac{|H_W + H_{MW}^-|^2}{H_W^2} \sim \mathrm{Re}\,\epsilon. \qquad (8.6)$$

- *Superweak*: We postulate the existence of a new $\Delta S = 2$ Hamiltonian that is highly suppressed:

$$H_{SW}^- \sim 10^{-9} \cdot H_W. \qquad (8.7)$$

Instead of Eq. (8.5) we have

$$\langle K^0 | H_{SW}^- | \overline{K}^0 \rangle \neq 0 \qquad (8.8)$$

to lowest order.

(i) Various *millistrong* models were proposed and analysed soon after the Fitch–Cronin experiment. Their phenomenology is rather accessible experimentally, for we expect **CP** and **T** asymmetries to be roughly of order 10^{-3} in every process unless forbidden by some other reasons. This class of models was quickly ruled out! (ii) The *superweak* scenario was first introduced by Wolfenstein [69] in August 1964. Historically and intellectually it had considerable impact, as we are going to describe. With the demise of the millistrong models it was the only ansatz left standing at that time. (iii) The KM ansatz that came later is of the *milliweak* type.

Apparently it was not realized for quite some time that **CP** violation cannot arise from the charged current interactions with u, d and s quarks only.[2] The community can be forgiven for not worrying about

---

[2] This no-go argument will be explained below.

a numerically tiny effect – $\mathrm{Br}(K_L \to \pi^+\pi^-) \simeq 2 \times 10^{-3}$ – when calculations of $\Gamma(K_L \to \mu^+\mu^-, \gamma\gamma)$ etc. yielded infinities in the absence of a renormalizable theory for the weak interactions.

Yet once the renormalizable Glashow–Salam–Weinberg theory had been formulated and scored a spectacular success in predicting neutral currents, the evaluation of the situation had to change significantly: it is quite remarkable that nobody noticed for some time that even the addition of charm quarks, rendering the theory renormalizable, did not allow for the implementation of **CP** violation, i.e. that New Physics was needed! We do not understand how such a blind spot could arise apart from offering the following observation. Most physicists (Gell-Mann included [70]) thought for a long time that quarks were merely mathematical objects. Furthermore, even if we viewed quarks as real objects, only three of them had been found. The existence of charm quarks to explain the absence of strangeness changing neutral currents – as suggested by Glashow, Illiopolous and Maiani (GIM) [71] – was not widely accepted; such a sentiment was expressed through the quote 'Nature must be smarter than Shelley'.[3] This general scepticism (or even agnosticism) may have been the reason for most to be content with a superweak model as explanation for **CP** violation. The situation at the physics department of Nagoya University was quite different, both on the theoretical and experimental side. Sakata, with his co-workers and students, believed in quarks as real objects. In cosmic ray experiments Niu [72] had found what he called X particles, now known as charm particles. In Nagoya there was thus no mental barrier in thinking about the four quark system. Once it was realized that phases of Yukawa couplings could be rotated away if there were only two families of quarks, it was natural to postulate more quarks [73].

8.2 The Standard Model

It is interesting to reflect upon the status of **CP** violation in particle physics now and compare it to that soon after its discovery. The cartoon of Fig. 8.1 was shown by Cabibbo [74] at the 1966 Berkeley Conference. At that time, physicists thought that the theory of the weak interactions was in good shape since they 'understood'[4] how to compute radiative correction to μ decay. In comparison, **CP** violation was like an atomic bomb blowing up in the background. We have made tremendous progress in understanding weak interactions since then. Similarly, we have a much better grasp on **CP** violation.

---

[3] The creator of this aphorism prefers to remain anonymous.
[4] The definition of 'understood' is clear from the cartoon.

Figure 8.1   This cartoon was presented by Cabibbo at the Berkeley conference in 1966. Today, we think that, indeed, the SM of electroweak interaction describes all of electromagnetic and weak interactions. But do we understand **CP** violation? We gained a tremendous amount of 'understanding' over the past 40 odd years. This is real progress. Yet the cartoon is relevant even today. This figure was reproduced from *Proceedings of the XIIth International Conference in High Energy Physics* by permission from N. N. Cabibbo.

The SM of the strong and electroweak forces is based on the gauge group $SU(3)_C \times SU(2)_L \times U(1)$; the electroweak part is referred to as the Glashow–Salam–Weinberg model.[5] We shall not attempt to discuss the basic principles which lead to the SM; instead we concentrate on those features relevant to our subsequent discussion. A more comprehensive and detailed description can be found in many textbooks [75].

8.2.1   QCD

QCD describes the strong forces as gauge interactions between quarks and gluons; its Lagrangian is given by

$$\mathcal{L}_{\mathrm{QCD}} = -\frac{1}{4} G^a_{\mu\nu} G^{\mu\nu,a} + \overline{Q} i \, \not{D} Q - \overline{Q}_R \mathcal{M} Q_L + \mathrm{h.c.} \qquad (8.9)$$

where $G^a_{\mu\nu}$ denotes the gluon field strength tensor

$$G^a_{\mu\nu} = \partial_\mu A^a_\nu - \partial_\nu A^a_\mu + g_S f^{abc} A^b_\mu A^c_\nu \qquad (8.10)$$

and D_μ the covariant derivative

$$D_\mu = \partial_\mu - i g_S t^a A^a_\mu \qquad (8.11)$$

---

[5] Calling it a model is actually a misnomer since it represents a self-consistent *theory*.

expressed with the gluon fields A_μ^a and $SU(3)$ generators t^a, $a = 1, ..., 8$; Q represents the column vector containing quark fields in all their colour and flavour reincarnations, with the couplings being flavour-diagonal. Note that once we define a set of Hermitian fields $t^a A_\mu^a$, g_S must be defined as real.

Thus it would seem that **CP** is naturally conserved by QCD, which obviously would be an attractive feature. Alas – upon further reflection it was realized that the situation is less clear-cut: there exists a gauge-invariant and renormalizable operator that was ignored in Eq. (8.9), namely $G_{\mu\nu}^a \tilde{G}^{\mu\nu,a}$, where $\tilde{G}^{\mu\nu,a}$ denotes the dual field strength tensor:

$$\tilde{G}_{\mu\nu}^a \equiv \frac{1}{2}\epsilon_{\mu\nu\alpha\beta}G^{\alpha\beta,a}. \tag{8.12}$$

The operator $G_{\mu\nu}^a \tilde{G}^{\mu\nu,a}$ violates **P**, **T** and **CP**. Even if its coefficient in $\mathcal{L}$ is set to zero at the tree-level, radiative corrections will resurrect it with a coefficient containing an ultraviolet divergence. On the other hand we deduce from the upper bound placed on the neutron electric dipole moment that this coefficient has to be 'unnaturally' tiny. This is called the 'strong **CP** problem'; we will address it in detail in Chapter 15.

8.2.2 The Glashow–Salam–Weinberg model

In the standard model, fermions come in families of left-handed doublets and right-handed singlets

$$Q_L = \begin{pmatrix} U \\ D \end{pmatrix}_L, \quad E_L = \begin{pmatrix} \nu_l \\ l^- \end{pmatrix}_L; \quad U_R, D_R, l_R^-, \tag{8.13}$$

where $U^{tr} = (u, c, ...)$, $D^{tr} = (d, s, ...)$ and $(l^-)^{tr} = (e^-, \mu^-, ...)$ denote the flavour eigenstates of up-type, down-type quarks and charged leptons, respectively, with the number of families unspecified at this point; the chiral fermion fields are defined by $\psi_{\binom{L}{R}} = \frac{1}{2}(1 \mp \gamma_5)\psi$.

The fermionic gauge interactions are expressed through charged current

$$J_\mu^+ = J_\mu^1 + iJ_\mu^2 = \overline{U}_L\gamma_\mu D_L + \bar{l}_L\gamma_\mu\nu_L, \tag{8.14}$$

neutral weak currents

$$J_\mu^3 = \frac{1}{2}(\overline{U}_L\gamma_\mu U_L - \overline{D}_L\gamma_\mu D_L + \overline{\nu}_L\gamma_\mu\nu_L - \bar{l}_L\gamma_\mu l_L) \tag{8.15}$$

and the electromagnetic current

$$J_\mu^{\text{em}} = \frac{2}{3}\overline{U}_L\gamma_\mu U_L - \frac{1}{3}\overline{D}_L\gamma_\mu D_L - \bar{l}_L\gamma_\mu l_L \tag{8.16}$$

in

$$\mathcal{L}_f = \mathcal{L}_{CC} + \mathcal{L}_{NC}$$

$$\mathcal{L}_{CC} = \frac{g}{\sqrt{2}} \left(J_\mu^+ W^{-,\mu} + J_\mu^- W^{+,\mu} \right)$$

$$\mathcal{L}_{NC} = e J_\mu^{\text{em}} A^\mu + \frac{g}{\cos\theta_W} \left(J_\mu^3 - \sin^2\theta_W J_\mu^{\text{em}} \right) Z^\mu \qquad (8.17)$$

where

$$\frac{g'}{g} = \tan\theta_W, \quad \frac{e}{g} = \sin\theta_W, \qquad (8.18)$$

with θ_W denoting the weak or Weinberg angle and

$$Z_\mu = \cos\theta_W A_\mu^3 - \sin\theta_W B_\mu, \quad A_\mu = \sin\theta_W A_\mu^3 + \cos\theta_W B_\mu. \qquad (8.19)$$

As discussed in the case of QCD, the gauge couplings must be defined as real. It would then seem that again no **CP** violation can emerge in the gauge forces. However, this would be a premature conclusion. For there is another dynamic sector in the theory, namely the one involving Higgs fields, which adds another layer of complexity.

The SM contains a single $SU(2)$ doublet of Higgs fields.[6] The Yukawa interactions of Higgs fields with quarks are described by:

$$L_{\text{Yukawa}} = - \sum_{i,j} (G_U)_{ij} (\overline{U}_{i,L}, \overline{D}_{i,L}) \begin{pmatrix} \phi^0 \\ -\phi^- \end{pmatrix} U_{j,R}$$

$$- \sum_{i,j} (G_D)_{ij} (\overline{U}_{i,L}, \overline{D}_{i,L}) \begin{pmatrix} \phi^+ \\ \phi^0 \end{pmatrix} D_{j,R} + \text{h.c.} \qquad (8.20)$$

The indices i and j run over 1 to n, the number of families. Once the neutral Higgs field acquires a vacuum expectation value (=VEV), $\langle \phi^0 \rangle = v$, fermion masses arise.

The mass matrices for the up-type and down-type quarks are then proportional to the corresponding Yukawa couplings with the scale set by v:

$$\mathcal{M}_U = v G_U \quad \text{and} \quad \mathcal{M}_D = v G_D. \qquad (8.21)$$

Since the Yukawa couplings are quite arbitrary, so are the mass matrices, and in general they will contain complex elements. In Eq. (4.90) we have seen that some coupling constants must be complex in order for **CP** to be violated. Complex Yukawa couplings raise the possibility of **CP** violation emerging from this sector – somehow. This question will be analysed next.

---

[6] Adding more doublets (which can easily be arranged) has phenomenological consequences like the existence of charged Higgs fields; such scenarios will be addressed in Chapter 17.

8.3 The KM ansatz

8.3.1 The mass matrices

Apart from a possible $G \cdot \tilde{G}$ term mentioned before, **CP** violation can enter SM dynamics only through the mass matrices $\mathcal{M}_U$ and $\mathcal{M}_D$. Their physical interpretation becomes more transparent when we write the Lagrangian in terms of the mass eigenstates of quarks. We diagonalize the mass matrices with the help of two unitary matrices each – $T_{U,L}$, $T_{U,R}$ and $T_{D,L}$, $T_{D,R}$ – acting in family space[7]

$$T_{U,L}\mathcal{M}_U T_{U,R}^{\dagger} = \mathcal{M}_U^{\text{diag}}, \qquad T_{D,L}\mathcal{M}_D T_{D,R}^{\dagger} = \mathcal{M}_D^{\text{diag}}; \qquad (8.22)$$

they transform the left- and right-handed quark fields, respectively, into their mass eigenstates:

$$U_{L[R]}^m = T_{U,L[R]}U_{L[R]}, \qquad D_{L[R]}^m = T_{D,L[R]}D_{L[R]}. \qquad (8.23)$$

T_L and T_R can be found by diagonalizing the Hermitian matrices $\mathcal{M}\mathcal{M}^{\dagger}$ and $\mathcal{M}^{\dagger}\mathcal{M}$.

$$\mathcal{M}_{\text{diag}}^2 = T_R \mathcal{M}^{\dagger}\mathcal{M} T_R^{\dagger} = T_L \mathcal{M}\mathcal{M}^{\dagger} T_L^{\dagger}. \qquad (8.24)$$

The expressions for the *neutral* currents – electroweak as well as strong – remain manifestly the same whether they are expressed in terms of *flavour* or *mass* eigenstates (see Problem 8.1) and *no* flavour-changing neutral currents arise on the tree-level; this is referred to as the GIM mechanism [71].

For the *charged* currents the situation is quite different; we find

$$\overline{U}_L\gamma_\mu D_L = \overline{U}_L^m T_{U,L}\gamma_\mu T_{D,L}^{\dagger}D_L^m = \overline{U}_L^m \gamma_\mu \mathbf{V} D_L^m, \qquad (8.25)$$

where

$$\mathbf{V} = T_{U,L}T_{D,L}^{\dagger} \qquad (8.26)$$

is called the CKM matrix. Note that $\mathbf{V}$ is unitary, which reflects weak universality, to be explained below.

8.3.2 Parameters of consequence

The mere fact that the quark mass matrices and thus also $\mathbf{V}$ contain complex phases does not necessarily mean that they generate observable

---

[7] There is a theorem called Singular Value Decomposition [76], which states: Any complex $m \times n$, $(m \geq n)$ matrix $\mathcal{M}$ can be diagonalized by $U\mathcal{M}V^{\dagger} = D$; D is a diagonal $n \times n$ matrix with positive diagonal elements, V an $n \times n$ unitary matrix and U an $n \times m$ matrix that for $n = m$ is unitary.

consequences like **CP** asymmetries; for we are allowed to redefine the phases of the quark fields. In their paper Kobayashi and Maskawa [73] analysed the question under which conditions *not all* of these phases can be rotated away.

We start by determining the number of independent physical parameters contained in **V**.

- A general $n \times n$ complex matrix contains $2n^2$ real parameters.
- Unitarity implies

$$\sum_j \mathbf{V}_{ij} \mathbf{V}_{jk}^* = \delta_{ik},\qquad (8.27)$$

yielding n constraints for $i = k$ and $2 \cdot \frac{1}{2} \cdot n \cdot (n-1) = n^2 - n$ for $i \neq k$. A unitary $n \times n$ matrix thus contains n^2 independent real parameters.
- The phases of the quark fields can be rotated freely:

$$U_i^m \to e^{i\phi_i^U} U_i^m, \quad D_j^m \to e^{i\phi_j^D} D_j^m,\qquad (8.28)$$

leading to

$$\mathbf{V} \to \begin{pmatrix} e^{-i\phi_1^U} & \cdots & 0 \\ \vdots & \ddots & \vdots \\ 0 & \cdots & e^{-i\phi_n^U} \end{pmatrix} \mathbf{V} \begin{pmatrix} e^{i\phi_1^D} & \cdots & 0 \\ \vdots & \ddots & \vdots \\ 0 & \cdots & e^{i\phi_n^D} \end{pmatrix},\qquad (8.29)$$

where **V** is multiplied by two diagonal matrices whose elements are pure phases. Since the overall phase is irrelevant, $2n-1$ *relative* phases can be removed from **V** in this way. Accordingly, **V** contains $(n-1)^2$ independent physical parameters.
- A general orthogonal $n \times n$ matrix is constructed from $\frac{1}{2}n(n-1)$ quantities describing the independent rotation angles:

$$N_{\text{angles}} = \frac{1}{2}n(n-1).\qquad (8.30)$$

- The number of independent phases in **V** is:

$$N_{\text{phases}} = (n-1)^2 - \frac{1}{2}n(n-1) = \frac{1}{2}(n-1)(n-2).\qquad (8.31)$$

- For *two* families we have $N_{\text{phase}} = 0$, $N_{\text{angles}} = 1$, i.e. just the Cabibbo angle. There is then *no* **CP** violation through **V**.
- For *three* families we have

$$N_{\text{angles}} = 3, \quad N_{\text{phases}} = 1,\qquad (8.32)$$

i.e. in addition to the three Euler mixing angles we encounter one irreducible phase that represents genuine **CP** violation.

- Beyond $n = 3$ there is a rapid proliferation of physical parameters: six angles and three phases arise for $n = 4$.
- If all up-type quarks were mass-degenerate, obviously we could not distinguish them by their masses. Any linear combination of mass eigenstates would still be a mass eigenstate, and we could make the transformation $\overline{U}_L^m \to \overline{U}_L^m \mathbf{V}^\dagger$ to remove the effect of $\mathbf{V}$. Similarly, if two up-type quarks were degenerate, we can make $\overline{U}_L^m \to \overline{U}_L^m A^\dagger$, where A is a block-diagonal matrix which mixes the two degenerate quarks. Clearly, the parameters in A can be adjusted to remove the complex phase in $\mathbf{V}$ (see Problem 8.2).

There is an infinity of ways to express the elements of $\mathbf{V}$ in terms of three rotation angles and one phase. Also the phase can be made to appear in many different elements; thus the reference to the Scarlet Pimpernel in the motto at the beginning of this chapter.

One representation has been sanctioned by the particle data group [11]:

$$
\mathbf{V} = \begin{pmatrix} c_{12}c_{13} & s_{12}c_{13} & s_{13}e^{-i\delta_{13}} \\ -s_{12}c_{23} - c_{12}s_{23}s_{13}e^{i\delta_{13}} & c_{12}c_{23} - s_{12}s_{23}s_{13}e^{i\delta_{13}} & s_{23}c_{13} \\ s_{12}s_{23} - c_{12}c_{23}s_{13}e^{i\delta_{13}} & -c_{12}s_{23} - s_{12}c_{23}s_{13}e^{i\delta_{13}} & c_{23}c_{13} \end{pmatrix}
$$

$$(8.33)$$

where $c_{ij} = \cos\theta_{ij}$ and $s_{ij} = \sin\theta_{ij}$ for the Euler angles θ_{ij} with i and j being family labels. Different parameterizations lead to the same physics, of course. Without some specific ideas about the mechanism for flavour generation, none is intrinsically superior for theoretical reasons; on the other hand, some can be more convenient on phenomenological grounds. In Section 8.4 we present one prime example, the Wolfenstein representation.

8.3.3 *Describing weak phases through unitarity triangles*

For the case of only three families, a geometric representation can greatly facilitate an intuitive understanding. The unitarity of the CKM matrix leads to two types of relations:

$$
\sum_{i=1}^{3} |V_{ij}|^2 = 1; \quad j = 1, \ldots, 3 \tag{8.34}
$$

$$
\sum_{i=1}^{3} V_{ji}V_{ki}^* = 0 = \sum_{i=1}^{3} V_{ij}V_{ik}^*; \quad j, \, k = 1, \ldots, 3, \quad j \neq k. \tag{8.35}
$$

The relations of Eq. (8.34) are of fundamental importance: they are often referred to as expressing *weak universality*, meaning that the overall charged current couplings of each quark – say an up-type quark – to all the down-type quarks is of universal strength. This concept first put forward in [74] deserves continuing experimental scrutiny. Within SM it is a consequence of the universality of non-abelian gauge couplings (of $SU(2)_L$). Yet Eq. (8.34) tells us nothing about the weak phases essential for **CP** violation. That information is contained in Eq. (8.35). Since the CKM matrix elements are complex, these relations imply that they form triangles in a complex plane. Two nice features of this representation can be stated right away.

- Changing the phase *convention* for one of the quark fields will (at most) rotate the whole triangle in the complex plane.
- The six triangles representing Eq. (8.35) are of very different shapes, see Fig. 8.2. *Yet they all possess the same area*, as can be seen quite easily. Consider the second equation of Eq. (8.35) with $j = d$ and $k = b$ and multiply it by the phase factor $\mathbf{V}_{ub}^*\mathbf{V}_{ud}$:

$$|\mathbf{V}_{ub}\mathbf{V}_{ud}|^2 + \mathbf{V}_{ub}^*\mathbf{V}_{ud}\mathbf{V}_{cd}^*\mathbf{V}_{cb} + \mathbf{V}_{ub}^*\mathbf{V}_{ud}\mathbf{V}_{td}^*\mathbf{V}_{tb} = 0,$$

and thus

$$\mathrm{Im}\mathbf{V}_{ub}^*\mathbf{V}_{ud}\mathbf{V}_{cd}^*\mathbf{V}_{cb} = -\mathrm{Im}\mathbf{V}_{ub}^*\mathbf{V}_{ud}\mathbf{V}_{td}^*\mathbf{V}_{tb}. \tag{8.36}$$

Multiplying by $\mathbf{V}_{cb}^*\mathbf{V}_{cd}$ instead shows that

$$\mathrm{Im}\mathbf{V}_{cb}^*\mathbf{V}_{cd}\mathbf{V}_{td}^*\mathbf{V}_{tb} = -\mathrm{Im}\mathbf{V}_{cb}^*\mathbf{V}_{cd}\mathbf{V}_{ud}^*\mathbf{V}_{ub}.$$

Repeating this procedure for the other relations of Eq. (8.35), we arrive at the following findings:

$$|\mathrm{Im}\mathbf{V}_{km}^*\mathbf{V}_{lm}\mathbf{V}_{kn}\mathbf{V}_{ln}^*| = |\mathrm{Im}\mathbf{V}_{mk}^*\mathbf{V}_{ml}\mathbf{V}_{nk}\mathbf{V}_{nl}^*| = J, \tag{8.37}$$

irrespective of the indices k, l, m, n! The quantity J is obviously invariant under changes in the phase conventions $|q\rangle \to e^{i\phi_q}|q\rangle$ for any of the quark fields (See Problems 8.4 and 8.5).

$$\mathrm{area\,(every\ triangle)} = \frac{1}{2}J. \tag{8.38}$$

This is the geometric translation of the algebraic result obtained before, that with three families there exists only a single irreducible phase [77].

Our discussion above shows that **CP** violation cannot be implemented unless

$$m_u \neq m_c \neq m_t, \; m_d \neq m_s \neq m_b \tag{8.39}$$

$$\theta_{12}, \theta_{13}, \theta_{23} \neq 0, \frac{\pi}{2}. \tag{8.40}$$

In addition

$$\delta \neq 0, \frac{\pi}{2} \tag{8.41}$$

has to hold. It should be noted that all these conditions, Eq. (8.39), Eq. (8.40) and Eq. (8.41) can be summarized in a compact manner. Define:

$$iC \equiv [\mathcal{M}_U \mathcal{M}_U^\dagger, \mathcal{M}_D \mathcal{M}_D^\dagger]. \tag{8.42}$$

Then we can put

$$\begin{aligned} \det C = -2J(m_t^2 - m_c^2)(m_c^2 - m_u^2)(m_u^2 - m_t^2) \\ \times (m_b^2 - m_s^2)(m_s^2 - m_d^2)(m_d^2 - m_b^2) \neq 0. \end{aligned} \tag{8.43}$$

A non-vanishing $\det C$ is a necessary condition for **CP** violation. This condition in general greatly suppresses **CP** violating observables. J is certainly a tiny dimensionless number, namely about 3×10^{-5} as obtained from present CKM phenomenology. Yet the interpretation of $\det C$ is not quite so straightforward, since it is a parameter carrying a high mass dimension: $\det C \propto M^{12}$! Depending on the relevant scale M for a certain process, $\det C$ can vary over many orders of magnitude. Arguing that baryogenesis at the electroweak scale is characterized by $M \sim 100\,\text{GeV}$ makes CKM dynamics utterly irrelevant for this process.

Then how can we observe **CP** violation at all? One has to keep two facts in mind: First, in many cases there are several relevant mass scales M, some of which are much smaller, like M_B, M_K, ΔM_B and ΔM_K. Second, the observable **CP** asymmetries in K and B decays are *ratios* of CKM parameters: while the numerators, being controlled by $\det C$, are certainly small, so are the denominators, and that is exactly what happens in B decays, as we shall see. The quantity

$$\tilde{C} \equiv J \cdot \log\left(\frac{m_t}{m_c}\right) \log\left(\frac{m_c}{m_u}\right) \log\left(\frac{m_u}{m_t}\right) \log\left(\frac{m_b}{m_s}\right) \log\left(\frac{m_s}{m_d}\right) \log\left(\frac{m_d}{m_b}\right) \tag{8.44}$$

also vanishes like Eq. (8.43), if any pair of the up-type or of down-type quarks were mass degenerate. In some cases it might reflect the possible impact of **CP** violation better than $\det C$.

8.4   A tool kit

Ten years after the KM paper the first evidence appeared – since confirmed – that the CKM matrix possesses a very peculiar structure. As described in Section 10.1.2 it was found that the lifetime of B mesons was longer than expected by an order of magnitude. Together with the observed preference of beauty to decay into charm rather than u quarks, this established the hierarchy

$$|\mathbf{V}_{ub}|^2 \ll |\mathbf{V}_{cb}|^2 \ll |\mathbf{V}_{us}|^2 \ll 1. \tag{8.45}$$

While we wish to collate historical events with our physics discussion, we shall use these experimental findings now as the simplification obtained by them focuses the theoretical discussion. As pointed out by Wolfenstein [78], we can express the CKM matrix through an expansion in powers of $\sin\theta_C = \lambda$; through $\mathcal{O}(\lambda^4)$, we have:

$$\mathbf{V} = \begin{pmatrix} 1 - \frac{1}{2}\lambda^2 & \lambda & A\lambda^3(\rho - i\eta + \frac{i}{2}\eta\lambda^2) \\ -\lambda & 1 - \frac{1}{2}\lambda^2 - i\eta A^2\lambda^4 & A\lambda^2(1 + i\eta\lambda^2) \\ A\lambda^3(1 - \rho - i\eta) & -A\lambda^2 & 1 \end{pmatrix}. \tag{8.46}$$

The three Euler angles and one KM phase of the PDG representation, Eq. (8.33), are replaced by the four real quantities λ, A, ρ and η. For such an expansion to be self-consistent, we have to require $|A|$, $|\rho|$ and $|\eta|$ to be all of order unity. Indeed, this will be shown in Section 10.10.2.

Let us look at the CKM matrix in a semi-quantitative way: Eq. (8.46) suggests one general observation while at the same time leading to a global prediction.

- This pattern goes well beyond the requirements of $\mathbf{V}$ merely being unitary: the matrix is almost diagonal and symmetric, and its elements get smaller the more we move away from the diagonal. Nature most certainly has encoded a profound message in this peculiar pattern. Alas – we have not (yet) succeeded in deciphering it!
- The six unitarity triangles [77, 79, 80] can now be characterized through their dependence on λ:

$$\underset{\mathcal{O}(\lambda)}{\mathbf{V}_{ud}^* \mathbf{V}_{us}} + \underset{\mathcal{O}(\lambda)}{\mathbf{V}_{cd}^* \mathbf{V}_{cs}} + \underset{\mathcal{O}(\lambda^5)}{\mathbf{V}_{td}^* \mathbf{V}_{ts}} = \delta_{ds} = 0 \tag{8.47}$$

$$\underset{\mathcal{O}(\lambda)}{\mathbf{V}_{ud} \mathbf{V}_{cd}^*} + \underset{\mathcal{O}(\lambda)}{\mathbf{V}_{us} \mathbf{V}_{cs}^*} + \underset{\mathcal{O}(\lambda^5)}{\mathbf{V}_{ub} \mathbf{V}_{cb}^*} = \delta_{uc} = 0 \tag{8.48}$$

$$\mathbf{V}_{us}^* \mathbf{V}_{ub} + \mathbf{V}_{cs}^* \mathbf{V}_{cb} + \mathbf{V}_{ts}^* \mathbf{V}_{tb} = \delta_{sb} = 0$$
$$\mathcal{O}(\lambda^4) \quad \mathcal{O}(\lambda^2) \quad \mathcal{O}(\lambda^2) \tag{8.49}$$

$$\mathbf{V}_{td} \mathbf{V}_{cd}^* + \mathbf{V}_{ts} \mathbf{V}_{cs}^* + \mathbf{V}_{tb} \mathbf{V}_{cb}^* = \delta_{ct} = 0$$
$$\mathcal{O}(\lambda^4) \quad \mathcal{O}(\lambda^2) \quad \mathcal{O}(\lambda^2) \tag{8.50}$$

$$\mathbf{V}_{td} \mathbf{V}_{ud}^* + \mathbf{V}_{ts} \mathbf{V}_{us}^* + \mathbf{V}_{tb} \mathbf{V}_{ub}^* = \delta_{ut} = 0$$
$$\mathcal{O}(\lambda^3) \quad \mathcal{O}(\lambda^3) \quad \mathcal{O}(\lambda^3) \tag{8.51}$$

$$\mathbf{V}_{ud} \mathbf{V}_{ub}^* + \mathbf{V}_{cd} \mathbf{V}_{cb}^* + \mathbf{V}_{td} \mathbf{V}_{tb}^* = \delta_{db} = 0,$$
$$\mathcal{O}(\lambda^3) \quad \mathcal{O}(\lambda^3) \quad \mathcal{O}(\lambda^3) \tag{8.52}$$

where below each product of matrix elements we have noted their size in powers of λ.

We see that the six triangles fall into three categories, as illustrated in Fig. 8.2:

(1) The first two triangles are extremely 'squashed': two sides are of order λ, the third one of order λ^5 and their ratio of order $\lambda^4 \simeq 2.3 \cdot 10^{-3}$; Equation (8.47) and Eq. (8.48) control the situation in strange and charm decays; the effective weak phases there are obviously tiny.

(2) The third and fourth triangles are still rather squashed, yet less so: two sides are of order λ^2 and the third one of order λ^4.

(3) The last two triangles have sides that are all of the same order, namely λ^3. All their angles are therefore naturally large, i.e. $\sim$ several $\times$ 10 degrees. To leading order in λ we have

$$\mathbf{V}_{ud} \simeq \mathbf{V}_{tb}, \ \mathbf{V}_{cd} \simeq -\mathbf{V}_{us}, \ \mathbf{V}_{ts} \simeq -\mathbf{V}_{cb}; \tag{8.53}$$

the triangles of Eq. (8.52) and Eq. (8.51) thus coincide to that order.

The sides of this triangle, having naturally large angles, are given by $\lambda \cdot \mathbf{V}_{cb}$, $\mathbf{V}_{ub}$ and $\mathbf{V}_{td}^*$; these are all quantities that control important aspects of B decays, namely CKM favoured and disfavoured B decays and $B_d - \overline{B}_d$ oscillations, as described in Chapter 10. Hence we can conclude: *the KM ansatz unequivocally predicts beauty transitions to contain relative weak phases of order unity!* To repeat once more: the crucial element in reaching this conclusion is the 'long' B lifetime.

We can actually go considerably further. With the $B - \overline{B}$ oscillation rate being quite similar to its decay rate, see Chapter 10, we have two different, yet *coherent* amplitudes of *comparable* size available in B^0 transitions. *Therefore there have to be large – or even huge –* **CP** *asymmetries in*

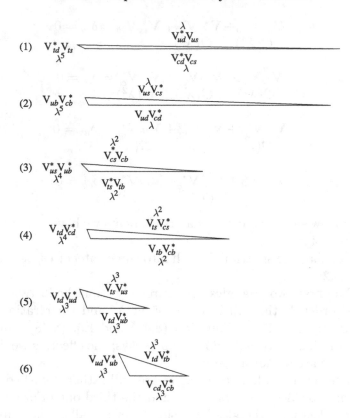

Figure 8.2   Each of six unitarity relations Eq. (8.47) to Eq. (8.52) can be represented by a triangle on the complex plane with equal area. The one corresponding to **CP** violation in K decay is given by (1). It is a spear rather than a triangle – this leads to small **CP** violation. In contrast, the one corresponding to B decay, (6), is expected to have three sides of $\mathcal{O}(\lambda^3)$ and decent angles – leading to large **CP** violation.

some decays of beauty hadrons! The challenge for theory is to figure out in which specific channels such asymmetries will surface and how reliably – parametrically as well as numerically – we can predict their size.

8.4.1   *The angles of the unitarity triangle*

We shall see that all six parameters of the unitarity triangle, three angles and three sides, defined by the unitarity relation Eq. (8.52), can be measured. The sides of the triangle are obtained by measuring decay rates, and the angles are measured by various asymmetries. When angles are determined from asymmetries, we must be careful about their signs. For this

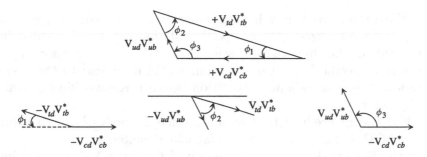

Figure 8.3 Angles of the unitarity triangle are related to the phases of the KM matrix. The right-hand rule gives the positive direction of the angle between two vectors.

reason, we shall record here the relationships between angles of the unitarity triangle and the KM phases. The angles are defined by the right-hand rule.

$$\phi_1 = \pi - \arg\left(\frac{-\mathbf{V}_{tb}^* \mathbf{V}_{td}}{-\mathbf{V}_{cb}^* \mathbf{V}_{cd}}\right),$$

$$\phi_2 = \arg\left(\frac{\mathbf{V}_{tb}^* \mathbf{V}_{td}}{-\mathbf{V}_{ub}^* \mathbf{V}_{ud}}\right),$$

$$\phi_3 = \arg\left(\frac{\mathbf{V}_{ub}^* \mathbf{V}_{ud}}{-\mathbf{V}_{cb}^* \mathbf{V}_{cd}}\right). \tag{8.54}$$

These results are obvious from Fig. 8.3.

8.5 The pundits' judgement

In summarizing the preceding discussion let us add some new twists.

- The KM analysis is a prime example for deducing the existence of New Physics *indirectly*: the transition $K_L \to \pi\pi$ involves known particles only, which are built from u, d and s quarks only. Yet it cannot proceed within a two-family SM. New degrees of freedom had to be postulated, namely the beauty and top quarks; it took 10 and 20 years, respectively, to find them.

  It is tempting to speculate what would have happened if history had proceeded somewhat differently, if the charm quark and τ leptons had been discovered a bit sooner, *before* the KM paper had been written. Or if pions were significantly lighter, in which case it would be hard to distinguish K_L from K_S and discover **CP** violation.

With the third family having made its first appearance, someone would have realized that now **CP** violation can be embedded into the SM. In that hypothetical scenario the direct observation of New Physics, specifically of the τ lepton, would have lead to the prediction of a *qualitatively* new transition between Known Physics states: $K_L \to \pi\pi$.

- CKM theory predicts that at least some B modes have to exhibit large **CP** asymmetries with *no plausible* deniability.

- In the SM quark masses are generated from the Yukawa couplings with a *single* Higgs doublet. This means that the Yukawa couplings are complex, i.e. that there is a *hard* breaking of **CP** symmetry, namely in dimension-four operators in the non-gauge sector of the Lagrangian. Yet the KM construction is more general than that: algebraically it works equally well if the 3×3 quark mass matrices arise from a more complex Higgs sector with multiple Yukawa couplings and vacuum expectation values; in such scenarios the breaking of **CP** invariance can occur also in a spontaneous fashion. We will return to these considerations when discussing alternative models for **CP** violation.

- As explained later, KM theory correctly describes the existing data. Does it mean that we have understood **CP** violation? The answer is *No!* In that case we have merely found an economical and phenomenologically successful way of embedding **CP** violation into the SM. **V** is derived from the quark mass matrices. To understand it we have to figure out the origin of quark masses.

Problems

8.1. Show that the neutral current interaction $\overline{U}_L \gamma_\mu U_L$ remains invariant when we express it in terms of mass eigenstates. By looking at the form of the weak neutral current given in Eq. (8.15), show that no flavour changing interaction can be generated.

8.2. Show that if any of the quark masses become degenerate, the CKM matrix can be made real. Show also that if any one of the CKM matrix elements vanishes, then the phases can be adjusted so that the CKM matrix is real.

8.3. Take a 2×2 unitary matrix and show explicitly that the phases can be rotated away by redefining the quark phases. Try the same procedure for a 3×3 unitary matrix and show that not all the phases can be

8.4. Show that J is rephasing invariant.

8.5. Show that all unitarity triangles have the same area.

8.6. Consider a representation of **V** with a 2×2 submatrix that is completely real. Show that such a matrix does *not* contain **CP** violation.

8.7. Derive Eq. (8.54) from Fig. 8.3, paying particular attention to the sign, as the SM will predict the sign as well as the magnitude of the angles. Verify that these three angles add up to π.

9

The theory of $K_L \to \pi\pi$ decays

In this chapter, we present theoretical expectations for various **CP** violating observables in $K \to \pi\pi$ decays, where we must face up to the challenge of explaining the $\Delta I = 1/2$ rule discussed in Section 5.3. A theoretical description of decay rates involves two main steps:

- evaluating the underlying Lagrangian at ordinary hadronic scales;
- calculating the matrix elements of the emerging operators.

Below we mainly state the results on operator renormalization and on the size of the relevant matrix elements giving few explanations. A more detailed treatment is found in [81].

9.1   The $\Delta S = 1$ non-leptonic Lagrangian

The electroweak Lagrangian given in Eq. (8.17) is defined at an energy scale M_W. At lower scales the W bosons are no longer real dynamic entities, yet their virtual exchange generates an effective coupling between charged currents:

$$\mathcal{H}^{CC}_{\text{weak}} = \frac{G_F}{\sqrt{2}} J^\dagger_\mu J^\mu, \ J^\mu = \begin{pmatrix} \overline{u} \\ \overline{c} \\ \overline{t} \end{pmatrix}^{tr} \gamma^\mu (1 - \gamma_5) \mathbf{V} \begin{pmatrix} d \\ s \\ b \end{pmatrix}. \qquad (9.1)$$

QCD corrections renormalize this interaction at scales below M_W. Evolving the Hamiltonian $\mathcal{H}$ to a lower scale μ means that physical degrees of freedom whose characteristic frequencies exceed μ are integrated out; i.e. they are eliminated as dynamic fields from the effective *operators* in $\mathcal{H}$; they can contribute only through quantum corrections which affect the *c-number coefficients* of those operators. 'Soft' degrees of freedom with characteristic frequencies below μ provide the fields from which the operators are built. The scale μ is introduced for separating short-distance effects to be included in the coefficient functions from long-distance dynamics in the matrix elements:

$$\text{long-distance} > \mu^{-1} > \text{short-distance.} \qquad (9.2)$$

This natural separation offers practical advantages as well: because QCD is asymptotically free, a perturbative treatment of it should yield a good approximation for short-distance dynamics. Accordingly we can calculate the QCD corrections to an operator in the short-distance domain and thus evaluate the dependence of its coefficient on μ for $\mu \gg \Lambda_{QCD}$. Thus we have the hierarchy

$$\Lambda_{QCD} \ll \mu \ll M_W. \qquad (9.3)$$

The leading QCD quantum corrections covering the range from M_W down to μ are evaluated in terms of $\alpha_S \cdot \log(M_W/\Lambda_{QCD})$ and $\alpha_S \cdot \log(\mu/\Lambda_{QCD})$ rather than merely α_S. These logarithmically enhanced numbers can be of order unity, making quantum corrections quite sizeable. For a numerically reliable treatment we have to sum them up; a rather elaborate theoretical tool box based on the renormalization group [82] allows us to achieve this goal in what is called the 'leading log' (and 'next-leading log' etc.) approximation (an excellent discussion of these points is presented in Ref. [83]).

What we have said so far is quite general. For the particular case under study here, namely strange decays, we have to note a qualitatively new feature: the original operator $\overline{u}_L\gamma_\mu s_L \overline{d}_L\gamma^\mu u_L$ which appears in Eq. (9.1) is *non*-multiplicatively renormalized: *additional* four-quark operators with different colour, flavour and even chirality structure emerge, as can be inferred on qualitative grounds already.

• Consider the scattering of left-handed quarks $s_L + u_L \rightarrow d_L + u_L$ which is given by

$$\mathcal{H} = \frac{G_F}{\sqrt{2}} \mathbf{V}_{ud}\mathbf{V}_{us}^* [O_- + O_+] + \text{h.c.}, \qquad (9.4)$$

$$O_\pm = 2[\overline{s}_L\gamma_\mu u_L \overline{u}_L\gamma^\mu d_L \pm \overline{s}_L\gamma_\mu d_L \overline{u}_L\gamma^\mu u_L]. \qquad (9.5)$$

$O_-[O_+]$ is odd [even] under the interchange of final state u and d quarks. Thus the final state is a pure $I_f = 0$ $[I_f = 1]$ state. With s being an iso-singlet, we see that O_- drives purely $\Delta I = 1/2$ transitions, whereas O_+ drives $\Delta I = 1/2$ and $3/2$ transitions. Since the strong interactions

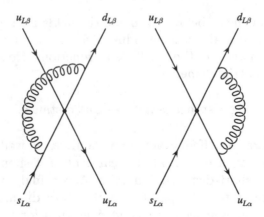

Figure 9.1   QCD corrections to four-quark operators. The blob represents the uncorrected four-Fermi interaction $\bar{u}_{L\alpha}\gamma_\mu s_{L\alpha}\bar{d}_{L\beta}\gamma^\mu u_{L\beta}$. Exchange of a gluon shown in these diagrams generates $\bar{u}_{L\beta}\gamma_\mu s_{L\alpha}\bar{d}_{L\alpha}\gamma^\mu u_{L\beta}$ which, after Fiertz transformation, becomes $\bar{u}_{L\beta}\gamma_\mu u_{L\beta}\bar{d}_{L\alpha}\gamma^\mu s_{L\alpha}$.

conserve I, they preserve the isospin structure and renormalize O_- and O_+ multiplicatively, yet by different amounts:

$$O_+(M_W) \xrightarrow{\text{QCD}} c_+(\mu)O_+(\mu)$$

$$O_-(M_W) \xrightarrow{\text{QCD}} c_-(\mu)O_-(\mu). \tag{9.6}$$

Figure 9.1 shows the radiative QCD correction to a four-quark operator. In the leading log approximation the coefficients $c_\pm$ are given by

$$c_-(\mu)|_{LL} = \left[\frac{\alpha_S(M_W^2)}{\alpha_S(\mu^2)}\right]^{-\frac{12}{33-2n_F}} , \qquad c_+(\mu)|_{LL} = \left[\frac{\alpha_S(M_W^2)}{\alpha_S(\mu^2)}\right]^{\frac{6}{33-2n_F}} \tag{9.7}$$

with n_F denoting the number of active flavours. Thus a new operator with a different flavour structure $(\bar{d}_L\gamma_\mu s_L)(\bar{u}_L\gamma^\mu u_L)$ appears.

Since $\alpha_S(M_W^2) < \alpha_S(\mu^2)$ for $M_W > \mu$ – QCD is asymptotically free – we have

$$c_-(\mu) > 1 > c_+(\mu) ; \tag{9.8}$$

i.e. radiative QCD corrections naturally enhance $\Delta I = 1/2$ over $3/2$ transitions [84, 85].

• Even the chirality structure can be modified by QCD corrections through the so-called penguin process introduced by the ITEP group [86] and shown in Fig. 9.2(a). The penguin operator is given by

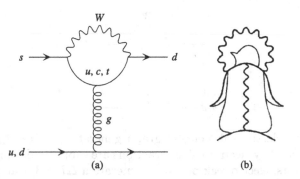

Figure 9.2 (a) Penguin graphs generated by QCD correction to the non-leptonic Lagrangian. This graph generates O_3 to O_6. (b) If we open up the blob in (a) and expose the W boson propagator, we can guess why (a) is called a penguin graph. This figure was reproduced from *Parity* by permission of T. Muta and T. Morozumi.

$$O_P = \sum_a \bar{s}_L \gamma_\mu t^a d_L (\bar{u} \gamma^\mu t^a u + \bar{d} \gamma^\mu t^a d). \tag{9.9}$$

The gluon field – being a vector – yields a $(V - A) \times V$ contribution; since it is an isoscalar, it contributes only to $\Delta I = 1/2$ transitions. Redrawing Fig. 9.2(a) as in Fig. 9.2(b) makes it more obvious why the name penguin has been coined for it.

• For a quantitative description we start with the three family charged current Hamiltonian defined at M_W:

$$\mathcal{H}(M_W) = \frac{G_F}{\sqrt{2}} \sum_{q=u,c,t} \mathbf{V}_{qs}^* \mathbf{V}_{qd}\, O_2^q + \text{h.c.}$$

$$= \frac{G_F}{\sqrt{2}} \mathbf{V}_{us}^* \mathbf{V}_{ud} [(1 - \tau)(O_2^u - O_2^c) + \tau(O_2^u - O_2^t)] + \text{h.c.}, \tag{9.10}$$

where $\tau = -\frac{\mathbf{V}_{ts}^* \mathbf{V}_{td}}{\mathbf{V}_{us}^* \mathbf{V}_{ud}}$ and $O_2^q = (\bar{s}q)_{V-A}(\bar{q}d)_{V-A}$ with $(\bar{s}q)_{V-A}(\bar{q}d)_{V-A} \equiv (\bar{s}\gamma_\mu(1-\gamma_5)q)(\bar{q}\gamma^\mu(1-\gamma_5)d)$. Running the renormalization group equation down from M_W to $\mu < m_c$ yields

$$O_2^u - O_2^c \quad \rightarrow \quad \sum_{i=1}^{10} z_i(\mu) O_i$$

$$O_2^u - O_2^t \quad \rightarrow \quad \sum_{i=1}^{10} v_i(\mu) O_i, \tag{9.11}$$

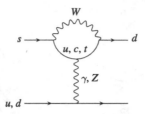

Figure 9.3　The electroweak penguin graph generating O_7 to O_{10}. While these diagrams generated by electroweak interactions are suppressed compared to gluon penguin operators, electroweak penguins generate a $\Delta I = 3/2$ amplitude. These contributions become relevant when $\Delta I = 3/2$ transitions are important – such as in ϵ'. The strength of electroweak penguins is enhanced by the large top quark mass.

where we have included the electromagnetic penguin operator shown in Fig. 9.3. Putting everything together – see [81, 86, 87, 88] for numerical evaluation of $z_i(\mu)$ and $v_i(\mu)$, – we obtain as low-energy Hamiltonian:

$$\mathcal{H}(\Delta S = 1) = \frac{G_F}{\sqrt{2}} \mathbf{V}_{us}^* \mathbf{V}_{ud} \left(\sum_{i=1}^{10} (z_i(\mu) + \tau y_i(\mu)) O_i \right) + \text{h.c.}, \qquad (9.12)$$

where $y_i(\mu) = v_i(\mu) - z_i(\mu)$, and

$$
\begin{aligned}
O_1^u &= (\bar{s}d)_{V-A}(\bar{u}u)_{V-A}, & O_2^u &= (\bar{s}u)_{V-A}(\bar{u}d)_{V-A} \\
O_3 &= (\bar{s}d)_{V-A}\sum_q(\bar{q}q)_{V-A}, & O_4 &= (\bar{s}_\alpha d_\beta)_{V-A}\sum_q(\bar{q}_\beta q_\alpha)_{V-A} \\
O_5 &= (\bar{s}d)_{V-A}\sum_q(\bar{q}q)_{V+A}, & O_6 &= (\bar{s}_\alpha d_\beta)_{V-A}\sum_q(\bar{q}_\beta q_\alpha)_{V+A} \\
O_7 &= \frac{3}{2}(\bar{s}d)_{V-A}\sum_q e_q(\bar{q}q)_{V+A}, & O_8 &= \frac{3}{2}(\bar{s}_\alpha d_\beta)_{V-A}\sum_q e_q(\bar{q}_\beta q_\alpha)_{V+A} \\
O_9 &= \frac{3}{2}(\bar{s}d)_{V-A}\sum_q e_q(\bar{q}q)_{V-A}, & O_{10} &= \frac{3}{2}(\bar{s}_\alpha d_\beta)_{V-A}\sum_q e_q(\bar{q}_\beta q_\alpha)_{V-A},
\end{aligned}
$$

$$(9.13)$$

with α and β being colour indicies.

9.2　Evaluating matrix elements

The $K \rightarrow (\pi\pi)_I$ amplitude is given by

$$\langle (\pi\pi)_I | \mathcal{H}(\Delta S = 1) | K \rangle = \frac{G_F}{\sqrt{2}} \mathbf{V}_{us}^* \mathbf{V}_{ud} \sum_{i=1}^{10} c_i(\mu)\langle (\pi\pi)_I | O_i | K \rangle_\mu, \qquad (9.14)$$

where $c_i(\mu) = z_i(\mu) + \tau y_i(\mu)$. Remember, the quantity μ was introduced to renormalize the operator $\mathcal{H}(\Delta S = 1)$ as demanded by quantum field theory; *in principle* the numerical value of such an auxiliary scale is irrelevant since the μ dependence of $c_i(\mu)$ is compensated for by a corresponding μ dependence in the matrix elements of $\langle (\pi\pi)_I | O_i | K \rangle_\mu$ so that it drops out from the transition amplitudes. However *in practice* its value has to be chosen judiciously for two reasons:

- We are faced with a 'Scylla and Charybdis' problem: on one hand we need $\mu \gg \Lambda_{QCD}$ for a perturbative treatment of the radiative corrections to be meaningful; on the other hand it has to be sufficiently low so that the non-perturbative algorithm we employ in calculating matrix elements can be applied.
- The situation is actually more dubious since we have to rely mainly on models or mere phenomenological prescriptions for computing matrix elements. Typically, those tools are not sufficiently refined to allow us to track the dependence on μ. We then rely on physical intuition to argue that μ should be matched to typical momenta in the hadronic wavefunctions, i.e. $\mu \sim 0.5 - 1 \, \text{GeV}$.

9.3 Chiral symmetry and vacuum saturation approximation

We have not learnt yet how to reliably compute matrix elements of O_i at low energies using analytical QCD algorithm. Often symmetries come to our rescue. We know that QCD possesses chiral symmetry; i.e. if quark masses are set to zero, left-handed and right-handed quarks do not talk to each other. We can write down an effective low energy theory with $U(3) \times U(3)$ chiral symmetry broken only by quark mass terms (see, for example [82] or [89]). Computing the Noether currents associated with chiral symmetry for both the QCD and the effective chiral Lagrangian [90], yields the following identities:

$$\bar{u}\gamma_\mu\gamma_5 u = \frac{F_\pi}{\sqrt{2}}\partial_\mu\pi^0, \qquad \bar{d}\gamma_\mu\gamma_5 u = F_\pi\partial_\mu\pi^+$$

$$\bar{s}\gamma_\mu\gamma_5 u = F_K\partial_\mu K^+, \qquad \bar{s}\gamma_\mu\gamma_5 d = F_K\partial_\mu K^0$$

$$\bar{s}\gamma_5 u = \frac{iM_K^2 F_K}{m_s + m_u}K^+, \qquad \bar{s}\gamma_5 d = \frac{iM_K^2 F_K}{m_s + m_d}K^0$$

$$\bar{s}u = \frac{M_K^2 - M_\pi^2}{m_s - m_u}\left(\pi^+ K^0 + \frac{1}{\sqrt{2}}\pi^0 K^+\right)$$

$$\bar{s}d = \frac{M_K^2 - M_\pi^2}{m_s - m_d}\left(\pi^- K^+ - \frac{1}{\sqrt{2}}\pi^0 K^0\right)$$

$$\bar{d}d = \frac{M_\pi^2}{2m_d}\left(\pi^+\pi^- + \frac{1}{2}\pi^0\pi^0\right)$$

$$\bar{s}\gamma_\mu u = -i\left(\frac{1}{\sqrt{2}}(\pi^0\partial_\mu K^+ - \partial_\mu\pi^0 K^+) + \pi^+\partial_\mu K^0 - \partial_\mu\pi^+ K^0\right)$$

$$\bar{s}\gamma_\mu d = -i\left(\frac{1}{\sqrt{2}}(K^0\partial_\mu\pi^0 - \partial_\mu K^0\pi^0) + \pi^-\partial_\mu K^+ - \partial_\mu\pi^- K^+\right).$$

$$(9.15)$$

Using these identities we can write the operators O_i to leading order in chiral perturbation theory as products of pseudo-Nambu–Goldstone bosons which arise from chiral symmetry breaking. The evaluation of $\langle(\pi\pi)_I|O_i|K\rangle_\mu$ to this lowest order is trivial (see Problem 9.5).

$$\langle\pi^+\pi^-|O_2^u|K^0\rangle = -\langle\pi^0\pi^0|O_1^u|K^0\rangle = X$$

$$\langle\pi^+\pi^0|O_1|K^+\rangle = \langle\pi^+\pi^0|O_2|K^+\rangle = \frac{1}{\sqrt{2}}X$$

$$\langle\pi^+\pi^-|O_6|K^0\rangle = \langle\pi^0\pi^0|O_6|K^0\rangle = Y, \qquad (9.16)$$

where

$$X = iF_\pi(M_K^2 - M_\pi^2) = 0.036i \text{ GeV}^3$$

$$Y = -2i(F_K - F_\pi)\left(\frac{M_K^2}{m_s + m_d}\right)^2$$

$$= -0.114i\left(\frac{175MeV}{m_s + m_d}\right)^2 \text{ GeV}^3. \qquad (9.17)$$

This ansatz is called *factorization* or the *vacuum saturation approximation*. Vacuum saturation cannot be an identity. For example, in this order matrix elements are independent of μ; thus the above cancellation of μ dependences cannot happen. We can of course compute higher order terms which require renormalization, and thus introduce a μ dependence. But we do not expect this path to be fruitful for reasons which will become obvious soon. Instead, we shall introduce fudge factors $B_i^I(\mu)$ with their deviation from unity representing the error in the approximation for a given operator, where I is the isospin of the two-pion system. For an energy scale $\mu \sim 0.7\,\text{GeV}$, where we expect chiral symmetry to be reasonably intact, we expect $B_i^I(\mu) = \mathcal{O}(1)$.

The operator O_6 has a mixed chirality structure of the form $(V - A) \times (V + A)$, see Eq. (9.13), which is due to the fact that the gluon couples to left- as well as right-handed quarks. Under a Fierz transformation it changes into $(S - P) \times (S + P)$; we might then suspect that it will have enhanced matrix elements for pseudoscalar hadrons [86]. This is indeed the case – see the Problems section – and it arises explicitly in Eq. (9.17) through the factor $[M_K/(m_s + m_d)]^2 \sim 10$.

9.4 $K \to \pi\pi$ decays

We are now ready to evaluate amplitudes for $K \to \pi\pi$ decays given by

$$\langle \pi^+\pi^0|\mathcal{H}(\Delta S = 1, \mu)|K^+\rangle = \frac{G_F}{\sqrt{2}}\mathbf{V}_{us}^*\mathbf{V}_{ud}\frac{1}{\sqrt{2}}(z_1(\mu) + z_2(\mu))X. \quad (9.18)$$

For $\mu \sim 0.7\,\text{GeV}$ (beyond this our chiral purturbation result is certainly questionable), $z_1 = -0.87$, $z_2 \sim 1.51$ [81]. Using these results,

$$|\langle \pi^+\pi^0|\mathcal{H}(\Delta S = 1, \mu)|K^+\rangle| \simeq 3 \times 10^{-8}\,\text{GeV}, \quad (9.19)$$

to be compared with an experimental number $1.8 \times 10^{-8}\,\text{GeV}$ – not bad for such a crude estimate!

Now, for the real part of the $K^0 \to \pi^+\pi^-$ decay amplitude, numerical studies show that all contributions except for those from O_2 and O_6 can be neglected:

$$\langle \pi^+\pi^-|\mathcal{H}(\Delta S = 1, \mu)|K^0\rangle = \frac{G_F}{\sqrt{2}}\mathbf{V}_{us}^*\mathbf{V}_{ud}(z_2(\mu)X + z_6(\mu)Y). \quad (9.20)$$

We thus obtain, with $z_6 = -0.082$,

$$\frac{\langle \pi^+\pi^-|\mathcal{H}(\Delta S = 1, \mu)|K^0\rangle}{\langle \pi^+\pi^0|\mathcal{H}(\Delta S = 1, \mu)|K^+\rangle} = \sqrt{2}\frac{z_2(\mu)X(\mu) + z_6(\mu)Y(\mu)}{(z_1(\mu) + z_2(\mu))X(\mu)} \sim 4. \quad (9.21)$$

This should be compared to the experimental value of

$$\frac{\langle \pi^+\pi^-|\mathcal{H}(\Delta S = 1, \mu)|K^0\rangle}{\langle \pi^+\pi^0|\mathcal{H}(\Delta S = 1, \mu)|K^+\rangle} = 15, \quad (9.22)$$

obtained from Eq. (5.9). Do we understand the $\Delta I = 1/2$ rule? The answer is obvious – and embarrassing. While we more or less understand the $\Delta I = 3/2$ amplitude, the precise origin of the $\Delta I = 1/2$ enhancement still has not been identified 50 years later. Various possible culprits have been lined up; we should note that QCD renormalization accelerates quite considerably the $\Delta I = 1/2$ rate through the enhanced operator O_-, as does the penguin operator through its large matrix element as inferred from chiral invariance. Yet there does not seem to exist a concise dynamical explanation for what had appeared to be an elegant rule. Instead Eq. (5.9) probably represents the cumulative result of various effects all working in the same direction,[1] yet no conclusive quantitative explanation has been given; this is not surprising since we are dealing here with predominantly nonperturbative effects. On the other hand it would be unfair to claim that this rule represents more than an embarrassment for theory.

9.5 ♠ Computation of ϵ'/ϵ ♠

Before we begin, a few general remarks might be useful:

- Since the interplay of three families is required for a **CP** asymmetry to become observable, direct **CP** violation cannot be generated from tree level diagrams – a one-loop operator is needed. Direct **CP** violation thus represents a pure quantum effect and can be expected to be reduced in the KM model. This can be seen by writing in the notation of Section 7.2

$$\frac{\epsilon'}{\epsilon} \sim \omega \frac{\xi_0 - \xi_2}{\epsilon_m}, \quad \omega \simeq 1/20, \tag{9.23}$$

  where $\epsilon_m = -\frac{\operatorname{Im} M_{12}}{\Delta M}$. The phase $\xi_0 - \xi_2$ of the loop diagrams is reduced even compared to ϵ_m.

- In the SM phases for the amplitudes A_0 and A_2 arise from penguin operators. With isospin symmetry penguin operators only produce $\Delta I = \frac{1}{2}$ transitions; i.e. A_2 is real. Then the phase ξ_2 arises from electroweak penguins dependent on quark charges, and matrix elements of interactions sensitive to $m_u - m_d$.

- Therefore we must include electroweak penguin operators $O_7 - O_{10}$ [88]. Once isospin breaking is introduced, the $\eta - \eta'$ system will mix with π^0. This will modify matrix elements associated with QCD penguin operators $O_3 - O_6$.

- Since ϵ' depends on $\xi_2 = \frac{1}{\omega} \frac{\operatorname{Im} A_2}{\operatorname{Re} A_0}$, these isospin breaking effects are enhanced by a factor of $1/\omega \sim 20$. Thus we need to take them seriously.

- Reflecting on the history of predicting ϵ'/ϵ, we note that ϵ is enhanced much more by a large top mass than ϵ'. This has to be kept in mind when judging older predictions on ϵ'/ϵ; they tend to be based on assuming a range for m_t that is well below the now measured value. Consider the 'old' analysis of Franzini [92]: it yields numbers very similar to present estimates once the now known value for m_t is inserted.

---

[1] For example, there is a strong attractive force in the $(\pi\pi)_0$ channel at around 400–800 MeV which further enhances the $\Delta I = 1/2$ matrix elements [91].

Unfortunately this opens the back door for $\epsilon'/\epsilon \simeq 0$. The strong penguin is not greatly enhanced by a large top mass; however, the electroweak penguin is – due to the longitudinal component of the virtual Z boson which is the reincarnation of one of the original Higgs fields – and it contributes with the opposite sign! Even $\epsilon'/\epsilon = 0$ can then occur for m_t large enough. This happens for $m_t \geq 200\,\text{GeV}$, which is not far above the observed mass.

To evaluate ϵ' using Eq. (7.34), we have to know the phases of A_0 and A_2, ξ_0 and ξ_2, respectively. Our lack of success in deriving the $\Delta I = 1/2$ rule from theory suggests that we should obtain the values of the various hadronic matrix elements from experimental data.

9.5.1 *Determining matrix elements from data*

We sketch the method used for obtaining matrix elements of these four quark operators to show you the complexities involved. Those who are interested in the actual determination are referred to Ref. [81].

At what value of μ shall we determine the matrix elements? It should be large enough so that the coefficient functions are reliably obtained using renormalization group equations. It will be clear below that $\mu = m_c$ is a good choice. But then we cannot rely on chiral perturbation theory to obtain the matrix element. This is just as well. For the region in which the coefficient functions are known, $\mu \gg \Lambda_{\text{QCD}}$, and the region where chiral perturbation theory determines the matrix elements, $\mu \sim m_\pi$, never overlap. Then we have nothing but experiments to help us determine the matrix elements. The experimental values we use are (see Problem 9.2):

$$\text{Re}\, A_0 = 3.38 \times 10^{-7}\,\text{GeV}$$
$$\text{Re}\, A_2 = 1.49 \times 10^{-8}\,\text{GeV}. \tag{9.24}$$

In the region $\mu > m_c$, the charm degree of freedom is not frozen and the Lagrangian is written as

$$\mathcal{H}(\Delta S = 1) = \frac{G_F}{\sqrt{2}} \mathbf{V}_{us}^* \mathbf{V}_{ud} \left((1 - \tau) \sum_{i=1,2} z_i(\mu)(O_i^u - O_i^c) + \tau \sum_{i=1}^{10} v_i(\mu) O_i \right). \tag{9.25}$$

With ϵ' so small, we have to include higher order terms in $\mathcal{H}(\Delta S = 1)$, including, in particular, *electroweak* penguin operators. Terms multiplied by τ can safely be neglected for the real parts of the matrix elements:

$$\text{Re}\, A_0 = \frac{G_F}{\sqrt{2}} \mathbf{V}_{us}^* \mathbf{V}_{ud} [z_1(\mu)\langle Q_1^u - Q_1^c \rangle_0 + z_2(\mu)\langle Q_2^u - Q_2^c \rangle_0]$$

$$\text{Re}\, A_2 = \frac{G_F}{\sqrt{2}} \mathbf{V}_{us}^* \mathbf{V}_{ud} [z_1(\mu)\langle Q_1^u\rangle_2 + z_2(\mu)\langle Q_2^u\rangle_2], \qquad (9.26)$$

with

$$\langle Q_i^q(\mu)\rangle_I \equiv \langle (\pi\pi)_I | Q_i^q | K \rangle. \qquad (9.27)$$

For illustration we use $z_1(m_c) = -0.459$ and $z_2(m_c) = 1.244$, which correspond to $\Lambda_{\overline{MS}} = 0.3\,\text{GeV}$ in the HV renormalization scheme. Noting that $\langle Q_1^u\rangle_2 = \langle Q_2^u\rangle_2$, and the renormalization group equation preserves this relation, we obtain

$$\langle Q_2\rangle_2 \simeq 0.010\,(\text{GeV})^3. \qquad (9.28)$$

Determining $\langle Q_i\rangle_0$ is more complicated. Imposing $\langle Q_-(m_c)\rangle_0 \geq \langle Q_+(m_c)\rangle_0 \geq 0$, where $Q_\pm$ is a matrix element of $O_\pm$ defined in Eq. (9.5), which holds in most non-perturbative approaches, we obtain

$$\langle Q_2\rangle_0 = (0.13 \pm 0.02)\,(\text{GeV})^3, \quad \langle Q_1\rangle_0 = (-0.06 \mp 0.06)\,(\text{GeV})^3. \quad (9.29)$$

Note that $\langle Q_2\rangle_0$ is much larger than $\langle Q_2\rangle_2$: for $\mu \geq m_c$ the charm penguin effect is included in $\langle Q_2\rangle$ rather than appearing as a separate operator, as is the case with $\mu \ll m_c$, discussed in the previous section.

$\langle Q_1\rangle_{0,2}$ and $\langle Q_2\rangle_{0,2}$ are thus known. How about the other matrix elements? $\text{Im}A_i$ arises from terms proportional to $\text{Im}\,\tau$ in Eq. (9.25). The matrix elements $\langle Q_5\rangle_0$, $\langle Q_6\rangle_0$, $\langle Q_7\rangle_2$ and $\langle Q_8\rangle_2$ have been studied in lattice QCD with the result that the vacuum saturation approximation seems to be reliable for these matrix elements. Finally, there are operator identities which allow us to write $\langle Q_4\rangle_0$, $\langle Q_9\rangle_0$, $\langle Q_{10}\rangle_0$, $\langle Q_9\rangle_2$, $\langle Q_{10}\rangle_2$ in terms of $\langle Q_1\rangle_I$, $\langle Q_2\rangle_I$, and $\langle Q_3\rangle_0$. So, if we know $\langle Q_3\rangle_0$, we are all set. So far there is no way to improve the reliability of the vacuum saturation approximation. But we note that vacuum saturation leads to $\langle Q_3\rangle_0 \sim X/3$; i.e. it is about a factor of 5 smaller than $\langle Q_2\rangle_0$. So, the result is not sensitive to $\langle Q_3\rangle_I$, and we shall use the vacuum saturation approximation.

9.5.2 Numerical estimates

It is by now quite clear that the theoretical status of the KM prediction for direct **CP** violation as expressed through ϵ'/ϵ is even more frustrating than the experimental situation used to be – and with less relief in sight. Inserting experimental numbers for ϵ and $\text{Re}\,A_0$, the result for ϵ'/ϵ given in Eq. (7.23) can be written as:

$$\frac{\epsilon'}{\epsilon} = -ie^{i(\delta_2 - \delta_0)} \times 10^{-4} \left(\frac{\text{Im}\,\lambda_t}{10^{-4}}\right) r \left(\sum_{i=3}^{6} y_i \langle Q_i\rangle_0 (1 - \Omega_{\eta\eta'}) - \frac{1}{\omega} \sum_{i=7}^{10} y_i \langle Q_i\rangle_2\right) \qquad (9.30)$$

where $r = \frac{G_F \omega}{|2\epsilon| \text{Re} \, A_0} = 336 \, \text{GeV}^{-3}$, and we used the fact that $y_1 = y_2 = 0$. We have written

$$\Omega_{\eta\eta'} = \frac{1}{\omega} \frac{\text{Im} \, A_2^{IB}}{\text{Im} \, A_0}, \tag{9.31}$$

where $\text{Im} \, A_2$ arises from the isospin breaking contribution of the penguin operators [93]. They give rise to $K \to \eta\pi^0 \to \pi^0\pi^0$ and $K \to \eta'\pi^0 \to \pi^0\pi^0$. Over the past 10 years, this number has gone down from 0.25 to 0.06 $\pm$ 0.08.

The numerical value of each contribution to Eq. (9.30), for $\Omega_{\eta\eta'} = 0$, is given as follows.

$$r \sum_{i=3}^{6} y_i \langle Q_i \rangle_0 \sim 0.133 - 3.93 + (-0.5 + 7.56) R_s$$

$$-\frac{r}{\omega} \sum_{i=7}^{10} y_i \langle Q_i \rangle_2 \sim (+0.19 - 3.49) R_s + 1.29 - 0.48, \tag{9.32}$$

where

$$R_s = \left(\frac{175 \, \text{MeV}}{m_s + m_d} \right)^2, \tag{9.33}$$

and we used $\Delta S = 0$ Wilson coefficients at $\mu = m_c = 1.3 \, \text{GeV}$, $\Lambda_{\text{QCD}} = 325 \, \text{MeV}$ and $m_t = 170 \, \text{GeV}$ as given by Buchalla *et al.* [94].

A few comments can provide orientation.

- The contribution from $y_3 \langle Q_3 \rangle_0$ is indeed small.
- $y_4 \langle Q_4 \rangle_0$ and $y_6 \langle Q_6 \rangle_0$ terms are from QCD charm and top quark penguin contributions, respectively. For $m_s = (170 \, \text{MeV} \sim 100 \, \text{MeV})$ they amount to $(4 \sim 19) \times 10^{-4}$.
- Electromagnetic penguins yield $-(2 \sim 9) \times 10^{-4}$ for the same range of m_s.
- QCD and electroweak penguins contribute with the opposite sign!
- Our very rough computation gives $\epsilon'/\epsilon = (4 \sim 10) \times 10^{-4}$.

There are considerable cancellations between various terms – especially between QCD and electroweak penguins. The theoretical prediction for ϵ'/ϵ thus suffers from a large uncertainty. It changes for different allowed input values of μ, Λ_{QCD}, m_c, $m_s(\mu)$, and these variations are magnified due to the cancellations. In the presentation above, we have retained the leading $1/N_C$ terms only mainly to illustrate the procedure in a more transparent way.

Going beyound our rough sketch of the theoretical analysis, one concludes that the standard model and the recent experimental result are consistent with each other [95].

9.6 $\Delta S = 2$ amplitudes

This is one of very few places in particle physics where second order weak amplitudes are accessible to experiment. It is therefore an ideal place to look for New Physics beyond the SM.

It also illustrates a point we have made before: proper treatment of **CP** phenomenology involves applying the full machinery of theoretical tools available in quantum field theory and an appreciation of its subtleties; thus it can be seen as a somewhat unconventional introduction into quantum field theory.

When the **CP** violating effect is small, we have from Eq. (6.39):

$$\Delta M_K = -2\overline{M}_{12}. \tag{9.34}$$

Without an elementary $\Delta S = 2$ interaction in the SM, we obtain the $\Delta S = 2$ amplitude by iterating the basic $\Delta S = 1$ coupling, i.e. we are dealing with a second-order weak process: $\mathcal{H}_{\text{eff}}(\Delta S = 2) = \mathcal{H}_{\text{eff}}(\Delta S = 1) \otimes \mathcal{H}_{\text{eff}}(\Delta S = 1)$. In doing so we get the celebrated box diagram shown in Fig. 9.4. There are local (c and t quark) and non-local (u quark loop) contributions. We shall concentrate on local contributions first and come back to non-local ones later. The contributions that do *not* depend on the mass of the internal quark cancel against each other due to the GIM mechanism [71]. Integrating over the internal lines then yields a convergent result. With three families we arrive at [96]

$$\mathcal{H}_{\text{eff}}^{\text{box}}(\Delta S = 2, \mu) = \left(\frac{G_F}{4\pi}\right)^2 M_W^2$$

$$\cdot \left[\eta_{cc}(\mu)\lambda_c^2 E(x_c) + \eta_{tt}(\mu)\lambda_t^2 E(x_t) + 2\eta_{ct}(\mu)\lambda_c\lambda_t E(x_c, x_t)\right] [\bar{d}\gamma_\mu(1 - \gamma_5)s]^2 \tag{9.35}$$

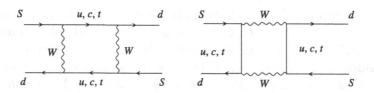

Figure 9.4   The box diagrams giving rise to a $\Delta S = 2$ operator.

with λ_i denoting combinations of KM parameters,

$$\lambda_i = \mathbf{V}_{is}\mathbf{V}_{id}^*, \ i = c, t, \tag{9.36}$$

$E(x_i)$ and $E(x_i, x_j)$ reflecting the box loops with equal and different internal quarks (charm or top), respectively:

$$E(x_i) = x_i \left(\frac{1}{4} + \frac{9}{4(1 - x_i)} - \frac{3}{2(1 - x_i)^2} \right) - \frac{3}{2} \left(\frac{x_i}{1 - x_i} \right)^3 \log x_i \tag{9.37}$$

$$E(x_c, x_t) = x_c x_t \left[\left(\frac{1}{4} + \frac{3}{2} \frac{1}{1 - x_t} - \frac{3}{4} \frac{1}{(1 - x_t)^2} \right) \frac{\log x_t}{x_t - x_c} + (x_c \leftrightarrow x_t) \right.$$

$$\left. - \frac{3}{4} \frac{1}{(1 - x_c)(1 - x_t)} \right] \ ; \ x_i = \frac{m_i^2}{M_W^2}, \tag{9.38}$$

and $\eta_{qq'}$ containing the QCD radiative corrections from evolving the effective Lagrangian from M_W down below m_c. The next-to-leading log corrections have been studied in order to understand the theoretical errors. We shall not go into the scale dependence as well as errors associated with uncertainties in Λ_{QCD}, m_t, etc. Such discussion can be found in Ref. [81]; the result is

$$\eta_{cc} \simeq 1.38 \pm 0.20, \ \ \eta_{tt} \simeq 0.57 \pm 0.01, \ \ \eta_{ct} \simeq 0.47 \pm 0.04. \tag{9.39}$$

As the last step, we have to evaluate matrix elements of the $\Delta S = 2$ transition operator. Even for a local four-fermion operator this is far from trivial since on-shell matrix elements are controlled by long-distance dynamics. Its size can be parametrized as follows (See Problem 9.6):

$$\langle K^0 | (\bar{d}\gamma_\mu(1 - \gamma_5)s)(\bar{d}\gamma_\mu(1 - \gamma_5)s) | \overline{K}^0 \rangle = -\frac{4}{3} B_K F_K^2 M_K. \tag{9.40}$$

The fudge factor B_K is called the bag factor for historical reasons. Let us see the origin of the minus sign. From Table 4.2 we note that $\mathbf{CP}\bar{s}(\vec{x})\gamma^\mu\gamma_5 d(\vec{x})\mathbf{CP}^\dagger = -\bar{d}(-\vec{x})\gamma^\mu\gamma_5 s(-\vec{x})$, while we have defined $\mathbf{CP}|K\rangle = +|\overline{K}\rangle$ by convention.

Several theoretical techniques have been employed to estimate the size of B_K. For a comprehensive review on hadronic matrix elements see Ref. [97]. The findings of several phenomenological studies can be summarized as follows:

$$B_K \simeq 0.8 \pm 0.2. \tag{9.41}$$

At which value of μ is Eq. (9.41) evaluated? Within the error stated in Eq. (9.41), it is reasonable to set $\mu \simeq 0.5$–$1\,\text{GeV}$ as the appropriate scale.

9.6.1 ΔM_K

Let us start with a numerical estimate of the short-distance contribution to ΔM_K. Equation (9.34) gives

$$\Delta M|_{SD} = -2\langle K|\mathcal{H}_{\text{eff}}^{\text{box}}(\Delta S = 2, \mu)|\overline{K}\rangle$$
$$= \frac{G_F^2}{6\pi^2} F_K^2 B_K M_K M_W^2 \left[\lambda_c^2 E(x_c)\eta_{cc}(\mu)\right.$$
$$\left. + \lambda_t^2 E(x_t)\eta_{tt}(\mu) + 2\lambda_c\lambda_t E(x_c, x_t)\eta_{ct}(\mu)\right]. \qquad (9.42)$$

For numerical estimates, we use the following numbers:

$$E(x_c) \sim 3 \times 10^{-4}, \quad E(x_t) \sim 2.7, \quad E(x_c, x_t) \sim 2.7 \times 10^{-3}$$
$$m_t = 180\,\text{GeV}, \quad m_c = 1.4\,\text{GeV}, \quad \lambda = 0.221. \qquad (9.43)$$

It is clear that since $\lambda_t \sim 3.5 \times 10^{-4} \ll \lambda_c$, the first term corresponding to the $\bar{c}c$ intermediate states dominates. Numerically, we have

$$\Delta M|_{SD} = 2 \times 10^{-12}\,\text{MeV}. \qquad (9.44)$$

Using the experimental value given in Eq. (5.42), we see that

$$\Delta M|_{SD} \sim \left(\frac{1}{3} \sim \frac{1}{2}\right) \cdot \Delta M_K|_{\text{exp}}. \qquad (9.45)$$

The deficit in $\Delta M|_{SD}$, viz. $\Delta M|_{\text{exp}}$, can – with the present reliability of our theoretical tools – be assigned to contributions from *non-local* operators corresponding to the u quark loop contribution which describes $K^0 \rightarrow \pi, \eta, \eta', \pi\pi, \ldots \rightarrow \overline{K}^0$ [98] intermediate states. These contributions can be summarized as follows:

$$\Delta M|_{SM} = \Delta M|_{SD} + \Delta M|_{LD},$$
$$\Delta M|_{LD} \sim \left(\frac{1}{2} \sim \frac{2}{3}\right) \cdot \Delta M|_{\text{exp}}. \qquad (9.46)$$

Obviously we wish the theoretical uncertainties were smaller. While we can expect the parameter B_K and thus $\Delta M|_{SD}$ to be known better in the foreseeable future, there is much less reason for such optimism with respect to $\Delta M|_{LD}$. Yet for proper perspective we should note that the quark box contribution in the *absence* of charm (and top) quarks amounts to (see Problem 9.7)

$$\Delta M|_{u,d,\prime}^{\text{box}} \simeq \frac{G_F^2}{3\pi^2} M_W^2 \cos^2\theta_C \sin^2\theta_C B_K F_K^2 M_K$$
$$\sim 10^{-8}\,\text{MeV} \sim 4000 \cdot \Delta M|_{\text{exp}}\,! \qquad (9.47)$$

While theory gives the right order of magnitude for ΔM_K, we should not overlook the fact that the theory also implies K_L must be heavier than K_S to give the correct sign for ΔM_K. Both of these conclusions are highly non-trivial.

9.6.2 ϵ

As shown in Eq. (7.42), we have

$$\epsilon \simeq \frac{1}{\sqrt{2}} e^{i\phi_{sw}} (\epsilon_m + \xi_0) \tag{9.48}$$

where

$$\epsilon_m = -\frac{\mathrm{Im}\, M_{12}}{\Delta M_K}. \tag{9.49}$$

The ξ_0 term comes from long-distance dynamics. The expression for ϵ'/ϵ given in Eq. (7.34) together with the experimental fact that $\epsilon' \ll \epsilon$ shows that the long-distance dynamics *cannot* generate ϵ by themselves. Within the KM ansatz, the interplay of all three families including top quarks is essential; since $m_t, m_c > M_K$ the relevant effective $\Delta S = 2$ operator is local; i.e. $\mathcal{H}_{\mathrm{eff}}^{\mathrm{box}}(\Delta S = 2)$ can be used in evaluating ImM_{12}:

$$\begin{aligned}
\epsilon_{KM} &\simeq \epsilon_{KM}^{\mathrm{box}} \\
&\simeq \frac{G_F^2}{12\sqrt{2}\pi^2} e^{i\phi_{sw}} \frac{M_W^2 M_K F_K^2 B_K}{\Delta M} \Big[\mathrm{Im}(\lambda_c^2) E(x_c)\eta_{cc} \\
&\quad + \mathrm{Im}(\lambda_t^2) E(x_t)\eta_{tt} + 2\mathrm{Im}(\lambda_c\lambda_t) E(x_c, x_t)\eta_{ct} \Big] \\
&\sim 1.9 \times 10^4 \, B_K e^{i\phi_{sw}} [\mathrm{Im}(\lambda_c^2) E(x_c)\eta_{cc} + \mathrm{Im}(\lambda_t^2) E(x_t)\eta_{tt} \\
&\quad + 2\mathrm{Im}(\lambda_c\lambda_t) E(x_c, x_t)\eta_{ct}].
\end{aligned} \tag{9.50}$$

Noting that $\lambda_t + \lambda_c + \lambda_u = 0$, we see that ϵ is proportional to J, defined in Eq. (8.38). Note also that the factors which depend on the KM matrix elements must provide about 10^{-8} suppression – a critical test of the KM ansatz. While ϵ has been measured very accurately, no tight bounds can be derived on these KM parameters due to the theoretical uncertainty in B_K, see Eq. (9.41). Yet, the constraint from ϵ will figure prominently in our discussion of **CP** asymmetries in B_d decays. All this will be discussed in detail later.

9.7   ♠ SM expectations for $\langle P_\perp \rangle$ in K_{l3} decays ♠

Let us see how a transversal polarization gets generated within the SM. The lowest order radiative correction to the $\overline{u}\gamma_\mu(1 - \gamma_5)sW^\mu$ vertex is

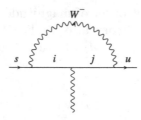

Figure 9.5 Feynman graph representing the vertex correction for the $\bar{u}\gamma_\mu$ $(1-\gamma_5)sW^\mu$ vertex. Here $i = u,\ c,\ t$; $j = d,\ s,\ b$ quarks.

shown in Fig. 9.5. What we want to do is to investigate how we get $\mathrm{Re}\,(F_P^* F_A) - \mathrm{Im}\,(F_S^* F_V)$, where the formfactors F_P, F_A, F_S, and F_V are defined in Eq. (7.57). Let us analyse Fig. 9.5. The loop integral I has the following form in the ultraviolet region:

$$\mathcal{I}_{UV} \propto \sum_{i,j} \mathrm{Im}\left(\mathbf{V}_{is}\mathbf{V}_{ij}^*\mathbf{V}_{uj}\right)$$

$$\times g^3 \int \frac{d^4l}{(2\pi)^4}\, \bar{u}\gamma_+\gamma_\alpha \frac{i}{\not{l}-m_i}\gamma_\mu\gamma_- \frac{i}{\not{l}-m_j}\gamma^\alpha\gamma_- s\, \frac{-i}{l^2-m_W^2}. \qquad (9.51)$$

It can be seen that the imaginary part of the integral has double GIM cancellations making the integral highly convergent. It is also seen that the most prominent part of Eq. (9.51) has exactly the $\bar{u}\gamma_\mu(1-\gamma_5)sW^\mu$ structure and it just changes the overall phase. Such terms do not contribute to the polarization. We estimate that the part which contributes to the polarization is bounded by:

$$\langle P_\perp \rangle < \frac{\alpha}{\pi \sin^2 \theta_W} A^2 \lambda^4 \eta \frac{m_b^2}{M_W^2} \sim 10^{-7}. \qquad (9.52)$$

9.8   Résumé

The full machinery of quantum field theory has been brought to bear on the theoretical treatment of the non-leptonic decays of strange hadrons: operator renormalization, even of the non-multiplicative variety based on the renormalization group, separation of perturbative and non-perturbative dynamics, simplifying hadronic matrix elements with the help of chiral symmetry, etc. The returns for our efforts have been mixed.

- We have not found a simple elegant reason underlying the $\Delta I = 1/2$ rule. We have identified several dynamical effects all enhancing

$\Delta I = 1/2$ amplitudes; yet we have not arrived at a quantitative dynamical understanding. It is a sore thumb – but not worse than that.

- We can understand why **CP** violation – and in particular its direct manifestations – are so delicate in strange decays. In addition to the required interplay of the three quark families, the $\Delta I = 1/2$ enhancement as well as the large top mass suppress quantities like ϵ'/ϵ.

- It naturally yields a small value for ϵ'/ϵ. Predicting a reliable numerical value for it is, however, quite a different story: the history of such predictions does not represent one of the glory stories of theoretical high energy physics. Yet measuring direct **CP** violation is of crucial importance, no matter what theory might – or might not – say.

- A semi-quantitative prediction for the $K_L - K_S$ mass difference ΔM has been obtained that agrees with its measured value – including the sign! The GIM mechanism is crucial for this success.

- We have established very good theoretical control over the **CP** breaking effective $\Delta S = 2$ Lagrangian; i.e. we can reliably express $\mathcal{L}(\Delta S = 2)$ in terms of fundamental SM parameters. Even the impact of long-distance dynamics is being brought under control, see Eq. (9.40) and Eq. (9.41).

- We will see in the next chapter that the observed value of ϵ can be reproduced without forcing any of the parameters, and that this is a highly non-trivial success that could easily have been an abject failure!

In any case, we have to be eager to tackle a new adventure where we can redeem ourselves and atone for past shortcomings.

Problems

9.1. Check Eq. (9.21) and try to understand Eq. (5.9)

$$\frac{\Gamma(K_S \to \pi^+\pi^-)}{\Gamma(K^+ \to \pi^+\pi^0)} = 4\left(\frac{z_2}{z_1 + z_2}\right)^2. \qquad (9.53)$$

Does it explain the $\Delta I = \frac{1}{2}$ rule?

9.2. First, remind yourself that

$$\Gamma(K^+ \to \pi^+\pi^0) = \frac{p_\pi}{8\pi M_K^2}|A(K^+ \to \pi^+\pi^0)|^2, \qquad (9.54)$$

then from the lifetimes for K^+, K_S, and branching ratios for the decays $K_S \to \pi^+\pi^-$ and $K^+ \to \pi^+\pi^0$, obtain

$$A(K^+ \to \pi^+\pi^0) = 1.8 \times 10^{-8} \,\text{GeV}$$
$$A(K^0 \to \pi^+\pi^-) = 27.5 \times 10^{-8} \,\text{GeV}$$
$$A(K^0 \to \pi^0\pi^0) = 26.4 \times 10^{-8} \,\text{GeV}. \tag{9.55}$$

9.3. Using identities given in Eq. (9.15), and remembering the normalization of states given in Section 6.1, derive

$$\langle 0|A_\mu^+(0)|P^+\rangle = -iF_P p_\mu, \qquad \langle 0|A_\mu^0(0)|P^0\rangle = \frac{-iF_P}{\sqrt{2}} p_\mu,$$

$$\langle \pi^0|V_\mu^+(0)|P^+\rangle = \frac{-1}{\sqrt{2}}(p_P + p_\pi)_\mu,$$

$$\langle \pi^+|V_\mu^0(0)|P^+\rangle = -(p_P + p_\pi)_\mu. \tag{9.56}$$

In our notation $F_{\pi^+} = 130 \,\text{MeV}$, $F_K = 160 \,\text{MeV}$.

9.4. Using $\mathbf{CP}A_\mu(t, \vec{x})\mathbf{CP}^\dagger = -A^{\dagger\mu}(t, -\vec{x})$ as in Table 4.2, show that

$$\int d^4x e^{ipx}\langle 0|A_\mu(t, \vec{x})|P^0\rangle = -\int d^4x e^{ipx}\langle 0|A^{\mu\dagger}(t, -\vec{x})|\overline{P}^0\rangle. \tag{9.57}$$

for $\mathbf{CP}|P^0\rangle = |\overline{P}^0\rangle$.

Note that the minus sign comes from our definition of $\mathbf{C}|P^0\rangle = -|\overline{P}^0\rangle$ – which also influences the definition of $\mathbf{CP}$ eigenstates P_1 and P_2.

9.5. We have evaluated the matrix elements using chiral perturbation theory. Historically, matrix elements were evaluated by inserting the vacuum in all possible ways. This shows that

$$\langle \pi^+\pi^-|O_2|K^0\rangle = \langle \pi^-|(\bar{s}u)_{V-A}|K^0\rangle\langle \pi^+|(\bar{u}d)_{V-A}|0\rangle$$
$$\langle \pi^+\pi^-|O_6|K^0\rangle = -2[\langle \pi^-|(\bar{s}u)_S|K^0\rangle\langle \pi^+| - (\bar{u}d)_P|0\rangle$$
$$+ \langle \pi^+\pi^-|(\bar{d}d)_S|0\rangle\langle 0|(\bar{s}u)_P|K^0\rangle]. \tag{9.58}$$

Using Eq. (9.15), derive Eq. (9.16).

9.6. To derive Eq. (9.40), show that

$$\langle \overline{K}^0|(\bar{s}_\alpha\gamma_\mu(1-\gamma_5)d_\alpha)(\bar{s}_\beta\gamma_\mu(1-\gamma_5)d_\beta)|K^0\rangle$$
$$= 2[\langle \overline{K}^0|\bar{s}_\alpha\gamma_\mu(1-\gamma_5)d_\alpha|0\rangle\langle 0|\bar{s}_\beta\gamma_\mu(1-\gamma_5)d_\beta|K^0\rangle$$
$$+ \langle \overline{K}^0|\bar{s}_\alpha\gamma_\mu(1-\gamma_5)d_\beta|0\rangle\langle 0|\bar{s}_\beta\gamma_\mu(1-\gamma_5)d_\alpha|K^0\rangle]. \tag{9.59}$$

Now, using Eq. (9.56), derive Eq. (9.40).

9.7. Ignoring QCD radiative corrections, calculate the quark box contribution to ΔM assuming there are neither charm nor top quarks. Is it finite? By dimensional argument convince yourself that Eq. (9.47) gives the right order of magnitude for this contribution.

9.8. Let us simplify the situation and consider a two family case. Write an expression corresponding to the Feynman graph shown in Fig. 9.4. Show that the u and c quark propagators combine to give

$$\frac{m_u - m_c}{(\not{p} - m_u)(\not{p} - m_c)} \tag{9.60}$$

a GIM cancellation.

9.9. Look at Eq. (9.50): setting $\lambda_t = 0$ or $x_t = 0$ would remove the top quark contribution to $|\epsilon|^{\text{box}}_{KM}$, yet it would seem that $|\epsilon|^{\text{box}}_{KM} \neq 0$ still holds – in conflict with the result that **CP** violation à la KM requires the interplay of three families! Resolve the apparent paradox.

10

Paradigmatic discoveries in
B physics

It's Beauty – Not Bottom![1]

In this chapter we describe the discoveries that have produced the B decay paradigm, namely the 'long' B lifetime, commensurate $B_d - \overline{B}_d$ oscillations and large **CP** asymmetries in $B_d \to \psi K_S$, $\pi\pi$, $K\pi$, $\eta' K_S$. The long lifetime was quite unexpected for lack of imagination, the $B - \overline{B}$ oscillation was unexpected since hardly anybody entertained the idea of top quarks being remotey as heavy as they turned out to be. As for the **CP** violation, they involve channels that are **Classes (C)** as well as **(B)** in the language of Section 6.7, and those of $B_d \to \psi K_S$, $\pi\pi$ decays have been predicted in 1980.

Here we want to emphasize the main points and illustrate the essential assistance obtained from subtle quantum mechanical effects and even from hadronization. Technical details and theoretical evaluations will be postponed to the two subsequent chapters.

10.1   The emerging beauty of B hadrons

In June 1997 an international conference entitled 'b20: Twenty Beautiful Years of Bottom Physics' was held at the Illinois Institute of Technology. This was a joint celebration of the twentieth anniversary of the discovery of Υ – the bound state of the $b\bar{b}$ quarks – and the seventy-fifth birthday of Leon Lederman, who led the discovery of the Υ. Following are brief excerpts from the discussions at the conference [99].

---

[1] Two sets of names had been suggested for the two quarks of the third family: 'truth' & 'beauty' or 'top' & 'bottom'. However, as formulated by M.K. Gaillard, while one set is pretentious, the other is vulgar. Believing that linguistic consistency can be overemphasized, we will use the names of 'top' & 'beauty'.

10.1.1 The discovery of beauty

Lederman and his group had performed experiments measuring

$$p + p \rightarrow X + \text{anything}$$
$$\hookrightarrow \mu^+ \mu^-. \tag{10.1}$$

This is considered to be an excellent way to detect a long lived new particle decaying into $\mu^+\mu^-$ [100]. Muons can penetrate a lot more material than any other particle except neutrinos; this property makes it easy to identify muons. Iron blocks from, say, old battleships placed in front of a detector can filter out particles other than muons (and neutrinos).

Sanda in his concluding talk reminisced:

> Back in 1970, I was a post-doc at the physics department of Columbia University, where Leon was a professor. Three post-docs shared an office on the 8th floor of Pupin. Leon came in one day and showed me the muon mass spectrum from the $p\bar{p}$ collision at the ISR, (Fig. 10.1). A curious thing about this spectrum is the shoulder at around 3 GeV. Leon used to tell us, jokingly, that there was a huge resonance at 6 GeV, and that the shoulder was its tail. It appears like a shoulder because there is not enough energy to actually produce a peak. He suggested that we work on computing the mass spectrum.

In the meantime, Lederman's younger competitor, Sam Ting, a professor at MIT, decided to look at this shoulder carefully. He performed an experiment at Brookhaven National Laboratory with much better resolution to measure the invariant mass of e^+e^- pairs at around 3 GeV. In 1974, he and his collaborators announced the discovery of the J particle, which showed up as a big peak at an invariant mass of 3 GeV [101]. The excess events which had made up Lederman's shoulder were coming from a very narrow bound state of a new quark – the charm quark – and its antiquark. Leon had been looking at the integrated version of the peak, as his resolution was much larger than the width of the resonance. This resonance was discovered simultaneously at the e^+e^- storage ring SPEAR of the Stanford Linear Accelerator Center [102], where it was named ψ. It is now listed in the Particle Data Book as J/ψ. For this discovery, Sam Ting and Burt Richter were awarded the Nobel Prize for physics in 1976.

Lederman had to keep on looking, and he did. His persistence paid off – as it often does, at least for some people – in 1976, when he eventually bumped into yet another narrow resonance, the Υ, around 10 GeV [103].

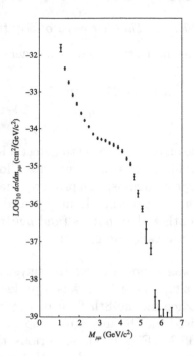

Figure 10.1   A curious shoulder in the muon mass spectrum of $p +$ Uranium $\rightarrow$ $\mu^+\mu^- +$ anything, which turned out to be J/ψ. This figure was reproduced from *Physical Review Letters* by permission of the American Physical Society.

Since the Υ had been discovered through its decay into a e^+e^- pair, it could be studied quite cleanly at an e^+e^- storage ring. The DORIS ring at DESY in Hamburg already had a successful career behind it analysing the dynamics of the charmonium family – the $\bar{c}c$ resonances just mentioned; in a perforce ride its energy was boosted threefold to reach the Υ regime [104].

Just around that time, the CESR ring at Cornell University in Ithaca was set up to study e^+e^- collisions in the appropriate energy region.[2] Now the race was on. Before long five resonances $\Upsilon(1S), \ldots, \Upsilon(5S)$ were identified. The important discovery for us was that of the $\Upsilon(4S)$ as shown in Fig. 10.2: its mass places it just slightly above the production threshold for B mesons, i.e. mesons containing a b quark together with a $\bar{d}$ or $\bar{u}$ antiquark. That the $\Upsilon(4S)$ decays into a $B\bar{B}$ pair almost all the time was observed by the CLEO [105] and CUSP [106] collaborations working at CESR.

---

[2] It was actually optimized for a c.m. energy of 16 GeV, but it was never operated there.

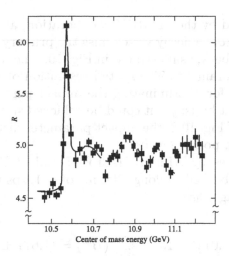

Figure 10.2   The hadronic cross-section for $e^+e^- \rightarrow$ hadrons near $\Upsilon(4S)$. By sitting on top of $\Upsilon(4S)$ the signal-to-noise ratio for $B^0\overline{B}^0$ production can be increased to 1 : 2.5. Not only that, $B^0\overline{B}^0$ is nearly at rest in the rest frame of $\Upsilon(4S)$ and also there is no $B^0\overline{B}^{*0}$ production as it is below the threshold. This figure was reproduced from *Physical Review Letters* by permission of the American Physical Society.

10.1.2   The longevity of B mesons

Aside from those well known gifted individuals who make major contributions in their teens, most of us get wiser as we get older. There are many things in life which can be obtained only by living longer. Likewise, the key discovery at the base of many interesting phenomena in B physics is that B mesons live for a 'long' time. This is remarkable also in two other respects, namely what 'long' means and how the discovery was achieved. Measuring the lifetimes of beauty hadrons became feasible through microvertex detectors; those had actually been developed with the goal of measuring the lifetimes of *charm* hadrons in hadronic collisions which just happen to have rather similar decay lengths. The basic idea is to track the time through space, that is, to infer the (proper) time of a decay from locating its decay vertex – i.e. a secondary vertex separated from the production point – in a tracking device. This is easier said than done as we are dealing with flight paths of only a few hundred microns (if that much) in an environment of numerous tracks belonging to the underlying event.

It was the MAC collaboration [107] working at the PEP ring of SLAC that found the first evidence of a long beauty life time. The discovery was quickly followed by the MARKII collaboration, also at PEPII [108]. Tracks emanating from a decay vertex miss the primary production vertex when extrapolated back, as illustrated in Fig. 10.3; i.e. they exhibit a non-vanishing impact parameter. The spatial resolution of the MAC detector was still too coarse for discriminating the impact parameter against zero on an event-by-event basis; yet it could be achieved statistically, as shown by the histogram of Fig. 10.3: the impact parameter distribution is shifted away from zero to a positive value.

Subsequent studies by ALEPH, DELPHI and OPAL at LEP and CDF at FNAL established[3] a 'long' lifetime of $\sim 1.5\,\mathrm{ps}$ for B mesons. The present world averages are

$$\tau(B_d) = (1.536 \pm 0.014) \times 10^{-12}\,\mathrm{s}., \quad \tau(B^{\pm}) = (1.671 \pm 0.018) \times 10^{-12}\,\mathrm{s}.$$

$$(10.2)$$

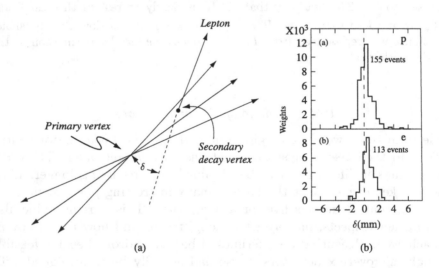

(a)   (b)

Figure 10.3   A B hadron is created at the primary vertex. It travels to the secondary decay vertex where it decays. The lifetime can be deduced by measuring the distance between the primary and secondary vertices, but with technology available at that time it was impossible to measure this distance. Therefore, a statistical analysis was performed. (a) δ is defined as an impact parameter of the lepton computed at the primary vertex. In (b) the distribution of δ is given. A non-vanishing δ was seen first by the MAC collaboration. This figure was reproduced from *Physical Review Letters* by permission of the American Physical Society.

---

[3] The history and up-to-date status of measurements of the lifetimes of charm and beauty hadrons is reviewed in Ref. [109].

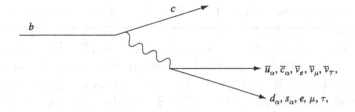

Figure 10.4 Feynman diagram responsible for b quark decay. If the final state quark masses are neglected, each channel contributes equally to the decay, and its contribution can be computed from μ decay. Note that there are nine channels as quarks come in three colours.

This implies one of the inequalities given in Eq. (8.45). To appreciate the fact that this is a long life time for a B meson, let us give a naive estimate. The total width of b quarks can be guessed at by scaling the expression for the muon width, accounting for the larger number of decay channels with quarks coming in three colours, see Fig. 10.4:

$$\Gamma_\mu = \frac{G_F^2}{192\pi^3}m_\mu^5 \implies \Gamma_b \sim \frac{G_F^2}{192\pi^3}m_b^5|\mathbf{V}_{bc}|^2 \times (2 \times 3 + 3). \quad (10.3)$$

The difference due to finite quark and lepton masses is ignored. For $|\mathbf{V}_{bc}| \sim 1$ we would have $\tau_b \sim \tau_\mu(m_\mu/m_b)^5/9 \simeq 10^{-15}$ s, i.e. a factor of 1000 shorter than observed. Actually at that time one typically estimated $\mathbf{V}_{cb} \sim \sin\theta_c$, since it described an inter-family transition:

$$\tau_b \sim \tau_\mu \left(\frac{m_\mu}{m_b}\right)^5 \frac{1}{9}\frac{1}{|\mathbf{V}_{cb}|^2} \sim 3 \cdot 10^{-14}\left|\frac{\sin\theta_C}{\mathbf{V}_{cb}}\right|^2 \text{ s.} \quad (10.4)$$

Compared with this expectation Eq. (10.2) represents a *long* lifetime; it shows that:

$$|\mathbf{V}_{cb}| \sim \frac{1}{30} \sim \mathcal{O}(\sin^2\theta_C) \ll \mathcal{O}(\sin\theta_C). \quad (10.5)$$

10.1.3 *The fluctuating identity of neutral B mesons*

Knowing B mesons to live a long time, we can hope that they would do something interesting while they are alive. Indeed, $B - \overline{B}$ oscillations were first discovered by the ARGUS collaboration through establishing the existence of same-sign dilepton events [110]:

$$e^+e^- \to \Upsilon(4S) \to B^0\overline{B}^0 \to \mu^\pm\mu^\pm + X, \quad (10.6)$$

where X stands for any final state. For without oscillations the two leptons from B^0 decays have to carry opposite charges. We must be careful, though, not to be fooled by leptons from $B^0\overline{B}^0 \to [l^+X_1][D+X_2]$ followed by $D \to l^+ + X_D$. Such secondary leptons can be separated from primary semileptonic decays $\overline{B}^0 \to l^- + X$ through various kinematic criteria like transverse momentum cuts, etc. ARGUS actually found a special beauty, namely the fully reconstructed event shown in Fig. 10.5.

To interpret oscillation data properly we have to apply quantum mechanical reasoning. The B meson pair in $\Upsilon(4S) \to B^0\overline{B}^0$ is produced into a **C** odd configuration with the two mesons flying apart from each other with momenta $\vec{k}$ and $-\vec{k}$ at time of production $t = 0$. Subsequently oscillations set in that are highly correlated for **C** $= -$: Bose statistics tells us that if one of the mesons is a B^0 at some time t, the other one cannot be a B^0 as well at *that* time, since the state must be odd under exchange of the two mesons. The time evolution of the pair is then simply[4]

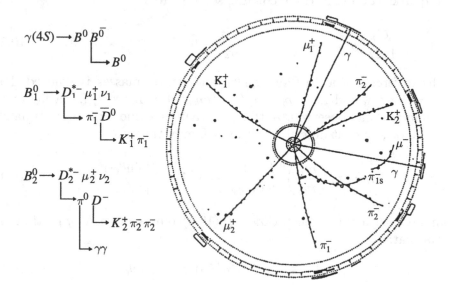

Figure 10.5   The gold-plated event discovered by the ARGUS collaboration. By knowing the momenta of all charged particles and the presence of π^0 from measuring the energy of $\gamma\gamma$, each decay chain can be reproduced. We can afford to have one neutrino for each B^0 meson decay as $B^0\overline{B}^0$ are produced nearly at rest. Reconstructed decay chains show that both are B^0 mesons which in turn shows that $B^0 - \overline{B}^0$ mixing must exist. This figure was reproduced from *Physics Letters* [110] by permission of Elsevier Science.

---

[4] It will be shown later that $\Delta\Gamma_B \ll \Delta M_B$ and thus we neglect $\Delta\Gamma$ in our argument below.

$$|(B^0\overline{B}^0)_{\mathbf{C}\,=-}(t)\rangle = e^{-\Gamma_B t}\frac{1}{\sqrt{2}}\left[|B^0(\vec{k})\overline{B}^0(-\vec{k})\rangle - |B^0(-\vec{k})\overline{B}^0(\vec{k})\rangle\right],$$

$$(10.7)$$

and we can never have like-sign primary dileptons emerge at a single time t. Once one of the B^0 hadrons has decayed, the coherence is lost, and the surviving $\overline{B}^0$ meson will oscillate without production constraint. From Eq. (10.7) it is straightforward to compute the time (t_2) dependence of the remaining B^0 once the first one has decayed at t_1 see (Problem 10.1). The probability that one charged lepton emerges at time t_1 and another one with the same charge at t_2 is given by [111]:

$$\Gamma((B^0\overline{B}^0)_{\mathbf{C}=-} \to l^+_{t_1} l^+_{t_2} X) \propto e^{-\Gamma(t_1+t_2)}\left|\frac{p}{q}\right|^2 \sin^2\frac{\Delta M_B}{2}(t_1 - t_2);$$

$$\Gamma((B^0\overline{B}^0)_{\mathbf{C}=-} \to l^-_{t_1} l^-_{t_2} X) \propto e^{-\Gamma(t_1+t_2)}\left|\frac{q}{p}\right|^2 \sin^2\frac{\Delta M_B}{2}(t_1 - t_2), \quad (10.8)$$

These rates indeed vanish for $t_1 = t_2$. Integrating over times of decay t_1, t_2 we obtain for the general case of a $B^0\overline{B}^0$ pair (Problem 10.2):

$$N_{++} = N([B^0\overline{B}^0]_{\mathbf{C}=\mp} \to l^+ l^+ + X) \sim \left|\frac{p}{q}\right|^2\left(1 - \frac{1 \pm x^2}{(1+x^2)^2}\right)$$

$$N_{--} = N([B^0\overline{B}^0]_{\mathbf{C}=\mp} \to l^- l^- + X) \sim \left|\frac{q}{p}\right|^2\left(1 - \frac{1 \pm x^2}{(1+x^2)^2}\right)$$

$$N_{+-} = N([B^0\overline{B}^0]_{\mathbf{C}=\mp} \to l^\pm l^\mp + X) \sim \left(1 + \frac{1 \pm x^2}{(1+x^2)^2}\right), \quad (10.9)$$

with $\left|\frac{q}{p}\right|^2 \simeq 1$. The rate for production of same sign dileptons compared to opposite sign dileptons is shown in Fig. 10.6. Interpretating the difference in the number of same-sign dilepton events for **C** even and odd is left as an exercise.

An observable which is classified as **Class (B)** in Section 6.7 is:

$$A_{SL} \equiv \frac{N_{--} - N_{++}}{N_{--} + N_{++}} = \frac{1 - \left|\frac{p}{q}\right|^4}{1 + \left|\frac{p}{q}\right|^4} \quad (10.10)$$

as shown in Eq. (6.72).

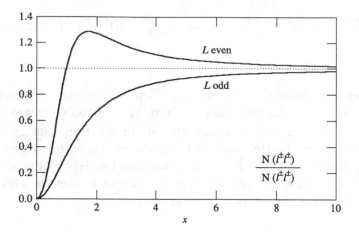

Figure 10.6 Rate for producing dilepton events for even and odd angular momentum $B^0\overline{B}^0$ states. The fact that there are fewer same sign dilepton events when the $B^0\overline{B}^0$ state is an odd angular momentum state than an even one can be understood by the fact that Bose statistics does not allow $B^0 B^0$ or $\overline{B}^0\overline{B}^0$ to exist at the same time.

Now let us look at the data. Starting with

$$\left.\frac{N_{++} + N_{--}}{N_{+-}}\right|_{C=-} = \frac{x^2}{2 + x^2}, \tag{10.11}$$

present data yield for B_d mesons (they actually go beyond the time-integrated result of Eq. (10.11) the following:

$$x_d = \frac{\Delta M_{B_d}}{\overline{\Gamma}_{B_d}} = 0.776 \pm 0.008, \quad \Delta M_{B_d} \simeq (3.34 \pm 0.03) \cdot 10^{-10}\,\mathrm{MeV}. \tag{10.12}$$

The physical meaning of x is very simple. Write:

$$\Delta M_B \cdot t = \frac{\Delta M_B}{\Gamma} \cdot \frac{t}{\tau}, \tag{10.13}$$

where Γ and τ are the total width and lifetime, respectively. What is important for the *observability* (rather than the occasional occurrence) of oscillations is that x be not too small, with $x \sim 1$ being optimal, since

$$\frac{\Delta M_B}{\Gamma} \sim \frac{\text{lifetime}}{\text{oscillation time}}. \tag{10.14}$$

The experimental sensitivity for the semileptonic **CP** asymmetry A_{SL} in B_d oscillations is approaching the 1% level [112]:

$$A_{SL}(B_d) = 0.0080 \pm 0.0090 \pm 0.0068 \tag{10.15}$$

Nature has provided us with an encore: there is a second neutral B meson that can oscillate, namely the one carrying strangeness: $B_s = [\bar{b}s]$. Within the SM ΔM_{B_s} (or $x_s = \Delta M_{B_s}/\bar{\Gamma}_{B_s}$) can rather reliably be related to ΔM_{B_d} (or $x_d = \Delta M_{B_d}/\bar{\Gamma}_{B_d}$), as explained below, with the unequivocal conclusion: $\Delta M_{B_s} \gg \Delta M_{B_d}$.

Both D0 and CDF at Fermilab have observed $B_s - \bar{B}_s$ oscillations with CDF obtaining [113]

$$x_s = 25.5 \pm 0.8, \quad \Delta M_{B_s} = (116.9 \pm 0.7 \pm 0.5) \times 10^{-10}\,\text{MeV}. \tag{10.16}$$

The measured value of ΔM_{B_s} agrees with the SM prediction as discussed below. For the semileptonic asymmetry one finds [112]

$$A_{SL}(B_s) = 0.020 \pm 0.021 \pm 0.009. \tag{10.17}$$

10.1.4 Another triumph for CKM dynamics

Let us focus on the two **CP** *in*sensitive observables $|V_{ub}/V_{cb}|$ and $\Delta M_{B_d}/\Delta M_{B_s}$ – i.e. quantities that do not require **CP** violation to acquire non-zero values. As explained in the next section also the second ratio represents a determination of CKM parameters with moderate uncertainty, namely $|V_{td}/V_{ts}|$. With $|V_{ub}/V_{cb}|$ and $|V_{td}/V_{ts}|$ one can construct the CKM triangle, see the plot on the left in Fig. 10.7.

It exhibits the following two facts.

- The measured values of these two quantities produce a *non*-flat CKM triangle: i.e. these two **CP** *in*sensitive observables tell us that **CP** violation has to exist in weak dynamics. This conclusion could not be drawn from the measurement of $|V_{ub}/V_{cb}|$ and ΔM_{B_d} due to the larger theoretical uncertainty in interpreting the latter.
- The CKM triangle thus constructed is even quantitatively consistent with the observed **CP** sensitive observables ϵ_K and sin $2\phi_1$, as shown by the plot on the right of Fig. 10.7.

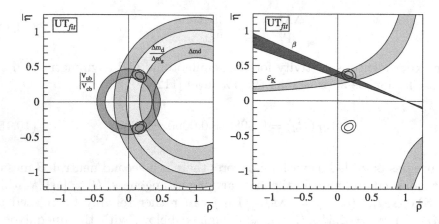

Figure 10.7   CKM unitarity triangle from $|V_{ub}/V_{cb}|$ and $|V_{td}/V_{ts}|$ on the left and compared to constraints from ϵ_K and $\sin 2\phi_1$ on the right (courtesy of M. Pierini).

10.2   What does the SM say about oscillations?

The off-diagonal element of the mass matrix is given by

$$\langle B^0|\mathcal{H}_{\text{eff}}(\Delta B = 2; \mu)^\dagger|\overline{B}^0\rangle_{(\mu)} \equiv M_{12} - \frac{i}{2}\Gamma_{12}. \tag{10.18}$$

While there are many similarities in the computation for ΔM_B and $\Delta\Gamma_B$, it is better to present them separately.

10.2.1   Computation of ΔM

The major contribution to ΔM_B comes from the top quark box graphs shown in Fig. 10.8. To compute M_{12}, start from Eq. (9.35) with $P = B$. Take the matrix element $\langle B^0|\mathcal{H}_{\text{eff}}^{\text{box}}|\overline{B}^0\rangle$ and keep only the top quark contribution proportional to $E(x_t)$. The result is:

$$M_{12} = \left(\frac{G_F}{4\pi}\right)^2 \xi_{tq}^2 M_W^2 E\left(\frac{m_t^2(\mu)}{M_W^2}\right) \eta_t^B(\mu)\langle B^0|(\overline{q}b)_{V-A}(\overline{q}b)_{V-A}|\overline{B}^0\rangle_{(\mu)},$$
$$\tag{10.19}$$

where $\eta_t^B(\mu)$ denotes QCD radiative corrections and

$$\xi_{tq} = V_{tb}V_{tq}^*, \ \ q = d, s. \tag{10.20}$$

Analogous to the K meson case, we express the matrix element with the help of the decay constant:

$$\langle B^0|(\overline{q}b)_{V-A}(\overline{q}b)_{V-A}|\overline{B}^0\rangle_{(\mu)} = -\frac{4}{3}B_B(\mu)F_B^2(\mu)M_B. \tag{10.21}$$

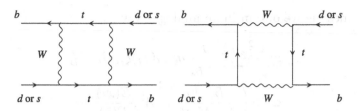

Figure 10.8 Feynman diagrams which are responsible for $\Delta B = 2$ transition.

$B_B = 1$ is again referred to as 'vacuum saturation' (VS) or factorization with F_B representing the decay constant for B mesons.

The oscillation parameter x_d can now be computed:

$$x_d \simeq -\frac{2\overline{M}_{12}}{\Gamma_{B_d}} \simeq \frac{G_F^2 F_B^2 M_B M_W^2}{\Gamma_{B_d}} \frac{\eta_t^B E(x_t)\, \bar{\xi}_{td}^2 B_B}{6\pi^2}$$

$$\simeq 0.63 \left(\frac{F_B}{150 MeV}\right)^2 B_B \left(\frac{E(x_t)}{E(5)}\right) A^2[(1-\rho)^2 + \eta^2]$$

$$\times \operatorname{sign}(\operatorname{Re} \xi_{td}^2) \tag{10.22}$$

where $\bar{\xi}_{td}^2 = |\xi_{td}^2| \times \operatorname{sign}(\operatorname{Re}\xi_{td}^2)$; we have calibrated the expression to $x_t = 5$ which corresponds to $m_t = 180\,\text{GeV}$. Also, $\tau_B \sim 1.5\,\text{ps}$ and $\eta_t^B = 0.55$ [114] have been used.

Note that the probability of observing like-sign dilepton events is proportional to x_d^2, see Eq. (10.11), and thus nearly proportional to $E(x_t)^2 \sim (m_t/M_W)^4$. While it is trivial to extend the oscillation analysis from the K to the B meson system, it takes considerable daring to entertain the possibility of the top quark being as heavy as $180\,\text{GeV}$ – which is needed to generate $x_d \sim 1$ – when data yielded a lower bound of a mere $20\,\text{GeV}$. Premature claims of the discovery of top quarks with $m_t \sim 40\,\text{GeV}$ did not help theorists much, either, at that time. The attentive reader will discern a slightly defensive tone here; yet in fairness this should be kept in mind when wondering why theorists did not predict $x_d \sim \mathcal{O}(1)$. On the other hand the ARGUS [110] discovery, confirmed speedily by CLEO [115], was readily perceived as the first clear evidence for an 'unusually heavy' top with m_t exceeding $100\,\text{GeV}$.

With a top quark mass of $20\,\text{GeV}$ – and no New Physics – the B_d meson would decay before it could exhibit a significant oscillation; likewise if τ_B had been of order $10^{-14}\,\text{s}$ – in either case we would not have written this book.

For B_s mesons we obtain in an analogous way:

$$x_s \simeq \frac{G_F^2 M_W^2}{6\pi^2} \frac{M_{B_s}}{\Gamma_{B_s}} \eta_t^B E(x_t) \bar{\xi}_{ts}^2 B_{B_s} F_{B_s}^2$$

$$\simeq 19 \left(\frac{\sqrt{B_{B_s}} F_{B_s}}{180 \text{ MeV}} \right)^2 A^2 \cdot \frac{E(x_t)}{E(5)}. \tag{10.23}$$

The new feature is that here all microscopic parameters are known, i.e. also $\lambda_t'^2 \simeq A^2 \lambda^4$ (at least in a three-family SM) in addition to $\eta_t^B E(x_t)$. The only unknown is the size of the hadronic matrix element expressed by $B_{B_s} F_{B_s}^2$. Using estimates of the latter in particular from lattice QCD one arrives at the following SM prediction:

$$x_s = 26.3^{+9.3}_{-2.2}, \quad \Delta M_{B_s} = \left(120^{+43}_{-10} \right) \cdot 10^{-10} \text{ MeV} \tag{10.24}$$

which is in good agreement with the experimental findings, see Eq. (10.16).

While the SM can thus account for at least most of the rapid B_s oscillation rate, one should note the sizable uncertainty in the SM prediction. Furthermore, while New Physics can make no more than a sub-dominant contribution to ΔM_{B_s}, it can still emerge as the dominant source of **CP** violation in $B_s \to l^- X$, $\psi\phi$ as discussed later.

10.3 ♠ On the sign of ΔM_B ♠

In considering the sign of ΔM_B we have to remember a few subtleties explained in Section 6.5 in particular.

- Whereas $\bar{\xi}_{ts}^2$ is necessarily positive within the SM, $\text{Re}\,\xi_{td}^2$ a priori could be either positive or negative; yet constraints from existing data tell us that it has to be positive as well.
- Thus we have $\overline{M}_{12} < 0$ for both B_d and B_s mesons with the convention $\mathbf{CP}|B\rangle = |\overline{B}\rangle$.
- However, the sign of ΔM_B by itself becomes an observable only if the two mass eigenstates can be distinguished by a difference in lifetimes or by their **CP** quantum numbers (if **CP** symmetry holds approximately).
- Neither of these criteria can be utilized in the B_d complex:
  - As stated below we predict the lifetimes of the two B_d mass eigenstates to differ by no more than about 1%. It is quite unlikely that such a small difference can be identified in the foreseeable future.
  - The KM ansatz unequivocally predicts large **CP** violation here.

- The situation is quite different for the B_s system.
  - As explained below lifetime differences here might amount to about 20% or so and become observable.
  - The $\Delta B = 2$ and the CKM favoured $\Delta B = 1$ effective operators for B_s transitions are **CP** conserving.
- Even if the sign of ΔM_B is not observable, the sign of the **CP** asymmetry $\sin\Delta M_B \cdot \mathrm{Im}\frac{q}{p}\bar{\rho}(f)$ is and can be predicted within a given theory of $\Delta B = 2$ dynamics.

10.4   CP violation in B decays – like in K decays, only different

Around 1976, A. Pais gave a seminar at Rockefeller University entitled 'CP Violation in Charmed Particle Decays' [116]. It represents the sentiment about **CP** violation at that time. Sanda's recollection of how Pais started out his seminar is as follows:

> There is good news and bad news. The good news is that **CP** violation in a heavy meson system is quite similar to that of the K meson system. The bad news is that there is little distinction like K_L and K_S mass eigenstates. For heavy meson systems, both lifetimes are short.

We strongly disagree with the good news. We claim that the good news is even better! The **CP** violation in B decay is three orders of magnitude larger than that of the K meson system. We agree with the bad news. Little did we know that it will take *only 20 years* to overcome the difficulties. Although $\Gamma_{K_S} \gg \Gamma_{K_L}$, for B_d meson, we know that $\Delta\Gamma_{B_d} \equiv \Gamma_{B_1} - \Gamma_{B_2} \ll \Gamma_{B_1} + \Gamma_{B_2}$ i.e. $\Gamma_{B_1} \simeq \Gamma_{B_2}$. While an initial beam of K^0 and $\overline{K}^0$ mesons is transformed through patience – i.e. waiting for the K_S component to decay away – into a practically pure K_L beam, this does not happen with $B_d/\overline{B}_d$ beams in any appreciable way, since the $B_{d,1}$ and $B_{d,2}$ lifetimes practically coincide. It has been stressed before that $\Gamma_{K_S} \gg \Gamma_{K_L}$ is due to a very exceptional situation, namely that only a handful of channels make up K^0 decays and the 'freak' accident of nature posting the mass of K_L meson barely above the 3π threshold.

So, we *must* look for some new feature in B decays. Not too long afterwards came the paper by Bander, Silverman and Soni [117], where they discussed **CP** asymmetries in b quark decay generated by penguin amplitudes.[5]

---

[5] This effect will be discussed later.

The new feature in B^0 decays which leads to large **CP** violation is that q/p is predicted to have a large phase for B_d mesons,

$$\frac{q}{p}\bigg|_{B_d} = \sqrt{\frac{M_{12}^*}{M_{12}} + \mathcal{O}\left(\frac{\Gamma_{12}}{M_{12}}\right)}\bigg|_{B_d} \simeq \frac{\mathbf{V}_{tb}^*\mathbf{V}_{td}}{\mathbf{V}_{tb}\mathbf{V}_{td}^*} \simeq \frac{(1-\rho)^2 - \eta^2 - 2i\eta(1-\rho)}{(1-\rho)^2 + \eta^2}.$$

$$(10.25)$$

The **CP** asymmetry in *semileptonic* B^0 decays due to $|p/q| \neq 1$ (see Eq. (10.10)), which Pais and Treiman had discussed is indeed quite tiny. Is there a **CP** violating observable which exposes this phase?

We need a novel idea to find an observable depending on q/p rather than its modulus. This was the impetus behind analysing flavour-nonspecific modes [118]. The **Class (C1)**, as discussed in Section 6.7, comes to our rescue. There exist final states f *common* to B^0 and $\overline{B}^0$ decays. $A(f) \cdot \overline{A}(f) \neq 0$. Then the **CP** asymmetry is proportional to Im $\frac{q}{p}\overline{\rho}(f)$.

The two decays shown in Fig. 10.9 are, however, different at the quark level, and it would appear we do not have a common final state f. One contains $(\overline{s}d)$ and the other $(\overline{d}s)$. Yet once we realize experimentalists see hadrons K_S and K_L, not quark states $(\overline{s}d)$ or $(\overline{d}s)$

$$(\overline{s}d) \to K_S + X \leftarrow (\overline{d}s) \qquad (10.26)$$

we have opened up a whole new territory to explore. *Hadronization thus provides essential help in making a* **CP** *asymmetry observable.*

The authors started their collaboration in 1980. At that time not a single B decay mode had been identified, let alone the lifetime measured, not even to mention oscillations. While there is a priori a host of such states – many of them of great practical relevance as discussed in the following chapters – we realized [119] that

$$B^0 \to \psi K_S \qquad (10.27)$$

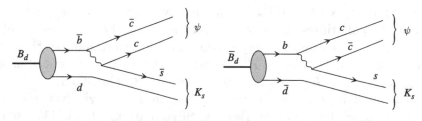

Figure 10.9   This is one of the very few places in which we can take advantage of the fact that we can only detect hadrons instead of quarks. Quark pairs $(\overline{s}d)$ and $(\overline{d}s)$ can be considered identical if we detect hadronic final states $K_S + X$ or $K_L + X$.

is an optimal channel for detecting a **CP** asymmetry for both experimental and theoretical reasons:

- The decays $\psi \to l^+l^-$ provide a striking signature; the accompanying $K_S \to \pi^+\pi^-$ is also relatively easy to detect.
- It is not KM suppressed. It was estimated to command a small branching ratio of order 10^{-3}. The often repeated comment that all channels with large **CP** asymmetries suffer from small branching ratios, however, overstates the facts: since a 5 GeV hadron has so many exclusive (non-leptonic) channels available for decay, they will all have smallish branching ratios. Semi-inclusive modes such as $B^0 \to \psi K_S + X$ will occur more frequently; yet the asymmetry will get washed out, unless X is chosen wisely, in particular if the final state is not a pure **CP** eigenstate.
- The statistical error on an asymmetry is

$$\text{Error}\left(\frac{N_+ - N_-}{N_+ + N_-}\right) \propto \frac{1}{\sqrt{N}}, \qquad (10.28)$$

where N is the smaller of $N_\pm$. To establish an asymmetry in a perfect detector, we need the ratio of asymmetry to error to be large.

$$\text{Asym} \times \sqrt{N} \gg 1. \qquad (10.29)$$

By choosing an inclusive decay mode with large branching ratio, such as that shown in Eq. (10.26), we gain in N but lose in the size of the asymmetry. The inclusive final state is not likely to be a **CP** eigenstate. The fact that the N dependence enters through a square root means that it is often much more effective to choose a decay channel with a large asymmetry at the expense of a smaller branching ratio than to choose a decay with larger branching ratio but smaller asymmetry.

- The final state is an odd **CP** eigenstate (ignoring the tiny **CP** odd component in the K_S state). With this mode being driven by the isoscalar coupling $b \to c\bar{c}s$, it is described by a *single* isospin transition amplitude. Therefore it is quite unlikely even in the presence of New Physics to exhibit *direct* **CP** violation; thus

$$|A(\psi K_S)| = |\overline{A}(\psi K_S)| \quad \text{or} \quad |\overline{\rho}(\psi K_S)| = 1. \qquad (10.30)$$

Taken together with $|q|^2 \simeq |p|^2$, we see that the observable $(q/p)\overline{\rho}(\psi K_S)$ represents practically a unit vector in the complex plane.

- Using $\Delta\Gamma_{B_d} \ll \Delta M_{B_d} \sim \Gamma_{B_d}$, the expressions derived in Section 6.7 give:

$$\Gamma(B_d[\overline{B}_d](t) \to \psi K_S) \propto e^{-\Gamma_B t} G_{\psi K_S}[\overline{G}_{\psi K_S}](t) \qquad (10.31)$$

$$\frac{G_{\psi K_S}(t)}{2} = |A(\psi K_S)|^2 \left[1 - \mathrm{Im}\left(\frac{q}{p}\overline{\rho}(\psi K_S)\right) \sin \Delta M_{B_d} t\right] \qquad (10.32)$$

$$\frac{\overline{G}_{\psi K_S}(t)}{2} = |A(\psi K_S)|^2 \left[1 + \mathrm{Im}\left(\frac{q}{p}\overline{\rho}(\psi K_S)\right) \sin \Delta M_{B_d} t\right]; \qquad (10.33)$$

i.e. the **CP** violation represented by $\mathrm{Im}((q/p)\overline{\rho}(\psi K_S)$ becomes observable due to $\Delta M_{B_d} \neq 0$ and depends on the time of decay:

$$\frac{d}{dt}\frac{G_{\psi K_S}(t)}{\overline{G}_{\psi K_S}(t)} \neq 0. \qquad (10.34)$$

- The asymmetry is reliably expressed in terms of fundamental CKM parameters only – i.e. with great parametric reliability – and is numerically large.[6] For this transition is driven by the effective *isoscalar* Hamiltonian

$$\mathcal{H}_{\mathrm{eff}}(\Delta B = 1) = \frac{G_F}{\sqrt{2}} \mathbf{V}_{cb}\mathbf{V}_{cs}^* \cdot [C_1 \overline{s}\gamma_\mu(1-\gamma_5)b\overline{c}\gamma_\mu(1-\gamma_5)c$$
$$+ C_2 \overline{c}\gamma_\mu(1-\gamma_5)b\overline{s}\gamma_\mu(1-\gamma_5)c] + \mathrm{h.c.}, \quad (10.35)$$

and we find

$$\frac{q}{p}\overline{\rho}(\psi K_S) \simeq -\frac{\mathbf{V}_{tb}^*\mathbf{V}_{td}}{\mathbf{V}_{tb}\mathbf{V}_{td}^*} \cdot \frac{\mathbf{V}_{cb}\mathbf{V}_{cs}^*}{\mathbf{V}_{cb}^*\mathbf{V}_{cs}} \cdot \frac{\mathbf{V}_{us}\mathbf{V}_{ud}^*}{\mathbf{V}_{us}^*\mathbf{V}_{ud}} = -e^{-2i\phi_1},$$

$$\mathrm{Im}\frac{q}{p}\overline{\rho}(\psi K_S) \simeq \sin 2\phi_1, \qquad (10.36)$$

where the minus sign is due to ψK_S being a **CP** odd state. There are three important points to Eq. (10.36).

- The factor $\frac{\mathbf{V}_{us}\mathbf{V}_{ud}^*}{\mathbf{V}_{us}^*\mathbf{V}_{ud}}$ will be omitted in discussions below. It comes from $K^0 \to \pi^+\pi^-$ decay which is used to identify K_S. This factor makes this physical observable independent of quark phases – as it must be.
- The hadronic matrix element (including its strong phase shift) drops out from the ratio since only a single isospin amplitude contributes in this channel [119].

---

[6] Figure 2 of Ref. [119] contains a prediction for the asymmetry that could be close to 100% for $|\mathbf{V}_{td}| \ll |\mathbf{V}_{cb}| \ll |\mathbf{V}_{us}|$ together with vacuum saturation.

    – The asymmetry is predicted to be naturally large, since ϕ_1 is an angle in a triangle with similar length sides, see Fig. 8.2(6).

• Finally as discussed in Section 6.8, we remind the reader that the sign of $\sin(\phi_1)$ can also be extracted irrespective of the sign of ΔM_B. This can be expressed as follows:

$$\frac{\Gamma(\overline{B}_d(t) \to \psi K_S) - \Gamma(B_d(t) \to \psi K_S)}{\Gamma(\overline{B}_d(t) \to \psi K_S) + \Gamma(B_d(t) \to \psi K_S)} = \sin(\Delta M_{B_d} t)\mathrm{Im}\left(\frac{q}{p}\bar{\rho}(\psi K_S)\right)$$

$$= \sin(|\Delta M_{B_d}|t)\sin(2\phi_1).$$

$$(10.37)$$

10.5 From sweatshops to beauty factories

Due to a high signal-to-noise ratio and low associated multiplicities, e^+e^- annihilation provides a clean laboratory for studying beauty physics; in addition, $\Upsilon(4S) \to B^0\overline{B}^0$ offers the considerable advantage that beam-energy constraints are particularly powerful for this two-body final state.

In $e^+e^- \to B^0\overline{B}^0$, however, the situation is more involved since both mesons oscillate – in a correlated fashion! Consider the event shown in Fig. 10.10, where a $B^0\overline{B}^0$ pair is produced in a **C** odd configuration, as in $\Upsilon(4S) \to B^0\overline{B}^0$. The time evolution of the *pair* is given by Eq. (10.7). Once one of the beauty mesons is tagged as a $\overline{B}^0$ by, for example, its semileptonic decay $\overline{B}^0 \to l^- + X$ at time t, we know that its pair produced partner has to be a B^0 at *that* time; its amplitude for decaying an interval Δt *later* into a **CP** eigenstate f depends – due to EPR correlations as explained in Section 10.10.4 – only on Δt, the time elapsed since the first decay, and is given by [118]:

$$A(f, t_2, t_1) = e^{-\frac{1}{2}\Gamma(t+\Delta t)}e^{-i\frac{1}{2}(M_1+M_2)(t+\Delta t)}$$

$$\times \left[A(f)\cos\left(\frac{1}{2}\Delta M\Delta t\right) + i\frac{q}{p}\overline{A}(f)\sin\left(\frac{1}{2}\Delta M_B\Delta t\right)\right]. \quad (10.38)$$

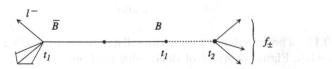

Figure 10.10 Configuration for a $B^0\overline{B}^0|_{\mathbf{C}=-}$ event in which $\overline{B}^0$ is identified at t_1 through its semileptonic decay with l^-, and its partner, a B^0 at t_1, decays to a **CP** eigenstate f at $t_2 > t_1$.

The same expression applies when $B^0 \to f$ actually occurs before $\overline{B}^0 \to l^- + X$ by an amount Δt; i.e. Eq. (10.38) holds for Δt both positive and negative. The probabilities for $(B^0\overline{B}^0)_{C=-} \to [l^\pm + X]_t + [f]_{t+\Delta t}$ are then given by (see Problem 10.1):

$$\Gamma(B^0\overline{B}^0|_{\mathbf{C}=-} \to [l^+X]_{t_l} f_{t_f}) \propto e^{-\Gamma(t_l+t_f)} |A(l^+)|^2 |A(f)|^2$$
$$\cdot \left[1 + \mathrm{Im}\left(\frac{q}{p}\overline{\rho}(f)\right) \sin\Delta M_B(t_f - t_l)\right]$$

$$\Gamma(B^0\overline{B}^0|_{\mathbf{C}=-} \to [l^-X]_{t_l} f_{t_f}) \propto e^{-\Gamma(t_l+t_f)} |\overline{A}(l^-)|^2 |A(f)|^2$$
$$\cdot \left[1 - \mathrm{Im}\,\frac{q}{p}\overline{\rho}(f)\sin\Delta M_B(t_f - t_l)\right], (10.39)$$

where we have assumed $|\overline{\rho}(f)| = 1$. This correlated time evolution is shown in Fig. 10.11 for $f = \psi K_S$ with $\mathrm{Im}\,\frac{q}{p}\overline{\rho}(\psi K_S) \simeq \sin 2\phi_1 = +0.6$.[7] Later we will see that the distributions that can actually be measured look a bit different for small values of $|t_f - t_l|$ for well-understood technical reasons.

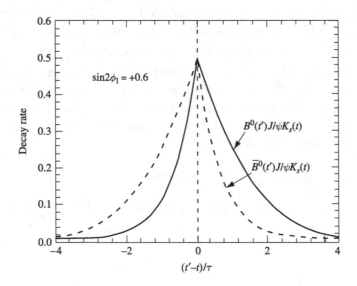

Figure 10.11　The calculated proper time distribution of $J/\psi K_S$ final states for $\sin 2\phi_1 = +0.6$. Figure courtesy of Belle collaboration.

---

[7] This figure was already included in the first edition of our book; i.e. it was drawn before 1999 and thus well before accurate data existed. Nevertheless it anticipated the experimental findings!

10.5.1 Disappointment at a symmetric machine

There arises a new obstacle. The asymmetry in Eq. (10.39) is an *odd* function of the difference between the two decay times t_f and t_l. When integrating over them, the asymmetry gets averaged to zero. Unfortunately the B mesons travel a mere $30\,\mu$m in the $\Upsilon(4S)$ rest-frame before they decay. Presently available technology for microvertex detectors does not provide us with the necessary resolution.

For a while there was hope that we could harness the seemingly magic powers of quantum mechanics in another way (rather than detector technology) to overcome this hurdle. Consider

$$e^+e^- \to B^{0*}\overline{B}^0 \to B^0\overline{B}^0 + \gamma; \tag{10.40}$$

the $B\overline{B}$ pair is now produced in a **C** *even* configuration.[8] For a **C** even $B^0\overline{B}^0$ state, the **CP** asymmetry depends on $\sin \Delta M_B(t_f + t_l)$ rather than $\sin \Delta M_B(t_f - t_l)$; in that case integrating over all times of decay would *not* average out the asymmetry. CLEO and CUSB have searched for the reaction of Eq. (10.40) at $\Upsilon(4S)$; alas, they did not find a sufficiently strong signal!

10.5.2 A crazy idea

It was a seemingly crazy idea that provided the breakthrough out of this impasse, namely to build an *asymmetric* e^+e^- collider, as first suggested by P. Oddone during our conversations with him on the difficulty mentioned above. If $\Upsilon(4S)$ was moving so that B^0 mesons have a large enough boost, they would leave a track that could be measured by existing vertex detectors.

Now, how much luminosity do we need? The cross section for B^0 production just above threshold is shown in Fig. 10.2. At the $\Upsilon(4S)$ we have a 1 nb cross section available. Let us aim for 100 events of the type $B^0\overline{B}^0 \to [l^{\pm}X][\psi K_S]$ to establish the asymmetry during one year of running. With $\text{BR}(B_d \to \psi K_S) \simeq 5 \times 10^{-4}$, $\text{BR}(K_S \to \pi^+\pi^-) \simeq 0.7$, $\text{BR}(\psi \to \mu^+\mu^-) \simeq 0.12$ and $\text{BR}(B_d \to l^+ + X) \simeq 0.1$ we need a luminosity of

$$L \sim \frac{3 \times 10^7 \text{ events}}{10^7 \text{ s } 10^{-33} \text{ cm}^2} = 3 \times 10^{33}\,\text{s}^{-1}\,\text{cm}^{-2} \tag{10.41}$$

for a perfect detector, where we have assumed that we can run 10^7 s in a calendar year with $\pi \cdot 10^7$ s. This was nearly 1000 times higher than

---

[8] The $B\overline{B}$ pair produced in $\Upsilon(4S)$ decays is equivalently labelled as a P wave or a **C** odd pair; like in this case, the **C** label often allows for a more concise argument.

the highest luminosity achieved *at that time* at CESR, the world record holder. With this kind of luminosity, it deserves to be hailed as a factory rather than a sweatshop.

At first, and for some time, some accelerator physicists let it be known that an asymmetric collider could not be built with a luminosity $\sim 10^{34}\,\mathrm{cm}^{-2}\,\mathrm{s}^{-1}$, let alone more. There were sound technical reasons for that scepticism – yet those challenges were overcome by the two teams building the B factories at KEK and SLAC. Their e^+e^- rings colliding positrons with about $3\,\mathrm{GeV}$ on electrons of about $9\,\mathrm{GeV}$, which became operational before the turn of the millenium, have achieved the peak luminosities of $17 \times 10^{33}\,\mathrm{s}^{-1}\,\mathrm{cm}^{-2}$ and $12 \times 10^{33}\,\mathrm{s}^{-1}\,\mathrm{cm}^{-2}$, respectively! Which is to show that truly formidable obstacles can be overcome through human ingenuity coupled with persistence – if the prize is attractive enough. This is certainly the case here.

10.6   First reward – $B_d \rightarrow \psi K_S$

The decay rate evolution of a beam initially made up of B_d $[\overline{B}_d]$ mesons is described by Eq. (10.32) and Eq. (10.33). There are several ways – usually called 'flavour tagging' – in which one can determine the *initial* flavour as a B_d or $\overline{B}_d$. One typical one is to observe a flavour *specific* transition. For example, Fig. 10.12 shows a semileptonic decay of the other beauty hadron produced electromagnetically or strongly in conjunction with the B_d or $\overline{B}_d$:

$$B_d\overline{B} \rightarrow B_d(l^-X)_B \quad vs. \quad \overline{B}_dB \rightarrow \overline{B}_d(l^+X)_B. \tag{10.42}$$

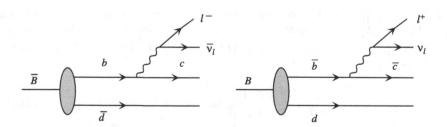

Figure 10.12   Feynman diagrams for right-sign semileptonic decays. These decays obey the $\Delta B = \Delta Q$ rule, i.e. only $\overline{B}^0 \rightarrow l^- + X$ and $B^0 \rightarrow l^+ + X$ occur. Thus leptonic decays can be used to identify the decaying objects as either B^0 or $\overline{B}^0$.

The *existence* of the asymmetries in $B^0 \to K_S + (\psi,\ \psi',\ \cdots)$ were established by both groups independently and basically simultaneously in 2001:

$$\sin(2\phi_1) = \begin{cases} 0.99 \pm 0.14 \pm 0.06 & \text{Belle} & [120] \\ 0.59 \pm 0.14 \pm 0.05 & \text{BaBar} & [121] \\ 0.79 \pm 0.10 & \text{world average.} \end{cases}$$

By the summer of 2007 accurate measurements of the $B^0 \to \psi K_S$ and $B^0 \to \psi K_L$ asymmetries had been obtained:

$$\sin(2\phi_1) = \begin{cases} 0.642 \pm 0.031 \pm 0.017 & \text{Belle} & [122] \\ 0.697 \pm 0.035 \pm 0.016 & \text{BaBar} & [123]. \end{cases}$$

In the notation of Eq. (6.81), we also have $C_{\psi K_S} = 0.018 \pm 0.025 \simeq 0$.

The observation of the **CP** asymmetry in $B_d \to \psi K_S$ represents an important turning point in several respects.

- Unlike the establishment of $\epsilon' \neq 0$, which marked the completion of an era, it signals the beginning of a new one being the first **CP** asymmetry observed outside the $K^0 - \overline{K}^0$ complex.
- It suggests the 'floodgates' for **CP** violation will open; i.e. the ingredients are there for **CP** asymmetries to emerge in many B^0 channels.
- **CP** violation has been 'demystified': if the dynamics are sufficiently complex for supporting **CP** violation, the latter does not have to be small; i.e. the observable phases can be large. This 'demystification' will be completed, once one finds **CP** violation anywhere in the lepton sector.
- Having passed this decisive test, the CKM paradigm has been promoted from a mere ansatz to a tested theory – albeit one we suspect to be incomplete.

10.7 The second reward – $B_d \to \pi^+\pi^-$

The $B_d \to \pi^+\pi^-$ decay mode has become well established in the last few years. The HFAG [112] average of BaBar, Belle, CLEO, CDF results is

$$\text{BR}(B_d \to \pi^+\pi^-) = (5.16 \pm 0.22) \cdot 10^{-6}, \tag{10.43}$$

which is somewhat smaller than the original prediction of $(1 \sim 2) \times 10^{-5}$.

As will be discussed in detail below, due to penguin diagrams we expect

$$|A(\pi^+\pi^-)| \neq |\overline{A}(\pi^+\pi^-)|. \tag{10.44}$$

Thus we expect **Class (C2) CP** violation, discussed in Section 6.7, to surface here. The **CP** asymmetry can be expressed as follows:

$$\frac{G_{\pi^+\pi^-}(t) - \overline{G}_{\pi^+\pi^-}(t)}{G_{\pi^+\pi^-}(t) + \overline{G}_{\pi^+\pi^-}(t)} = C_{\pi^+\pi^-}\cos\Delta M_{B_d}t - S_{\pi^+\pi^-}\sin\Delta M_{B_d}t \quad (10.45)$$

$$C_{\pi^+\pi^-} = \frac{1 - |\overline{\rho}(\pi^+\pi^-)|^2}{1 + |\overline{\rho}(\pi^+\pi^-)|^2}, \; S_{\pi^+\pi^-} = \frac{2\mathrm{Im}((q/p)\overline{\rho}(\pi^+\pi^-))}{1 + |\overline{\rho}(\pi^+\pi^-)|^2} \quad (10.46)$$

An interesting consistency check for experiments is

$$C^2_{\pi^+\pi^-} + S^2_{\pi^+\pi^-} \leq 1. \quad (10.47)$$

BaBar [124] and Belle [125] report:

$$S_{\pi^+\pi^-} = \begin{cases} -0.61 \pm 0.10 \pm 0.04 & \text{Belle} \\ -0.61 \pm 0.11 \pm 0.03 & \text{BaBar} \end{cases} \quad (10.48)$$

$$C_{\pi^+\pi^-} = \begin{cases} -0.55 \pm 0.08 \pm 0.05 & \text{Belle} \\ -0.21 \pm 0.09 \pm 0.02 & \text{BaBar} \end{cases} \quad (10.49)$$

$C_{\pi^+\pi^-} \neq 0$ – which evidently holds for the Belle measurement, yet less so for BaBar's analysis – represents *direct* **CP** violation. The latter is expected to arise here, since in contrast to $B_d \to \psi K_S$ we have $|\overline{\rho}(\pi\pi)| \neq 1$ due to the concurrence of the following two facts.

- The final state, which happens to be **CP** even, is made up of a combination of $I = 0$ and $I = 2$ configurations which can conceivably possess significantly different strong phase shifts.
- There are two quark-level diagrams with different topologies and different CKM parameters that contribute, namely the spectator process and the Cabibbo reduced penguin reaction, as shown in Fig. 10.13; the latter affects the $I = 0$ final state only.

Direct **CP** violation could also manifest itself through $S_{\pi^+\pi^-} \neq -S_{\psi K_S} \simeq \sin 2\phi_1$, yet presently there is no significant difference between $S_{\pi^+\pi^-}$ and $\sin 2\phi_1$.

CP violation has been established in $B^0 \to \pi\pi$ decay. Can we get some information about the unitarity triangle from this data? The answer is that much more work is needed, since the non-vanishing $C_{\pi^+\pi^-}$ implies that there are non-negligible penguin contributions.

If we could ignore the penguin contribution (which would represent a poor approximation of course) and consider only the tree level graph shown in Fig. 10.13, then we could proceed exactly as for $B^0 \to \psi K_S$, with the matrix elements being:

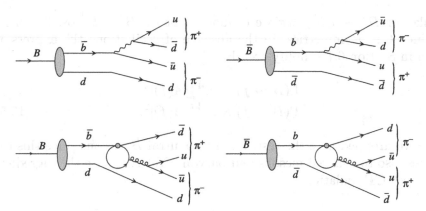

Figure 10.13 Tree and penguin quark diagrams which contribute to $B^0 \to \pi\pi$ decay. The penguin diagrams give us considerable problems in predicting **CP** violation in $B^0 \to \pi\pi$ decay.

$$\overline{A}(\pi\pi) = \frac{4G_F}{\sqrt{2}} \mathbf{V}_{ub}\mathbf{V}_{ud}^* \left[C_1 \langle \pi\pi | \overline{d}_L\gamma_\mu b_L \overline{u}_L\gamma_\mu u_L | \overline{B} \rangle \right.$$
$$\left. + C_2 \langle \pi\pi | \overline{u}_L\gamma_\mu b_L \overline{d}_L\gamma_\mu u_L | \overline{B} \rangle \right], \tag{10.50}$$

$$A(\pi\pi) = \frac{4G_F}{\sqrt{2}} \mathbf{V}_{ub}^*\mathbf{V}_{ud} \left[C_1 \langle \pi\pi | (\overline{d}_L\gamma_\mu b_L \overline{u}_L\gamma_\mu u_L)^\dagger | B \rangle \right.$$
$$\left. + C_2 \langle \pi\pi | (\overline{u}_L\gamma_\mu b_L \overline{d}_L\gamma_\mu u_L)^\dagger | B \rangle \right]. \tag{10.51}$$

In that simple scenario the hadronic matrix element still drops out from the ratio $\overline{\rho}(\pi^+\pi^-)$; i.e.

$$\frac{q}{p}\overline{\rho}(\pi\pi) \simeq \frac{\mathbf{V}_{tb}^*\mathbf{V}_{td}}{\mathbf{V}_{tb}\mathbf{V}_{td}^*} \frac{\mathbf{V}_{ub}\mathbf{V}_{ud}^*}{\mathbf{V}_{ub}^*\mathbf{V}_{ud}} = e^{2i\phi_2}, \quad \mathrm{Im}\frac{q}{p}\overline{\rho}(\pi\pi) \simeq \sin 2\phi_2. \tag{10.52}$$

More work is needed to get at ϕ_2. We shall come back to this in Chapter 11.

10.8 More rewards – $B^0 \to K\pi$, $\eta'K_S$

10.8.1 B → Kπ

It had been recognized for a long time [117, 118] that all the dynamic ingredients are there for a sizeable **Class (B)** direct **CP** asymmetry to surface in the four modes:

$$\overline{B}_d \to K^-\pi^+, \ \overline{K}^0\pi^0, \quad B^- \to K^-\pi^0, \ \overline{K}^0\pi^-. \tag{10.53}$$

Unlike for $B^0 \to \psi K_S$, $\pi\pi$ we cannot count on $B^0 - \overline{B}^0$ oscillations to produce **CP** asymmetries. In the absence of oscillations the expressions given in Section 6.7 simplify greatly:

$$\Gamma(B \to f) \propto e^{-\Gamma t}|A(f)|^2$$
$$\Gamma(\overline{B} \to \overline{f}) \propto e^{-\Gamma t}|\overline{A}(\overline{f})|^2. \tag{10.54}$$

We have dropped here the restriction to neutral B mesons since this case applies also to charged mesons and baryons. Yet there are the ingredients for direct **CP** violation.

- Two isospin amplitudes drive $B^0 \to K\pi$, one with $\Delta I = 0$ and one with $\Delta I = 1$; their strong phase shifts have no reason to be the same.
- In addition to the spectator transition, there are penguin operators containing different CKM parameters.

  Let us denote the spectator amplitude, Fig. 10.14(a), as tree amplitude and write

$$T(K\pi) = \mathbf{V}_{ub}\mathbf{V}_{us}^* T. \tag{10.55}$$

The penguin amplitudes, Fig. 10.14(b), with top and charm contribution can be written with their CKM factors:

$$P^c(K\pi) = \mathbf{V}_{cb}\mathbf{V}_{cs}^* P^c, \qquad P^t(K\pi) = \mathbf{V}_{tb}\mathbf{V}_{ts}^* P^t. \tag{10.56}$$

A **CP** asymmetry is thus proportional to $\sin\phi_3$.
- The $b \to u$ spectator contribution is CKM suppressed, making it roughly comparable to the penguin term. This enhances the interference between them and thus improves the prospects for a sizeable **CP** asymmetry.
- The penguin process with an internal charm quark provides at least a qualitative model for final state interactions.

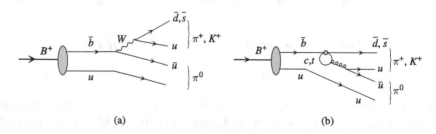

(a)   (b)

Figure 10.14   Quark diagrams for $B^+ \to K^+\pi^0$ and $B^+ \to \pi^+\pi^0$. (a) The tree graph contribution. (b) Penguin contribution.

Thus one expects and even predicted sizeable direct **CP** asymmetries to emerge in some of the $B^0 \rightarrow K + \pi$ channels. For example in 1988, well before $B^0 \rightarrow \pi\pi$, $K\pi$ had even been seen, it was predicted [44]:

$$\text{BR}(B_d \rightarrow K^+\pi^-) \sim \mathcal{O}(10^{-5}) \tag{10.57}$$

$$\frac{\Gamma(\overline{B}_d \rightarrow K^-\pi^+) - \Gamma(B_d \rightarrow K^+\pi^-)}{\Gamma(\overline{B}_d \rightarrow K^-\pi^+) + \Gamma(B_d \rightarrow K^+\pi^-)} \sim -0.1. \tag{10.58}$$

The world average [112] of BaBar, Belle, CLEO, CDF data reads

$$\frac{\Gamma(\overline{B}_d \rightarrow K^-\pi^+) - \Gamma(B_d \rightarrow K^+\pi^-)}{\Gamma(\overline{B}_d \rightarrow K^-\pi^+) + \Gamma(B_d \rightarrow K^+\pi^-)} = -0.095 \pm 0.013. \tag{10.59}$$

Significant hints for class **Class (B)** asymmetries have been observed by Belle and BaBar in ηK^{*0}, ηK^+, and $K^+\rho^0$ channels. To extract information on the unitarity triangle is more difficult, since accurate computations of strong decay amplitudes and their phases are needed.

10.8.2 $B_d \rightarrow \eta' K_S$

$B_d \rightarrow \eta' K_S$ has an unexpectedly large branching ratio:

$$\text{Br}(B_d \rightarrow \eta' K_S) = (3.4 \pm 0.2) \cdot 10^{-5}. \tag{10.60}$$

This makes it easier to search for a **CP** asymmetry. What range of **CP** asymmetry do we expect? First let us get some idea about the dynamics driving this decay. Figure 10.13(b) represents penguin transitions. Replacing the penguin transition $b \rightarrow d(\overline{u}u)$ by $b \rightarrow s(\overline{s}s)d$, we will get $B_d \rightarrow \eta' K_S$ with

$$\frac{q}{p}\overline{\rho}(\eta' K_S) = \frac{\mathbf{V}_{tb}^*\mathbf{V}_{td}}{\mathbf{V}_{tb}\mathbf{V}_{td}^*}\frac{\mathbf{V}_{tb}\mathbf{V}_{ts}^*}{\mathbf{V}_{tb}^*\mathbf{V}_{ts}} = e^{2i\phi_1}. \tag{10.61}$$

Thus we expect a large asymmetry like in $B_d \rightarrow \psi K_S$, though with the opposite sign. The data yield [113]

$$S_{\eta' K_S} = \begin{cases} 0.64 \pm 0.10 \pm 0.04 & \text{Belle} \\ 0.58 \pm 0.10 \pm 0.03 & \text{BaBar} \end{cases} \tag{10.62}$$

$$C_{\eta' K_S} = \begin{cases} 0.01 \pm 0.07 \pm 0.05 & \text{Belle} \\ -0.16 \pm 0.07 \pm 0.03 & \text{BaBar} \end{cases} \tag{10.63}$$

in full agreement with the SM predictions $S_{\eta' K_S} \sim \sin 2\phi_1$, $C_{\eta' K_S} \simeq 0$. This consistency check on the SM will become quite powerful with more statistics.

10.9   CPT invariance vs. T and CP violation

Let us analyse Fig. 10.15.[9] It shows that B^0 tag decays tend to happen with $\Delta t > 0$; i.e. after the transition to ψK_S. This by itself represents

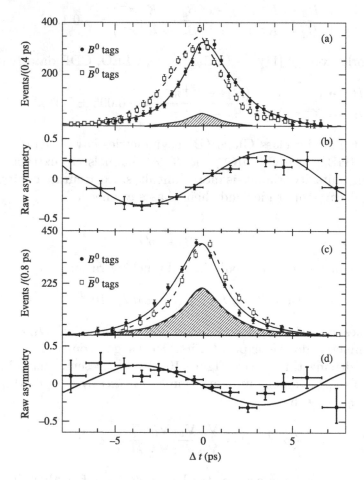

Figure 10.15   (a) The measured Δt distributions in ψK_S, $\psi(2S)K_S$, χK_S and $\eta_c K_S$ where the **CP** parity is -1. All these channels have exactly the same asymmetry shown in Eq. (10.37). (b) The **CP** asymmetry as a function of Δt. (c) and (d) are corresponding distributions for the **CP** even final state ψK_L. Note that the **CP** asymmetries shown in (b) and (d) are of opposite sign as it should be. This figure is obtained from Ref. [127].

---

[9] We are happy to find that Fig. 10.11, drawn back in 1999, turned out to be very close to the experimental findings. One qualitative difference between the theoretical and the experimental curves – Fig. 10.11 and Fig. 10.15, respectively – is easily understood: the detection efficiency for very 'early' B decays deteriorates leading to a smoothing out of the cusp at $\Delta t = 0$ in Fig. 10.11.

T violation. For **T** invariance implies that $\Delta t > 0$ and $\Delta t < 0$ events happen with equal probability. The **CP** conjugate of the B^0 tags events is given by the $\overline{B}^0$ tag curve. Those two curves are manifestly different. Both curves taken together show **T** as well as **CP** to be violated.

Yet we can learn more, namely that those two features are correlated, by considering a **CPT** transformation. Let us reverse the time for $\overline{B}^0$ tags $[\Delta t \to -\Delta t]$ and compare the resulting curve with that for B^0 tags. Within measurement errors those two curves coincide; i.e. the observed **T** and **CP** violations are *commensurate as required by* **CPT** *symmetry*. To look at it from an empirical point of view: detailed comparisons of the B^0 and $\overline{B}^0$ tag curves represent **CPT** tests.

It should be noted that the EPR correlation between B^0 and $\overline{B}^0$ imposed by Bose(–Einstein) statistics is essential for the argument, since it synchronizes the oscillations of the B^0 and $\overline{B}^0$ and thus leads to the dependence on Δt.[10]

10.10 Reflections[11]

10.10.1 On the virtue of 'over-designing'

In our estimate of the required luminosity given in Section 10.5.2, we assumed implicitly an asymmetry of about 30 %. If one had known that the **CP** asymmetry in $B_d \to \psi K_S$ is close to 70%, one might have argued that a luminosity of $\sim 10^{33}\,\mathrm{s}^{-1}\,\mathrm{cm}^{-2}$ would suffice rather than the design values of $10 \times 10^{33}\,\mathrm{s}^{-1}\,\mathrm{cm}^{-2}$ and $3 \times 10^{33}\,\mathrm{s}^{-1}\,\mathrm{cm}^{-2}$ of KEKB and PEP-II, respectively, let alone the actually achieved values. Yet we should appreciate that the machines are *not* over-designed – the higher luminosity is *not* a luxury.

- As explained later, our ability to make accurate predictions for a host of other **CP** asymmetries is greater than thought originally.
- Rare transitions like $B_d \to \phi K_S$, $\eta' K_S$ can now be analysed quantitatively.
- We will argue below that one cannot *count* on New Physics intervening 'massively' in B decays.
- As described later a 'Super-B' factory with a luminosity on the $10^{36}\,\mathrm{s}^{-1}\,\mathrm{cm}^{-2}$ level has an essential research programme fully complementary to the ones at the LHC and the anticipated ILC. It

---

[10] The B^0 curve *by itself* also implies **CP** violation. For its time dependence is *not* given by a simple exponential as required by the theorem given in Section 7.1.4. Yet the latter assumes the strict validity of the Wigner–Weisskopf approximation.

[11] Some readers might view it as 'Pontificating'.

would have seemed frivolous to contemplate such an undertaking when existing B factories were operating 'merely' at $10^{33}\,\mathrm{s}^{-1}\,\mathrm{cm}^{-2}$. We are actually convinced that even such a facility would be far from 'over-designed'.

10.10.2   The 'unreasonable' success of CKM theory

Before discussing B decays in more and quantitative detail we want to briefly recapitulate salient features of the theory of flavour-changing neutral currents and emphasize the major successes scored by the SM with the KM ansatz added.

As explained in Chapter 8 the constraints on the CKM parameters imposed by observables can be expressed through triangles, in particular the so-called unitarity triangle (UT). Figure 10.16 illustrates our knowledge in 1998, *before* the observation of **CP** violation in B decays. It shows three bands representing constraints due to measurements of $|\mathbf{V}_{ub}/\mathbf{V}_{cb}|$, ΔM_{B_d}, and ϵ_K. At first we might attribute little weight to the observation that these three bands do intersect, since they appear fairly broad, mainly due to theoretical uncertainties. However, such an evaluation – although often repeated – misses the profound message as we read it.

- The observables ϵ_K, ΔM_{B_d} and $\Gamma(B \to lX_u)$, which get connected by the KM framework, represent quite different dynamic regimes that proceed on very different time scales. A priori it would seem that the extracted value of parameters could have been quite different – by orders of magnitude! – from what they turned out to be. We are tempted to say that it borders on the miraculous that these quantities

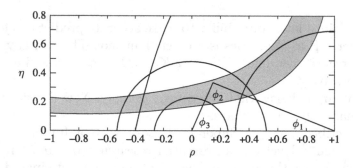

Figure 10.16   The allowed region of (ρ, η) of the unitarity triangle. The semicircular band with centre at $(0,0)$ represents the constraint from $|\mathbf{V}_{ub}/\mathbf{V}_{cb}|$. The semicircle with centre at $(1,0)$ represents the constraint from $B^0 - \overline{B}^0$ oscillations. Finally, the meshed band represents the constraint from ϵ_K.

can be described in terms of a unitary CKM matrix with $|\rho|, |\eta| \leq 1$. An uncertainty by a factor of 2 in the values of ρ and η should be viewed as puny if the related observables could have been different by orders of magnitude. In other words: the KM phenomenology can accommodate only a tiny corner of a general parameter space – yet nature has put the observables exactly into that 'neighbourhood'!

- Even after committing ourselves to the KM 'neighbourhood', phenomenological success is far from guaranteed. Keep in mind that in the 1970s and early 1980s values like $|\mathbf{V}_{cb}| \sim 0.04$ and $|\mathbf{V}_{ub}| \sim 0.004$ would have seemed quite unnatural; claiming the top mass having to be $\sim 180\,\text{GeV}$ would have been perceived as preposterous even in the 1980s! Consider a scenario with $|\mathbf{V}_{cb}| \simeq 0.04$ and $|\mathbf{V}_{ub}| \simeq 0.003$, yet $m_t = 40\,\text{GeV}$; in the mid-1980s this appeared to be quite natural (in particular in view of the claimed 'discovery' – later withdrawn – of top quarks with $m_t = 40 \pm 10\,\text{GeV}$). In that case we would need a value for η much larger than allowed by the $|V_{ub}/V_{cb}|$ constraints.

- The point of these considerations is to illustrate that the KM description could easily have suffered a phenomenological collapse over the last fifteen years – but it did not!

We humbly submit to the reader that the KM ansatz is quite unlikely to be merely a coincidence or – worse still – mirage: we see it as reflecting a significant aspect of **CP** violation – though presumably not the only one – even if we are ignorant about its dynamical foundation.

10.10.3 *Praising hadronization*

Hadronization and nonperturbative dynamics in general are usually viewed as a highly unwelcome complication – if not outright nuisance – for our description of fundamental dynamics. Based on our present theoretical control over these effects (or lack thereof) we cannot rule out large fractions of the observed ΔM_K, ϵ_K and ΔM_B and even most of ϵ'_K being due to New Physics.

However such a perspective misses the deeper truth. *Without* hadronization neutral kaons could not form as bound states of quarks and antiquarks, and thus there could be *no* $K^0 - \bar{K}^0$ oscillations. The latter involving quantum mechanical coherence over macroscopic distances allows to measure the truly tiny mass differences $\text{Im}M_{12} \sim \mathcal{O}(10^{-8}\,\text{eV})$ that drives indirect **CP** violation there.

Let us analyse the situation in a bit more detail. Hadronization helps significantly in three aspects.

- The rate for the **CP** conserving transition $K_L \to 3\pi$ is reduced by a factor ~ 500 due to *hadronic* phase space (this enhancement factor would be much less if the pion mass were half its real value).
- It awards 'patience'; i.e. one can 'wait' to obtain a pure K_L beam by letting the K_S component decay away.
- It is essential for generating the **CP** violating signal through the *existence* of a reaction – $K_L \to \pi\pi$ – rather than an asymmetry between two conjugate reactions – $s \to u\bar{u}d$ versus $\bar{s} \to \bar{u}u\bar{d}$. Fewer events are needed to establish the former than the latter.

Having different, yet *coherent* states – $|K^0\rangle$ and $|\overline{K}^0\rangle$ – and transition amplitudes – $T(K^0 \to f)$ and $T(K^0 \to \overline{K}^0 \to f)$ – is essential for quantum mechanical state mixing and **CP** asymmetries. Hadronization provides the portals for this to happen; it acts as a 'cooling' process enhancing the coherence of amplitudes, and its intrinsic strength is of great advantage in this context.

The situation is quite analogous for the beauty sector. It is only due to the efficient intervention of hadronization that $B^0 - \overline{B}^0$ oscillations can occur and the tiny mass difference $\Delta M_{B_d} \sim \mathcal{O}(10^{-4}\,\text{eV})$ be measured. Of course these minute quantities are very sensitive to the intervention of New Physics – yet that should be viewed as an extra bonus.

While hadronization presents serious challenges to our computational capabilities, we should not lose sight of the fact that without it many phenomena could not occur and thus not be studied in the first place. In later chapters we will give more examples illustrating hadronization's role as a difficult, yet indispensable ally. Hadronization should therefore be seen as a hero of the tale of **CP** violation rather than the villain it is usually portrayed.

10.10.4 *EPR correlations – a blessing in disguise*

The expressions derived in Eq. (10.39) state that $B^0\overline{B}^0$ pairs produced in

$$e^+e^- \to \Upsilon(4S) \to B_d\bar{B}_d, \tag{10.64}$$

may have an asymmetry depending on $t_f - t_l$ as:

$$Asym = \sin(\Delta M_B(t_f - t_l))\sin 2\phi_1. \tag{10.65}$$

Originally this caused big disappointment, as explained in Section 10.5.1. It lead us to the difficult path of building an asymmetric collider. In retrospect, it was truly a blessing in disguise.

- The time of decay cannot be measured directly, yet inferred from the length of the flight path of the B meson between its production point and decay vertex. While the decay vertex is usually well defined, the B meson production point is not due to finite beam sizes. The time interval $t_f - t_l$ can be inferred from the distance between two decay vertex points, while t_f and t_l cannot be determined separately with good accuracy.
- We had hoped, as discussed in Section 10.5.1, that a $B^0 \overline{B}^{*0}$ pair is produced, so that the asymmetry could be seen in a symmetric machine. Yet having a clean source of $B^0 \overline{B}^0$ pair without contamination of $B^0 \overline{B}^{0*}$ is a tremendous blessing – otherwise, the determination of ϕ_1 would have depended on the branching ratio for $\Upsilon(4S) \to B^0 \overline{B}^{0*}$ and our limited knowledge of the production vertex.
- **CP** violation implies, due to **CPT** invariance, violation of **T** invariance. One sign of it is the fact that the time dependence of the **CP** asymmetry is given by $\sin \Delta M_B (t_f - t_l)$, which flips sign under time reversal $t \to -t$; see the lower plot in Fig. 10.15. This dependence on Δt is caused by a EPR correlation as explained in Section 10.9.

The main point to be noted is that EPR correlations, which represent some of quantum mechanics' most puzzling features, serve as an essential precision tool, which is routinely used in these measurements. While we might not have an intuitive grasp of EPR correlations, we should appreciate them as a gift from nature rather than view them as a paradox.

10.11 Résumé

As it happens with the most intriguing cases in life, the conclusion of one chapter in the tale of **CP** violation is intertwined with the opening of a new one. In this case we have learnt the following lessons.

- The *first* **CP** violation *outside* the $K^0 - \overline{K}^0$ complex has been established by the Belle and BaBar experiments.
- The paradigm of truly large **CP** asymmetries unequivocally predicted by KM theory has indeed emerged in four transitions and counting.
- The numerical CKM prediction for the asymmetry in $B_d \to \psi K_S$ – as inferred from ΔM_{B_d}, $|V_{ub}/V_{cb}|$, $|\epsilon_K|$ and the lower bound on ΔM_{B_s} and refined through experimental as well as theoretical efforts over the last few years – has been confirmed in impressive fashion with $\sin 2\phi_1$ measured to within 5%.

- The promise of B^0 transitions as a highly sensitive probe for New Physics is being realized with the first credible hints for the latter emerging in a class of channels.

This seminal achievement is however just an intermediate station – and presumably even an early one – on a much longer journey. A comprehensive and multi-faceted programme can be pursued combining high sensitivity with high accuracy – as described in the next chapter.

Problems

10.1. Suppose that a $B^0\overline{B}^0$ pair is created at $t = 0$, with momenta $\vec{k}$ and $-\vec{k}$ in the centre of mass system with given **C** parity. Assuming $\Delta\Gamma = 0$, show that a state with B^0 or $\overline{B}^0$ at time t_1 with momentum $\vec{k}$ and B^0 or $\overline{B}^0$ at time t_2 with momentum $-\vec{k}$ can be written as:

$$|(P^0\overline{P}^0)_{C=\pm}(t_1, k), (t_2, -k)\rangle = \frac{1}{\sqrt{2}} e^{-\frac{1}{2}\Gamma(t_1+t_2)}$$

$$\left(-i\sin\frac{\Delta M(t_{-k} \pm t_k)}{2} \left[\frac{p}{q}|P_k P_{-k}\rangle \pm \frac{q}{p}|\overline{P}_k \overline{P}_{-k}\rangle\right]\right.$$

$$\left. + \cos\frac{\Delta M(t_{-k} \pm t_k)}{2} \left[|P_k \overline{P}_{-k}\rangle \pm |\overline{P}_k P_{-k}\rangle\right]\right). \qquad (10.66)$$

Consider the case in which one of the particles decays in a flavour-specific way, while the other one generates a flavour-non-specific final state:

$$\overline{P} \to l^- + X \not\leftarrow P$$
$$\overline{P} \not\to l^+ + X \leftarrow P$$
$$\overline{P} \to f \leftarrow P, \qquad (10.67)$$

where for simplicity we have assumed the final state f to be a CP eigenstate.

Derive the decay rates:

$$\Gamma\left(P^0\overline{P}^0\Big|_{C=\pm} \to (l^- X)_k(t_k) f_{-k}(t_{-k})\right)$$

$$\sim e^{-\Gamma(t_k+t_{-k})}|A(l)|^2|A(f)|^2 \left[(1 - \cos\Delta M(t_k \pm t_{-k}))\left|\frac{q}{p}\right|^2 |\overline{\rho}(f)|^2\right.$$

$$\left. + (1 + \cos\Delta M(t_k \pm t_{-k})) - 2\sin\Delta M(t_{-k} \pm t_k)\mathrm{Im}\frac{q}{p}\overline{\rho}(f)\right],$$

$$(10.68)$$

where

$$\bar{\rho}(f) \equiv \frac{\bar{A}(f)}{A(f)} \equiv \frac{1}{\rho(f)}. \tag{10.69}$$

Derive a similar expression for $\Gamma\left(P^0\bar{P}^0\big|_{C=\pm} \to (l^+X)_k(t_k)\, f_{-k}(t_{-k})\right)$.

Now for B^0 mesons with $|p/q| = 1$ and for special final states with $|\rho(f)| = 1$ derive Eq. (10.8).

10.2. Using Eq. (10.66), find the amplitude for same- and opposite-sign dilepton production

$$(B^0\bar{B}^0)_{C=\pm} \to l^{\pm}l^{\mp} + X$$
$$(B^0\bar{B}^0)_{C=\pm} \to l^{\pm}l^{\pm} + X, \tag{10.70}$$

and derive Eq. (10.9).

10.3. Under special circumstances the mere *existence* of a transition rather than an asymmetry between two reactions can establish **CP** violation. Consider

$$e^+e^- \to \gamma^* \to P^0\bar{P}^0 \to f_+^a f_+^b,\ f_-^a f_-^b \tag{10.71}$$

with

$$\mathbf{CP}|f_{\pm}^i\rangle = \pm|f_{\pm}^i\rangle,$$

i.e. both P^0 and $\bar{P}^0$ decay into **CP** eigenstates with the same **CP** parity. Convince yourself that this is a **CP** violating reaction.

For $|q/p| = 1$ and $\Delta\Gamma = 0$ verify the time integrated decay rate

$$\text{rate}\left(P^0\bar{P}^0\big|_{C=-} \to f_{\pm}^a f_{\pm}^b\right) \propto \left|A(f_{\pm}^a)A(f_{\pm}^b)\right|^2$$
$$\cdot\left[\left(1 + \frac{1}{1+x^2}\right)\left|\bar{\rho}(f_{\pm}^a) - \bar{\rho}(f_{\pm}^b)\right|^2 \right.$$
$$\left. + \frac{x^2}{1+x^2}\left|1 - \left(\frac{q}{p}\right)^2 \bar{\rho}(f_{\pm}^a)\bar{\rho}(f_{\pm}^b)\right|^2\right]. \tag{10.72}$$

10.4. If the two final states are identical

$$f_{\pm}^a = f_{\pm}^b \equiv f_{\pm},$$

then show that

$$\text{Br}\left(P^0\bar{P}^0\big|_{C=-} \to f_{\pm}f_{\pm}\right) \tag{10.73}$$
$$= [\text{Br}(P^0 \to f_{\pm})]^2 \cdot \frac{x^2}{1+x^2} \cdot \left(2\text{Im}\frac{q}{p}\bar{\rho}(f_{\pm})\right)^2. \tag{10.74}$$

Explain why

$$e^+e^- \to \gamma^* \to P^0\overline{P}^0 \to f_\pm f_\pm \qquad (10.75)$$

is not forbidden by Bose statistics. Hint: look at this decay in terms of mass eigenstates P_1 and P_2,

$$\gamma^* \nrightarrow P_1 P_1, \ P_2 P_2$$
$$\gamma^* \to P_1 P_2 \to (f_\pm)_1 (f_\pm)_2. \qquad (10.76)$$

10.5. Consider a decay in which the amplitude ratios for $P^0 \to f_\pm^a$ and $P^0 \to f_\pm^b$ exhibit a relative phase

$$\frac{\bar{\rho}(f_\pm^b)}{\bar{\rho}(f_\pm^a)} = e^{i\alpha_{ab}}, \quad \alpha_{ab} \neq 0, \pi, \qquad (10.77)$$

which means that $f_\pm^a$ and $f_\pm^b$ definitely have to be different states. Then show that $\gamma^* \to P^0\overline{P}^0 \to f_\pm^a f_\pm^b$ can proceed even in the absence of oscillations, and the decay rate is given by:

$$\mathrm{Br}\left(P^0\overline{P}^0\Big|_{C=-} \to f_\pm^a f_\pm^b\right) \sim \mathrm{Br}(P^0 \to f_\pm^a)\mathrm{Br}(P^0 \to f_\pm^b)\left(\sin\frac{\alpha_{ab}}{2}\right)^2 \qquad (10.78)$$

for $x = 0$.

10.6. Discuss why you expect more same-sign dilepton events from semi-leptonic decays of **C** even than **C** odd $B^0\overline{B}^0$ pairs.

10.7. Considering the fact that $M(\Upsilon(4S)) = 10.5800 \pm 0.0035\,\mathrm{GeV}$ and $m(B) = 5.2792 \pm 0.0018\,\mathrm{GeV}$, and $c\tau = 468\,\mu\mathrm{m}$, compute the distance a B^0 travels before it decays when the $B^0\overline{B}^0$ pair is produced at rest. Compute the typical track length for a $3\,\mathrm{GeV} \times 9\,\mathrm{GeV}$ asymmetric e^+e^- collider.

10.8. Prove Eq. (10.47) by using $|q/p| = 1$.

10.9. Show Eq. (10.36) and Eq. (10.52) by going back to the definition of ϕ_i given in Eq. (8.54).

11

Let the drama unfold – B
CP phenomenology

> **CP** *violation in B decays:*
> *exactly like in K decays –*
> *only different!*

While the observations of **CP** asymmetries in $B_d \to \psi K_S$, $\pi\pi$, $K\pi$ and $\eta' K_S$ are clearly seminal discoveries, it had been argued from the start that there will be many more channels exhibiting manifestations of **CP** violation; those could be of a different origin. Yet even we are amazed by how quickly the field is expanding and new 'actors' appear. Anticipating a 'flood' of experimental information on such asymmetries, we will focus more on 'strategic' than on 'tactical' considerations. At the time when the first edition was written, one of our objectives was to establish the KM ansatz. This has been accomplished. So we go on to our main goal, which is connected to the central challenge faced by high energy physics in general, namely to discover phenomena that cannot be explained by the existing theory; to find some defect in the theory so that, during the process of correcting it, we can get to the next layer of fundamental dynamics. In this chapter, we list ways to determine the fundamental parameters and point to future research in this vibrant field.

11.1 Pollution from water fowls and others

In Chapter 10 we have stated that asymmetries in $B \to \psi K_S$ and $B \to \pi\pi$ decays lead to information on ϕ_1, and ϕ_2, respectively, provided that only the tree level amplitudes contributed. To what extent is this valid? If not, how can we get to the angles of the unitarity triangle?

Complications arise whenever there are two amplitudes with different weak phases contributing to a decay. Consider two operators with different CKM parameters driving $B \to f$:

$$A(B \to f) = e^{i\xi_1}e^{i\delta_1}|\mathcal{A}_1| + e^{i\xi_2}e^{i\delta_2}|\mathcal{A}_2|, \qquad (11.1)$$

where δ_i and ξ_i are the strong interaction and weak phases, respectively; the moduli of the CKM parameters have been incorporated into $|\mathcal{A}_i|$. We then find

$$\overline{\rho}(f) = e^{-2i\xi_1} \frac{1 + e^{-i\Delta\xi}e^{i\Delta\delta}\left|\frac{\mathcal{A}_2}{\mathcal{A}_1}\right|}{1 + e^{i\Delta\xi}e^{i\Delta\delta}\left|\frac{\mathcal{A}_2}{\mathcal{A}_1}\right|} \ ; \quad \Delta\xi = \xi_2 - \xi_1, \ \Delta\delta = \delta_2 - \delta_1, \quad (11.2)$$

leading to

$$\overline{\rho}(f) \sim e^{-2i\xi_1} \cdot \left(1 - 2i\left|\frac{\mathcal{A}_2}{\mathcal{A}_1}\right|e^{i\Delta\delta}\sin\Delta\xi\right), \qquad (11.3)$$

for $|\mathcal{A}_2/\mathcal{A}_1| \ll 1$. $|\overline{\rho}(f)| \neq 1$ is an unambiguous sign for the presence of a second weak operator. Yet the inverse is not necessarily true: $|\overline{\rho}(f)| = 1$ does not mean that there is only one amplitude; $\Delta\xi = 0$ or $\Delta\delta = 0$ lead to $|\overline{\rho}(f)| = 1$ (see Problem 11.7).

From Eq. (11.3) we obtain [128, 129]:

$$\mathrm{Im}\frac{q}{p}\overline{\rho}(f) \sim \sin 2(\Phi_m - \xi_1) - 2\left|\frac{\mathcal{A}_2}{\mathcal{A}_1}\right|\sin\Delta\xi\cos(2\Phi_m - 2\xi_1 + \Delta\delta), \quad (11.4)$$

where $2\Phi_m = \arg(q/p)$. The presence of the second term on the right-hand side is often referred to as penguin pollution since the penguin mechanism often generates such a second amplitude, see Fig. 10.13 . Yet we find it unfair to single out penguins in such a negative way – in particular, since the guilty party is us, the theorists. It would be more appropriate to state that they enrich the phenomenology even when making it more complex. For example, we would not have seen **CP** violation in $B \to K\pi$ if not for the penguins. The presence of a second weak amplitude poses a challenge to our ability to extract weak parameters from the data. How this can best be achieved depends on the specifics of the channel under study.

One more general remark is in order: we have so far employed the terms *tree* and *penguin* operators in a somewhat un-reflected way, guided by simple Feynman diagrams. Yet we should keep in mind that if two operators contribute to the same final state they obviously mix under QCD renormalization; their relative weight depends on the scale at which the operators are evaluated (which in turn is compensated by the scale dependence of their matrix elements). *They should then be distinguished by their dependence on the CKM parameters, together with their colour and chiral structure.*

The $\Delta B = 1$ weak Hamiltonian can be obtained proceeding as for the $\Delta S = 1$ case. One difference is that at an energy scale $\sim M_B$, the charm degree of freedom is not frozen. Thus we have four types of operator: $(\bar{b}c)_{V-A}(\bar{c}q')_{V-A}$, $(\bar{b}u)_{V-A}(\bar{u}q')_{V-A}$, where $q' = s$, d. Let us define four-Fermi operators in a notation analogous to that of Eq. (9.13) which defined $O_1 - O_{10}$:

$$O^q_{q'1} = (\bar{b}q')_{V-A}(\bar{q}q)_{V-A}, \qquad O^q_{q'2} = (\bar{b}q)_{V-A}(\bar{q}q')_{V-A},$$

$$O_{q'3} = (\bar{b}q')_{V-A}\sum_q(\bar{q}q)_{V-A}, \qquad O_{q'4} = (\bar{b}_\alpha q'_\beta)_{V-A}\sum_q(\bar{q}_\beta q_\alpha)_{V-A},$$

$$O_{q'5} = (\bar{b}q')_{V-A}\sum_q(\bar{q}q)_{V+A}, \qquad O_{q'6} = (\bar{b}_\alpha q'_\beta)_{V-A}\sum_q(\bar{q}_\beta q_\alpha)_{V+A},$$

$$O_{q'7} = \tfrac{3}{2}(\bar{b}q')_{V-A}\sum_q e_q(\bar{q}q)_{V+A}, \quad O_{q'8} = \tfrac{3}{2}(\bar{b}_\alpha q'_\beta)_{V-A}\sum_q e_q(\bar{q}_\beta q_\alpha)_{V+A},$$

$$O_{q'9} = \tfrac{3}{2}(\bar{b}q')_{V-A}\sum_q e_q(\bar{q}q)_{V-A}, \quad O_{q'10} = \tfrac{3}{2}(\bar{b}_\alpha q'_\beta)_{V-A}\sum_q e_q(\bar{q}_\beta q_\alpha)_{V-A},$$

where α and β are colour indices. The $\Delta B = 1$ Hamiltonian is then given by

$$\mathcal{H}(\Delta B = 1) = \frac{G_F}{\sqrt{2}}\left[\mathbf{V}^*_{cb}\mathbf{V}_{cq'}[C_1(\mu)O^c_{q'1} + C_2(\mu)O^c_{q'2}]\right.$$
$$+ \mathbf{V}^*_{ub}\mathbf{V}_{uq'}[C_1(\mu)O^u_{q'1} + C_2(\mu)O^u_{q'2}]$$
$$\left. - \mathbf{V}^*_{tb}\mathbf{V}_{tq'}\sum_{i=3}^{10}C_i(\mu)O_{q'i} + \text{h.c.}\right]. \qquad (11.5)$$

The first two terms are called tree operators and the third is called a penguin operator. Finally, we shall denote hadronic matrix elements as

$$Q^q_{q'i}(F) = \langle F|O^q_{q'i}|B\rangle. \qquad (11.6)$$

Living with penguins

As seen from Eq. (6.81), $|\bar{\rho}(f)| \neq 1$ leads to $C_{\pi\pi} \neq 0$. This in turn implies that there is a second amplitude with a different weak phase, as seen in Eq. (11.3). We have seen in Eq. (10.49) and Eq. (10.48) that $C_{\pi\pi}$ is comparable to $S_{\pi\pi}$. This means that penguins – or diagrams other than the tree diagram – play a major role in $B \to \pi\pi$ decay. We have also stated in Section 10.8 that large direct **CP** violation has been observed in $B \to K\pi$ decay mode. This asymmetry is a direct consequence of interference between tree and penguin amplitudes. Therefore we have to live with the penguins and appreciate them, and at the same time clean up after them.

11.2   Determining ϕ_1

11.2.1   How clean is $B_d \to \psi K_S$?

In addition to the diagram shown in Fig. 10.9, the decay $B \to \psi K_S$ gets contributions from the penguin diagrams shown in Fig. 11.1. The resulting operators are

$$T(\psi K_S) = \frac{G_F}{\sqrt{2}} \mathbf{V}_{cb}^* \mathbf{V}_{cs} [C_1(\mu) Q_{s1}^c(\psi K_S) + C_2(\mu) Q_{s2}^c(\psi K_S)]$$

$$P_u(\psi K_S) = \frac{G_F}{\sqrt{2}} \mathbf{V}_{ub}^* \mathbf{V}_{us} [C_1(\mu) Q_{s1}^u(\psi K_S) + C_2(\mu) Q_{s2}^u(\psi K_S)]$$

$$P_t(\psi K_S) = \frac{G_F}{\sqrt{2}} (-\mathbf{V}_{tb}^* \mathbf{V}_{ts}) \sum_{i=3}^{10} C_i(\mu) Q_{si}(\psi K_S), \qquad (11.7)$$

where the letters $T(\psi K_S)$, $P_u(\psi K_S)$ and $P_t(\psi K_S)$ obviously refer to the diagrammatic origin of these operators.[1]

First note that the weak phases of $T(\psi K_S)$ and $P_t(\psi K_S)$ are the same to leading order in λ. Thus we see from Eq. (11.4) that the presence of $P_t(\psi K_S)$ does not change the asymmetry. It is different for $P_u(\psi K_S)$ – yet that operator is considerably suppressed by CKM parameters and the coefficient:

$$\left| \frac{P_u(\psi K_S)}{T(\psi K_S)} \right| \sim \mathcal{O}\left(\lambda^2 \cdot \frac{Q_{s1,s2}^u(\psi K_S)}{Q_{s1,s2}^c(\psi K_S)} \right) \leq 1\%. \qquad (11.8)$$

The penguin contributions of Fig. 11.1, while potentially significant for the width, are therefore limited to less than 1% for the **CP** asymmetry.

$$\left| \mathrm{Im} \frac{q}{p} \bar{\rho}(\psi K_S) - \sin 2\phi_1 \right| < 1\%. \qquad (11.9)$$

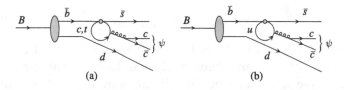

(a)   (b)

Figure 11.1   Penguin graphs for $B \to \psi K_S$ decay: (a) has the same weak phase as the tree graph; thus this diagram does not change the asymmetry. (b) introduces some theoretical uncertainties, but this contribution is expected to be less than 1%.

---

[1] Note that P_c and T cannot be distinguished from each other after renormalization.

11.2.2 ♠ *Other ways to get at* ϕ_1 ♠

$B \to D\overline{D}K_S$, $D\overline{D}$, $D^*\overline{D}$, ϕK_S, $\eta' K_S$, ωK_S are some of the channels which should exhibit **Class (C1)** type **CP** violation, with their asymmetries predicted to be $\sin 2\phi_1$. This can most easily be seen in the Wolfenstein representation of the CKM matrix, where the decay amplitudes are real, and the asymmetry is thus given by $\text{Im}\,(q/p)$. Present data are at least not in conflict with that. The C_f for these final states are still compatible with 0 and the S_f consistent with $\sin 2\phi_1$ from $B \to \psi K_S$ [112]. The rare modes $B_d \to \phi K_S$, $\eta' K_S$, $\pi^0 K_S$, etc. hold the 'promise' to show some deviations from the SM pattern, since they are generated by penguin processes; i.e. they represent pure quantum effects. We will address these channels in Section 11.5.6.

Transitions $B \to PP$, PV, with P and V denoting a pseudoscalar and vector meson, respectively, are described by only one partial wave $l = 0$ [1] for $B \to PP\,[PV]$. For $B \to VV$, on the other hand, is made up by $l = 0, 1, 2$ configurations – the total amplitude is a sum of **CP** even and odd amplitudes.

This can also be seen in the helicity analysis. If we denote the final state in terms of their helicities, angular momentum conservation requires the states to be $|++\rangle$, $|--\rangle$, and $|00\rangle$. Under parity

$$\mathbf{P}|++\rangle = |--\rangle, \quad \mathbf{P}|00\rangle = |00\rangle; \tag{11.10}$$

thus the parity eigenstates with eigenvalue ± 1 are:

$$\begin{aligned} |\,\|\,\rangle \\ |\perp\rangle \end{aligned} = \frac{1}{\sqrt{2}}(|++\rangle \pm |--\rangle). \tag{11.11}$$

The fact that there are final states with different **CP** eigenvalues, implies that an *indiscriminate* sum over the final states will tend to wash out the total asymmetry. Fortunately, final states with different **CP** quantum numbers can be separated by measuring angular distributions [130, 131, 132, 133, 134]. For definiteness in defining the angles in the transversity frame, let us consider

$$\begin{aligned} B \to \quad &\psi \quad K^{*0} \to K_S \pi^0 \\ &\hookrightarrow \mu^+\mu^-. \end{aligned} \tag{11.12}$$

Define θ and ϕ as in Fig. 11.2, and let θ_K be the direction of K_S relative to $-\vec{p}_\psi$ in the rest frame of K^{*0}. The amplitudes are denoted by: $A(f)_{\|,\perp,0} = \langle f_{\|,\perp,00}|H|B\rangle$ and $\overline{A}(f)_{\|,\perp,0} = \langle f_{\|,\perp,00}|H|\overline{B}\rangle$, where the subscripts in $f_{\|,\perp,00}$ imply that the state f has transversity $\|,\perp,00$. The angular distribution is given by

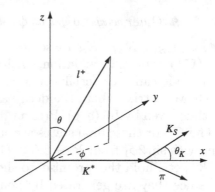

Figure 11.2   Angular distribution for $B \to \psi K_S^*$ based on the transversity variables. Let ψ be at rest. The x-axis is defined by the K^{*0} momentum and the x–y plane is defined by the momentum of K_S and π^0 from the K_S^* decay; θ and ϕ are defined by the momentum of l^+ from $\psi \to l^+l^-$. Finally θ_K is the direction of K_S with respect to the x axis. For $B \to D^{*-}D^{*+}$ decay distributions, particles labelled in this figure should be replaced as follows: $\psi \to D^{*+}$, $K^* \to D^{*-}$, $l^+ \to \pi^+$, $K \to \pi^-$. θ, ϕ are transversity angles for π^+, θ_K is replaced by the angle which π^- makes relative to the x axis.

$$\frac{d\Gamma}{d\cos\theta\, d\cos\theta_K\, d\phi\, dt} = \frac{9}{32\pi}\Big[|A(f,t)_0|^2 G_0 + |A(f,t)_\parallel|^2 G_\parallel + |A(f,t)_\perp|^2 G_\perp$$

$$+ \operatorname{Im}\left(A(f,t)_\parallel^* A(f,t)_\perp\right)G_{\parallel\perp} + \operatorname{Re}\left(A(f,t)_0^* A(f,t)_\parallel\right)G_{0\parallel}$$

$$+ \operatorname{Im}\left(A(f,t)_0^* A(f,t)_\perp\right)G_{0\perp}\Big], \tag{11.13}$$

where the angular functions $G_{\parallel,\perp,0}$ are given in Table 11.1.
   The angular distribution for

$$
\begin{aligned}
B \to \quad & D^{*-}\ \ D^{*+} \to D^0\pi^+ \\
& \hookrightarrow \overline{D}^0\pi^-
\end{aligned}
\tag{11.14}
$$

can be obtained in the same manner using Table 11.1.
   Oscillations introduce a time dependence into $A(f)_{\parallel,\perp,0}$. It can easily be computed using Eq. (6.48):

$$A(f,t)_\lambda \propto A(f)_\lambda\left[\cos(\tfrac{1}{2}\Delta M t) + i\sin(\tfrac{1}{2}\Delta M t)\frac{q}{p}\overline{\rho}(f)_\lambda\right]$$

$$\overline{A}(f,t)_\lambda \propto \overline{A}(f)_\lambda\left[\cos(\tfrac{1}{2}\Delta M t) + i\sin(\tfrac{1}{2}\Delta M t)\frac{p}{q}\rho(f)_\lambda\right] \tag{11.15}$$

Table 11.1 *Transeversity frame angular distribution functions*
[135]. The angles are defined in Fig. 11.2.

	ψK_S	$D^{+*}D^{-*}$
G_0	$2\cos^2\theta_K(1 - \sin^2\theta\cos^2\phi)$	$2\cos^2\theta_K\sin^2\theta\cos^2\phi$
$G_\parallel$	$\sin^2\theta_K(1 - \sin^2\theta\sin^2\phi)$	$\sin^2\theta_K\sin^2\theta\sin^2\phi$
$G_\perp$	$\sin^2\theta_K\sin^2\theta$	$\sin^2\theta_K\cos^2\theta$
$G_{\parallel\perp}$	$\sin^2\theta_K\sin 2\theta\sin\phi$	$\sin^2\theta_K\sin 2\theta\sin\phi$
$G_{0\parallel}$	$-\frac{1}{\sqrt{2}}\sin 2\theta_K\sin^2\theta\sin 2\phi$	$-\frac{1}{\sqrt{2}}\sin 2\theta_K\sin^2\theta\sin 2\phi$
$G_{0\perp}$	$\frac{1}{\sqrt{2}}\sin 2\theta_K\sin 2\theta\cos\phi$	$\frac{1}{\sqrt{2}}\sin 2\theta_K\sin 2\theta\cos\phi$

where $\bar{\rho}_\lambda(f) = \frac{\overline{A}(f)_\lambda}{A(f)_\lambda}$ and $\rho_\lambda(f) = \frac{A(f)_\lambda}{\overline{A}(f)_\lambda}$. The functions appearing in the angular distribution are:

$$|A(f,t)_\lambda|^2 \propto \frac{1}{2}|A(f)_\lambda|^2 \left[(1 + |\bar{\rho}_\lambda|^2) + (1 - |\bar{\rho}_\lambda|^2)\cos(\Delta Mt)\right.$$
$$\left. - 2\text{Im}\left(\frac{q}{p}\bar{\rho}_\lambda\right)\sin(\Delta Mt)\right]$$

$$\text{Re}\,(A(f,t)^*_{\lambda_1}A(f,t)_{\lambda_2}) \propto \text{Re}\,(A^*_{\lambda_1}A_{\lambda_2})\cos^2\left(\frac{\Delta Mt}{2}\right) + \text{Re}\,(\overline{A}^*_{\lambda_1}\overline{A}_{\lambda_2})\sin^2\left(\frac{\Delta Mt}{2}\right)$$
$$+ \text{Im}\left\{A^*_{\lambda_1}A_{\lambda_2}\left[\left(\frac{q}{p}\bar{\rho}_{\lambda_1}\right)^* - \frac{q}{p}\bar{\rho}_{\lambda_2}\right]\right\}\sin(\Delta Mt)$$

$$\text{Im}\,(A(f,t)^*_{\lambda_1}A(f,t)_{\lambda_2}) \propto \text{Im}\,(A^*_{\lambda_1}A_{\lambda_2})\cos^2\left(\frac{\Delta Mt}{2}\right) + \text{Im}\,(\overline{A}^*_{\lambda_1}\overline{A}_{\lambda_2})\sin^2\left(\frac{\Delta Mt}{2}\right)$$
$$- \text{Re}\left\{A^*_{\lambda_1}A_{\lambda_2}\left[\left(\frac{q}{p}\bar{\rho}_{\lambda_1}\right)^* - \frac{q}{p}\bar{\rho}_{\lambda_2}\right]\right\}\sin(\Delta Mt).$$

$$(11.16)$$

Corresponding formulae for $\overline{B}$ decay can be obtained by making the substitution $A \leftrightarrow \overline{A}$, $\rho \leftrightarrow \bar{\rho}$, and $p \leftrightarrow q$ in the expressions above.

B → ψK* *amplitudes*

For $B \to \psi K^*$, we set $\eta_0 = \eta_\parallel = -\eta_\perp = +1$, and $\bar{\rho}_\lambda = \eta_\lambda|\bar{\rho}_\lambda|$ in the Wolfenstein parameterization. So we have:

$$|A(t)_\lambda|^2 = |A_\lambda|^2[1 \pm \eta_\lambda\sin(2\phi_1)\sin(\Delta Mt)] \qquad \text{for } \lambda = 0,\ \parallel,\ \perp$$
$$\text{Re}\,(A(t)^*_\parallel A(t)_0) = \text{Re}\,(A^*_\parallel A_0)[1 \pm \sin(2\phi_1)\sin(\Delta Mt)]$$

$$\text{Im}\,(A(t)_{\|,0}A(t)_\perp) = \pm\text{Im}\,(A^*_{\|,0}A_\perp)\cos(\Delta Mt)$$
$$\mp\text{Re}\,(A^*_{\|,0}A_\perp)\cos(2\phi_1)\sin(\Delta Mt). \tag{11.17}$$

There is a two-fold ambiguity in the determination of ϕ_1 from $\sin(2\phi_1)$: the sine of ϕ_1 and $\frac{\pi}{2} - \phi_1$ are the same. But the cosine of these angles has the opposite sign. So, in order to remove the ambiguity, it is important to measure $\cos(2\phi_1)$.

The helicity analysis allows one to do just that. It is possible to obtain both $\sin(2\phi_1)$ and $\cos(2\phi_1)$ from observing polarizations of this decay. Both Belle and BaBar obtained results consistent with $\cos(2\phi_1) > 0$. This removes the two-fold ambiguity[136, 137].

11.3   Determining ϕ_2

11.3.1   Penguins in $B_d \to \pi\pi$

From the diagrams of Fig. 10.13 we infer the presence of the following transition operators.[2]

$$T(\pi\pi) = \frac{G_F}{\sqrt{2}}\mathbf{V}^*_{ub}\mathbf{V}_{ud}[C_1(\mu)Q^u_{d1}(\pi\pi) + C_2(\mu)Q^u_{d2}(\pi\pi)]$$

$$P_c(\pi\pi) = \frac{G_F}{\sqrt{2}}\mathbf{V}^*_{cb}\mathbf{V}_{cd}[C_1(\mu)Q^c_{d1}(\pi\pi) + C_2(\mu)Q^c_{d2}(\pi\pi)]$$

$$P_t(\pi\pi) = \frac{G_F}{\sqrt{2}}(-\mathbf{V}^*_{tb}\mathbf{V}_{td})\sum_{i=3}^{10} C_i(\mu)Q_{di}(\pi\pi). \tag{11.18}$$

The qualitatively new element is that now $T(\pi\pi)$, $P_c(\pi\pi)$ and $P_t(\pi\pi)$ are of roughly comparable strength in terms of their CKM parameters:

$$|\mathbf{V}^*_{ub}\mathbf{V}_{ud}| \sim |\mathbf{V}^*_{cb}\mathbf{V}_{cd}| \sim |\mathbf{V}^*_{tb}\mathbf{V}_{td}| \sim \mathcal{O}(\lambda^3), \tag{11.19}$$

yet with different weak phases.

From the discussion above, penguins are large and we have to discuss methods to obtain ϕ_2 in a clean manner.

11.3.2   Overcoming pollution

We have to take on the challenge of reducing the considerable uncertainties in relating the observable **CP** asymmetry in, say, $B_d \to \pi\pi$ to the microscopic quantities, namely the CKM parameters. While our understanding

---

[2] As before it should be noted that while the penguin diagram with an internal *top* quark generates a *local* operator, the other two do not.

of non-leptonic B decays will certainly improve, we think an accurate *ab initio* theoretical calculation of the relevant transition amplitudes will not become available soon. Instead we have to harness additional experimental information in a judicious way. The concrete procedure depends on the specifics of the mode and will involve a learning curve. It is therefore not our intent to discuss it in exhaustive detail; instead we want to outline important features of possible strategies. An essential element is employing isospin. Some authors use $SU(3)_F$ relations [138, 139, 140]. We discourage the reader from relying on $SU(3)_F$. It is hard to convince anyone that you have found evidence for New Physics, if your result depended on $SU(3)_F$ symmetry. The latter is useful, however, in obtaining qualitative results which can be used as a guide in evaluating various methods.

Here we will focus on the task of extracting ϕ_2 from $B_d \to \pi's$; application of these ideas to determining ϕ_3 from $B_s \to K\pi's$ is quite analogous, although probably much more difficult quantitatively.

11.3.3 B → ππ

There are six channels, namely $B^+ \to \pi^+\pi^0$, $B^0 \to \pi^+\pi^-$, $\pi^0\pi^0$ plus their charge conjugate ones. The isospin decomposition of these amplitudes is exactly the same as that of $K \to \pi\pi$ decays given in Eq. (7.18). The only difference is that we can no longer use Watson's theorem to relate the final state interaction phase to the $\pi\pi$ phase shift. We shall absorb the final state interaction phase into A_i. With two isospin amplitudes, A_0 and A_2, and three measurable rates for B mesons (and likewise for $\overline{B}$), there is a constraint among the latter [141]:

$$A^{+-} = -A^{00} + \sqrt{2}A^{+0}. \tag{11.20}$$

Now we can write

$$\rho(\pi^+\pi^-) = \frac{A(\pi^+\pi^-)}{\overline{A}(\pi^+\pi^-)} = \frac{A_2}{\overline{A}_2}\frac{1+z}{1+\overline{z}}, \tag{11.21}$$

where $z = \sqrt{2}A_0/A_2$, $\overline{z} = \sqrt{2}\overline{A}_0/\overline{A}_2$ and $A(\overline{B} \to \pi\pi) \equiv \overline{A}(\pi\pi)$. Since the $P_{c,t}(\pi\pi)$ operators for $b \to d$ can generate $\Delta I = 1/2$ modes only, the $I = 2$ amplitude is given by the T operator alone, and therefore:

$$\mathrm{Im}\frac{q}{p}\overline{\rho}(\pi^+\pi^-) = \left|\frac{1+\overline{z}}{1+z}\right| \sin\left[2\phi_2 + \arg\left(\frac{1+\overline{z}}{1+z}\right)\right]. \tag{11.22}$$

Figure 11.3(a) and (b) illustrate how we can obtain z and $\overline{z}$ from branching ratios. We can thus extract $\sin 2\phi_2$. This is a theoretically clean method.

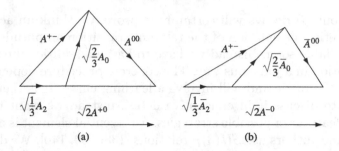

Figure 11.3   (a) Isospin relations between $B^+ \to \pi^+\pi^0$, $B^0 \to \pi^+\pi^-$, $B^0 \to \pi^0\pi^0$ modes. We see that both the magnitudes of A_0, A_2 and the relative phase between these two amplitudes can be obtained. (b) Isospin relations between the charge conjugate decays.

11.3.4   B → πρ, ρρ

The quark level diagrams shown in Fig. 10.13 drive these modes as well. With ρ and π being distinct hadrons, the final states become more numerous – $B_d \to \rho^+\pi^-$, $\rho^-\pi^+$, $\rho^0\pi^0$, $B^- \to \rho^-\pi^0$, $\rho^0\pi^-$ plus their charge conjugate versions – and more complex: only $\rho^0\pi^0$ constitutes a **CP** eigenstate and final states can carry isospin 0, 1 or 2. There are two reasons why we look at these transitions, nevertheless.

- The branching ratios for the $B \to \pi\rho$ modes are significantly larger and easier to measure than for $B \to \pi\pi$, where $B \to \pi^0\pi^0$ represents a bottleneck.
- The CKM parameters are of course the same for $B \to \pi\pi$ and $B \to \rho\pi$; analysing $B \to \pi\rho$, including their time evolution, can then provide us with important information on the impact of the strong forces.

Details can be found in Refs. [142] and [143].

Experimentalists have gone even further and included $B_d \to \rho\rho$ channels. They argued that because $B \to \rho^0\rho^0$ branching ratio is small compared to $B \to \rho^+\rho^-$ and $B^\pm \to \rho^\pm\rho^0$, the decay is dominated by two ρ mesons in a longitudinally polarized state. Thus the final state $\rho\rho$ are **CP** $= +$ eigenstate. They proceeded to do the isospin analysis. At the present level of accuracy, this is sufficient. Belle's and BaBar's results are summarized in Ref. [112].

A general caveat should be mentioned here. Going from the experimental starting point $B \to 3\pi$ to $B \to \pi\rho$ configurations is quite non-trivial. There are other contributions to the three-pion final state such as $\sigma\pi$ (with

σ denoting a scalar $\pi\pi$ compound), and cutting on the dipion mass provides a rather imperfect filter due to the large ρ width. It hardly matters in this context whether the σ is a bona fide resonance or some other dynamical enhancement. This actually leads to a further complication, namely that the σ structure cannot be described adequately by a Breit-Wigner shape. As analysed first in Ref. [144], and then in more detail in Ref. [145], ignoring such complications can induce a significant systematic uncertainty in the extracted value of ϕ_2. The modes $B_{d,u} \to \rho\rho$ contain even more theoretical complexities, since they have to be extracted from $B \to 4\pi$ final states, where one has to allow for $\sigma\rho$, 2σ, $\rho 2\pi$ etc. in addition to 2ρ.

Our point here is one of caution rather than of agnosticism. The concerns sketched above might well be more academic than practical with the present statistics. Yet they have to be addressed when reducing the uncertainties in ϕ_2 down to the 5% level, as has to be our goal. In the end we will need:

- to perform time-*dependent* Dalitz plot analyses; and
- to involve the expertise that already exists or can be obtained concerning low-energy hadronization processes like final state interactions among low energy pions and kaons; valuable information can be gained on those issues from $D_{(s)} \to \pi$'s, kaons, etc. as well as $D_{(s)} \to l\nu K\pi/\pi\pi/KK$, in particular when analysed with state-of-the-art tools of chiral dynamics.

11.4 Determining ϕ_3

Measuring ϕ_3 directly and comparing it with the value inferred from CKM trigonometry represents a high sensitivity probe for the intervention of New Physics. Many observables depend on ϕ_3 – and, alas, on other and ill-known quantities as well, making an accurate determination of ϕ_3 a formidable task theoretically as well as experimentally. Here we want to address some of the nitty-gritty details in such endeavours.

Consider the two modes [119, 147, 148, 149]

$$B^- \to D^0 K^-, \quad B^- \to \overline{D}^0 K^- \tag{11.23}$$

for which the quark level diagrams are shown in Fig. 11.4. The transition amplitudes are

$$A(D^0 K^-) = |\mathcal{A}_D|\ e^{i[\arg(\mathbf{V}_{cb}\mathbf{V}_{us}^*)+\delta_D]}$$
$$A(\overline{D}^0 K^-) = |\mathcal{A}_{\overline{D}}|\ e^{i[\arg(\mathbf{V}_{ub}\mathbf{V}_{cs}^*)+\delta_{\overline{D}}]} \tag{11.24}$$

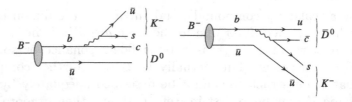

Figure 11.4   Feynman diagrams for $B^- \to D^0 K^-$ and $B^- \to \overline{D}^0 K^-$.

$$\overline{A}(\overline{D}^0 K^+) = |\mathcal{A}_D| \, e^{i[\arg(\mathbf{V}_{cb}^* \mathbf{V}_{us}) + \delta_D]}$$
$$\overline{A}(D^0 K^+) = |\mathcal{A}_{\overline{D}}| \, e^{i[\arg(\mathbf{V}_{ub}^* \mathbf{V}_{cs}) + \delta_{\overline{D}}]}. \qquad (11.25)$$

At first we might think that they are clearly distinct and do not interfere. That would imply that only **Class(B) CP** violation can be seen. However, that is not necessarily the case. For the flavour identity of the neutral D meson has to reveal itself through its weak decay. Semileptonic transitions, being flavour-specific, tell us unambiguously whether the decaying meson is a D^0 or $\overline{D}^0$. Yet final states that are **CP** eigenstates reveal nothing about the flavour of the D meson:

$$D^0 \to K^+ K^-, \pi^+ \pi^-, K_S \pi^0, K_S \rho^0, K_S \omega, K_S \eta \leftarrow \overline{D}^0. \qquad (11.26)$$

Thus a **Class(C)** asymmetry can be seen by restricting the decay modes of D and $\overline{D}$. Let us consider six types of channel:

$$B^- \to D^0 K^-, \overline{D}^0 K^-, D_{1,2} K^-$$
$$B^+ \to \overline{D}^0 K^+, D^0 K^+, D_{1,2} K^+, \qquad (11.27)$$

where the amplitude for $B \to D_{1,2} K^-$ is given by

$$A(D_{1,2} K^-) = \frac{1}{\sqrt{2}} \left[A(D^0 K^-) \pm A(\overline{D}^0 K^-) \right]. \qquad (11.28)$$

CPT invariance tells us that

$$|A(B^- \to D^0 K^-)| = |A(B^+ \to \overline{D}^0 K^+)|$$
$$|A(B^- \to \overline{D}^0 K^-)| = |A(B^+ \to D^0 K^+)|, \qquad (11.29)$$

since a single quark level diagram shown in Fig. 11.4 drives these flavour-specific decays. In contrast, for $B^\pm \to D_{1,2} K^\pm$ both diagrams contribute, and they do it coherently. Their interference will generate a direct **CP** asymmetry that depends on $\sin(\delta_D - \delta_{\overline{D}})$ and $\sin\phi_3$:

$$\frac{|A(B^- \to D_{1,2}K^-)|^2 - |A(B^+ \to D_{1,2}K^+)|^2}{|A(B^- \to (D^0/\overline{D}^0)K^-)|^2 + |A(B^+ \to (D^0/\overline{D}^0)K^+)|^2}$$

$$= \pm \frac{2|R|\sin(\delta_D - \delta_{\overline{D}})}{1 + |R|^2 + 2|R|\cos\phi_3\cos(\delta_D - \delta_{\overline{D}})} \cdot \sin\phi_3, \qquad (11.30)$$

where $R = A(B^- \to \overline{D}^0 K^-)/A(B^- \to D^0 K^-)$. The exciting new element here is that the final state phase shift $(\delta_D - \delta_{\overline{D}})$ – instrumental for the asymmetry to materialize in the first place – can be extracted from the data.

Measuring the widths for $B^- \to D^0 K^-$, $\overline{D}^0 K^-$ and $D_{1,2}K^-$ defines a triangle in the complex plane [147] as shown in Fig. 11.5 (a). Likewise for the triangle formed by the **CP** conjugate amplitudes, see Fig. 11.5(b). These two triangles allow us to determine the phases.

$$\angle[A(D^0 K^-), A(\overline{D}^0 K^-)] \equiv |-\phi_3 + \delta_{\overline{D}} - \delta_D|$$
$$\angle[\overline{A}(\overline{D}^0 K^+), \overline{A}(D^0 K^+)] \equiv |+\phi_3 + \delta_{\overline{D}} - \delta_D|. \qquad (11.31)$$

Comparing those two angles allows us to deduce the size of ϕ_3 up to a binary ambiguity. Note that ϕ_3 can be determined this way even if $\delta_{\overline{D}} - \delta_D = 0$! There is, however, a major practical drawback to this method: we expect $\mathrm{Br}(B^- \to \overline{D}^0 K^-)$ to be tiny since it represents a colour-suppressed mode, see Fig. 11.5.

$$\frac{\mathrm{Br}(B^- \to \overline{D}^0 K^-)}{\mathrm{Br}(B^- \to D^0 K^-)} \simeq \left|\frac{\mathbf{V}_{ub}\mathbf{V}_{cs}^*}{\mathbf{V}_{cb}\mathbf{V}_{us}^*}\right|^2 \left|\frac{a_2}{a_1}\right|^2 \sim \mathcal{O}(0.01), \qquad (11.32)$$

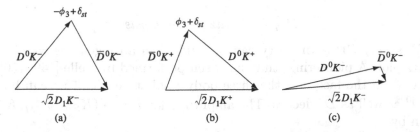

Figure 11.5 (a) Triangle representing the relationship between $A(D_1 K^-)$, $A(D^0 K^-)$ and $A(\overline{D}^0 K^-)$. (b) The triangle representing the charged conjugate version. (c) A more realistic triangle (spear). Here we have denoted $\delta_{st} = \delta_{\overline{D}} - \delta_D$.

where a_2/a_1 is the usual phenomenological colour suppression factor. The triangle under discussion will be quite squashed, with one of its sides being a mere ~10% of the others. The resulting error on ϕ_3 may be too large for it to be useful. It is possible that we are too pessimistic here; yet we had better have an alternative plan.

11.4.1 *Using doubly Cabibbo-suppressed decays*

Consider again the overall reaction chain [148]:

$$B^- \to \begin{pmatrix} K^- D^0 \\ K^- \overline{D}^0 \end{pmatrix} \to K^- f. \qquad (11.33)$$

The practical problem of having a squashed triangle, stated in the previous section, can be avoided by choosing f such that the two decay rates are of the same order of magnitude. i.e. we compensate $|A(B^- \to D^0 K^-)| \gg |A(B^- \to \overline{D}^0 K^-)|$ through $|A(D^0 \to f)| \ll |A(\overline{D}^0 \to f)|$. Then the situation becomes much more favourable: although we lose in statistics, the triangles can be constructed with more quantitative precision.

Final states f with $S = +1$ will do the trick for us since they are doubly Cabibbo suppressed for D^0, yet Cabibbo allowed for $\overline{D}^0$: $|A(D^0 \to [S = +1])/A(\overline{D}^0 \to [S = +1])| \sim \mathcal{O}(\theta_C^2)$. More specifically we observe

$$\frac{\mathrm{Br}(D^0 \to K^+\pi^-)}{\mathrm{Br}(\overline{D}^0 \to K^+\pi^-)} = 0.0077 \pm 0.0025 \pm 0.0025. \qquad (11.34)$$

All together we estimate

$$\frac{\mathrm{Br}(B^- \to K^- D^0)}{\mathrm{Br}(B^- \to K^- \overline{D}^0)} \cdot \frac{\mathrm{Br}(D^0 \to [S = +1])}{\mathrm{Br}(\overline{D}^0 \to [S = +1])} \sim 100 \cdot \mathcal{O}(\tan^4\theta_C) \sim 1 \quad (11.35)$$

as desired! For details we refer you to Ref. [148].

11.4.2 *Dalitz plot analysis*

A **Class(C) CP** asymmetry can be studied in an ingenious way in the Dalitz plot. A pioneering study has been performed by Belle [149, 150]. It is based on the fact that the three-body final state $K_S\pi^+\pi^-$ can be fed by D^0 as well as $\overline{D}^0$ decays. The amplitude for $B^\pm \to (K_S\pi^+\pi^-)_D K^\pm$ is given by

$$M^\pm = f(m_+^2, m_-^2) + re^{\pm i\phi_3 + i\delta} f(m_-^2, m_+^2) \qquad (11.36)$$

where δ is the difference of strong phases:

$$\delta = \delta_{D^0 K_S} - \delta_{\overline{D}^0 K_S}. \qquad (11.37)$$

The angle ϕ_3 can be determined by fitting over the entire Dalitz plot. Using this method Belle and BaBar find [112]:

$$\phi_3 = \begin{cases} (53 \, {}^{+15}_{-18}\big|_{\text{stat}} \pm 3|_{\text{syst}} \pm 9|_{\text{model}})^\circ & \text{Belle} \\ (92 \pm 41|_{\text{stat}} \pm 11|_{\text{syst}} \pm 12|_{\text{model}})^\circ & \text{BaBar.} \end{cases} \quad (11.38)$$

The uncertainties are obviously very substantial, yet this method carries the promise of providing accurate results in the future, statistically as well as systematically. One obvious challenge one has to overcome in this method is to determine the complex function $f(m_+^2, m_-^2)$. This introduces some model dependence into the analysis. The redeeming feature is that a fit to the full Dalitz plot provides a number of over-constraints. The 'vice' of having to choose a certain parameterization can actually be turned into a 'virtue': for the Dalitz description of $D \to K_S \pi^+ \pi^-$ can be tested independently in $e^- e^+ \to \psi(3770) \to D^0 \overline{D}^0$ – a task being done by CLEO-c and BESIII.

11.5 Search for New Physics

As stated in the begining of this chapter, our main goal is to search for New Physics. Seaching for New Physics in **CP** violating observable makes much sense for the following reasons.

(1) **CP** violation is a higher order weak effect. The SM background and the impact of long-distance dynamics are reduced.

(2) As we see later, the **CP** violation from the KM mechanism cannot explain the baryon excess of the universe. There must be **CP** violating New Physics.

(3) Many B decay amplitudes involve loop diagrams. New Physics can reveal itself in these decays.

With this in mind, looking for deviations from the SM predictions in

- direct **CP** violation;
- the unitarity triangle by:
  - measuring ϕ_1, ϕ_2, and ϕ_3 accurately in different ways to test the trigonometric relation:

$$\phi_1 + \phi_2 + \phi_3 = 180^\circ; \quad (11.39)$$

  - measuring the sides of the triangle.

are promising ways to look for New Physics. In this section we shall discuss some methods known at present.

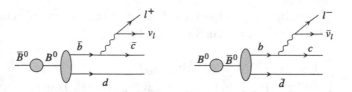

Figure 11.6　Because of the $\Delta B = \Delta Q$ rule, existence of wrong-sign leptonic decay implies $B - \overline{B}$ mixing.

11.5.1　*Wrong-sign semileptonic decays:* **Class(B)**

These decays proceed through Fig. 11.6; i.e. they require the intervention of $B - \overline{B}$ oscillations. The asymmetry is given in Eq. (6.72):

$$A_{SL} = \frac{\Gamma(B^0(t) \to l^- \nu X) - \Gamma(\overline{B}^0(t) \to l^+ \nu X)}{\Gamma(B^0(t) \to l^- \nu X) + \Gamma(\overline{B}^0(t) \to l^+ \nu X)} = \frac{1 - |p/q|^4}{1 + |p/q|^4}. \quad (11.40)$$

11.5.2　♠ *Theoretical estimate of* A_{SL} ♠

We expect the quark box diagram shown in Fig. 11.7 to provide us with a decent approximation to $\Delta \Gamma_B$.[3] Yet the internal quarks are now charm and up quarks which are considerably lighter than M_B; it is then the latter that sets the scale for $\Delta \Gamma_B$: $\Delta \Gamma_B \propto M_B$. Since $\Delta M_{B_q} \propto m_t$, we expect on rather general grounds

$$\Delta \Gamma_B \ll \Delta M_B. \quad (11.41)$$

Even so, $\Delta \Gamma_{B_s}$ could still be sizeable, as we see in a rough guestimate of $\Delta \Gamma_B$: the transition amplitude $T(B_s \to \overline{D}_s^{(*)} D_s^{(*)} \to \overline{B}_s)$ which contributes prominently to $\Delta \Gamma_{B_s}$ is of second order in $\mathbf{V}_{cb}$ – as is $\overline{\Gamma}(B_s \to [c\overline{c}s\overline{s}])$. With the latter making up a smallish yet significant part of $\overline{\Gamma}_{B_s}$, we can state an order of magnitude estimate: $\Delta \Gamma_{B_s}/\overline{\Gamma}_{B_s} \sim \mathcal{O}(10\%)$ [151] with $\overline{\Gamma}_{B_s}$ denoting the average B_s width. On the other hand, channels common to B_d

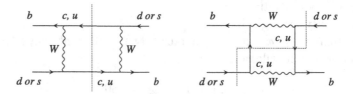

Figure 11.7　Diagrams contributing to Γ_{12}.

---

[3] Such an expectation would be utterly absurd for $\Delta \Gamma_K$ and highly miraculous for $\Delta \Gamma_D$.

and $\overline{B}_d$ decays are CKM suppressed; e.g. $B_d \to \overline{D}^{(*)}D^{(*)} / \pi\pi / \rho\rho \to \overline{B}_d$; or $B_d \to \overline{D}^{(*)}\pi + D^{(*)}\pi \to \overline{B}_d$ which are CKM favoured for B_d and doubly CKM suppressed for $\overline{B}_d$ decays or vice versa. Accordingly, we guestimate $\Delta\Gamma_{B_d}/\Gamma_{B_d} \sim \mathcal{O}(\text{few} \times 10^{-3})$, where cancellations can take place between the two classes of common channels exemplified above.

To arrive at a less hand waving estimate we evaluate the quark box diagram; we have to keep in mind that here it does *not* yield a *local* operator: while the W boson fields can be integrated out as for ΔM_B, the internal charm quark fields *cannot* since m_c and even $2m_c$ fall below m_b; this gives rise to absorptive parts at $m_c + m_u$ and $2m_c$. Evaluating the quark box diagram [44, 152],

$$(\Gamma_{B_q})_{12} \simeq -\frac{G_F^2 M_{B_q}^3}{8\pi} B_B F_B^2\Big|_{B_q} \left(\xi_{cq}^2 P(cc) + \xi_{uq}^2 P(uu) + 2\xi_{u,q}\xi_{cq}P(uc)\right). \tag{11.42}$$

The functions $P(cc)$, $P(uu)$ and $P(uc)$ denote the weight of the configurations with $\bar{c}c$, $\bar{u}u$ and $\bar{u}c$ or $\bar{c}u$ intermediate states, respectively, see Fig. 11.7. With $m_u^2 \ll m_c^2 \ll m_b^2$ they are given by very simple expressions:

$$P(uu) \simeq 1, \quad P(uc) \simeq 1 - \frac{4}{3}\frac{m_c^2}{m_b^2}, \quad P(cc) \simeq 1 - \frac{8}{3}\frac{m_c^2}{m_b^2}. \tag{11.43}$$

Hence

$$(\Gamma_{B_q})_{12} \simeq -\frac{G_F^2 M_{B_q}^3}{8\pi} B_B F_B^2|_{B_q} \left[(\xi_{uq} + \xi_{cq})^2 - \frac{8}{3}\frac{m_c^2}{m_b^2}\xi_{cq}(\xi_{uq} + \xi_{cq})\right]. \tag{11.44}$$

Unitarity of the 3×3 CKM matrix requires

$$\xi_{uq} + \xi_{cq} = -\xi_{tq}, \tag{11.45}$$

and thus

$$\left(\frac{(\Gamma_{B_q})_{12}}{(M_{B_q})_{12}}\right)_{B_q} \simeq \frac{3\pi}{2}\frac{M_{B_q}^2}{M_W^2}\frac{1}{\eta_t^B(\mu)E(x_t)}\left(1 + \frac{8}{3}\frac{m_c^2}{m_b^2}\frac{\xi_{cq}}{\xi_{tq}}\right) \sim \mathcal{O}(\text{few} \times 10^{-3}), \tag{11.46}$$

with $(M_{B_q})_{12}$ as defined in Eq. (10.19). This is in line with the general expectations stated above, although we have to take the number in Eq. (11.46) with a grain of salt: for it reflects cancellations taking place between $P(uu)$, $P(cc)$ and $P(ud)$.

The above result can be turned into

$$\frac{\Delta\Gamma}{\overline{\Gamma}} = \left(\frac{\Gamma_{12}}{M_{12}}\right)_{B_q}\frac{x_q}{2}$$
$$= \mathcal{O}(10^{-3})x_q. \qquad (11.47)$$

The measured value of $x_s = 26.5 \pm 1.3$ [113] leads to

$$\frac{\Delta\Gamma}{\overline{\Gamma}}|_{B_s} \simeq \text{several } \%. \qquad (11.48)$$

Similar numbers are obtained when we employ phenomenological models to describe $B_s \to \overline{D}_s^{(*)}D_s^{(*)} \to \overline{B}_s$. A more careful evaluation of the quark box diagrams yields [153]:

$$\frac{\Delta\Gamma}{\overline{\Gamma}}|_{B_s} = 0.147 \pm 0.060. \qquad (11.49)$$

Using the expression for $\frac{q}{p}$ given in Eq. (6.43), we obtain

$$A_{SL}(B_q) = \frac{1 - |p/q|^4}{1 + |p/q|^4} = -r\sin\zeta_B, \qquad (11.50)$$

where $r\sin\zeta_B = \text{Im}\,\Gamma_{12}/M_{12}$ as defined in Eq. (6.37). The quantity $r\sin\zeta_B$ can be predicted in the KM ansatz within some uncertainties. Using Eq. (11.46), we have

$$r \simeq \frac{3\pi}{2}\frac{M_B^2}{M_W^2}\frac{1}{\eta_t^B(\mu)E(x_t)} \qquad (11.51)$$

$$\sin\zeta_{B_q} \simeq \tan\zeta_{B_q} \simeq \frac{8}{3}\frac{m_c^2}{m_b^2}\text{Im}\frac{\mathbf{V}_{cb}\mathbf{V}_{cq}^*}{\mathbf{V}_{tb}\mathbf{V}_{tq}^*}. \qquad (11.52)$$

Thus

$$\sin\zeta_{B_d} \simeq -\frac{8}{3}\frac{m_c^2}{m_b^2}\frac{\eta}{(1-\rho)^2 + \eta^2}, \quad \sin\zeta_{B_s} \simeq -\frac{8}{3}\frac{m_c^2}{m_b^2}\lambda^2\eta. \qquad (11.53)$$

Several features of Eq. (11.52) and Eq. (11.53) are to be noted:
(a) On the positive side:
⊕ The quantity $\sin\zeta_B$ measures the intrinsic strength of **CP** violation in $B \leftrightarrow \overline{B}$ oscillations in an unambiguous way.
⊕ It can be expressed with reasonable accuracy within KM theory in terms of its basic parameters in an intuitively understandable way:

• In addition to the suppression from $r = \mathcal{O}(\text{few} \times 10^{-3})$, there is a suppression from $\sin \zeta_B$. For $\sin \zeta_B$, the reduction factor m_c^2/m_b^2 is produced by the GIM mechanism (and actually reads $(m_c^2 - m_u^2)/m_b^2$, for there can be no **CP** violation in the KM scheme when $m_c = m_u$).

We have to remember, however, that the extent of this cancellation depends on taking the quark diagrams seriously. See the discussion leading to Eq. (11.46).

• We find $\zeta_{B_d} \gg \zeta_{B_s}$ since ζ_{B_s} is CKM suppressed: for on the leading level, only quarks of the second and third family contribute here, namely t, b, c and s; accordingly no **CP** violation can arise to leading order in the KM parameters.

(b) On the negative side:

⊖ The observable $1 - |q/p|$ provides a numerically poor handle on $\sin \zeta_{B_s}$. For the latter's coefficient $[(\Gamma_B)_{12}/(M_B)_{12} \simeq \Delta\Gamma_B/\Delta M_B]$ is, as discussed in the preceding section, expected to be quite small on rather general grounds and independently of the strength of **CP** violation. To be more specific: $\Delta\Gamma_B/\Delta M_B \leq 0.01$ and thus

$$\left| 1 - \left| \frac{q}{p} \right|^2 \right| < 0.01. \tag{11.54}$$

The situation is quite different for neutral kaons where – due to purely accidental reasons – we have $\Gamma_{12} \simeq M_{12}$!

⊖ The prediction for $\Delta\Gamma_B/\Delta M_B$ is not 'gold-plated'; it could conceivably represent an underestimate. In principle $\Delta\Gamma_B$ could be measured in addition to ΔM_B; yet this would go beyond a merely academic proposition only if $\Delta\Gamma_B$ were much larger than anticipated – at least for B_d mesons. ⊖Altogether we arrive at the predictions [153]

$$A_{SL}(B_d)|_{KM} \sim 5 \times 10^{-4}, \tag{11.55}$$

$$A_{SL}(B_s)|_{KM} \sim 2 \times 10^{-5}. \tag{11.56}$$

These are very small asymmetries, in particular for $B_s \leftrightarrow \overline{B}_s$ oscillations, and searching for them clearly poses challenging demands on the control over the systematics an experiment has to maintain. Nevertheless such searches are highly worthwhile:

• They can be executed once we obtain a sample of 'wrong-sign' leptons from B^0 decays. This has indeed been done, most recently at the B factories through

$$e^+ e^- \rightarrow B_d \overline{B}_d \rightarrow (l^+ X)_B (l^+ X)_B \;\; vs. \;\; (l^- X)_B (l^- X)_B. \tag{11.57}$$

The world average reads [112]:

$$\left|\frac{q}{p}\right|_{B_d} = 1.0025 \pm 0.0019, \quad \left|\frac{q}{p}\right|_{B_s} = 1.0015 \pm 0.0051. \qquad (11.58)$$

With $|A_{SL}| \simeq 2||q/p| - 1|$ one thus has

$$|A_{SL}|_{B_d} \simeq (4.9 \pm 3.8) \cdot 10^{-3}, \quad |A_{SL}|_{B_s} \simeq (3.0 \pm 10.1) \cdot 10^{-3} ; \qquad (11.59)$$

i.e. one is starting to probe the 1% regime, where New Physics might realistically manifest itself. Yet one has to reduce the systematic uncertainties, before the sensitivity can be improved significantly.

• Establishing an unambiguous signal would constitute a discovery of the very first rank. Comparing the present experimental number on $|q/p|$, Eq. (11.58), with the general bound of Eq. (11.54) tells us that the experimental sensitivity is only now entering a regime where one can realistically hope for a signal.

• New Physics could manifest itself here in unequivocal ways:

(1) It is conceivable that data would reveal $A_{SL}(B_d)$ to be around 0.01, i.e. around the present upper bound and thus considerably larger than suggested by Eq. (11.55) while at the same time the SM expectations for $\Delta\Gamma_B/\Delta M_B$ were satisfied. If, on the other hand, both $A_{SL}(B_d)$ and $\Delta\Gamma_B/\Delta M_B$ were found to be enhanced over the KM and SM estimates, respectively, then no clear conclusion could be drawn – beyond conceding that we had presumably overestimated our computational powers!

(2) The prediction for $A_{SL}(B_s)$ in Eq. (11.56) is suppressed relative to the natural scale of $\sim \mathcal{O}(10^{-3})$, as set by $\Delta\Gamma_B/\Delta M_B$, due to the very specific mechanism in which **CP** violation is implemented in the KM scheme: as for $B_s \to J/\psi\phi$ (see below) quarks of only the second and third family contribute on the leading level; therefore the *relative* phase between $(M_{B_s})_{12}$ and $(\Gamma_{B_s})_{12}$ is CKM suppressed, Eq. (11.53). New Physics can quite conceivably contribute to M_{12}, but is unlikely to do so for Γ_{12}; it would then vitiate the CKM suppression of ζ_{B_s}, possibly leading to

$$A_{SL}(B_s)|_{NP} \sim \mathcal{O}(0.01). \qquad (11.60)$$

We have already stated that B–$\overline{B}$ oscillations occur with considerable speed. For good measure nature has actually provided us with two such systems, namely B_d and B_s mesons. There are then two mass eigenstates, defined in Eq. (6.19), with different lifetimes for neutral beauty mesons without and with strangeness: $B_{d,2}$ versus $B_{d,1}$ and $B_{s,2}$ versus $B_{s,1}$.

B_s mesons pose an experimental challenge for a e^+e^- collider: First, it is almost impossible to trace the time dependence which goes like $\sin(x_s t/\tau_s)$; Second, if a non-vanishing y_s is established, it may lead to a certain preponderance of $B_{s,2}$ mesons in a $B_s/\overline{B}_s$ beam which might be exploitable through modern detection devices.

The fact that $|(\Gamma_K)_{12}| \sim |(M_K)_{12}|$ whereas $|(\Gamma_B)_{12}| \ll |(M_B)_{12}|$ has also a direct impact on **CP** observables. For ϵ_K, it generates a complex phase of $\simeq 45°$; concentrating on the observable $\mathrm{Re}\,\epsilon_K$ rather than $|\epsilon_K|$ thus leads to only a slight reduction in sensitivity. In the $B_d - \overline{B}_d$ system on the other hand with $|\Gamma_{12}| \ll |M_{12}|$ we have $\mathrm{Re}\,\epsilon_B \ll |\epsilon_B|$ (see Problem 11.1).

These differences and the fact that **CP** violation is so feeble in K decays have a considerable impact on how to go about searching for **CP** asymmetries.

11.5.3   *What can oscillations tell us about New Physics?*

Top quarks have been discovered with $m_t \sim 180\,\mathrm{GeV}$; their mass is expected to be measured with an accuracy of $\mathcal{O}(1\%)$ in the foreseeable future. Yet even then no accurate SM prediction can be given for ΔM_B since neither $|\mathbf{V}_{td}|$ nor $B_B F_B^2$ are well known numerically. Fortunately we can go beyond this agnostic statement in four respects.

(i) Nature has been generous enough to provide us with two kinds of neutral B states that can exhibit oscillations, namely B_d and B_s, i.e. beauty mesons without and with strangeness, respectively. The dependence on m_t drops out from the ratio of their oscillation rates:[4]

$$\frac{\Delta M_{B_s}}{\Delta M_{B_d}} = \frac{[B_B F_B^2]_{B_s}}{[B_B F_B^2]_{B_d}} \frac{\overline{\mathbf{V}}_{ts}^2}{\overline{\mathbf{V}}_{td}^2}; \qquad (11.61)$$

with $|\overline{\mathbf{V}}_{ts}^2/\overline{\mathbf{V}}_{td}^2| \sim \mathcal{O}(1/\lambda^2) \gg 1$ we predict $\Delta M_{B_s} \gg \Delta M_{B_d}$. We can say more. The ratio of the hadronic parameters $B_B F_B^2$ represents a quantity characterizing flavour $SU(3)$ breaking; therefore we should be able to predict it with decent accuracy. The best available theoretical technologies, at present, yield:

$$\frac{F_{B_s}}{F_{B_d}} \simeq 1.1 \sim 1.2, \qquad (11.62)$$

---

[4] We ignore here small and calculable differences due to $M_{B_d} \neq M_{B_s}$.

and therefore

$$\frac{\Delta M_{B_s}}{\Delta M_{B_d}} = (1.2 \sim 1.5) \times \frac{|\mathbf{V}_{ts}|^2}{|\mathbf{V}_{td}|^2}. \tag{11.63}$$

The measurement of $\Delta M_{B_s}/\Delta M_{B_d}$ then allows us to extract $|\mathbf{V}_{td}|$ since $|\mathbf{V}_{ts}| \simeq |\mathbf{V}_{cb}|$ holds (for three families). The data are fully compatible with the SM prediction of Eq. (11.63), while still allowing for significant contributions from New Physics, as discussed in Section 10.2. Possible scenarios for such New Physics are described in Sections 17.1.5, 17.2.3 and 19.4.

(ii) Once a theoretical prediction for F_D, the decay constant for D mesons, has been confirmed in $D^+, D_s^+ \to \mu^+\nu$ and the underlying method thus validated, we can be confident of arriving at a reliable estimate for F_B by extrapolating from m_c to m_b. Measuring both F_{D_s} and F_D with, say, about 10% accuracy would obviously be quite beneficial in this context. We can also hold out the hope of measuring F_B directly through $\Gamma(B \to \tau\nu)$.

(iii) We might succeed in extracting $|\mathbf{V}_{td}|$ also from other decay rates – either directly from $\mathrm{Br}(K^+ \to \pi^+\nu\bar{\nu})$ or as $|\mathbf{V}_{td}/\mathbf{V}_{ts}|$ from $\mathrm{Br}(B \to \gamma\rho/\omega)/\mathrm{Br}(B \to \gamma K^*)$. Once this has been achieved we can – limited by the numerical accuracy to which $|\mathbf{V}_{td}|$ is known –

- extract F_B from ΔM_{B_d} and compare that value with other determinations of it;
- scrutinize $\Delta M_{B_s}/\Delta M_{B_d}$ for manifestations of New Physics.

(iv) Measuring **CP** asymmetries in B^0 decays will shed further light on F_B, $|\mathbf{V}_{td}|$ and $|\mathbf{V}_{ts}|$, as discussed in detail later.

11.5.4 $B_s \to \psi\phi$, $\psi\eta^{(\prime)}$, $D_s^+D_s^-$: *Class (C2)*

The channels $B_s \to \psi\phi$, $\psi\eta^{(\prime)}$, $D_s^+D_s^-$ occurring on the *leading* KM level fit the bill as sensitive probes for New Physics. In particular $B_s \to \psi\phi$ is well established with $\mathrm{BR}(B_s \to \psi\phi) = (9.3 \pm 3.3) \cdot 10^{-4}$. It is being used the measure the average B_s lifetime as well as $\Delta\Gamma_s$ as sketched above. While $\psi\eta^{(\prime)}$ and $D_s^+D_s^-$ are even **CP** eigenstates, $\psi\phi$ is a combination of even [odd] eigenstates for S and D [P] waves. S waves dominate, yet the two components can be separated out quite effectively in the data.

Although the present HFGAG average of CDF, D0, ALEPH and DELPHI data [112] – $\Delta\Gamma_s = 0.138^{+0.068}_{-0.074}\,\mathrm{ps}^{-1}$ – does not provide a clear verdict on $\Delta\Gamma_s$, it is probably not insignificant for B_s mesons. Therefore we retain it (while still using $|p| = |q|$). For the final states $f = \psi\phi$, $\psi\eta$, $D_s^+D_s^-$ we have – certainly within the SM – $|\bar{\rho}(f)| = 1 = |\rho(f)|$. Thus:

$$G_f(t) = |A(B \to f)|^2 \cdot \left[1 + e^{\Delta\Gamma_{B_s}t} + (1 - e^{\Delta\Gamma_{B_s}t})\mathrm{Re}\left(\frac{q}{p}\bar{\rho}(f)\right) \right.$$

$$-2\text{Im}\left(\frac{q}{p}\overline{\rho}(f)\right)e^{\Delta\Gamma_{B_s}t/2}\sin\Delta M_{B_s}t\bigg], \tag{11.64}$$

$$\overline{G}_f(t) = |\overline{A}(B\to f)|^2\cdot\left[1 + e^{\Delta\Gamma_{B_s}t} + (1 - e^{\Delta\Gamma_{B_s}t})\text{Re}\left(\frac{p}{q}\rho(f)\right)\right.$$

$$\left.-2\text{Im}\left(\frac{p}{q}\rho(f)\right)e^{\Delta\Gamma_{B_s}t/2}\sin\Delta M_{B_s}t\right]. \tag{11.65}$$

The diagrams responsible for $B_s \to \psi\phi$ (or $B_s \to \psi\eta^{(\prime)}$) are shown in Fig. 11.8; those for $B_s \to D_s^+ D_s^-$ are obtained by rearranging the quark lines. The crucial point here is the observation that to leading order in λ only four quarks and antiquarks participate in $B_s \to \psi\phi$ and $B_s \to \overline{B}_s \to \psi\phi$: t, b, c and s, i.e. members of the third and second family only. It is the peculiar feature of the KM ansatz that no **CP** violation can arise with only two active families. That means that the **CP** asymmetry in this channel has to be CKM suppressed, since it enters through the effective violation of weak universality due to the presence of the first family [119].

To be more specific, we note that these transitions are described by single isospin amplitudes each; therefore – as for $B_d \to \psi K_S$ – no *direct* **CP** violation can arise: $|A(B_s \to f)| = |A(\overline{B}_s \to f)|$ and within KM we find

$$\frac{q}{p}\overline{\rho}(B_s \to f) \simeq \text{Im}\left[\frac{(\mathbf{V}_{tb}^*\mathbf{V}_{ts})^2}{|\mathbf{V}_{tb}\mathbf{V}_{ts}|^2}\frac{(\mathbf{V}_{cb}\mathbf{V}_{cs}^*)^2}{|\mathbf{V}_{cb}\mathbf{V}_{cs}|^2}\right] \simeq 2\lambda^2\eta < 0.05. \tag{11.66}$$

In calling such asymmetries small, we might note that they are still an order of magnitude larger than what has been observed in K_L decays. On the other hand New Physics scenarios to be sketched in later chapters will generate asymmetries that can well reach the few $\times 10$ percentage level. This holds true despite the observed value of ΔM_{B_s} agreeing with the SM prediction – within the latter's sizable uncertainty.

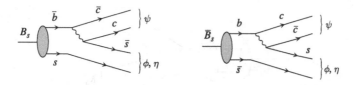

Figure 11.8 Feynman diagrams for $B \to \psi\phi$ and $\overline{B} \to \psi\phi$ decays. Note that there are only second and third generation quarks involved in these graphs. Thus, in the SM, we expect a suppressed asymmetry. Observation of sizeable asymmetry implies existence of New Physics [119].

The striking experimental signature of $B_s \to \psi\phi$ might turn out to be essential in overcoming the experimental challenge of resolving the rapid $B_s \to \overline{B}_s$ oscillations driven by the large value of ΔM_{B_s}.

11.5.5 $B_s \to K_S\rho^0$: *Class (C1, C2)*

The tree level quark diagram shown in Fig. 11.9 generates this transition. If it were the only underlying mechanism $(q/p)\overline{\rho}$ could be expressed merely in terms of CKM parameters:

$$\frac{q}{p}\overline{\rho}(B_s \to K_S\rho) \simeq -\frac{\mathbf{V}_{tb}^*\mathbf{V}_{ts}}{\mathbf{V}_{tb}\mathbf{V}_{ts}^*}\frac{\mathbf{V}_{ub}\mathbf{V}_{ud}^*}{\mathbf{V}_{ub}^*\mathbf{V}_{ud}} \equiv -e^{-2i\phi_3}. \tag{11.67}$$

Yet this channel also receives contributions from a CKM reduced penguin operator which introduces additional weak phases, giving rise to direct **CP** violation and making the extraction of ϕ_3 murky. This difficulty is compounded by $B_s - \overline{B}_s$ oscillations being very speedy and the transition being CKM suppressed with a less than outstanding signature for the final state!

11.5.6 $B_d \to \phi K_S, \eta K_S$: *Class(C2)*

The modes discussed so far can be produced already by tree level processes. This is no longer true for $B \to \phi K$ with $\phi = [\overline{s}s]$. It requires a loop process as is readily provided by the Penguin operator $b \to s\overline{s}s$. The rate should thus be suppressed, and indeed one finds $\mathrm{BR}(B \to \phi K) \sim 10^{-5}$, roughly in line with expectations. Within the SM the $b \to s\overline{s}s$ operator does not contain a weak phase; accordingly one expects $B_d \to \phi K_S$ to exhibit the same **CP** asymmetry as $B_d \to \psi K_S$, i.e. in the notation of Eq. (6.81)

$$S_{\phi K_S}^{SM} = 0.726 \pm 0.037, \quad C_{\phi K_S}^{SM} = 0. \tag{11.68}$$

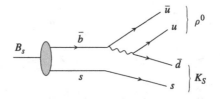

Figure 11.9   Tree diagram responsible for $B_s \to \rho K_S$. In addition, there is a penguin graph which makes precise prediction of this asymmetry difficult.

The branching ratios for the two channels are about as suppressed as expected [112]:

$$\mathrm{BR}(B_d \to \phi K^0) = (8.6^{+1.3}_{-1.1}) \cdot 10^{-6}, \ \mathrm{BR}(B^+ \to \phi K^+) = (9.3 \pm 1.0) \cdot 10^{-6}.$$
$$(11.69)$$

The measured values for the **CP** sensitive observables $S_{\phi K_s}$ and $C_{\phi K_s}$ have wandered considerably over the years; the 2007 'snap shot' reads [112]:

$$S_{\phi K_S} = \begin{cases} 0.50 \pm 0.21 \pm 0.06 & \text{Belle} \\ 0.21 \pm 0.26 \pm 0.11 & \text{BaBar} \end{cases} \tag{11.70}$$

$$C_{\phi K_S} = \begin{cases} -0.07 \pm 0.15 \pm 0.05 & \text{Belle} \\ 0.08 \pm 0.18 \pm 0.04 & \text{BaBar.} \end{cases} \tag{11.71}$$

While the two data sets are not incompatible, they tend to paint a different picture. Whereas Belle's numbers are within one sigma of the prediction, BaBar's value for $S_{\phi K_s}$ is a bit on the low side. More data will clarify the experimental issue. While there is no conclusive verdict at present, it offers much motivation to continue the search for New Physics, since

- the SM prediction of Eq. (11.68) is quite reliable;
- this channel had been recognized even before it was observed as one with high potential to reveal the intervention of New Physics; and
- SUSY scenarios could accommodate deviations in an unforced way.

One can also point out that, from the result summarized by HFAG [112], all channels driven by a transition operator $b \to s\bar{q}q$, $q = u, d, s - B_d \to \phi K_S, \eta^{(\prime)} K_S, f_0 K_S, \pi^0 K_S, K^+ K^- K_S$ – on average seem to exhibit signals for a **CP** asymmetry *below* SM expectations. Since these channels provide such natural portals to New Physics, it is mandatory to obtain significantly more data to clarify the situation.

One should, however, keep the following practical complication in mind: what one observes directly is $B \to K\overline{K}K$. This final state is then analysed and interpreted in terms of different components, namely the resonant channels ϕK, $f_0 K$ plus a non-resonant component. Applying merely mass cuts on the $K\overline{K}$ pair is insufficient to obtain an unambiguous interpretation. This ambiguity is enhanced by the following fact: since ϕK_S and $f_0 K_S$ carry opposite **CP** parity, their **CP** asymmetries are opposite in sign as well (if produced by the same underlying quark transition operator). Thus even a small admixture of $f_0 K_S$ in the ϕK_S can reduce the inferred asymmetry for the latter significantly.

In Chapter 10 we have already discussed a mode that belongs in this class of penguin-driven reactions, namely $B \to \eta' K_S$. **CP** violation has actually been established that at present is consistent with the SM prediction.

11.5.7 $B_s \to D_s^{\pm} K^{\mp}$: *Class(C1,C2)*

For flavour-non-specific final states that are not **CP** eigenstates the situation becomes more complex. Since $\bar{f} \neq f$, we have to deal with four possibly distinct decay modes, namely $\overline{B}^0 \to f, \bar{f}$ and $B^0 \to f, \bar{f}$. Yet they are of great practical value, as our subsequent discussion will show. Here we will describe only one such example.

The four channels

$$\overline{B}_s \to D_s^+ K^-, \ D_s^- K^+$$
$$B_s \to D_s^- K^+ , \ D_s^+ K^- \tag{11.72}$$

might allow us a relatively decent direct extraction of ϕ_3. They proceed through the diagrams shown in Fig. 11.10 and their **CP** conjugates.

In $\overline{B}_s \to D_s^+ K^-$ $[D_s^- K^+]$ driven by $b \to c(\bar{u}s)$ $[b \to u(\bar{c}s)]$ isospin changes by $I = (1/2, -1/2)$ $[I = (1/2, 1/2)]$; i.e. both transitions are described by a single isospin amplitude:

$$A(\overline{B}_s \to D_s^+ K^-) = e^{i\delta_-} \mathbf{V}_{cb} \mathbf{V}_{us}^* \mathcal{A}_-$$
$$A(\overline{B}_s \to D_s^- K^+) = e^{i\delta_+} \mathbf{V}_{ub} \mathbf{V}_{cs}^* \mathcal{A}_+$$
$$A(B_s \to D_s^- K^+) = e^{i\delta_-} \mathbf{V}_{cb}^* \mathbf{V}_{us} \mathcal{A}_-$$
$$A(B_s \to D_s^+ K^-) = e^{i\delta_+} \mathbf{V}_{ub}^* \mathbf{V}_{cs} \mathcal{A}_+. \tag{11.73}$$

Therefore

$$|A(\overline{B}_s \to D_s^+ K^-)| = |A(B_s \to D_s^- K^+)|$$
$$|A(\overline{B}_s \to D_s^- K^+)| = |A(B_s \to D_s^+ K^-)|. \tag{11.74}$$

Yet $|A(\overline{B}_s \to D_s^+ K^-)| = |A(\overline{B}_s \to D_s^- K^+)|$ obviously does *not* hold as an identity. Setting $\Delta\Gamma_B = 0$ for simplicity we find

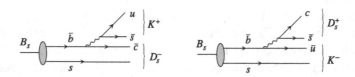

Figure 11.10 These two decay channels and their charge conjugate decays allow us to determine ϕ_3.

$$G_{D_s^+ K^-} = |A(D_s^+ K^-)|^2 \Big[1 + |\bar{\rho}(D_s^+ K^-)|^2$$

$$+ (1 - |\bar{\rho}(D_s^+ K^-)|^2)\cos\Delta M_B t - 2\mathrm{Im}\frac{q}{p}\bar{\rho}(D_s^+ K^-)\sin\Delta M_B t\Big]$$

$$(11.75)$$

$$\overline{G}_{D_s^- K^+} = |\overline{A}(D_s^- K^+)|^2 \Big[1 + |\rho(D_s^- K^+)|^2$$

$$+ (1 - |\rho(D_s^- K^+)|^2)\cos\Delta M_B t - 2\mathrm{Im}\left(\frac{p}{q}\rho(D_s^- K^+)\right)\sin\Delta M_B t\Big]$$

$$(11.76)$$

$$G_{D_s^- K^+} = |A(D_s^- K^+)|^2 \Big[1 + |\bar{\rho}(D_s^- K^+)|^2$$

$$+ (1 - |\bar{\rho}(D_s^- K^+)|^2)\cos\Delta M_B t - 2\mathrm{Im}\frac{q}{p}\bar{\rho}(D_s^- K^+)\sin\Delta M_B t\Big]$$

$$(11.77)$$

$$\overline{G}_{D_s^+ K^-} = |\overline{A}(D_s^+ K^-)|^2 \Big[1 + |\rho(D_s^+ K^-)|^2$$

$$+ (1 - |\rho(D_s^+ K^-)|^2)\cos\Delta M_B t - 2\mathrm{Im}\left(\frac{p}{q}\rho(D_s^+ K^-)\right)\sin\Delta M_B t\Big]$$

$$(11.78)$$

$$\bar{\rho}(D_s^+ K^-) = e^{i(\delta_- - \delta_+)}\frac{\mathbf{V}_{cb}\mathbf{V}_{us}^*}{\mathbf{V}_{ub}^*\mathbf{V}_{cs}}\frac{A_-}{A_+}$$

$$\simeq e^{i(\delta_- - \delta_+)}e^{-i\phi_3}\frac{1}{\mathcal{R}}\frac{A_-}{A_+}$$

$$\bar{\rho}(D_s^- K^+) = e^{-i(\delta_- - \delta_+)}\frac{\mathbf{V}_{ub}\mathbf{V}_{cs}^*}{\mathbf{V}_{cb}^*\mathbf{V}_{us}}\frac{A_+}{A_-}$$

$$\simeq e^{-i(\delta_- - \delta_+)}e^{-i\phi_3}\mathcal{R}\frac{A_+}{A_-},\qquad (11.79)$$

where $\mathcal{R} = \sqrt{\rho^2 + \eta^2}$. From the four observables $G_{D_s^\pm K^\mp}$, $\overline{G}_{D_s^\pm K^\mp}$ we can reliably extract the magnitudes and phases of $\bar{\rho}(D_s^+ K^-)$ and $\bar{\rho}(D_s^- k^+)$. This leads to the determination of ϕ_3 and $\delta_- - \delta_+$. The **CP** asymmetries in these channels depend on $\sin\phi_3$ rather than $\sin 2\phi_3$ as in $B_s \to K_S \rho^0$; it is thus maximal for $\phi_3 = 90°$, whereas $\sin 2\phi_3$ would vanish then.

11.6 Resumé

The aim of this chapter was to illuminate how rich a **CP** phenomenology we can expect in the weak decays of beauty hadrons. While we can use the same classification as in kaon decays, the numbers and even the pattern are quite different, which is to a large degree due to $\Gamma_{K_L} \ll \Gamma_{K_S}$ vs $\Gamma_{B_L} \sim \Gamma_{B_S}$, $\Delta\Gamma_K \sim \Delta M_K$ vs $\Delta\Gamma_B \ll \Delta M_B$. The situation is summarized in the following list.

- The flavour-specific semileptonic B^0 decays unambiguously probe for **CP** violation in $B - \overline{B}$ oscillations. The asymmetry expressed through $|q/p| \neq 1$ is small, namely less than 0.1%. While in absolute terms this might not be much smaller than what happens in K_L decays, it is a poor measure of the strength of **CP** violation in B decays.
- We can summarize the landscape for non-leptonic transitions as shown in Table 11.2. The third column might require one comment: only if the final state f is a **CP** eigenstate, does $|\overline{\rho}_{B^0 \to f}| \equiv |A(\overline{B}^0 \to f)/A(B^0 \to f)| \neq 1$ necessarily constitute a direct **CP** asymmetry.
- The predictions for the asymmetries in $B_d \to \psi K_S$ and $B_s \to \psi\phi$ are done with *high parametric* reliability. This is however not true for those in the other decay classes: apart from stating that they can exhibit direct **CP** asymmetries, we have ignored here the impact of penguin operators that can muddle the extraction of ϕ_1 and ϕ_2. This is merely indicated by the symbol $\sim$.
- While in these channels penguin operators are thus a nuisance we could easily do without, they become an essential agent in other modes like $B \to K\pi$, for they can provide the second weak amplitude required for a direct **CP** asymmetry to become observable. Of course they remain annoying in the sense that their evaluation is beset with considerable theoretical uncertainties.
- The predicted plethora of large **CP** asymmetries in B decays is beginning to emerge in the data.
- Experiments at hadronic colliders have begun making significant contributions to this field.
- There are three classes of **CP** violating observables in the sector of beauty hadrons.
  (1) **CP** violation *involving* $B - \overline{B}$ oscillations: It has been established in $B_d(t) \to \psi K_S$ since 2001.
  (2) *Direct* **CP** violation: It has been found in $B_d \to K\pi$ in 2004.
  (3) **CP** violation *in* $B - \overline{B}$ oscillations: It is predicted to occur in the SM on a reduced level and is being searched for through $B^0(t) \to l^- X$ vs. $\overline{B}^0(t) \to l^+ X$.

Table 11.2 *KM predictions on* **CP** *asymmetries involving* $B - \overline{B}$ *oscillations. Here* $\mathcal{R} = \sqrt{\rho^2 + \eta^2}$. *For the analysis of* $B \to D^0 K_S$ *and* $B \to \overline{D}^0 K_S$ *decays see Problem* 11.8.

Quark level transition	Example of hadronic channel f	$\lvert\bar{\rho}_{B^0 \to f}\rvert$	KM prediction for **CP** asymmetry	**CP** parity
$b \to c c \bar{s}$	$B_d \to \psi K_S$	1	$\sin 2\phi_1 + \mathcal{O}(\lambda^2)$	odd
	$B_s \to \psi \phi$	1	$2\lambda^2 \eta$	~even
$b \to u \bar{u} d$	$B_d \to \pi^+ \pi^-$	$\simeq 1$	$\sim \sin 2\phi_2$	even
	$B_s \to K_S \rho^0$	$\simeq 1$	$\sim \sin 2\phi_3$	odd
	$B_d \to \rho^\pm \pi^\mp$	$\mathcal{O}(1)$	$\sim \sin 2\phi_2$	none
$b \to c(\bar{u}s)$	$B_d \to D^0 K_S$	$\mathcal{O}(1)$	$\sim -\frac{1}{\mathcal{R}}\sin(2\phi_1 + \phi_3)$	none
$b \to u(\bar{c}s)$	$B_d \to \overline{D}^0 K_S$	$\mathcal{O}(1)$	$\sim -\mathcal{R}\sin(2\phi_1 + \phi_3)$	none
	$B_s \to D_s^+ K^-$	$\mathcal{O}(1)$	$\sim -\frac{1}{\mathcal{R}}\sin\phi_3$	none
	$B_s \to D_s^- K^+$	$\mathcal{O}(1)$	$\sim -\mathcal{R}\sin\phi_3$	none

- There is actually a fourth class that deserves separate attention, namely *direct* **CP** violation in *semileptonic B* decays – in particular in $B \to \tau \nu X$. Even with New Physics it is very unlikely one could obtain an observable **CP** asymmetry in the *integrated rates*. However in analogy to $K_{\mu 3}$ decays one could search for a transverse polarization of the final state τ lepton or better still for **T**-*odd* moments in $B \to \tau \nu D \pi$.

In this chapter we have presented a painting in rather broad (theoretical) brush strokes. In the next chapter we will address more precisely various elements of our theoretical treatment, their subtleties and uncertainties and how to deal with them.

A final thought

As we all have experienced in hiking or climbing a mountain, each step up is often tough and tiring. Therefore on occasion we stop, rest and enjoy the view. When we look down at the path we have just climbed, we are amazed at our own achievements. Let us rest and look back. Figure 11.11 shows what we know about the unitarity triangle as the book is going into press. Compare it with Fig. 10.16. In 1998 the only data we could use were measurements of ΔM_{B_d}, ϵ_K and $\mathbf{V}_{ub}/\mathbf{V}_{cb}$. With two new entries – (i) the determination of ϕ_1, and (ii) $\Delta M_{B_s}/\Delta M_{B_d}$ – the allowed region

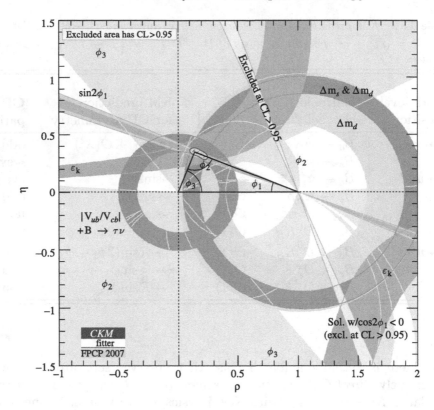

Figure 11.11 New measurements of ϕ_1 and $\Delta M_{B_s}/\Delta M_{B_d}$ reduced the allowed region on the $\rho - \eta$ plane [154]; similar results can be found in [155]. This should be compared to Fig. 10.16 which was drawn in 1998.

is restricted to a small area. In the meantime the circular band around $(\rho, \eta) = (0, 0)$ has shrunk due to better data on $\mathbf{V}_{ub}/\mathbf{V}_{cb}$. The figure also shows the constraint from the determination of ϕ_2 and ϕ_3. Note that all of these constraints can simultaneously be satisfied with values for ρ and η from a rather small area.

We urge the reader to reread Section 10.10.2. The discussion made still applies. Also, while this agreement with the SM is very impressive, this is only an intermediate stage. It convinces us that the **CP** violation observed so far is generated by KM dynamics. However, we know that these dynamics do not explain the baryon asymmetry of the universe. There must be more surprises waiting to be uncovered by us. We want to establish small discrepancies presumably hidden in the data. To do this, we need to make the allowed region in Fig. 11.11 as small as the tip of a needle. Lots of work remains; i.e. we have to continue climbing the mountain!

Problems

11.1. We have repeatedly emphasized that strong final state interaction phases are important in exhibiting a **CP** asymmetry in the width. At the same time, we have insisted that the asymmetry in $B \to \psi K_S$ can be expected entirely with CKM parameters. Resolve the apparent paradox.

11.2. What do we mean by ϵ_B and is this phase convention dependent?

11.3. Show that $|q/p| \simeq 1$ for $\Delta M \gg \Delta \Gamma$. Under what condition can we write

$$\text{Im}\left(\frac{q}{p}\bar{\rho}\right) = -\text{Im}\left(\frac{p}{q}\rho\right)? \qquad (11.80)$$

11.4. What is the relationship between $\phi_{\psi K_s}$ and $\phi_{\pi\pi}$ if the phase enters only through q/p? This is the case for the superweak theory.

11.5. Consider $B \to f_i$ decay where f_i is a **CP** eigenstate, and assume that there is only one amplitude contributing. Define

$$\rho(f_i) = \frac{\langle f_i | \mathcal{H} | B \rangle}{\langle \bar{f}_i | \mathcal{H} | \bar{B} \rangle}. \qquad (11.81)$$

If **CP** is conserved, $\rho(f_i) = $ **CP** eigenvalue of f_i. Now if **CP** is not conserved, $\mathcal{H}$ contains a complex phase. If decays to f_1 and f_2 are both caused by the same Feynman graph, show that $\rho(f_1) = \pm \rho(f_2)$ where the $\pm$ sign is given by the relative **CP** eigenvalue.

11.6. Show that

$$\text{Im}\frac{q}{p}\bar{\rho}(\psi K_S) = \sin 2\phi_1$$

$$\text{Im}\frac{q}{p}\bar{\rho}(\pi\pi) = \sin 2\phi_2. \qquad (11.82)$$

11.7. Show that **CP** symmetry implies:

$$\langle \psi K_S | \bar{b}_L \gamma_\mu c_L \bar{c}_L \gamma^\mu s_L | B \rangle = -\langle \psi K_S | (\bar{b}_L \gamma_\mu c_L \bar{c}_L \gamma^\mu s_L)^\dagger | \bar{B} \rangle$$

$$\langle \pi\pi | \bar{b}_L \gamma_\mu c_L \bar{c}_L \gamma^\mu s_L | B \rangle = \langle \pi\pi | (\bar{b}_L \gamma_\mu c_L \bar{c}_L \gamma^\mu s_L)^\dagger | \bar{B} \rangle. \qquad (11.83)$$

11.8. Consider $B \to D^0 K_S$ and $B \to \overline{D}^0 K_S$ decays with strong final state interaction phases δ_{DK_s} and $\delta_{\overline{D}K_s}$, respectively. Show that

$$\rho(D^0 K_S) = e^{i(\delta_{\overline{D}K_s} - \delta_{DK_s})} \frac{A_{\overline{D}}}{A_D} e^{i\phi_3}$$

$$\rho(\overline{D}^0 K_S) = e^{-i(\delta_{\overline{D}K_s} - \delta_{DK_s})} \frac{A_D}{A_{\overline{D}}} e^{i\phi_3}. \qquad (11.84)$$

Can we use these to extract ϕ_3?

11.9. Show that

$$A(f,t)^*_{\lambda_1} A(f,t)_{\lambda_2}$$
$$= |A(f,t)^*_{\lambda_1} A(f,t)_{\lambda_2}| e^{i(\delta_{\lambda_2} - \delta_{\lambda_1})} \left[e^{i(\phi_{\lambda_2} - \phi_{\lambda_1})} \cos^2 \frac{\Delta M t}{2} \right.$$
$$+ |\bar{\rho}_{\lambda_1} \bar{\rho}_{\lambda_2}| \eta_{\lambda_1} \eta_{\lambda_2} e^{i(\phi_{\lambda_1} - \phi_{\lambda_2})} \sin^2 \frac{\Delta M t}{2}$$
$$+ \frac{i}{2} \sin(\Delta M t) \left(|\bar{\rho}_{\lambda_2}| \eta_{\lambda_2} e^{i(\xi - \phi_{\lambda_1} - \phi_{\lambda_2})} - |\bar{\rho}_{\lambda_1}| \eta_{\lambda_1} e^{i(-\xi + \phi_{\lambda_1} + \phi_{\lambda_2})} \right) \Big]$$

$$\overline{A}(f,t)^*_{\lambda_1} \overline{A}(f,t)_{\lambda_2}$$
$$= |A(f,t)^*_{\lambda_1} A(f,t)_{\lambda_2}| |\bar{\rho}_{\lambda_1} \bar{\rho}_{\lambda_2}| \eta_{\lambda_1} \eta_{\lambda_2} e^{i(\delta_{\lambda_2} - \delta_{\lambda_1})} \left[e^{i(\phi_{\lambda_1} - \phi_{\lambda_2})} \cos^2 \frac{\Delta M t}{2} \right.$$
$$+ |\rho_{\lambda_1} \rho_{\lambda_2}| \eta_{\lambda_1} \eta_{\lambda_2} e^{i(\phi_{\lambda_2} - \phi_{\lambda_1})} \sin^2 \frac{\Delta M t}{2}$$
$$+ \frac{i}{2} \sin(\Delta M t) \left(|\rho_{\lambda_2}| \eta_{\lambda_2} e^{i(-\xi + \phi_{\lambda_1} + \phi_{\lambda_2})} - |\rho_{\lambda_1}| \eta_{\lambda_1} e^{i(\xi - \phi_{\lambda_1} - \phi_{\lambda_2})} \right) \Big]$$

$$(11.85)$$

11.10. Derive Eq. (11.4).

11.11. From Eq. (11.24) and Eq. (11.25) we see that the imaginary parts of these amplitudes do not vanish in the limit of the CKM matrix elements being real. Does this mean that **CP** violation exists even for a real CKM matrix?

11.12. Show that in Eq. (11.24) and Eq. (11.25) the strong final state interaction phase for $A(DK^-)$ is equal to that of $A(\overline{D}K^+)$.

11.13. Consider

$$B_s \rightarrow D^0 \phi, \ \overline{D}^0 \phi, \ D_{\pm} \phi \qquad (11.86)$$

with $|D_{\pm}\rangle = \frac{1}{\sqrt{2}}(|D^0\rangle \pm |\overline{D}^0\rangle)$. Show that

$$\frac{q}{p} \bar{\rho}(D_{\pm}\phi) = \frac{(q/p)\bar{\rho}(D^0 \phi) \pm 1}{1 \pm (p/q)\bar{\rho}(\overline{D}^0 \phi)^{-1}} \qquad (11.87)$$

holds in the KM ansatz. Check $(q/p)\bar{\rho}(D^0\phi)$ and $(q/p)\bar{\rho}(\overline{D}^0\phi)$ are rephasing invariant.

Hint: first show that

$$\frac{p}{q}\rho(D^0_{1,2}\phi) = \frac{\frac{q}{p}\rho(D^0\phi) \pm \frac{p}{q}e^{2i\arg[\mathbf{V}^*_{cb}\mathbf{V}_{us}]}}{1 \pm \left(\frac{q}{p}\rho(\overline{D}^0\phi)\right)^{-1}\frac{p}{q}e^{2i\arg[\mathbf{V}^*_{cb}\mathbf{V}_{us}]}}. \tag{11.88}$$

Then show that $\lambda(D^0_{1,2}\phi)$ is rephasing invariant. If you concluded that $\frac{p}{q}e^{-2i\arg[\mathbf{V}^*_{cb}\mathbf{V}_{us}]}$ is not a rephasing invariant, you are correct. Then think about the definition of $D_{1,2}$. How does this definition change as we change the phases of the u and c quarks?

12

Rare K and B decays – almost perfect laboratories

If a meson lives long enough, it has a chance to show us many interesting phenomena. With ordinary transitions slowed down, quantum effects due to virtual intermediate states become relevant, and those can involve dynamical entities beyond the SM operating at much higher energy scales than M_K or M_B. Search for new phenomena where uninteresting ordinary decays are suppressed has yielded many hints in the past, and we can expect it to do so again in the future.

A voluminous tool chest has been created over the years. This field was pioneered by Gaillard and Lee [156]; the renormalization group analysis including penguins was introduced by Shifman, Vainshtein, Voloshin and Zakharov [86]; Gilman and Wise [87, 157] and also Vysotskii [158] refined the analysis; electromagnetic penguins were introduced by Flynn and Randall [88]; the heavy quark mass corrections were computed by Inami and Lim [96]; finally the QCD corrections to two loops have been evaluated by Buras and co-workers [81, 83, 94, 159]. In order not to divert too much from our main theme we just sketch the procedure and quote numerical results, and refer the truly committed student to the aforementioned literature.

12.1 Rare K decays

12.1.1 $K_L \to \mu^+\mu^-$ and $K^+ \to \pi^+ e^+ e^-$

The phenomenological landscape can be characterized by two tiny ratios:

$$\frac{\Gamma(K^+ \to \pi^+ e^+ e^-)}{\Gamma(K^+ \to \pi^0 e^+ \nu)} \sim 6 \times 10^{-6}, \tag{12.1}$$

248

$$\frac{\Gamma(K_L \to \mu^+\mu^-)}{\Gamma(K^+ \to \mu^+\nu)} \sim 3 \times 10^{-9}. \tag{12.2}$$

These ratios concern the strength of strangeness-changing neutral currents. They are greatly suppressed; this near-absence of strangeness-changing currents provided one of the guiding principles in constructing the SM. Yet the suppression is of quite different strength – by roughly three orders of magnitude – in leptonic and in semileptonic transitions. Such a huge numerical difference cannot be laid at the doorstep of non-perturbative corrections due to the strong interations. There has to be a structural explanation. Let us see how the SM addresses this challenge. Both $K_L \to \mu^+\mu^-$ and $K^+ \to \pi^+ e^+ e^-$ are driven by the process depicted in Fig. 12.1. The effective interaction $\mathcal{H}(s \to d\gamma)$ which is responsible for Fig. 12.1(a) is of special interest [160]:

$$\mathcal{H}(s \to d\gamma) = \frac{1}{6}\langle r_{sd}\rangle \bar{s}(-\partial_\lambda\partial^\lambda g_{\mu\nu} + \partial_\mu\partial_\nu)\gamma^\nu(1 - \gamma_5)dA^\mu$$
$$+ \langle \mu_{sd}\rangle \bar{s}\sigma_{\mu\nu}[(m_s + m_d) + \gamma_5(m_s - m_d)]dF^{\mu\nu},$$

$$\frac{1}{6}\langle r_{sd}\rangle = \frac{G_F}{\sqrt{2}}\sin\theta_c\cos\theta_c\left(-\frac{e}{9\pi^2}\right)\ln\frac{m_c}{m_u},$$

$$\langle \mu_{sd}\rangle = \frac{G_F}{\sqrt{2}}\sin\theta_c\cos\theta_c\left(\frac{7e}{192\pi^2}\right)\frac{m_c^2 - m_u^2}{M_W^2}, \tag{12.3}$$

where A^μ denotes the photon field and $F^{\mu\nu}$ its field strength; the top quark contribution has been ignored for simplicity. The first term in Eq. (12.3) is called the charge radius term and the second one the magnetic moment term.

The GIM mechanism requires these amplitudes to vanish if $m_u = m_c$. This is realized in two different ways: while the charge radius term has a GIM suppression $\sim \log m_c/m_u$ – which actually provides no suppression – it is of the form $(m_c^2 - m_u^2)/M_W^2$ for the magnetic moment term as well as the other diagram, Fig. 12.1(b). The fact that the charge radius interaction

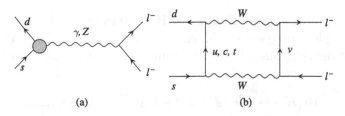

(a) (b)

Figure 12.1 The diagrams contributing to $s\bar{d} \to l^+l^-$. The key contribution is the $s\bar{d}\gamma$ vertex in (a) which has a GIM suppression of the form $\alpha_s \log(m_c/m_u)$, and it contributes only to $K \to \pi e^+ e^-$.

cannot contribute to $K \to \mu^+\mu^-$ (see Problem 13.2) explains the factor ~ 1000 difference shown in Eq. (12.2). A word of caution: QCD corrections change the effective Hamiltonian considerably. For example, the magnetic moment term gets an additional $\alpha_s \log(m_c/m_u)$ term [86].

12.1.2 $K_L \to \pi^0 l^+ l^-$

The final state $\pi^0 l^+ l^-$ is **CP** even to lowest order in the electroweak coupling. (Problem 13.3). It can be reached from the **CP** even component in K_L

$$K_L \to K_S \to \pi^0 l^+ l^- \tag{12.4}$$

or through direct **CP** violation in $\mathcal{H}(\Delta S)$, see Fig. 12.1. In the case of $K_L \to \pi\pi$, the direct **CP** violation parameter, ϵ'/ϵ, is suppressed by an additional factor of ω due to the $\Delta I = \frac{1}{2}$ rule. For many rare decays, such non-leptonic dynamical effects are not expected to play a role. So, we expect the strength of *direct* **CP** violation to be roughly comparable to that via $K_L \to K_S$. Detailed calculations indeed yield [159, 161]

$$\mathrm{Br}(K_L \to \pi^0 e^+ e^-) = (4.5 \pm 2.6) \times 10^{-12}$$
$$\mathrm{Br}(K_L \to K_S \to \pi^0 e^+ e^-) < 1.5 \times 10^{-12}. \tag{12.5}$$

Unfortunately, there are other contributions [157], the chain reaction

$$K_L \to \pi^0 \gamma^* \gamma^* \to \pi^0 l^+ l^-, \tag{12.6}$$

which is neither necessarily short-distance dominated nor **CP** violating. Using chiral perturbation theory, we deduce from the data on $K_L \to \pi^0 \gamma\gamma$ [162, 163] that

$$\mathrm{Br}(K_L \to \pi^0 e^+ e^-) = (0.3 - 1.8) \times 10^{-12}, \tag{12.7}$$

which is not much smaller than the **CP** breaking contributions.

A Dalitz plot analysis would enable us to extract the **CP** violating component unambigously [164]. Since the experimental limit for this decay is [11]

$$\mathrm{Br}(K_L \to \pi^0 e^+ e^-) < 2.8 \times 10^{-9}, \quad 90\% \text{ C.L.}, \tag{12.8}$$

this is not likely to happen soon – unless unexpected New Physics generates a signal to be observed by those who dare venture in this direction.

12.1.3 $K \to \pi\nu\bar{\nu}$

The measurement of the branching ratios for this class of decays truly tests experimentalists' patience as well as expertise. Yet the theoretical predictions are rather clean thus allowing a more conclusive interpretation of experimental findings than for most of the rare K decays. Recently, there has been much progress theoretically as well as experimentally. We also look forward to even more experimental progress with the advent of J-PARC. It is a facility being built in Tokai-mura, Japan, which will provide a 50 GeV proton beam with 3×10^{14} particles per pulse. The matrix element for this decay is given by [96]:

$$\langle \pi\nu\bar{\nu}|H(d\bar{s} \to \nu\bar{\nu})|K\rangle = \frac{G_F}{\sqrt{2}} \frac{\alpha}{2\pi \sin^2 \theta_W} \sum_{i=e,\mu\tau} D\langle \pi|(\bar{s}d)_{V-A}|K\rangle(\bar{\nu}\nu)_{V-A},$$
(12.9)

where

$$D = \mathbf{V}_{td}\mathbf{V}_{ts}^* X(x_t) + \mathbf{V}_{cd}\mathbf{V}_{cs}^* X(x_c),$$

$$X(x) = \frac{x}{8}\left[3\frac{x-2}{(1-x)^2}\ln x + \frac{x+2}{x-1}\right], \quad x_i = \frac{m_i^2}{M_W^2}.$$
(12.10)

We have ignored QCD corrections here as well as terms of $\mathcal{O}(\frac{m_i^2}{M_W^2})$. For these corrections and details of the computation, as well as a more detailed review of the subject, we refer to Ref. [161], and references therein.

In evaluating the hadronic matrix element $\langle \pi|J_\mu^{\text{had}}|K\rangle$ we can no longer ignore long-distance dynamics. What saves the day is the following: isospin invariance implies the equality of the quantities $\langle \pi^+|J_\mu^{\text{had,neut}}|K^+\rangle$ and $\langle \pi^0|J_\mu^{\text{had,neut}}|K^0\rangle$ with $\langle \pi^0|J_\mu^{\text{had,ch}}|K^+\rangle$; the latter is known from the data on the ordinary semileptonic transition $K^+ \to \pi^0 l^+ \nu$.

$K^+ \to \pi^+ \nu\bar{\nu}$

A new three-loop SM prediction [165] reduces the uncertainty considerably:

$$\text{Br}(K^+ \to \pi^+ \nu_i \bar{\nu}_i) = (8.0 \pm 1.2) \times 10^{-11}.$$
(12.11)

Virtually identical numbers are given in Ref. [166]. There is reasonable hope that the present 15% uncertainty in this SM prediction can be reduced to 4%.

One measures this branching ratio by looking for events of the type

$$K^+ \to \pi^+ + \text{nothing}.$$
(12.12)

The AGS experiment E787 at Brookhaven National Laboratory has been searching for events of this type. Just to give you some idea how formidable the search is, note that what we are searching for is an event – with less than sterling signature! – occurring only once in about 10^{10} decays. For example, we must make sure that other K^+ decays such as

$$K^+ \to \mu^+ \nu_\mu, \ \mu^+ \nu_\mu \gamma, \ \pi^+ \pi^0, \tag{12.13}$$

do not mimic the signal at the level of once every 10^{10} decays. The presence of π^0 must be excluded by vetoing γ's. The misidentification of μ^+ for π^+ is avoided by observing the chain decay

$$\pi^+ \to \mu^+ \nu$$
$$\hookrightarrow e^+ \nu_e \bar{\nu}_\mu. \tag{12.14}$$

E787 and its newer version E949 have found two candidates for $K^+ \to \pi^+ \nu \bar{\nu}$ [167]. Interpreting them as a signal, the branching ratio is:

$$\mathrm{Br}(K^+ \to \pi^+ \nu_i \bar{\nu}_i) = 1.47\ ^{+1.3}_{-0.89} \times 10^{-10}. \tag{12.15}$$

Comparing Eq. (12.11) and Eq. (12.15), we see that the SM will be tested in the near future.

What justifies the heroic effort required to measure such a tiny and hard to observe branching ratio? As stated above one can expect the uncertainty in the predicted SM value to be reduced to about 4%. This points to a dual motivation: (i) one can extract from the measurement an accurate value for the CKM parameter $|\mathbf{V}_{td} \mathbf{V}^*_{ts}|$, which provides another constraint on the unitarity triangle; (ii) inserting the best value for $|\mathbf{V}_{td} \mathbf{V}^*_{ts}|$ as inferred from, say, ΔM_{B_d} and ΔM_{B_s} one can use the great accuracy of the SM prediction to search for an even smaller New Physics contribution in $K^+ \to \pi^+ \nu \bar{\nu}$. The existence of additional Higgs bosons, right-handed gauge bosons or SUSY particles could change the branching ratio significantly.

A rather exotic example for physics beyond the SM may affect this process *directly*:

$$K^+ \to \pi^+ + X^0, \tag{12.16}$$

where X^0 is a massless particle, will lead to a striking signature. The final state π^+ is monoenergetic in the rest frame of K^+. The 90% confidence limit for such a decay is [167]

$$\mathrm{Br}(K^+ \to \pi^+ + X^0) < 0.73 \times 10^{-10}. \tag{12.17}$$

In Ref. [167], the upper limit is given as a function of X^0 mass.

$K_L \rightarrow \pi^0 \nu \bar{\nu}$

This decay can occur only due to **CP** violation since there is no two-photon and practically no two-Z^0 intermediate state. To go beyond these qualitative remarks we follow the usual and by now familiar procedure, namely to first derive the effective $\Delta S = 1$, $\Delta Q = 0$ coupling on the quark level and then evaluate the appropriate matrix elements.

It can formally be expressed through an effective low energy Hamiltonian involving hadronic fields:

$$\mathcal{H}_{\text{eff}} = c_{CPV} \phi_\pi \partial_\mu \phi_{K_L} \bar{\nu}_L \gamma^\mu \nu_L. \tag{12.18}$$

Under **CP** we have $\bar{\nu}\gamma^\mu\nu \rightarrow -\bar{\nu}\gamma_\mu\nu$, $\partial_\mu\phi_{K_L} \rightarrow -\partial^\mu\phi_{K_L}$ and $\phi_{\pi^0} \rightarrow -\phi_{\pi^0}$, i.e. $\mathcal{H}_{\text{eff}}$ is odd under **CP**; therefore c_{CPV} has to be proportional to $\text{Im}\mathbf{V}_{td}\mathbf{V}_{ts}^*$. Putting everything together, we arrive at

$$\frac{\text{Br}(K_L \rightarrow \pi^0 \nu \bar{\nu})}{\text{Br}(K^+ \rightarrow \pi^0 e^+ \nu_e)} = \frac{3}{2}\frac{\tau(K_L)}{\tau(K^+)}\frac{\alpha^2}{\pi^2 \sin^4\theta_W}\left|\frac{\text{Im}(\mathbf{V}_{td}\mathbf{V}_{ts}^*)}{\mathbf{V}_{us}}X(x_t)\right|^2. \tag{12.19}$$

Again using presently available information on the KM parameters, we arrive at the numerical prediction [166]:

$$\text{Br}(K_L \rightarrow \pi^0 \nu \bar{\nu}) = (2.85 \pm 0.39) \times 10^{-11}. \tag{12.20}$$

One can expect the present uncertainty of 14% to get reduced to maybe as low as 2%. The irreducible error on the SM prediction for $K_L \rightarrow \pi^0 \nu \bar{\nu}$ is lower than for $K^+ \rightarrow \pi^+ \nu \bar{\nu}$, since the former is less sensitive to m_c and long-distance dynamics.

This decay is dominated by *direct* **CP** violation. If for some reason the direct **CP** violation computed above is absent, the **CP** even component in the K_L wavefunction can give rise to this decay. The decay rate for $K_L \rightarrow \pi^0 \nu \bar{\nu}$ is then reduced relative to that for $K^+ \rightarrow \pi^+ \nu \bar{\nu}$ by the impurity parameter ϵ_K^2.

Of course, the present experimental bound is above the expectation by several orders of magnitude [168]:

$$\text{Br}(K_L \rightarrow \pi^0 \nu \bar{\nu}) < 2.1 \times 10^{-7}, \quad 90\% \text{ CL}. \tag{12.21}$$

The J-PARC facility in Japan might eventually provide about 1000 events. An simple argument gives an experimentally relevant bound [169]

$$\langle \pi^0 \nu \bar{\nu} | H | K_L \rangle = \frac{1}{\sqrt{2}}(pA - q\bar{A}) \tag{12.22}$$

where A, and $\overline{A}$ are the amplitudes for $K^0 \to \pi^0 \nu\overline{\nu}$ and $\overline{K}^0 \to \pi^0 \nu\overline{\nu}$, respectively. Defining $\lambda = \frac{q}{p}\frac{\overline{A}}{A} = e^{2i\theta}$ we obtain

$$\Gamma(K_L \to \pi^0 \nu\overline{\nu}) = 2\Gamma(K^0 \to \pi^0 \nu\overline{\nu})\sin^2\theta. \qquad (12.23)$$

Using the isospin relation $\Gamma(K^0 \to \pi^0 \nu\overline{\nu}) = \frac{1}{2}\Gamma(K^+ \to \pi^+ \nu\overline{\nu})$ we obtain

$$\frac{\Gamma(K_L \to \pi^0 \nu\overline{\nu})}{\Gamma(K^+ \to \pi^+ \nu\overline{\nu})} = \sin^2\theta. \qquad (12.24)$$

Since $\sin\theta \leq 1$, we have a model-independent bound:

$$\mathrm{Br}(K_L \to \pi^0 \nu\overline{\nu}) < 6.1 \times 10^{-10}. \qquad (12.25)$$

Observing $K_L \to \pi^0 \nu\overline{\nu}$ – let alone measuring its branching ratio – represents an even stiffer experimental challenge than for $K^+ \to \pi^+ \nu\overline{\nu}$. The dual goals to extract the value of the CKM parameter V_{td} and search for manifestations of New Physics can be pursued more forcefully by measuring the widths of *both* $K^+ \to \pi^+ \nu\overline{\nu}$ *and* $K_L \to \pi^0 \nu\overline{\nu}$: While the former is controlled by $|V_{td}|$, the latter is by $\mathrm{Im}V_{td}$; those two observables thus allow the construction of the unitarity triangle, since the base line is normalized to unity. The version of the triangle inferred from $\Delta S = 1, 2$ dynamics can then be compared with its version as it emerges from B oscillations and decays. Any significant difference in the two versions points to the intervention of New Physics, presumably in the $\Delta S = 2$ and/or $\Delta B = 2$ sectors.

12.1.4 ♠ K $\to \pi\pi\gamma^{(*)}$ ♠

$\mathrm{K}^\pm \to \pi^\pm \pi^0 \gamma$

CPT symmetry enforces $\Gamma(K^+ \to \pi^+\pi^0) = \Gamma(K^- \to \pi^-\pi^0)$ to the degree that electromagnetic corrections can be ignored; the latter allows a tiny asymmetry to sneak in. It would then appear more promising to search for direct **CP** violation in radiative K decays – $K^+ \to \pi^+\pi^0\gamma$ vs. $K^- \to \pi^-\pi^0\gamma$ – thus trading in a larger branching ratio for a larger asymmetry. QCD corrections actually enhance its rate. Upon closer examination, however, we have encountered [170] a strong kinematic suppresion. The decay is dominated by the bremsstrahlung amplitude and a magnetic multipole amplitude. **CP** violation is generated by the interference of the bremsstrahlung and a small electric multipole amplitude. Theoretical estimates yield

$$a(\pi\pi\gamma) = \frac{\Gamma(K^+ \to \pi^+\pi^0\gamma) - \Gamma(K^- \to \pi^-\pi^0\gamma)}{\Gamma(K^+ \to \pi^+\pi^0\gamma) + \Gamma(K^- \to \pi^-\pi^0\gamma)} \sim 10^{-4} - 10^{-5}, \quad (12.26)$$

at best, which is discouraging.

The difficulty can be traced to the fact that the dominating magnetic multipole amplitude cannot interfere with the bremsstrahlung amplitude, once we sum over the photon polarization.

$$\mathbf{K} \to \pi^+\pi^-\gamma^{(*)}.$$

The modes $K_{L,S} \to \pi^+\pi^-\gamma$ have been observed with [171]

$$\text{Br}(K_L \to \pi^+\pi^-\gamma) = (4.66 \pm 0.15) \cdot 10^{-5}, \qquad (12.27)$$
$$\text{Br}(K_S \to \pi^+\pi^-\gamma) = (4.87 \pm 0.11) \cdot 10^{-3}, \qquad (12.28)$$

for $E_\gamma > 20\,\text{MeV}$. Two mechanisms can drive these channels and an analysis of the photon spectra indeed reveals the intervention of both:

- bremsstrahlung off the pions through an E1 transition:

$$K_L \xrightarrow{\Delta S=1} \pi^+\pi^- \xrightarrow{E1} \pi^+\pi^-\gamma, \ \ K_S \xrightarrow{\Delta S=1} \pi^+\pi^- \xrightarrow{E1} \pi^+\pi^-\gamma, \tag{12.29}$$

  where only the first step in the K_L decay is **CP** violating.
- Direct photon emission of the M1 type

$$K_L \xrightarrow{M1\&\Delta S=1} \pi^+\pi^-\gamma, \ \ K_S \xrightarrow{M1\&\Delta S=1} \pi^+\pi^-\gamma, \tag{12.30}$$

  which is **CP** conserving [violating] for the K_L [K_S] process.

In analogy to ϵ_K we define a ratio of E1 amplitudes

$$\eta_{+-\gamma} = \frac{A(K_L \to \pi^+\pi^-\gamma, E1)}{A(K_S \to \pi^+\pi^-\gamma, E1)}, \tag{12.31}$$

that measures **CP** violation. In the absence of direct **CP** violation we have $\eta_{+-\gamma} = \eta_{+-}$.

We can proceed in close analogy to the $K_L \to \pi^+\pi^-$, $\pi^0\pi^0$ case and through the use of a regenerator in a K_L beam compare the decay rate evolution of $K_L \to \pi^+\pi^-\gamma$ and $K_S \to \pi^+\pi^-\gamma$ as a function of the time of decay. This has been done in Ref. [171] where it has been demonstrated that $K_L - K_S$ interference occurs in $K \to \pi^+\pi^-\gamma$ and that the **CP** violating parameters of these modes are quite consistent with those of $K_L \to \pi\pi$.

We want to focus on a special aspect of it. The interference of the **CP** violating E1 and conserving M1 amplitudes will yield a circularly polarized photon. This polarization represents **CP** violation. To be more specific: the *interference* yields a triple correlation between the pion momenta and the photon polarization

$$P_\perp^\gamma = \langle \vec{\epsilon}_\gamma \cdot (\vec{p}_{\pi^+} \times \vec{p}_{\pi^-}) \rangle, \tag{12.32}$$

which is **CP** odd; its leading contribution is proportional to η_{+-} entering in the E1 amplitude.

$\mathbf{K_L \rightarrow \pi^+ \pi^- e^+ e^-}$

The photon polarization which constitutes the **CP** signal can be probed best for off-shell photons

$$K_L \rightarrow \pi^+ \pi^- \gamma^* \rightarrow \pi^+ \pi^- e^+ e^-. \tag{12.33}$$

A general discussion which includes direct E1 transition as well as the charge radius term $K_L \rightarrow K_S e^+ e^- \rightarrow \pi^+ \pi^- e^+ e^-$ can be performed [172, 173]. Actually these give new direct **CP** violating effects, but as discussed for the asymmetry in $K \rightarrow \pi\pi\gamma$ decay, these new effects are rather small. Here we restrict ourselves to the leading contributions for the asymmetry. The amplitude for this decay can be written as

$$T(K_L \rightarrow \pi^+ \pi^- e^+ e^-) = e|T(K_S \rightarrow \pi^+ \pi^-)|\cdot$$

$$\cdot \left[\frac{g_{BR}}{M_K^4} \left(\frac{p_+^\mu}{p_+ \cdot k} - \frac{p_-^\mu}{p_- \cdot k} \right) + \frac{g_{M1}}{M_K^4} \epsilon_{\mu\nu\alpha\beta} k^\nu p_+^\alpha p_-^\beta \right] \frac{e}{k^2} \bar{u}(k_-) \gamma_\mu v(k_+) \tag{12.34}$$

(with $k = k_+ + k_-$) in terms of two couplings $g_{BR,M1}$ for the bremsstrahlung and $M1$ transitions, respectively.

From the observed M1 rate, we infer

$$g_{BR} = \eta_{+-} e^{i\delta_0(m_K^2)}, \quad g_{M1} = 0.76i \, e^{i\delta_1(s_\pi)}, \tag{12.35}$$

where $\delta_{0,1}$ denote the S- and P-wave $\pi\pi$ phase shifts.

The **CP** violating effect appears as a correlation between the $e^+ e^-$ and $\pi^+ \pi^-$ planes. Denoting by Φ the angle between the $e^+ e^-$ and $\pi^+ \pi^-$ planes, we obtain

$$\frac{d}{d\Phi} \Gamma(K_L \rightarrow \pi^+ \pi^- e^+ e^-) = \Gamma_1 \cos^2\Phi + \Gamma_2 \sin^2\Phi + \Gamma_3 \cos\Phi \sin\Phi. \tag{12.36}$$

It is easy to see that the term $\cos\Phi \sin\Phi$ changes sign under **CP** and **T** (see Problem 13.4); Γ_3 thus represents **CP** violation. It can be projected out by comparing the Φ distribution integrated over two quadrants:

$$A = \frac{\int_0^{\pi/2} d\Phi \frac{d\Gamma}{d\Phi} - \int_{\pi/2}^{\pi} d\Phi \frac{d\Gamma}{d\Phi}}{\int_0^{\pi} d\Phi \frac{d\Gamma}{d\Phi}} = \frac{2\Gamma_3}{\pi(\Gamma_1 + \Gamma_2)}. \tag{12.37}$$

Including all terms the result reads [172]:

$$A \simeq (15 \cos \Theta_1)\% + \left(38 \left| \frac{g_{E1}}{g_{M1}} \right| \cos \Theta_2 \right) \%, \qquad (12.38)$$

where $\Theta_1 = \phi_{+-} + \delta_0 - \bar{\delta}_1 - \frac{\pi}{2}$ (mod π); $\Theta_2 = \phi_{+-} + \delta_0 - \frac{\pi}{2}$ (mod π); and g_{E1} is a coupling for the E1 transition. Here $\bar{\delta}_1$ is the P wave $\pi\pi$ phase shift averaged over the kinematical region. Experimentally, $\delta_0 - \bar{\delta}_1 \sim 30°$. For $\phi_{+-} = \phi_{SW}$, and $g_{E1}/g_{M1} = 0.05$, the asymmetry becomes:

$$A = (14.3 \pm 1.3)\%. \qquad (12.39)$$

The main theoretical uncertainty resides in what we assume for the hadronic form factors. In the above analysis a phenomenological ansatz was employed for them; chiral perturbation theory yields similar numbers [174].

The KTeV experiment at FNAL [175] and subsequently NA48 at CERN measured the rate and asymmetry for this mode in full agreement with the predictions; their average yields [11]

$$A = (13.7 \pm 1.5)\% . \qquad (12.40)$$

The discovery of such a spectacularly large **CP** asymmetry is a significant result, be it only to show that **CP** violation is not uniformly tiny in K_L decays. We should note, though, that A is driven by η_{+-} entering through $K_L \to \pi^+\pi^- \to \pi^+\pi^-\gamma^* \to \pi^+\pi^- e^+ e^-$ and not the prediction of a specific model.

Some comments are in order.

- This **CP** asymmetry is so large because the **CP** violating amplitude is enhanced by kinematic bremsstrahlung factors.
- It is often stated that this asymmetry represents the observation of *direct* **T** violation. It is tempting to think that for the first time **T** violation is clearly seen, independent of **CPT** conservation. Yet with the time reversal operator being *anti*-unitary a **T** odd correlation can arise, as discussed in Section 4.10.3, even with **T** invariant dynamics, if complex phases are present; final state interactions can generate such phases.
- *Direct* **CP** violation can contribute as well to A and its size depends on the details of the dynamics underlying **CP** violation. Yet its contributions to the observable A averaged over all final states are tiny, namely $< 10^{-3}$ for the KM ansatz [176]; it is hard to see how they could be significantly larger for other models. While they can be significantly larger in certain parts of phase space, no promising avenue has been pointed out yet.

12.2 Beauty decays

Transitions driven by flavour-changing neutral currents represent pure quantum effects within the SM, where they are generated from penguin graphs on the one-loop level. Such modes have unequivocally been observed in various radiative transitions. The first such example was the exclusive channel $B^+ \to K^* \gamma$ [177], for which present data yield:

$$\text{Br}(B^+ \to K^* \gamma) = (4.03 \pm 0.26) \times 10^{-5}. \qquad (12.41)$$

Despite its delicate origin as a pure quantum effect, this channel commands a branching ratio very similar to that for the tree graph mode $B \to \psi K$ once the decay of the ψ resonance is included: $\text{BR}(B_d \to \psi K_S \to [l^+ l^-]_\psi K_S) \sim 4 \times 10^{-5}$. This is a reflection of the general rule of thumb that all B decays are on the small side, once one includes the decays of heavy final state particles into pions, kaons, etc. In that sense loop-driven B decays, which are presumably more sensitive to the presence of New Physics, are not truly rare, in particular with the data sets available from the B factories.

Another feature enhances our theoretical control over the impact of strong dynamics in B decays and thus sharpens our probes for New Physics. With the b quark mass m_b much heavier than typical QCD scales $\bar{\Lambda} \lesssim 1\,\text{GeV}$, one can treat *non*perturbative QCD corrections to nonleptonic, and semileptonic and radiative B decays through an expansion in powers of $m_b/\bar{\Lambda}$. That formalism is typically referred to as Heavy Quark Theory or Heavy Quark Expansion [178].

In this section we investigate B decays which only occur through loop effects: $B \to \gamma X$, $B \to l^+ l^- X$, and $B \to l^+ l^-$. The form of the effective Hamiltonian without QCD corrections is quite similar to that of corresponding reactions in K decays. However, the top quark contributions enter in full strength, and the GIM mechanism no longer leads to simple suppressions.

12.2.1 B → X$_s$γ

This is an inclusive transition; i.e. one in which we sum over all possible final states. Using a heavy quark expansion we can show that the leading term is given by the decay of a b quark. The non-perturbative corrections are suppressed by factors $\sim \mathcal{O}\left(\frac{\bar{\Lambda}}{m_b^2}, \frac{\bar{\Lambda}}{m_c^2}\right)$. While we will not address those explicitly, their contributions are included in the SM prediction given below.

The weak Hamiltonian for the underlying b quark transition is given by

$$\mathcal{H} = -\frac{G_F}{\sqrt{2}} \mathbf{V}_{ts}^* \mathbf{V}_{tb} C_{7\gamma}(\mu) Q_{7\gamma}(\mu), \tag{12.42}$$

where

$$Q_{7\gamma} = \frac{e}{8\pi^2} m_b \bar{s} \sigma^{\mu\nu} (1 + \gamma_5) b F_{\mu\nu}. \tag{12.43}$$

The value for $C_{7\gamma}(M_W)$ can be read off from Ref. [96], and $C_{7\gamma}(\mu)$ with the two loop QCD corrections can be found in Ref. [94]. Very recently the SM prediction has been evaluated [179] even on the next-to-next-to-leading log level, where terms of order $G_F \alpha_S^2 (\alpha_S \log)^n$ are included and summed over n. How formidable the required computations were can be read off from the fact that the paper lists 17 (!) theorists as authors – a number that a mere generation ago would mark a sizable experimental collaboration in high energy physics – with one of them singled out as a spokesperson. They find:

$$\mathrm{BR}(B \to X_s \gamma)|_{E_\gamma > 1.6\,\mathrm{GeV}} = (3.15 \pm 0.23) \times 10^{-4}, \tag{12.44}$$

where the various theoretical uncertainties have been added in quadrature.

Both Belle [180] and BaBar [181] have measured this rate in a fully inclusive fashion. Their average reads

$$\mathrm{BR}(B \to X_s \gamma)|_{E_\gamma > 1.6\,\mathrm{GeV}}$$
$$= (3.55 \pm 0.24|_{\mathrm{stat}} \pm 0.10|_{\mathrm{syst}} \pm 0.03|_{\mathrm{shape}}) \times 10^{-4}; \tag{12.45}$$

the last error stands for the mild uncertainty in extrapolating the measured photon energy spectrum below 1.9 GeV.

When comparing the predicted with the measured branching ratio one is reminded of the question whether a glass is half empty or half full; i.e, depending on one's inclination one could conclude that the data are in impressive agreement with the SM prediction – or that a deficit is opening up. Yet we can already conclude that such a deficit can be at most of moderate size. We would like to remind the attentive reader that we had predicted already in the first edition of this book that we cannot *count* on numerically large deviations from SM predictions in most B decay rates. This in turn implies that the experimental sensitivity is only now reaching a level, where one can 'realistically hope' for New Physics leaving its footprint.

12.2.2 B → μ⁺μ⁻

This decay is unlikely to teach us much about **CP** violation – $B \to \tau^+ \tau^-$ might have a better chance there. On the other hand it has such a clear signature that it can be probed also in hadronic collisions, which allows for a very high sensitivity. Within the SM the rate for $B_q \to \mu^+\mu^-$ is controlled by $|\mathbf{V}_{tb}^*\mathbf{V}_{tq}f_{B_q}|$, $q = d, s$, which therefore can be determined by measuring the branching ratio.

The effective interaction Hamiltonian is given by [96]

$$\mathcal{H} = -\frac{G_F}{\sqrt{2}} \frac{\alpha}{2\pi \sin\theta_w^2} \mathbf{V}_{tb}^*\mathbf{V}_{tq}C(x_t)(\bar{b}q)_{V-A}(\bar{\mu}\mu)_{V-A}, \quad q = d, s \qquad (12.46)$$

with

$$C(x) = \frac{x}{8}\left[\frac{3x}{(1-x)^2}\ln x + \frac{4-x}{1-x}\right]. \qquad (12.47)$$

This leads to [94]

$$\frac{\mathrm{Br}(B_d \to \bar{\mu}\mu)}{\mathrm{Br}(B_s \to \bar{\mu}\mu)} = \frac{\tau(B_d)M_{B_d}F_{B_d}^2|\mathbf{V}_{td}|^2}{\tau(B_s)M_{B_s}F_{B_s}^2|\mathbf{V}_{ts}|^2}. \qquad (12.48)$$

Measuring this ratio of branching ratios thus allows to extract $|\mathbf{V}_{td}|^2/|\mathbf{V}_{ts}|^2$. Yet this poses a great experimental challenge, since these transitions are extremely rare:

$$\mathrm{Br}(B_s \to \mu^+\mu^-)$$
$$= 4.18\times10^{-9}\left(\frac{\tau(B_d)}{1.6\,\mathrm{ps}}\right)\left(\frac{F_B}{230\,\mathrm{MeV}}\right)^2\left(\frac{|\mathbf{V}_{ts}|^2}{0.040}\right)\left(\frac{m_t(m_t)}{170\,\mathrm{GeV}}\right)^{3.12}. \qquad (12.49)$$

The present bounds coming mainly from the CDF and D0 experiments at FNAL read as follows [112]:

$$\mathrm{Br}(B_d \to \mu^+\mu^-) \leq 1.5 \times 10^{-8} \qquad (12.50)$$
$$\mathrm{Br}(B_s \to \mu^+\mu^-) \leq 4.7 \times 10^{-8}\;; \qquad (12.51)$$

i.e. while the present bound for B_d is still two orders of magnitude away from the SM expectation, that for B_s is within one order of magnitude. We should also keep in mind that these rates could be greatly enhanced in particular in some SUSY models. Yet to measure the B_d rate on the SM level of $\mathcal{O}(10^{-10})$ with 1% accuracy requires about 10^{14} B_d mesons!

12.2.3 B → X + $\nu\bar{\nu}$

The SM prediction for the inclusive transition $B \to X + \nu\bar{\nu}$ is theoretically clean, and its rate should be much more abundant than for $B \to \mu^+\mu^-$ – if we could only identify them!

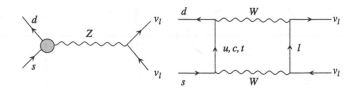

Figure 12.2 The diagrams driving $K \to l\bar{l}$ and $s\bar{d} \to \nu\bar{\nu}$.

Evaluating the diagrams shown in Fig. 12.2, yet with replacing s and d with b and s, respectively, one finds for the decay rate [94]:

$$\frac{\text{Br}(B \to X_s \nu\bar{\nu})}{\text{Br}(B \to X_c e\bar{\nu})} = \frac{3\alpha^2}{4\pi^2 \sin^4 \theta_W} \frac{|V_{ts}|^2}{|V_{cb}|^2} \frac{X^2(x_t)}{f(z)} \frac{\bar{\eta}}{\kappa(z)}, \tag{12.52}$$

$\bar{\eta} \simeq 0.83$ and $\kappa(z) \simeq 0.88$ are QCD corrections. Numerically we obtain

$$\text{Br}(B \to X_s \nu\bar{\nu}) = 4.1 \times 10^{-5} \frac{|V_{ts}|^2}{|V_{cb}|^2} \left(\frac{m_t(m_t)}{170 \, \text{MeV}} \right)^{2.3}. \tag{12.53}$$

The ALEPH collaboration has searched for this decay and obtained an upper limit [182]:

$$\text{Br}(B \to X_s + \nu\bar{\nu}) < 7.7 \times 10^{-4} \quad (90\% \, \text{CL}). \tag{12.54}$$

Let us wish the best of luck to experimentalists taking up this challenge.

12.2.4 $B \to X_s + \mu^+\mu^-$

This decay gets a contribution from the Feynman diagram in Fig. 12.3. Note that the final state in the KM favoured $b \to c\bar{c}s$ contains a $(c\bar{c})$ pair which may form a resonance J/ψ, ψ', ... [183, 184]. These non-perturbative corrections will introduce inherent uncertainties. An estimate

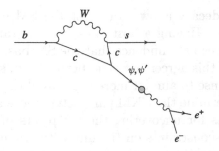

Figure 12.3 The long-distance contribution to $b \to se^+e^-$.

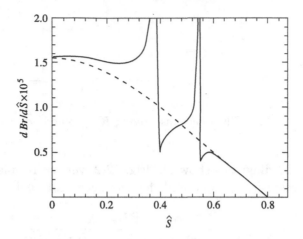

Figure 12.4 We see that due to the interference of long-distance and short-distance contributions, the shapes of Ψ and Ψ' resonances are altered. New Physics contributions may change the interference pattern. This figure was reproduced from *Physics Letters* by permission of Elsevier Science.

of the branching ratio as a function of $\hat{s} = Q^2/m_b^2$, where Q^2 is the invariant mass square of the l^+l^- system, is shown in Fig. 12.4.

In this figure, QCD corrections are not fully considered. For detailed QCD corrections to this decay, see Ref. [185]. Note that the width of J/ψ and ψ' is much larger than their actual width. This is because the diagram shown in Fig. 12.3 interferes with the penguin graph with internal top quark. Similarly, if there is New Physics, such a contribution can also interfere with the diagram shown in Fig. 12.3. Such interference will also affect the angular distributions [186]. It is a fertile searching ground for New Physics.

12.3 Résumé

Various K and B decays allow us to test the SM even on the level of quantum corrections. Having a host of rare decay rates agree with such SM expectations gives us confidence that the SM basically paints the correct picture, even if this agreement is no better than, say, 20%. Even then it makes eminent sense to aim for more accurate data. At the very least it will allow us to determine the CKM parameters more accurately and thus improve the chances for uncovering the footprints of New Physics indirectly through over-constraints on the unitarity triangle. One example is to construct the unitarity triangle separately from $\Delta S \neq 0$ processes – $K^+ \to \pi^+\nu\bar{\nu}$ and $K_L \to \pi^0\nu\bar{\nu}$ – and compare it with the version arising from $\Delta B \neq 0$ dynamics. It is also possible that we might hit the 'jackpot' of discovering New Physics more directly, namely when the disagreement between experiment and the SM prediction is sufficiently large to be conclusive already in a single channel!

The study of K and B physics along these lines will undoubtably continue for many decades. Don't forget, rare kaon decays of the K meson have been under scrutiny for over 50 years. Who knows, some bright young experimentalist might come up with a way to do a dedicated experiment to collect huge numbers of B decays – say 10^{14}?

Problems

12.1. Show that the charge radius term can only contribute to a virtual photon process $s \to de^+e^-$ and not to a real photon emission process $s \to d\gamma$.

12.2. Show that $\langle \mu\bar{\mu}|\mathcal{H}|K\rangle = 0$, where $\mathcal{H}$ is a charge radius contribution to the effective Hamiltonian for $s \to d\mu\bar{\mu}$. Show also that the magnetic moment term cannot contribute to $K \to \mu\bar{\mu}$ either. Hint: consider $\langle 0|\sigma_{\mu\nu}|K\rangle$.

12.3. Show that the contribution from Fig. 12.1 to $K_L \to Z^*\pi \to l^+l^-\pi$ and $K_L \to \gamma^*\pi \to l^+l^-\pi$ must violate **CP** symmetry. First, a physical argument: consider the **CP** quantum number of the final states $(Z^*\pi)$ and $(\gamma^*\pi)$. Noting that $\mathbf{CP}|Z^*\rangle = |Z^*\rangle$, $\mathbf{CP}|\gamma^*\rangle = |\gamma^*\rangle$ and two particles are in a relative P wave state, deduce the **CP** eigenvalues of the final states.

Second, a more formal argument: note that Fig. 12.1 gives an operator of the form $\bar{d}\gamma_\mu\gamma_- s\bar{l}\gamma^\mu\gamma_- l$. Show that the matrix element of this operator between the initial and the final states vanishes.

12.4. Consider $K_L \to \pi^+\pi^-e^+e^-$; let p_+, p_-, k_+ and k_- denote the four-momenta of π^+, π^-, e^+ and e^-, respectively.

(a) How does the triple correlation

$$P = (\vec{p}_+ \times \vec{p}_-) \cdot (\vec{k}_+ - \vec{k}_-), \qquad (12.55)$$

transform under **CP** and **T**? What does a non-vanishing expectation value of P in K_L decays teach you?

(b) Define the normal vectors of the $\pi^+ - \pi^-$ and $e^+ - e^-$ planes:

$$\vec{n}_\pi \equiv \frac{\vec{p}_+ \times \vec{p}_-}{|\vec{p}_+ \times \vec{p}_-|}, \quad \vec{n}_e \equiv \frac{\vec{k}_+ \times \vec{k}_-}{|\vec{k}_+ \times \vec{k}_-|}. \qquad (12.56)$$

Show that $\vec{n}_\pi$ and $\vec{n}_e$ change sign under **CP**. How does

$$\vec{n}_e \times \vec{n}_\pi \cdot \frac{(\vec{p}_+ + \vec{p}_-)}{|\vec{p}_+ + \vec{p}_-|} \qquad (12.57)$$

transform under **CP** and **T**? Denote by Φ the angle between $\vec{n}_\pi$ and $\vec{n}_e$; derive the identity

$$\sin\Phi \cos\Phi = \vec{n}_e \times \vec{n}_\pi \cdot \frac{(\vec{p}_+ + \vec{p}_-)}{|\vec{p}_+ + \vec{p}_-|} \vec{n}_e \cdot \vec{n}_\pi. \qquad (12.58)$$

How does $\sin\Phi \cos\Phi$ behave under **CP** and **T**?

(c) Show that the interference between the bremsstrahlung and M1 amplitudes generates a term in the angular distribution which is proportional to $\sin\Phi \cos\Phi$.

12.5. With 10^{14} identified B's, what is the sensitivity on the branching ratio of rare B decay? To what extent is it sensitive to New Physics?

13

♠ CPT violation – could it be in K and B decays? ♠

As we have seen, **CPT** symmetry is a very general property of quantum field theories: it can be derived by invoking little more than locality and Lorentz invariance [187]. Despite this impeccable pedigree, it makes sense to ask whether limitations exist.

- Precisely because the **CPT** theorem rests on such essential pillars of our present paradigm, we have to make every reasonable effort to probe its universal validity.
- Although no *appealing* theory of **CPT** violation exists, we should keep in mind that superstring theories – suggested as more fundamental than quantum field theories and by some seen as TOE, the 'Theory of Everything' – are intrinsically *non*-local and thus do not satisfy one of the basic axioms of the **CPT** theorem.
- An intriguing phenomenon has been suggested by Hawking [188]. Near a black hole *pure* quantum states can evolve into *mixed* ones, since some of the information carried by them gets funnelled into the black hole due to the latter's overpowering gravitational pull, and thus is lost for the 'outside world'. This sequence violates both conventional quantum mechanics and **CPT** invariance. Hawking used the density formalism to specifically discuss $K^0 - \overline{K}^0$ oscillations in the presence of a black hole. An experimental test of Hawking's idea was proposed in Ref. [189].

While the intrinsic non-locality of superstring theories allows for limitations to **CPT** invariance, it does *not* demand it. Our attitude can then be described as one of educated curiosity: while we know of no benchmark for where **CPT** invariance might be violated, we like to emphasize two points: (i) **CPT** studies have in the past played a role in strengthening the community's belief in the relevance of quantum field theories. (ii) Particle–antiparticle oscillations allow measurements of spectacular sensitivity by tracking coherent effects building up over *macroscopic* distances. Thus they enable us to probe **CPT** invariance – and the foundations it is based on, namely Lorentz invariance and linear quantum mechanics – with unparallelled accuracy. The scope of such tests is greatly extended through detailed studies of $B^0 - \overline{B}^0$ oscillations and by analysing EPR correlations in the $K^0\overline{K}^0$ and $B^0\overline{B}^0$ states produced at the Φ and $\Upsilon(4S)$ factories, respectively.

For this purpose we apply the expressions derived in Chapter 6 from linear quantum mechanics, yet *without* imposing **CPT** invariance. First we sketch relevant existing data and then make brief remarks concerning possible scenarios for limitations to **CPT** and Lorentz invariance and linear quantum mechanics.

13.1 Equality of masses and lifetimes

The 'classical' tests of **CPT** symmetry concern the equality of masses and lifetimes for particles and antiparticles (see Chapter 4). The experimental situation can be characterized by the following [11] set of upper bounds:

$$\frac{m_{e^+} - m_{e^-}}{m_e} < 8 \times 10^{-9}, \quad 90\% \text{ CL}, \qquad \frac{\frac{|q_{\bar{p}}|}{M_{\bar{p}}} - \frac{q_p}{M_p}}{\frac{q_p}{M_p}} = (-9 \pm 9) \times 10^{-11},$$

$$\frac{M_{\bar{n}} - M_n}{M_n} < 1.0 \times 10^{-8}, \quad 90\% CL, \qquad \frac{M_{\pi^+} - M_{\pi^-}}{M_\pi} = (2 \pm 5) \times 10^{-4},$$

$$\frac{M_{K^+} - M_{K^-}}{M_K} = (-0.6 \pm 1.8) \times 10^{-4}, \qquad \frac{M_\Lambda - M_{\overline{\Lambda}}}{M_\Lambda} = (-1.0 \pm 0.9) \times 10^{-5},$$

$$\frac{M_{W^+} - M_{W^-}}{M_W} = (-2 \pm 7) \times 10^{-3}, \qquad \frac{\tau_{\mu^+} - \tau_{\mu^-}}{\tau_{\mu^+}} < (2 \pm 8) \times 10^{-5},$$

$$\frac{\tau_{\pi^+} - \tau_{\pi^-}}{\tau_{\pi^+}} = (6 \pm 7) \times 10^{-4}, \qquad \frac{\tau_{K^+} - \tau_{K^-}}{\tau_{K^+}} = (1.1 \pm 0.9) \times 10^{-3}.$$

$$(13.1)$$

For proper evaluation we should keep the following in mind.

- The masses provide a very ad hoc calibration stick for a *difference* in particle–antiparticle masses, since the overwhelming bulk of the masses is not generated by the weak forces.
- While the decay widths in Eq. (13.1) are of weak origin, bounds $\sim \mathcal{O}(\text{few} \times 10^{-4})$ can hardly be called decisive *empirical* verifications of a fundamental symmetry.

Another bound is usually quoted as best test:

$$\frac{|M_{\overline{K}^0} - M_{K^0}|}{M_K} < 9 \times 10^{-19} \,\hat{=}\, \frac{|M_{\overline{K}^0} - M_{K^0}|}{\Gamma_{K_S}} < 7 \times 10^{-5}. \qquad (13.2)$$

Later in this chapter we will derive it and state the assumptions that go into it. For now we want to state that even with the caveat just mentioned, Eq. (13.2) appears impressive – as also indicated by the more meaningful comparison to Γ_{K_S}. It demonstrates that the interferometry in particle–antiparticle oscillations allows the measurement of truly tiny energy differences.

13.2 Theoretical scenarios

Proofs of the **CPT** theorem are based on natural axioms like Lorentz invariance, the existence of a unique ground state and local fields obeying the 'correct' spin-statistics connection, see Chapter 4. They also involve some more technical assumptions, like the fields containing a *finite* number of components only – i.e. belonging to a *finite* dimensional representation of the Lorentz group. In 1968 Oksak and Todorov gave two explicit examples for a Lorentz invariant free field theory that does *not* obey **CPT** invariance. The fields form unitary representations of $SL(2, C)$ – the special linear group in two dimensions with complex coefficients – and contain an *infinite* number of components.[1]

While this might be viewed as an academic curiosity, we can face the foundations of the **CPT** theorem more squarely by casting doubts on the absolute validity of *Lorentz invariance*. This had been done in Ref. [190] and was revived in the context of so-called tumbling gauge theories in Ref. [191]: Lorentz breaking is engineered by postulating a *vector* condensate to form in the vacuum: $\langle 0|\overline{\psi}\gamma_\mu\psi|0\rangle \neq 0$. This apparent Lorentz violation could still get relegated to an unphysical sector if $\langle 0|\overline{\psi}\gamma_\mu\psi|0\rangle$ takes on the role of a gauge fixing vector [190, 191]. Alternatively, it could lead to physical effects [192]. Since $(\mathbf{CPT})^{-1}\overline{\psi}\gamma_\mu\psi(x)\mathbf{CPT} = -\overline{\psi}\gamma_\mu\psi(-x)$, its VEV would break **CPT** symmetry. Fermion masses can then obtain a contribution that is not a Lorentz scalar, but the zeroth component of a vector, which would lift the degeneracy in the masses of particles and antiparticles [193].

The even more radical suggestion has been made that Lorentz symmetry is merely an approximate infrared phenomenon [194]. Unlike the previous scenario, the invariance is now broken at the highest energy scales like

---

[1] One of the examples satisfies the spin-statistics theorem, the other one does not. We should note that this theorem does in general not hold for infinite-component fields.

the Planck mass. Yet there is an attractive infrared fixed point in the renormalization flow that is Lorentz invariant; i.e. the coefficients of *non-covariant* operators, which are assumed to be of order unity at the Planck scale, are reduced when the scales are lowered and get renormalized down to zero in the infrared limit.

As mentioned above, recent theoretical developments have provided some fresh impetus to raising these seemingly esoteric questions. Super-string theories are defined in 10 dimensions in a (generalized) Lorentz invariant way. It has been argued [195] that in the compactification down to four space–time dimensions *spontaneous* violation of Lorentz invariance can arise, due to intrinsically non-local string interactions which lead to the emergence of VEVs for some non-scalar Lorentz operators. This is somewhat similar to the scenario sketched before and generates **CPT** violation.

These speculations are certainly on the more exotic side, yet should not be dismissed completely out-of-hand. The basic idea behind them can be formulated as follows: it is quite conceivable that at the Planck scale – the domain of quantum gravity – **CPT** and Lorentz invariance are of limited validity. It is likewise conceivable that such effects can get communicated to lower energies. The real and most difficult question – to which we do not know the answer at all – is whether and where they can retain an observable magnitude. In the absence of any reliable guidance from theory, we will concentrate on those processes that

- promise us the highest experimental sensitivity; and
- differentiate between different flavour sectors.

13.3 CPT phenomenology for neutral kaons

Without **CPT** constraint, the two neutral kaon mass eigenstates are still described by linear combinations of K^0 and $\overline{K}^0$, yet with two coefficients q_1/p_1 and q_2/p_2 rather than the single term q/p [56, 196, 197, 198], as described in Chapter 6. Consequently, even a perfect determination of the K_L state would *not* tell us the exact composition of the K_S state; i.e. there are more independent ways for **CP** violation to manifest itself.

In Section 6 we used spherical coordinates to diagonalize the effective mass matrix $\mathbf{M} - \frac{i}{2}\mathbf{\Gamma}$ given in Eq. (6.7). The **CPT** violation is characterized by non-vanishing of $\cos\theta = E_3/E$. Noting that $E = \sqrt{(M_{12} - \frac{i}{2}\Gamma_{12})(M_{12}^* - \frac{i}{2}\Gamma_{12}^*)}$, using Eq. (6.23), Eq. (6.24) and Eq. (6.29) we obtain an expression for $\cos\theta$ which shows explicitly that it is a parameter which characterizes **CPT** violation. The **CP** violation is characterized by E_2/E_1 given in Eq. (7.32):

$$\cos\theta = \frac{M_{22} - M_{11} - \frac{i}{2}(\Gamma_{22} - \Gamma_{11})}{\Delta M + \frac{i}{2}\Delta\Gamma}$$

$$\phi = 2\frac{\text{Im }M_{12} - \frac{i}{2}\text{Im }\Gamma_{12}}{\Delta M + \frac{i}{2}\Delta\Gamma}. \tag{13.3}$$

While we have no interest in introducing superfluous parameters, we also want to touch base with PDG notations: $\delta = \frac{1}{2}\cos\theta$, $\epsilon_0 = -\frac{i}{2}\phi$.

13.3.1 Semileptonic decays

Rather than q_1/p_1 and q_2/p_2, we will use here two other parameters to describe the kaon states, namely $\cos\theta$ and $i\phi$ [196]. For the convenience of the reader, we present here a translation of notations:

$$|K^0(t)\rangle = \frac{1}{2}\left[(1 + 2\delta)|K^0\rangle + (1 - 2\epsilon_0)|\overline{K}^0\rangle\right]e^{-iM_St}e^{-\frac{1}{2}\Gamma_St}$$

$$+ \frac{1}{2}\left[(1 - 2\delta)|K^0\rangle - (1 - 2\epsilon_0)|\overline{K}^0\rangle\right]e^{-iM_Lt}e^{-\frac{1}{2}\Gamma_Lt},$$

$$|\overline{K}^0(t)\rangle = \frac{1}{2}\left[(1 - 2\delta)|\overline{K}^0\rangle + (1 + 2\epsilon_0)|K^0\rangle\right]e^{-iM_St}e^{-\frac{1}{2}\Gamma_St}$$

$$+ \frac{1}{2}\left[(1 + 2\delta)|\overline{K}^0\rangle - (1 + 2\epsilon_0)|K^0\rangle\right]e^{-iM_Lt}e^{-\frac{1}{2}\Gamma_Lt}. \tag{13.4}$$

In the notation of Eq. (7.49) for the semileptonic decay amplitudes, we can compute the following decay rates:

$$\Gamma(K^0(t) \to \pi^-l^+\nu) = \left|\frac{F_l}{2}\right|^2\left[\{1 - 2\text{Re }y_l - 2\text{Re }(2\delta + x_l)\}e^{-\Gamma_Lt}\right.$$

$$+ \{1 - 2\text{Re }y_l + 2\text{Re }(2\delta + x_l)\}e^{-\Gamma_St}$$

$$+ 2e^{-\overline{\Gamma}t}\{(1 - 2\text{Re }y_l)\cos(\Delta Mt)$$

$$\left. - 2\text{Im }(2\delta + x_l)\sin(\Delta Mt)\}\right], \tag{13.5}$$

$$\Gamma(\overline{K}^0(t) \to \pi^-l^+\nu) = \left|\frac{F_l}{2}\right|^2\left[\{1 + 2\text{Re }(2\epsilon_0 - y_l) - 2\text{Re }x_l\}e^{-\Gamma_Lt}\right.$$

$$+ \{1 + 2\text{Re }(2\epsilon_0 - y_l) + 2\text{Re }x_l\}e^{-\Gamma_St}$$

$$- 2e^{-\overline{\Gamma}t}\{[1 + 2\text{Re }(2\epsilon_0 - y_l)]\cos(\Delta Mt)$$

$$\left. - 2\text{Im }x_l\sin(\Delta Mt)\}\right], \tag{13.6}$$

$$\Gamma(K^0(t) \to \pi^+ l^- \overline{\nu}) = \left|\frac{F_l}{2}\right|^2 \Big[\{1 - 2\mathrm{Re}\,(2\epsilon_0 - y_l) - 2\mathrm{Re}\,\overline{x}_l)\}\, e^{-\Gamma_L t}$$
$$+ \{1 - 2\mathrm{Re}\,(2\epsilon_0 - y_l) + 2\mathrm{Re}\,\overline{x}_l\}\, e^{-\Gamma_S t}$$
$$- 2e^{-\overline{\Gamma} t}\{[1 - 2\mathrm{Re}\,(2\epsilon_0 - y_l)]\cos(\Delta M t)$$
$$+ 2\mathrm{Im}\,\overline{x}_l \sin(\Delta M t)\}\Big], \tag{13.7}$$

$$\Gamma(\overline{K}^0(t) \to \pi^+ l^- \overline{\nu}) = \left|\frac{F_l}{2}\right|^2 \Big[\{1 + 2\mathrm{Re}\,y_l + 2\mathrm{Re}\,(2\delta - \overline{x}_l)\}\, e^{-\Gamma_L t}$$
$$+ \{1 + 2\mathrm{Re}\,y_l - 2\mathrm{Re}\,(2\delta - \overline{x}_l)\}\, e^{-\Gamma_S t}$$
$$+ 2e^{-\overline{\Gamma} t}\{[1 + 2\mathrm{Re}\,y_l]\cos(\Delta M t)$$
$$+ 2\mathrm{Im}\,(2\delta + \overline{x}_l)\sin(\Delta M t)\}\Big], \tag{13.8}$$

with $\overline{\Gamma} = (\Gamma_L + \Gamma_S)/2$ and where we recall the following constraints:

$\Delta S = \Delta Q$ rule:	$x_l = \overline{x}_l = 0$
CP invariance:	$x_l = \overline{x}_l^*$; $F_l = F_l^*$; $y_l = -y_l^*$; $\delta = 0$; $\epsilon_0 = 0$
T invariance:	$\mathrm{Im}\,F = \mathrm{Im}\,y_l = \mathrm{Im}\,x_l = \mathrm{Im}\,\overline{x}_l = 0$; $\epsilon_0 = 0$
CPT invariance:	$y_l = 0$; $x_l = \overline{x}_l$; $\delta = 0$.

Equations (13.5) to (13.8), which hold to first order in $\cos\theta$, ϵ_0, x_l, and y_l, are the master equations from which we form various combinations to test these different selection rules.

13.3.2 Asymmetries

We have listed four leptonic decays. With them we can form asymmetries sensitive to the complex parameters: x_l, $\overline{x}_l$, ϵ_0, δ, y_l.

Charge asymmetry in $K_L \to l^{\pm}\nu\pi^{\mp}$

The charge asymmetry in semileptonic K_L decays:

$$A_l = \frac{\Gamma(K_L \to l^+\nu\pi^-) - \Gamma(K_L \to l^-\nu\pi^+)}{\Gamma(K_L \to l^+\nu\pi^-) + \Gamma(K_L \to l^-\nu\pi^+)}$$
$$= 2\mathrm{Re}\,\epsilon_0 - 2\delta - \mathrm{Re}\,(x_l - \overline{x}_l) - 2\mathrm{Re}\,y_l \tag{13.9}$$

has already appeared in Eq. (7.14), where we have used it to determine the value of ϵ assuming **CPT** invariance. Now we see that it can tell us about **CPT** violation.

Test of the ΔS $= \Delta$Q *rule and measurement of* ΔM

Semileptonic decays allow us to track the strangeness of the meson *at the time of decay* – if the $\Delta Q = \Delta S$ rule holds. While this rule had been discovered empirically, it is a natural consequence of the SM, where it holds with high accuracy: $x_l|_{SM}, \overline{x}_l|_{SM} \approx 10^{-6} \sim 10^{-7}$ [199]. Finding a non-vanishing x_l at the level presently reachable implies the presence of New Physics.

This rule can be tested by probing

$$
\begin{aligned}
A_{\Delta M} &= \frac{\Gamma(K^0(t) \to e^+) + \Gamma(\overline{K}^0(t) \to e^-) - \Gamma(K^0(t) \to e^-) - \Gamma(\overline{K}^0(t) \to e^+)}{\Gamma(K^0(t) \to e^+) + \Gamma(\overline{K}^0(t) \to e^-) + \Gamma(K^0(t) \to e^-) + \Gamma(\overline{K}^0(t) \to e^+)} \\
&= \frac{2e^{-\overline{\Gamma}t}[\cos \Delta Mt + \mathrm{Im}\,(\overline{x}_l - x_l)\sin \Delta Mt]}{[1 + \mathrm{Re}\,(x_l + \overline{x}_l)]e^{-\Gamma_S t} + [1 - \mathrm{Re}\,(x_l + \overline{x}_l)]e^{-\Gamma_L t}},
\end{aligned} \tag{13.10}
$$

where the short-hand notation $\Gamma(K^0(t) \to e^+) \equiv \Gamma(K^0(t) \to e^+ \nu \pi^-)$ etc. has been employed. Given enough statistics, we can separate out $\mathrm{Re}\,(x_l + \overline{x}_l)$, $\mathrm{Im}\,(x_l - \overline{x}_l)$ and ΔM.

Tests of T *and* CPT *invariance*

Since

$$
\begin{aligned}
\Gamma(\overline{K}^0 \to K^0) &\xRightarrow{\mathbf{T}} \Gamma(K^0 \to \overline{K}^0) \\
\Gamma(\overline{K}^0 \to \overline{K}^0) &\xRightarrow{\mathbf{CPT}} \Gamma(K^0 \to K^0),
\end{aligned} \tag{13.11}
$$

we can probe **T** and **CPT** symmetry through the following ratios:

$$
\begin{aligned}
A_T(t) &= \frac{\Gamma(\overline{K}^0 \to K^0(t)) - \Gamma(K^0 \to \overline{K}^0(t))}{\Gamma(\overline{K}^0 \to K^0(t)) + \Gamma(K^0 \to \overline{K}^0(t))} \\
A_{CPT}(t) &= \frac{\Gamma(\overline{K}^0 \to \overline{K}^0(t)) - \Gamma(K^0 \to K^0(t))}{\Gamma(\overline{K}^0 \to \overline{K}^0(t)) + \Gamma(K^0 \to K^0(t))}.
\end{aligned} \tag{13.12}
$$

Those asymmetries can be re-expressed through directly observable quantities:

$$
\begin{aligned}
A_T(t) &= \frac{\Gamma(\overline{K}^0(t) \to e^+) - \Gamma(K^0(t) \to e^-)}{\Gamma(\overline{K}^0 \to e^+) + \Gamma(K^0 \to e^-)} \\
A_{CPT}(t) &= \frac{\Gamma(\overline{K}^0(t) \to e^-) - \Gamma(K^0(t) \to e^+)}{\Gamma(\overline{K}^0 \to e^-) + \Gamma(K^0 \to e^+)}.
\end{aligned} \tag{13.13}
$$

Both of these asymmetries are given by

$$A = \frac{ae^{-\Gamma_L t} + be^{-\Gamma_s t} + ce^{-\bar{\Gamma} t}\cos\Delta Mt + de^{-\bar{\Gamma} t}\sin\Delta Mt}{e^{-\Gamma_L t} + e^{-\Gamma_s t} - 2e^{-\bar{\Gamma} t}\cos\Delta Mt}$$
$$\simeq a \quad \text{for } t \gg \Gamma_S^{-1}. \tag{13.14}$$

The coefficients for $A_T(t)$ are:

$$a = 4\mathrm{Re}\,\epsilon_0 - 2\mathrm{Re}\,y_l - \mathrm{Re}\,(x_l - \bar{x}_l)$$
$$b = 4\mathrm{Re}\,\epsilon_0 - 2\mathrm{Re}\,y_l + \mathrm{Re}\,(x_l - \bar{x}_l)$$
$$c = 4(-2\mathrm{Re}\,\epsilon_0 + \mathrm{Re}\,y_l)$$
$$d = 2\mathrm{Im}\,(x_l + \bar{x}_l); \tag{13.15}$$

and for $A_{CPT}(t)$:

$$a = 2\mathrm{Re}\,y_l + 4\mathrm{Re}\,\delta + \mathrm{Re}\,(x_l - \bar{x}_l)$$
$$b = 2\mathrm{Re}\,y_l - 4\mathrm{Re}\,\delta - \mathrm{Re}\,(x_l - \bar{x}_l)$$
$$c = 4\mathrm{Re}\,y_l$$
$$d = 8\mathrm{Im}\,\delta + 2\mathrm{Im}\,(x_l + \bar{x}_l). \tag{13.16}$$

In going from Eq. (13.12) to Eq. (13.14), the interpretation of A_T has changed significantly: $A_T \neq 0$ *now does not necessarily establish* **T** *violation!* If $\epsilon_0 = \mathrm{Im}x_l = \mathrm{Im}\bar{x}_l = \mathrm{Im}y_l = 0$ – as required by **T** invariance (see above) – we have

$$A_T(t) \simeq -2\mathrm{Re}y_l - \mathrm{Re}(x_l - \bar{x}_l) = -2y_l - (x_l - \bar{x}_l) \quad \text{for } t \gg \Gamma_S^{-1} ; \tag{13.17}$$

i.e. $A_T \neq 0$ can be due to **CPT** not being conserved in two variants: (i) $y_l \neq 0$, which in turn corresponds to **CP** violation as well, since $y_l = -y_l^*$ cannot be satisfied for Re $y_l \neq 0$. (ii) $x_l \neq \bar{x}_l$, i.e. a limitation of the $\Delta S = \Delta Q$ rule. Since this rule is used to track the oscillating strangeness of the kaons through their semileptonic decays, it is easy to understand how its violation can *fake* a **T** asymmetry. This ambiguity can be avoided by summing the two asymmetries:

$$A_T(t) + A_{CPT}(t) \simeq 4(\mathrm{Re}\,\epsilon_0 + \mathrm{Re}\,\delta) \quad \text{for } t \gg \Gamma_S^{-1} ; \tag{13.18}$$

i.e. the sum of these two asymmetries is independent of y_l, x_l and $\bar{x}_l$; if it does not vanish, then **T** and/or **CPT** are violated.

Experimental results

There have been considerable new data on this topic since the first edition coming from the CPLEAR, KLOE, KTEV and NA48 collaborations. We list their combined results, as analysed by the PDG, in Table 13.1.

Table 13.1 *PDG values for* **CPT** *and* $\Delta S = \Delta Q$ *violating parameters in units of* 10^{-3}: $\mathrm{x}_\pm = \frac{1}{2}(\mathrm{x}_l \pm \overline{\mathrm{x}}_l)$

	x_+	x_-	δ	y
Real part	-0.8 ± 3.1	-0.8 ± 2.5	0.29 ± 0.27	0.4 ± 2.5
Imag. part	1.2 ± 2.1		-0.002 ± 0.02	

13.3.3 Non-leptonic neutral K decays

Proceeding in close analogy (including the notation) to the derivation for ϵ and ϵ' given in Chapter 7 for the **CPT** symmetric case, we obtain (see Problem 13.2):

$$\epsilon \simeq \frac{1}{2}\left(1 - \frac{q_2}{p_2}\frac{\overline{A}_0}{A_0}\right)$$

$$\epsilon' \simeq \frac{1}{2\sqrt{2}}\omega e^{i(\delta_2 - \delta_0)}\frac{q_2}{p_2}\left(\frac{\overline{A}_0}{A_0} - \frac{\overline{A}_2}{A_2}\right), \tag{13.19}$$

which illustrate some important general features:

- **CPT** violation in the $\Delta S = 2$ sector leads to the two mass eigenstates being described by two different mixing parameters q_1/p_1 and q_2/p_2.
- ϵ and ϵ' measured in $K_L \to \pi\pi$ depend on $\Delta_I = 1 - (q_2\overline{A}_I)/(p_2 A_I)$, yet are *not* sensitive to q_1/p_1. In Section 7.5 we have discussed $K^0, \overline{K}^0 \to \pi^+\pi^-\pi^0$ decays; from Eq. (7.78) we read off

$$A_{+-0}(\infty) = |A_L(3\pi)|^2\left(1 - \left|\frac{q_1}{p_1}\right|\right); \tag{13.20}$$

  i.e. we can extract $|q_1/p_1|$ from the **CP** asymmetry in $K_L \to \pi^+\pi^-\pi^0$ and compare it with $|q_2/p_2|$ from $K_L \to \pi^+\pi^-$. Nature seems to be kind to us. By looking at 2π and 3π decay modes of K_L we can extract both $\frac{q_1}{p_1}$ and $\frac{q_2}{p_2}$.
- *Direct* **CPT** violation in $\Delta S = 1$ amplitudes leads to $|\overline{A}_i| \neq |A_i|$. The amount of **CPT** – and **CP** violation as well – can thus depend on the isospin I of the final state. Equation (13.19) shows there are then three gateways through which a direct asymmetry can emerge in $K \to \pi\pi$: $|A_0| \neq |\overline{A}_0|$ and $|A_2| \neq |\overline{A}_2|$ in addition to $\arg(A_0\overline{A}_2/\overline{A}_0 A_2) \neq 0$, which can occur already within **CPT** symmetry.

- **CPT** symmetry implies, as discussed in Chapter 7, that the relative phase between ϵ and ϵ' is either close to $0°$ or $180°$, since the phase of $\epsilon - \arg\epsilon \simeq \phi_{SW} = \tan^{-1}(\Delta M/\Delta\Gamma) = 43.46° \pm 0.08°$ – and the strong phase shift basically coincide – $\delta_0 - \delta_2 = 42° \pm 4°$. A component of ϵ' *orthogonal* to ϵ thus breaks **CPT** invariance.

CPT *Tests in the phases of* η_{+-} *and* η_{00}

The most stringent test of **CPT** is provided by the phases of η_{+-} and η_{00}, ϕ_{+-} and ϕ_{00}, respectively. In Problem 13.4 we are going to derive [200]

$$\frac{|\eta_{+-}|\left(\frac{2}{3}\phi_{+-} + \frac{1}{3}\phi_{00} - \phi_{SW}\right)}{\sin\phi_{SW}} = \left(\frac{M_{\overline{K}} - M_K}{2\Delta M_K} + R_{\text{direct}}\right). \qquad (13.21)$$

Here

$$R_{\text{direct}} = \frac{-ie^{-i\phi_{SW}}}{\sin\phi_{SW}} \sum_{f\neq(2\pi)_{I=0}} \epsilon(f) + \frac{1}{2}\left(\text{Re}\,\frac{\overline{A}_0}{A_0} - 1\right) \qquad (13.22)$$

where $\epsilon(f)$ is defined by Eq. (13.29). Using the result from an analysis without assuming **CPT** symmetry: $\phi_{+-} = (43.4 \pm 0.7)°$ and $\Delta\phi = \phi_{00} - \phi_{+-} = (0.2 \pm 0.4)°$ [11], we obtain

$$\frac{2}{3}\phi_{+-} + \frac{1}{3}\phi_{00} - \phi_{SW} \simeq (0.04 \pm 0.70)° \qquad (13.23)$$

and thus find [200]

$$\left(\frac{M_{\overline{K}} - M_K}{2\Delta M_K} + R_{\text{direct}}\right) = (0.23 \pm 4.0) \times 10^{-5}, \qquad (13.24)$$

which is completely consistent with being zero. A few comments might help in evaluating Eq. (13.24).

- Assuming $R_{\text{direct}} = 0$ Eq. (13.24) can be re-expressed as follows:

$$\frac{M_{\overline{K}} - M_K}{M_K} = (0.31 \pm 5.6) \times 10^{-19}, \qquad (13.25)$$

which is the updated version of an often quoted bound.
- Eq. (13.25) looks numerically truly spectacular – yet it overstates the case since M_K is largely generated by the strong interactions! A more reasonable scale for the $\Delta S = 2$ difference $M_{\overline{K}} - M_K$ is provided by noting that **CP** invariance would yield $M_{\overline{K}} - M_K = 0$ even if **CPT** symmetry were invalid. So the most natural calibrator

for $M_{\overline{K}} - M_K$ might be $|\eta|\Gamma_S$; Eq. (13.25) is then re-expressed as follows:

$$\frac{M_{\overline{K}} - M_K}{|\eta|\Gamma_S} = (0.1 \pm 1.6) \times 10^{-2}. \tag{13.26}$$

The numerical accuracy here is good, though not overwhelming!

Estimating Im Γ_{12}

In Section 7.2.4, we derived:

$$\epsilon \simeq \frac{1}{\sqrt{1 + (\frac{\Delta\Gamma}{2\Delta M})^2}} e^{i\phi_{sw}} \left(-\frac{\text{Im}M_{12}}{\Delta M_K} + \xi_0 \right). \tag{13.27}$$

as seen in Eq. (7.42). This result is obtained by making two assumptions: (1) **CPT** symmetry; (2) $\text{Im}\,\Gamma_{12}$ gets contribution only from $f = (2\pi)_{I=0}$ and all other possible states were neglected. As seen in Eq. (13.27), these assumptions lead to a prediction of the phase of ϵ, namely ϕ_{SW}.

Relaxing these assumptions leads to Eq. (13.21). This is often stated as a **CPT** test. As it stands, however, Eq. (13.21) gets contribution from **CP** violation as well as **CPT** violation. We now want to estimate the contribution to R_{direct} which comes from **CP** violation. In doing so, we can formulate a genuine test for **CPT**.

The expression for ϵ in Eq. (7.36) contains a term coming from:

$$\Gamma_{12} = \sum_f \Gamma_{12}^f \,, \quad \Gamma_{12}^f \equiv 2\pi\rho_f \langle K^0|H_W|f\rangle\langle f|H_W|\overline{K}^0\rangle, \tag{13.28}$$

Putting it in the form of ϵ, the additional contribution is:

$$\epsilon(f) = i \cos \phi_{SW} e^{i\phi_{sw}} \frac{\text{Im}\Gamma_{12}^f}{\Delta\Gamma}. \tag{13.29}$$

Note that this term appears in Eq. (13.22). Assuming **CPT** symmetry Problem 13.5 helps you derive:

$$\text{Im}\,\Gamma_{12}^f = i\eta_f 2\pi\rho_f A(f)^2 \left(1 - \eta_f \frac{\overline{A}(f)}{A(f)} \right). \tag{13.30}$$

Putting Eq. (13.30) into Eq. (13.29), we obtain:

$$\left| \frac{\epsilon(f)}{\epsilon} \right| = \frac{1}{\sqrt{1 + (\frac{\Delta\Gamma}{2\Delta M})^2}} \frac{\Delta\Gamma}{2\Delta M} \frac{\Gamma(K_{S,L} \to f)}{2\Gamma_S} \frac{\Delta_f}{\epsilon} \tag{13.31}$$

Table 13.2 *Estimates of* $\epsilon(f)$ *where we have used:* $Br(K_S \to (3\pi)^{\mathbf{CP}+}) \sim 3.4 \times 10^{-7}$; $Br(K_L \to (3\pi)^{\mathbf{CP}-}) \sim 12\%$; $Br(K_L \to (3\pi)^0) \sim 19\%$; $\Gamma_L/\Gamma_S = 571$.

f	$(\pi\pi)_2$	$\mathbf{CP}+:(\pi^+\pi^-\pi^0)$	$\mathbf{CP}-:(\pi^+\pi^-\pi^0)$	$3\pi^0$
$\|\epsilon(f)\|$	2×10^{-7}	2×10^{-11}	2×10^{-8}	3×10^{-8}

where $\Gamma(K_{S,L} \to f)$ stands for $\Gamma(K_S \to f)$ for $\mathbf{CP}=+1$ final state f, and $\Gamma(K_L \to f)$ for $\mathbf{CP}=-1$ final state f. Finally, we denoted $\Delta_f = 1 - \eta_f \frac{\overline{A}(f)}{A(f)}$. We shall now analyse $|\epsilon(f)|$ for the different possible final states.

Remembering that the major contribution to the direct **CP** violation in K decays comes from the penguin amplitude, which contributes only to Δ_0, we expect $|\Delta_2| \ll |\Delta_0|$, and it is quite reasonable to expect

$$\epsilon' \sim \frac{1}{2\sqrt{2}}\omega \, e^{i(\delta_2-\delta_0)} \Delta_0, \tag{13.32}$$

Let us first estimate Δ_f from the value of ϵ'. One reasonable assumption we make is that $|\Delta_f| = \mathcal{O}(|\Delta_0|)$. We then conclude that:

$$\left|\frac{\Delta_f}{\epsilon}\right| \sim \mathcal{O}\left(\frac{2\sqrt{2}}{\omega}\frac{\epsilon'}{\epsilon}\right). \tag{13.33}$$

Using this estimate,

$$\left|\frac{\epsilon(f)}{\epsilon}\right| \sim \left(\frac{\Gamma(K_{S,L} \to f)}{\Gamma_S}\frac{1}{\omega}\frac{\epsilon'}{\epsilon}\right), \tag{13.34}$$

we obtain the results given in Table 13.2.

Now consider $f = \pi l \nu$. Non-vanishing $\epsilon(\pi l \nu)$ implies violation of the $\Delta Q = \Delta S$ rule as expressed by the complex parameter x defined in Eq. (7.49). Since

$$\mathrm{Im}\,\Gamma_{12}^{\pi l \nu} \simeq \mathrm{Im}\,x\Gamma(K \to \pi\mu\nu), \tag{13.35}$$

we find

$$|\epsilon(\pi l \nu)| \le 4 \cdot 10^{-7}, \tag{13.36}$$

where we have used the bound $\mathrm{Im}\,x = (0.5 \pm 2.5) \times 10^{-3}$ from Ref. [11] and $Br(K_S \to \pi^\mp\mu^\pm\nu) \sim 5 \times 10^{-4}$.

Now what about the second term in the right-hand side of Eq. (13.22)? In evaluating $\rho_0 = \frac{\overline{A}_0}{A_0}$, we note that there are tree and penguin contributions:

$$\rho_0 = \frac{Te^{i(-\phi_T+\delta_T)} + Pe^{i(-\phi_P+\delta_P)}}{Te^{i(\phi_T+\delta_T)} + Pe^{i(\phi_P+\delta_P)}} \tag{13.37}$$

where ϕ^T and ϕ^P are tree and penguin phases for A, and $\delta_{T.P}$ are strong phases. Using Watson's theorem we conclude that $\delta^P = \delta^T$ so that $\rho_0 = e^{2i\xi_0}$ where ξ is defined in Eq. (7.33). Then we obtain

$$\frac{1}{2}\left(\mathrm{Re}\,\frac{\overline{A}_0}{A_0} - 1\right) \sim \frac{1}{2}(\cos(2\xi_0) - 1) \sim \mathcal{O}(\xi_0^2) \tag{13.38}$$

so that contribution of **CP** violation to ρ_0 is negligible.

In summary, we can conclude[2]

$$\left|\sum_f \epsilon^f\right| \leq \sum_f |\epsilon^f| \leq 10^{-6}. \tag{13.39}$$

and

$$R_{\mathrm{direct}} = \mathcal{O}(10^{-6}) + \textbf{CPT}\ \text{violating terms.} \tag{13.40}$$

- We have shown that the prediction for arg ϵ where we had included only the contribution of $(2\pi)_0$ to Im Γ_{12} holds indeed to within more than one part in 10^3.
- Using

$$R_{\mathrm{direct}} < 10^{-6}, \tag{13.41}$$

we obtain:

$$\left(\frac{M_{\overline{K}} - M_K}{2\Delta M_K} + \frac{1}{2}\left(\mathrm{Re}\,\frac{\overline{A}_0}{A_0} - 1\right)_{\textbf{CPT}}\right) = (0.23 \pm 4.0_{\mathrm{exp.}} \pm 0.1_{\mathrm{theory}}) \times 10^{-5}, \tag{13.42}$$

which is completely consistent with being zero.

-

$$\left(\frac{M_{\overline{K}} - M_K}{2\Delta M_K}\right) = (0.23 \pm 4.0_{\mathrm{exp.}} \pm 0.1_{\mathrm{theory}}) \times 10^{-5}, \tag{13.43}$$

is valid if we assume that there is no **CPT** violation in *decay* amplitudes.

- **CPT** symmetry implies vanishing

$$\left|\frac{2}{3}\phi_{+-} + \frac{1}{3}\phi_{00} - \phi_{SW}\right| = (0.04 \pm 0.70_{\mathrm{exp.}} \pm 0.2_{\mathrm{th.}})^\circ. \tag{13.44}$$

To judge the accuracy of the test for **CPT**, we can compare the error with respect to ϕ_{SW}. Comparing the right hand side to ϕ_{SW}, we have tested **CPT** at the level of 1.5%. This test is limited by the experimental error. If the statistics improves, we have a potential to test **CPT** to the level of 0.04%.

---

[2] We follow the arguments of [42]. Although updated numbers are used, the final results hardly differ from the result obtained by Lee and Wu 40 years ago.

13.4 Harnessing EPR correlations

In Chapter 10 we saw that the reaction

$$e^+e^- \to \Upsilon(4S) \to B_d\overline{B}_d \qquad (13.45)$$

enables us to perform novel measurements of high sensitivity.

- Both beauty mesons can oscillate into each other.
- Since the $B\overline{B}$ pair is produced into a *single coherent quantum mechanical state* their oscillations are highly *correlated* with each other. We have actually an unusual EPR situation at hand [201]: irrespective of **CPT** invariance and of how fast the $B_d - \overline{B}_d$ oscillations proceed, the **C** *odd* $B_d\overline{B}_d$ pair in Eq. (13.45) can never transmogrify itself into a B_dB_d or $\overline{B}_d\overline{B}_d$ pair; this is a consequence of Bose statistics. Equivalently, we can say that the $\Upsilon(4S)$ resonance always decays into two *different* neutral *mass* eigenstates B_1 and B_2:

$$e^+e^- \to \Upsilon(4S) \to B_1B_2 \; ; \; \nrightarrow B_1B_1 \, , \, B_2B_2. \qquad (13.46)$$

Yet we cannot predict whether the particular neutral meson reaching the detector will turn out to be B_1 or B_2 any more than whether it will be a B_d or $\overline{B}_d$.
- Studying

$$e^+e^- \to \Upsilon(4S) \to B_d\overline{B}_d \to f_1f_2 \qquad (13.47)$$

for various channels $B_d/\overline{B}_d \to f_{1,2}$ thus represents double interferometry, which typically allows us to measure tiny quantities like $\Delta M(B_d)$ and phases that otherwise are unaccessible. Some examples are discussed in the Problems section of Chapter 10.

Now we want to go beyond these statements.

- Qualitatively, we encounter an analogous quantum mechanical situation at ϕ factories[3]

$$e^+e^- \to \phi(1020) \to K_SK_L. \qquad (13.48)$$

Yet big quantitative differences arise due to $\Gamma(K_S) \gg \Gamma(K_L)$ vs $\Gamma(B_1) \simeq \Gamma(B_2)$.
- The high sensitivity of the double interferometry referred to above can be harnessed to probe for **CPT** violation.

---

[3] Likewise for the reaction $e^+e^- \to \psi'' \to D^0\overline{D}^0/D_1D_2$.

- The specific form predicted for the EPR correlations is based on the superposition principle. A detailed analysis of the observed correlations will probe for the presence of *nonlinear* effects beyond ordinary quantum mechanics.

The formalism presented in Chapter 6 is completely general and applicable even without **CPT** symmetry. The amplitude for finding a final state f_1 emerging at time t_k and f_2 at time t_{-k} is given by [202]

$$
\begin{aligned}
A\left(P^0\overline{P}^0|_{C=\pm} \to f_1(t_k)f_2(t_{-k})\right) \\
= \left(g_+(t_k)\overline{g}_-(t_{-k}) + (-1)^C\overline{g}_-(t_k)g_+(t_{-k})\right) A(f_1)A(f_2) \\
+ \left(g_+(t_k)\overline{g}_+(t_{-k}) + (-1)^C\overline{g}_-(t_k)g_-(t_{-k})\right) A(f_1)\overline{A}(f_2) \\
+ \left(g_-(t_k)\overline{g}_-(t_{-k}) + (-1)^C\overline{g}_+(t_k)g_+(t_{-k})\right) \overline{A}(f_1)A(f_2) \\
+ \left(g_-(t_k)\overline{g}_+(t_{-k}) + (-1)^C\overline{g}_+(t_k)g_-(t_{-k})\right) \overline{A}(f_1)\overline{A}(f_2), \quad (13.49)
\end{aligned}
$$

where

$$
\begin{aligned}
g_+(t) &= f_+(t) + 2\delta f_-(t) \\
g_-(t) &= (1 - 2\epsilon_0)f_-(t) \\
\overline{g}_+(t) &= f_+(t) - 2\delta f_-(t) \\
\overline{g}_-(t) &= (1 + 2\epsilon_0)f_-(t), \quad (13.50)
\end{aligned}
$$

and the functions $f_\pm(t)$ are as defined by Eq. (6.49):

$$
f_\pm(t) = \frac{1}{2}e^{-iM_1 t}e^{-\frac{1}{2}\Gamma_1 t}\left[1 \pm e^{-i\Delta M t}e^{\frac{1}{2}\Delta\Gamma t}\right].
$$

As stated in Eq. (6.18), $\delta \neq 0$ [$\epsilon_0 \neq 0$] requires simultaneous **CPT** [**T**] and **CP** violation.

It is obvious from Eq. (13.49) that by studying the rate for a certain final state f_1 appearing at some time and f_2 an interval Δt later, we can – at least in principle – extract δ, ϵ_0 and the transformation properties of the matrix elements. We will sketch the procedure for ϵ_0 and beauty factories.

13.4.1 φ factory

It might be of help to repeat some statements on the quantum mechanical analysis of

$$
e^+e^- \to \phi(1020) \to K^0\overline{K}^0. \quad (13.51)
$$

The kaon pair is produced in a coherent **C** *odd* quantum state,[4] the evolution of which is described in terms of a single time variable – until one of the kaons decays at time $t = t_1$. We infer from Bose statistics that *up to that moment* the two kaons have to remain *distinct*; this is expressed most concisely through their mass eigenstates:

$$|K^0\overline{K}^0; t = 0\rangle = |K_S K_L; t = 0\rangle \rightarrow |K_S K_L; t > 0\rangle \quad \text{for } t \leq t_1. \quad (13.52)$$

Once the first kaon decays as, say, a K^0 or K_S, the other keeps evolving as a $\overline{K}^0$ or K_L, respectively, till it meets its decay at time t_2.

Even at a symmetric ϕ factory like DAΦNE the vertices for the kaon decays can be resolved with existing technology. Yet the *primary* vertex where the kaons were produced is known very poorly. We can therefore deduce quite reliably the *difference* in the (proper) time of decay – $\Delta t = t_2 - t_1$ – but not t_1 and t_2 themselves. In consequence we have to integrate the predicted distributions over $t_1 + t_2$ to obtain observable effects:

$$|A(\Delta t; f_1; f_2)|^2 \equiv \frac{1}{2} \int_{|\Delta t|}^{\infty} d(t_1 + t_2)|A(f_1, t_1; f_2, t_2)|^2. \quad (13.53)$$

We can then consider two types of asymmetries, namely

- those where we compare the rates *integrated* over $\Delta t > 0$ and $\Delta t < 0$, respectively; and
- those that are given as a function of Δt, see Eq. (13.53).

As final states we can employ $f = \pi\pi$, $l^{\pm}\nu\pi^{\mp}$. In this way all the quantities of interest – $\cos\theta$, ϕ, y_l, x_l and $\overline{x}_l$ can be measured. Since a detailed discussion can be found in Ref. [196], we limit ourselves to stating the main results.

Time-integrated asymmetries

With **CPT** symmetry requiring $|A(\Delta t; \pi^- l^+ \nu_l, \pi^+ l^- \overline{\nu}_l)|^2$ to be an *even* function of Δt, the following asymmetry measures **CPT** breaking with or without the $\Delta S = \Delta Q$ rule:

$$\langle A_{l+l-} \rangle$$

$$\equiv \frac{\int_0^\infty d(\Delta t)|A(\Delta t; \pi^- l^+ \nu_l, \pi^+ l^- \overline{\nu}_l)|^2 - \int_{-\infty}^0 d(\Delta t)|A(\Delta t; \pi^- l^+ \nu_l, \pi^+ l^- \overline{\nu}_l)|^2}{\int_0^\infty d(\Delta t)|A(\Delta t; \pi^- l^+ \nu_l, \pi^+ l^- \overline{\nu}_l)|^2 + \int_{-\infty}^0 d(\Delta t)|A(\Delta t; \pi^- l^+ \nu_l, \pi^+ l^- \overline{\nu}_l)|^2}$$

$$= -2\text{Re}(2\delta + x_l - \overline{x}_l^*). \quad (13.54)$$

---

[4] The radiative process $\phi \rightarrow K\overline{K}\gamma$ provides a negligible background.

If both kaons decay through the same semileptonic mode, we can compare *like*-sign dilepton final states

$$\langle A_{l\pm l\pm} \rangle$$
$$\equiv \frac{\int_{-\infty}^{\infty} d(\Delta t) |A(\Delta t; \pi^+ l^- \bar{\nu}_l, \pi^+ l^- \bar{\nu}_l)|^2 - \int_{-\infty}^{\infty} d(\Delta t) |A(\Delta t; \pi^- l^+ \nu_l, \pi^- l^+ \nu_l)|^2}{\int_{-\infty}^{\infty} d(\Delta t) |A(\Delta t; \pi^+ l^- \bar{\nu}_l, \pi^+ l^- \bar{\nu}_l)|^2 + \int_{-\infty}^{\infty} d(\Delta t) |A(\Delta t; \pi^- l^+ \nu_l, \pi^- l^+ \nu_l)|^2}$$
$$= 4(\operatorname{Re} y_l - \operatorname{Re} \epsilon_0); \tag{13.55}$$

i.e. $\langle A_{l\pm l\pm} \rangle \neq 0$ reveals **T** and **CP** breaking – $\phi \neq 0$ – and/or **CPT** and **CP** violation – $\operatorname{Re} y_l \neq 0$.

Δt-dependent asymmetries

Assuming both decays to be semileptonic, we find

$$A_{l+l-}(\Delta t)$$
$$\equiv \frac{|A(\Delta t > 0; \pi^- l^+ \nu_l, \pi^+ l^- \bar{\nu}_l)|^2 - |A(\Delta t < 0; \pi^- l^+ \nu_l, \pi^+ l^- \bar{\nu}_l)|^2}{|A(\Delta t > 0; \pi^- l^+ \nu_l, \pi^+ l^- \bar{\nu}_l)|^2 + |A(\Delta t < 0; \pi^- l^+ \nu_l, \pi^+ l^- \bar{\nu}_l)|^2}$$
$$= -\frac{2\operatorname{Re}(\Delta_l)\left(e^{-\Gamma_L \Delta t} - e^{-\Gamma_S \Delta t}\right) + 4\operatorname{Im}(\Delta_l) e^{-\frac{\bar{\Gamma}}{2}\Delta t} \sin(\Delta M \Delta t)}{e^{-\Gamma_L \Delta t} + e^{-\Gamma_S \Delta t} + 2e^{-\frac{\bar{\Gamma}}{2}\Delta t} \cos(\Delta M \Delta t)}, \tag{13.56}$$

where $\Delta_l \equiv 2\delta + x_l - \bar{x}_l^*$. Both the real and the imaginary part of Δ_l can be extracted by analysing this asymmetry for small and large time intervals, namely $\Delta t = (0 \sim 15)\tau_S$ and $\Delta t \geq 20\tau_S$, respectively [196].

The corresponding ratio for like-sign dileptons is actually *independent* of Δt:

$$A_{l\pm l\pm} \equiv \frac{|A(\Delta t; \pi^+ l^- \bar{\nu}_l, \pi^+ l^- \bar{\nu}_l)|^2 - |A(\Delta t; \pi^- l^+ \nu_l, \pi^- l^+ \nu_l)|^2}{|A(\Delta t; \pi^+ l^- \bar{\nu}_l, \pi^+ l^- \bar{\nu}_l)|^2 + |A(\Delta t; \pi^- l^+ \nu_l, \pi^- l^+ \nu_l)|^2}$$
$$= 4(\operatorname{Re} y_l - \operatorname{Re} \epsilon_0). \tag{13.57}$$

13.4.2   Tests of **CPT** symmetry in B decays

The completion of the programme for a ϕ factory outlined above does not free us from the obligation to perform an analogous detailed study for beauty decays.

- 'Exotic' physics could have a different effect on the decays of beauty than strange hadrons.
- The time-scale characterizing decays and oscillations is quite different in the two systems.

- Asymmetric beauty factories allowing one to track the time evolution of $B_d\overline{B}_d$ pairs are available!

We start from the same master equations as for the $K\overline{K}$ case, yet with $\Delta\Gamma_B \simeq 0$. Again we aim at illustrating salient features rather than giving an exhaustive discussion.

Defining

$$s = \cot\theta = \frac{1}{2}\left(\frac{q_2}{p_2} - \frac{q_1}{p_1}\right)e^{2\epsilon_0}, \tag{13.58}$$

we can express the time-integrated ratio of like-sign to opposite-sign dileptons and the charge asymmetry in it as follows:

$$R_{(\mathbf{C}=-1)} = \frac{\Gamma(B^0\overline{B}^0|_{\mathbf{C}=-} \to l^\pm l^\pm X)}{\Gamma(B^0\overline{B}^0|_{\mathbf{C}=-} \to l^\pm l^\mp X)} = \frac{1}{2}\frac{(|e^{2\epsilon_0}|^2 + |e^{-2\epsilon_0}|^2)x^2}{|1+s^2|(2+x^2) + |s|^2x^2} \tag{13.59}$$

$$a = \frac{\Gamma(B^0\overline{B}^0|_{\mathbf{C}=-} \to l^+l^+X) - \Gamma(B^0\overline{B}^0|_{\mathbf{C}=-} \to l^-l^-X)}{\Gamma(B^0\overline{B}^0|_{\mathbf{C}=-} \to l^\pm l^\mp X)}$$

$$= \frac{|e^{2\epsilon_0}|^2 - |e^{-2\epsilon_0}|^2}{|e^{2\epsilon_0}|^2 + |e^{-2\epsilon_0}|^2}, \tag{13.60}$$

where $x = \Delta M(B^0)/\Gamma(B^0)$ is to be distinguished from x_l. Note that while a is independent of s, its measurement will lead to a determination of Im ϕ. For the reasons discussed in Section 11.5.2, we expect Im ϕ to be small. Then R gives only a constraint on s and x. To determine x in the presence of **CPT** violation, we need to measure the time dependence of the dilepton rates:

$$N_{\mathbf{C}=-}^{\pm\pm} \sim e^{-\Gamma(t_1+t_2)}[1 - \cos\Delta M(t_1 - t_2)]$$

$$N_{\mathbf{C}=-}^{+-} \sim e^{-\Gamma(t_1+t_2)}[1 + 2s^2 + \cos\Delta M(t_1 - t_2)]. \tag{13.61}$$

Finally, let us now discuss how **CPT** violation might influence the ψK_S **CP** asymmetry. In the presence of small **CPT** violation, Eq. (10.39) becomes:

$$\Gamma([l^\pm X]_{t_l} f_{t_f}) \propto e^{-\Gamma(t_l+t_f)}|A(l)|^2|A(f)|^2 \left[1 \pm \mathrm{Im}\left(e^{-2\epsilon_0}\rho(f)\right)\sin\Delta M(t_f - t_l)\right.$$

$$\left. \pm s\mathrm{Re}\left(e^{-2\epsilon_0}\rho(f)\right)[\cos\Delta M(t_l + t_f) - 1]\right]. \tag{13.62}$$

where we have kept only the leading term in s. There is one non-trivial modification. **CPT** violation leads to a time integrated asymmetry,

$$\frac{\Gamma([l^+ X]_{t_l} f_{t_f}) - \Gamma([l^- X]_{t_l} f_{t_f})}{\Gamma([l^+ X]_{t_l} f_{t_f}) + \Gamma([l^- X]_{t_l} f_{t_f})} = -s\mathrm{Re}\left[e^{-2\epsilon_0}\rho(f)\right]\frac{x^2}{1+x^2}. \tag{13.63}$$

Note that this asymmetry depends linearly on s. So, for small s, it might be more sensitive to **CPT** violation than like-sign dilepton decays.

13.5 The moralist's view

We often hear the claim that **CPT** invariance has been tested at the $\mathcal{O}(10^{-18})$ level. However, this refers to a very specific observable only, namely $(M_{\overline{K}^0} - M_{K^0})/M_K$, and it has been inflated by relating the mass difference to the overall kaon mass. Calibrating this mass difference by the weak width Γ_S, we arrive at a bound $\sim \mathcal{O}(10^{-4})$.

Of course, the **CPT** *theorem* is based on noble assumptions about nature's basic structure. Yet these are assumptions nonetheless, and it behoves us to subject them to empirical scrutiny. We are not claiming that such an obligation would justify the construction of new facilities, in particular in the absence of meanigful benchmarks about the size and form of violations. However, the arrival of ϕ and beauty factories enables us to perform novel tests of **CPT** invariance; the high sensitivity that can be achieved by harnessing EPR correlations makes it even relevant as a probe for the possible presence of *non*-linear terms in the Schrödinger equation. It is conceivable that **CP** asymmetries in $e^+e^- \to B_d\overline{B}_d$ decays emerge, as predicted, but with a dependence on the time difference Δt that differs from expectations.

Problems

13.1. Ignore the effect of **CPT** and **CP** violation and consider a violation of the $\Delta S = \Delta Q$ rule. Then we can set $\phi = 2\delta = y_l = 0$. Compute the time dependence of the decay:

$$\Gamma(K \to \pi^- l^+ \nu) = \left|\frac{F_l}{2}\right| [(1 - 2\text{Re } x_l)e^{-\Gamma_L t} + (1 + 2\text{Re } x_l)e^{-\Gamma_s t}]$$
$$+ 2e^{\frac{1}{2}\Gamma t}\cos(\Delta Mt) - 4\text{Im } x_l e^{\frac{1}{2}\Gamma t}\sin(\Delta Mt),$$
$$\Gamma(K \to \pi^+ l^- \overline{\nu}) = \left|\frac{F_l}{2}\right| [(1 - 2\text{Re } x_l)e^{-\Gamma_L t} + (1 + 2\text{Re } x_l)e^{-\Gamma_s t}]$$
$$- 2e^{\frac{1}{2}\Gamma t}\cos(\Delta Mt) - 4\text{Im } x_l e^{\frac{1}{2}\Gamma t}\sin(\Delta Mt).$$
$$(13.64)$$

13.2. Repeat the derivation of ϵ and ϵ' given in Section 7.3 without assuming **CPT** symmetry, i.e. keeping track of label i on p_i and q_i.

13.3. What is the **CP** eigenvalue of the $n\pi^0$ state? Show that the **CP** violating $K_S \to \pi^+\pi^-\pi^0$ decay is kinematically enhanced over the **CP** conserving $K_S \to \pi^+\pi^-\pi^0$ decay.

13.4. Here is a problem to guide you through the derivation of Eq. (13.21).

(a) Using the expression for ϵ from Eq. (13.19) and for q_2/p_2 from Eq. (6.20) and Eq. (7.23), and assuming that **CP** and **CPT** violation is small, show that

$$\epsilon = \frac{1}{2}[1 - (1 + i\phi)(1 + 2\delta)(1 + r_A)]$$

$$= -\frac{1}{2}\left(i\frac{E_2}{E_1} + \frac{E_3}{E} + r_A\right) \qquad (13.65)$$

where $r_A = \frac{\bar{A}_0}{A_0} - 1$.

(b) Derive:

$$\epsilon = \frac{-\operatorname{Im} M_{12} + \frac{i}{2}\operatorname{Im}\Gamma_{12}}{\sqrt{(\Delta M)^2 + (\frac{1}{2}\Delta\Gamma)^2}}e^{i\phi_{sw}}$$

$$- \frac{i(M_{11} - M_{22}) + \frac{1}{2}(\Gamma_{11} - \Gamma_{22})}{2\sqrt{(\Delta M)^2 + (\frac{1}{2}\Delta\Gamma)^2}}e^{i\phi_{sw}} - \frac{1}{2}r_A. \qquad (13.66)$$

(c) Starting from

$$r_A = \operatorname{Re} r_A + 2i\frac{\operatorname{Im}\Gamma_{12}((2\pi)_0)}{\Delta\Gamma}, \qquad (13.67)$$

show that

$$-\frac{1}{2}r_A e^{-i\phi_{sw}} = -\frac{i}{2}\frac{\operatorname{Im}\Gamma_{12}((2\pi)_0)}{\sqrt{(\Delta M)^2 + (\frac{1}{2}\Delta\Gamma)^2}}$$

$$- \frac{\operatorname{Im}\Gamma_{12}((2\pi)_0)}{\Delta\Gamma}\sin\phi_{SW} - \frac{1}{2}\operatorname{Re} r_A e^{-i\phi_{sw}}. \qquad (13.68)$$

(d) Now, inserting the expression for r_A into that of ϵ, show that the $(2\pi)_0$ contribution to $\operatorname{Im}\Gamma_{12}$ cancels and derive

$$\epsilon = e^{i\phi_{SW}} \sin\phi_{SW} \frac{1}{\Delta M} \left(- \operatorname{Im} M_{12} + \frac{i}{2} \sum_{f \neq (2\pi)_0} \operatorname{Im}\Gamma_{12}(f) \right.$$

$$- \frac{i}{2}(M_{11} - M_{22}) - \frac{1}{4}(\Gamma_{11} - \Gamma_{22})$$

$$\left. - \frac{1}{2}\operatorname{Im}\Gamma_{12}((2\pi)_0)\tan\phi_{SW} \right) - \frac{1}{2}\operatorname{Re} r_A. \qquad (13.69)$$

(e) Multiplying by a phase factor $e^{-i2\phi_{sw}}$ and taking the imaginary part, derive:

$$|\eta_{+-}| \left(\frac{2}{3}\phi_{+-} + \frac{1}{3}\phi_{00} - \phi_{SW} \right)$$
$$= \left(\frac{M_{\overline{K}} - M_K}{2\Delta M_K} + R_{\text{direct}} \right) \sin\phi_{SW}, \qquad (13.70)$$

where

$$R_{\text{direct}} = \frac{1}{2}\left(\frac{\operatorname{Im}\Gamma_{12}}{\Delta M} + \operatorname{Re} r_A \right)$$
$$= \frac{-ie^{-i\phi_{sw}}}{\sin\phi_{SW}} \sum_{f \neq (2\pi)_0} \epsilon(f) + \frac{1}{2}\operatorname{Re} r_A \qquad (13.71)$$

and $\epsilon(f)$ is defined in Eq. (13.29).

13.5. Set $\langle f \, ; \text{out}|H|K\rangle = e^{i\delta_f} A_f$ and $\langle f \, ; \text{out}|H|\overline{K}\rangle = e^{i\delta_f} \overline{A}_f$, where f is a **CP** eigenstate, and δ_f is the final state strong interaction phase. Show that **CPT** symmetry implies $A_f = \eta_f \overline{A}_f^*$, and that

$$\operatorname{Im}\Gamma_{12}^f = -i\eta_f \pi \rho_f (\overline{A}_f + A_f)(\overline{A}_f - A_f), \qquad (13.72)$$

where the η_f corresponds to the **CP** eigenvalue of f. Now derive Eq. (13.30).

14

CP violation in charm decays – the dark horse

I know she invented fire –
but what has she done recently?

14.1 On the uniqueness of charm

The well travelled quote above concisely describes a widespread attitude towards charm. Charm dynamics is recognized as physics with a great past: it was instrumental in driving the paradigm shift from quarks as mathematical entities to physical objects and in providing essential support for accepting QCD as the theory of the strong interactions. Yet it is often viewed as one without a future. For the predicted SM electroweak phenomenology for $\Delta C \neq 0$ is on the 'dull' side: the CKM parameters are 'known' from other sources, $D - \overline{D}$ oscillations slow, **CP** asymmetries small and loop driven decays extremely rare.

Yet more thoughtful observers had realized that the very 'dullness' of charm's SM phenomenology provides us with a dual opportunity, namely to

(1) probe our *quantitative* understanding of QCD's non-perturbative dynamics thus calibrating our theoretical tools for describing B decays;

(2) perform almost 'zero-background' searches for New Physics, mainly in the area of **CP** violation.

The former item provides the central motivation for the CLEO-c programme at Cornell University to be followed up by the BESIII programme at Beijing's IHEP. The intent is to provide high precision measurements of observables that can be evaluated by a new generation of unquenched Monte Carlo simulations of QCD on the lattice. Validating theoretical control over non-perturbative dynamics will refine also our searches for New Physics in D as well as in B decays.

Concerning the second goal, honesty compels us to say that while New Physics signals can exceed SM predictions on **CP** asymmetries by orders of magnitude, they might not be large in absolute terms, as specified below. Even so charm transitions actually provide a unique and novel portal to flavour dynamics with many experimental features being a priori favourable. Furthermore baryogenesis requires **CP** violating dynamics from a New Physics source. The latter has to compete against much smaller SM contributions in charm than in beauty decays. In that sense the *theoretical* 'signal-to-noise' ratio for New Physics could be significantly larger in charm than in beauty transitions. All of this makes a dedicated programme of **CP** studies in charm transitions mandatory.

The prospects for finding New Physics in charm transitions have received a major boost through the strong evidence for $D - \overline{D}$ oscillations presented by BaBar and Belle in the spring of 2007. Before discussing this evidence and its consequences, we will make some general statements concerning the uniqueness of charm. Those will remain valid whether this evidence is confirmed or not.

The SM has been constructed such that there are no FCNC at the tree level and that they are GIM reduced at one-loop to accommodate the huge suppression observed for strangeness changing neutral currents. As a consequence charm and top changing neutral currents are suppressed even more strongly. It is the hallmark of New Physics scenarios in general to induce FCNC. A priori there seems to be no reason why those should follow the same pattern as in the SM. It is thus conceivable that New Physics FCNC are more noticeable in the transitions of up-type than down-type quarks. We want to emphasize that *charm is the only up-type quark allowing the full range of probes for flavour-changing neutral currents and New Physics in general*:

- Top quarks decay *before* they can hadronize [203]. Without top *hadrons* $T - \overline{T}$ oscillations cannot occur. This denies us the possibility to search for New Physics in $\Delta T = 2$ transitions that are most highly suppressed in the SM. A forteriori this limits our options to search for **CP** asymmetries in top decays, since one cannot call on oscillations to provide the required second amplitude.
- Hadrons built with u and $\overline{u}$ quarks like π^0 and η are there own antiparticle; thus there can be no $\pi^0 - \pi^0$ etc. oscillations as a matter of definition. Furthermore they possess so few decay channels that already **CPT** invariance basically rules out **CP** asymmetries in their decays.

None of this means one should not search for **CP** violation in the dynamics of top and up (as well as down) quarks. The former will be discussed in the next chapter, while searches for electric dipole moments, discussed in Section 3.6, probe the latter. The point is that charm dynamics offer unique phenomenological possibilities for discovering manifestations of New Physics, and we will argue that only very recently have experiments reached a range of sensitivity, where one can realistically expect the sought-after effects to show up.

14.2 $D^0 - \overline{D}^0$ oscillations

As stated before we do not believe that expressions in their most general form provide the best way to approach a new subject. We know, as explained below, that **CP** violation is at most of moderate size in charm decays. In discussing oscillations per se and the experimental evidence for it in this section we will assume **CP** invariance to hold. Later we will address **CP** asymmetries.

14.2.1 Experimental evidence
$D \to K^+K^-$

Finding two different lifetimes in the decays of neutral D mesons constitutes an unequivocal manifestation of oscillations. Then the mass eigenstates of neutral D mesons have to be **CP** eigenstates as well, and the decay rate evolution in time for $D^0 \to h^+h^-$, $h = K$, π is given by a single exponential function, controlled by the width for the **CP** even state:

$$\Gamma(D^0(t) \to h^+h^-) \propto e^{-\Gamma_+ t}|A(D^0 \to h^+h^-)|^2. \qquad (14.1)$$

The final state $K^-\pi^+$ on the other hand comes in equal measure from the **CP** even and odd states:

$$\Gamma(D^0(t) \to K^-\pi^+) \propto \frac{1}{2}\left(e^{-\Gamma_+ t} + e^{-\Gamma_- t}\right)|A(D^0 \to K^-\pi^+)|^2. \qquad (14.2)$$

The ratio between the two modes then exhibits the following time dependence:

$$\frac{\Gamma(D^0(t) \to K^-\pi^+)}{\Gamma(D^0(t) \to h^+h^-)} \propto \frac{1}{2}\left(1 + e^{(\Gamma_+ - \Gamma_-)t}\right) \simeq 1 + y_{CP} \cdot \frac{t}{\tau_D}, y_{CP} = \frac{\Gamma_+ - \Gamma_-}{2\overline{\Gamma}_D}.$$
$$(14.3)$$

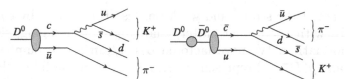

Figure 14.1   Feynman diagram which gives $D^0 \to K^+\pi^-$.

Belle has obtained a 3.2 σ signal for a difference in the effective lifetimes for $D^0 \to K^+K^-$ and $D^0 \to K^-\pi^+$ [204]:

$$y_{CP} = (1.31 \pm 0.32 \pm 0.25) \cdot 10^{-2}. \tag{14.4}$$

$\mathbf{D \to K^+\pi^-}$

In addition to the Cabibbo allowed transition $D^0 \to K^-\pi^+$ one can obtain also the **CP** conjugate final state: $D^0 \to K^+\pi^-$. There are two sources for 'wrong-sign' kaons in D^0 decays; i.e. the selection rule $\Delta C = \Delta S$ can effectively be violated by two different mechanisms, namely by doubly Cabibbo-suppressed $c \to d\bar{s}u$ transitions (DCSD) as well as by $D - \overline{D}$ oscillations. This is sketched in Fig. 14.1. By analysing the decay rate evolution as a function of (proper) time, one can disentangle the two sources for 'wrong-sign' kaons.

$$\frac{\Gamma(D^0(t) \to K^+\pi^-)}{\Gamma(D^0(t) \to K^-\pi^+)} = \frac{|A(D^0 \to K^+\pi^-)|^2}{|A(D^0 \to K^-\pi^+)|^2}$$

$$\times \left[1 + Y_{K\pi} \cdot \left(\frac{t}{\tau_D}\right) + Z_{K\pi} \cdot \left(\frac{t}{\tau_D}\right)^2\right]. \tag{14.5}$$

The time-independent term in the square brackets of Eq. (14.5) represents the pure DCSD transition. The third term $Z_{K\pi}$ quadratic in the normalized time of decay t/τ_D is the pure oscillation process. The second term $Y_{K\pi}$ linear in t/τ_D reflects the interference between the DCSD and oscillation amplitude. It requires a somewhat tedious calculation to obtain the explicit expressions for $Y_{K\pi}$ and $Z_{K\pi}$; in section 14.3.3 we provide guidance for the determined reader.

$$Y_{K\pi} = \frac{y_D'}{\tan^2\theta_C} \left|\frac{q}{p}\hat{\rho}(K^+\pi^-)\right|, \quad Z_{K\pi} = \left(\frac{x_D'^2 + y_D'^2}{4\tan^4\theta_C}\right)\left|\frac{q}{p}\hat{\rho}(K^+\pi^-)\right|^2,$$

$$\tag{14.6}$$

where we have used the following notation:

$$\frac{q}{p}\hat{\rho}(K^+\pi^-) \equiv -\frac{1}{\tan^2\theta_C}|\hat{\rho}(K^+\pi^-)|e^{-i(\delta - \phi_{K\pi})},$$

$$\frac{p}{q}\rho(K^-\pi^+) \equiv -\frac{1}{\tan^2\theta_C}|\hat{\rho}(K^-\pi^-)|e^{-i(\delta + \phi_{K\pi})}; \tag{14.7}$$

we have denoted the strong and weak phases of $\overline{\rho}(K^+\pi^-)$ by $\phi_{K\pi}$ and δ, respectively, with the former [latter] changing [not changing] sign under **CP** transformation. We have factored out $1/\tan^2\theta_C$ to emphasize that $\overline{\rho}(K^+\pi^-)$ and $\rho(K^-\pi^+)$ representing ratios between Cabibbo allowed and doubly suppressed amplitudes are greatly enhanced by $1/\tan^2\theta_C \sim 20$. Finally:

$$x'_D \equiv x_D\cos\delta + y_D\sin\delta, \; y'_D \equiv -x_D\sin\delta + y_D\cos\delta \qquad (14.8)$$

yielding $(x'_D)^2 + (y'_D)^2 = x_D^2 + y_D^2$. Since a priori there is no reason why δ should be particularly small, one should keep track of the difference between (x_D, y_D) and (x'_D, y'_D).

BaBar has found [205]

$$y'_D = (0.97 \pm 0.44 \pm 0.31) \cdot 10^{-2}, \; (x'_D)^2 = (-2.2 \pm 3.0 \pm 2.1) \cdot 10^{-4} \quad (14.9)$$

representing a 3.9 σ signal for $[y'_D, (x'_D)^2] \neq [0, 0]$ due to the correlations between y'_D and $(x'_D)^2$.

This reaction is a prime candidate for revealing **CP** violation due to New Physics, and later we will discuss it in more detail.

$\mathbf{D \to K_S\pi^+\pi^-}$

The final state in $D^0(t) \to K_S\pi^+\pi^-$ contains three resonant components: the **CP** eigenstate $K_S\rho^0$ common to D^0 and $\overline{D}^0$ decays and $K^{*-}\pi^+$ together with $K^{*+}\pi^-$. The second component is predominantly a Cabibbo favoured mode, whereas the third one can have a dual origin, namely coming from a DCSD or from an oscillation plus subsequent decay. Oscillations will lead to a variation in the relative weight of these three components as a function of decay time. A *time-dependent* Dalitz plot study will probe for such a variation. Having both the Cabibbo favoured and doubly suppressed transition in the final state allows to determine also the strong phase between their amplitudes. Belle finds [206]

$$x_D \equiv \frac{\Delta M_D}{\Gamma_D} = (0.80 \pm 0.29 \pm 0.17) \cdot 10^{-2}, y_D = (0.33 \pm 0.24 \pm 0.15) \cdot 10^{-2},$$

$$\qquad (14.10)$$

which amounts to a 2.4 σ signal for $x_D \neq 0$. Various cross checks can be applied to ensure the reliability of the specific Dalitz plot model used. The latter can be validated in $e^+e^- \to \psi(3770) \to D^0\overline{D}^0$ analysed by Cleo-c and BESIII.

$\mathbf{D^0 \to l^-X}$

Searching for 'wrong-sign' leptons in semileptonic decays of neutral D mesons represents the theoretically (yet not experimentally) cleanest probe of oscillations irrespective of **CP** invariance due to the SM selection rule in $c \to l^+ \nu q$, see Eq. (6.87) [112]:

$$r_D \equiv \frac{\Gamma(D^0 \to l^- X)}{\Gamma(D^0 \to l^+ X)} \simeq \frac{x_D^2 + y_D^2}{2} = (1.7 \pm 3.9) \cdot 10^{-4}. \qquad (14.11)$$

14.2.2 First résumé

Establishing $D^0 - \overline{D}^0$ oscillations would provide a novel insight into flavour dynamics. After having discovered oscillations in *all three* mesons built from *down*-type quarks – K^0, B_d and B_s – it would be the first observation of oscillations with *up*-type quarks; it would also remain the only one (at least for three-family scenarios), as explained above in Section 14.1. While the experimental evidence is not yet conclusive, it is most intriguing. An average over all data points to [112]:

$$x_D = 0.0100 \pm 0.0026 \,, \quad y_D = 0.0076 \pm 0.0018. \qquad (14.12)$$

14.2.3 Theoretical expectations on ΔM_D & $\Delta \Gamma_D$

Within the SM two structural reasons combine to make x_D and y_D small in contrast to the situation for $B^0 - \overline{B}^0$ and $K^0 - \overline{K}^0$ oscillations:

- The amplitude for $D^0 \leftrightarrow \overline{D}^0$ transitions is twice Cabibbo suppressed and therefore x_D, $y_D \propto \sin^2 \theta_C$. The amplitudes for $K^0 \leftrightarrow \overline{K}^0$ and $B^0 \leftrightarrow \overline{B}^0$ are also twice Cabibbo and KM suppressed – yet so are their decay widths.
- Due to the GIM mechanism one has $\Delta M = 0 = \Delta \Gamma$ in the limit of flavour symmetry. Yet the $K^0 \leftrightarrow \bar{K}^0$ amplitude is driven by $SU(4)_F$ breaking characterized by $m_c^2 \neq m_u^2$, which represents no suppression on the usual hadronic scales $\mu_{\text{had}} \leq 1 \, \text{GeV}$. In contrast $D^0 \leftrightarrow \overline{D}^0$ is controlled by $SU(3)_F$ breaking typified by $m_s^2 \neq m_d^2$ (or in terms of hadrons $M_K^2 \neq M_\pi^2$), which on the scale M_D^2 provides an effective suppression.

Thus we conclude for the overall oscillation strength:

$$\left. \frac{\Delta M_D}{\overline{\Gamma}_D} \right|_{\text{SM}}, \; \left. \frac{\Delta \Gamma_D}{\overline{\Gamma}_D} \right|_{\text{SM}} \sim SU(3)_F \; \text{breaking} \times \sin^2 \theta_C. \qquad (14.13)$$

If $SU(3)_F$ *breaking* is controlled by hadronic rather than quark masses – say, $M_K^2 / \mu_{\text{had}}^2$ instead of m_s^2 / m_c^2 – one cannot count on it to provide a large suppression; i.e.

$$\frac{\Delta M_D}{\overline{\Gamma}_D}\bigg|_{\text{SM}}, \ \frac{\Delta \Gamma_D}{\overline{\Gamma}_D}\bigg|_{\text{SM}} < \text{few} \times 0.01 \tag{14.14}$$

provides a conservative bound. To go beyond such a guestimate, we have to develop a (semi-)quantitative description of $SU(3)_F$ *breaking* as a central challenge. No conclusive answer has been found yet, as sketched in Section 14.2.5. Since a number of rather technical arguments are involved, that section is marked by ♠. Here we just summarize the findings. Concerning the predictions one has to distinguish carefully between two similar sounding questions.

- 'What are the *most likely* values for x_D and y_D *within the SM*?' The answer as explained below:

$$x_D, \ y_D \sim \mathcal{O}(10^{-3}). \tag{14.15}$$

- 'How large could x_D and y_D *conceivably* be within the SM?' The answer: One *cannot rule out* 10^{-2}.

In interpreting the experimental numbers given above, we have to 'sit on the fence': while the observed signals might be generated from SM dynamics alone, they could also be due to large or even dominant New Physics contributions. Yet according to the principle *'in dubio pro reo'* we cannot convict the SM of being incomplete.

Despite the ambiguity in the predictions one should make the utmost efforts to probe $D^0 - \overline{D}^0$ oscillations down to the $x_D, y_D \sim 10^{-3}$ level. After all, we might be only one theory breakthrough away from making a precise prediction. Yet even without that, we should note that D^0 oscillations represent an intriguing quantum mechanical phenomenon. Most importantly, they constitute an important ingredient for **CP** asymmetries arising in D^0 decays due to New Physics, as explained later.

14.2.4   *New Physics contributions to* ΔM_{D} *and* $\Delta \Gamma_{\text{D}}$?

While the expected numbers for x_D and y_D are similar, the dynamics driving these two $\Delta C = 2$ observables are not. Before going on, we remind the reader that $M_{12} - \frac{i}{2}\Gamma_{12}$ is an analytic function of M_D^2. It satisfies a dispersion relation in the complex M_D^2 plane.

- ΔM_D being generated by contributions from virtual states is sensitive to New Physics which could raise it to the percent level.
- $\Delta \Gamma_D$ being driven by on-shell transitions can hardly be sensitive to New Physics (for an dissenting view see Ref. [207]). At the same time,

however, it is very vulnerable to violations of local duality: a nearby narrow resonance could easily wreck any GIM cancellation and raise the value of $\Delta\Gamma_D$ by an order of magnitude.

If one had observed $x_D > 0.01 \gg y_D$, one would have a good prima facie case for New Physics. Yet such a scenario is hardly allowed by the new data. While on the other hand a scenario $\Delta M_D \ll \Delta\Gamma_D$ is conceivable, it is not a very natural one. For if one has an *on*-shell state f with $D^0 \to f \to \overline{D}^0$ that contributes to $\Delta\Gamma_D$, then a dispersion relation expresses its *off*-shell contribution to ΔM_D.

14.2.5 ♠ *Numerical predictions for* ΔM_D *and* $\Delta\Gamma_D$ ♠

The history of the predictions concerning $D^0 - \overline{D}^0$ oscillations does not represent one of the glory pages of theoretical high energy physics. On the other hand we are not clueless either. Before presenting quantitative SM predictions we will illustrate the general considerations given above by considering transitions to two pseudoscalar mesons, which are common to D^0 and $\bar{D}^0$ decays and can thus communicate between them:

$$D^0 \overset{CS}{\Rightarrow} K^+K^-, \pi^+\pi^- \overset{CS}{\Rightarrow} \overline{D}^0, \tag{14.16}$$

$$D^0 \overset{CA}{\Rightarrow} K^-\pi^+ \overset{DCS}{\Rightarrow} \overline{D}^0 \text{ or } D^0 \overset{DCS}{\Rightarrow} K^+\pi^- \overset{CA}{\Rightarrow} \overline{D}^0, \tag{14.17}$$

where CA, CS and DCS denote the channels as Cabibbo allowed, singly and doubly suppressed, respectively. We have used the symbol '$\Rightarrow$' to indicate that these transitions can be real on-shell ones – for $\Delta\Gamma_D$ – as well as virtual off-shell ones – for Δm_D. Since

$$A(D^0 \Rightarrow K^-\pi^+/K^+\pi^- \Rightarrow \overline{D}^0) \propto -\sin^2\theta_C\cos^2\theta_C$$
$$A(D^0 \Rightarrow K^-K^+/\pi^+\pi^- \Rightarrow \overline{D}^0) \propto \sin^2\theta_C\cos^2\theta_C \tag{14.18}$$

one obviously has in the $SU(3)_F$ limit $\Delta\Gamma(D^0 \to K\bar{K}, \pi\pi, K\pi, \pi\bar{K}) = 0$: the amplitudes for Eq. (14.17) would then be equal in size and opposite in sign to those of Eq. (14.16). Yet the measured branching ratios [11]

$$\text{BR}(D^0 \to K^+K^-) = (3.84 \pm 0.1) \cdot 10^{-3}$$
$$\text{BR}(D^0 \to \pi^+\pi^-) = (1.364 \pm 0.032) \cdot 10^{-3}$$
$$\text{BR}(D^0 \to K^-\pi^+) = (3.8 \pm 0.07) \cdot 10^{-2}$$
$$\text{BR}(D^0 \to K^+\pi^-) = (1.43 \pm 0.04) \cdot 10^{-4}$$

show very considerable $SU(3)$ breakings:

$$\frac{\text{BR}(D^0 \to K^+K^-)}{\text{BR}(D^0 \to \pi^+\pi^-)} \simeq 2.82 \pm 0.10, \quad \frac{\text{BR}(D^0 \to K^+\pi^-)}{\text{BR}(D^0 \to K^-\pi^+)} \simeq (1.6 \pm 0.05) \cdot \tan^4\theta_C$$

(14.19)

compared to ratios of unity and $\tan^4\theta_C$, respectively, in the symmetry limit. This suggests that the need for $SU(3)_F$ breaking might provide not that much suppression for $\Delta\Gamma_D$ and possibly also ΔM_D.

Yet when one compares more inclusive rates, one finds considerably less deviations from $SU(3)_F$ invariance; e.g.

$$\frac{\text{BR}(D^0 \to K^+K^-) + \text{BR}(D^0 \to K^+K^-\pi^+\pi^-)}{\text{BR}(D^0 \to \pi^+\pi^-) + \text{BR}(D^0 \to 2\pi^+2\pi^-)} \simeq 0.71.$$

(14.20)

This type of SM analysis in terms of hadronic channels can and has been extended by the authors of Refs. [208, 209]: (i) they estimate the amount of $SU(3)_F$ breaking for $\Delta\Gamma_D$ from phase space differences alone for classes of two-, three- and four-body D modes and conclude that y_D could be as 'large' as 0.01; and (ii) they infer x_D from y_D via a dispersion relation they arrive at $0.001 \leq |x_D| \leq 0.01$ with x_D and y_D being of opposite sign.

It is natural to ask whether one could not give also a quark-level description of $D - \overline{D}$ oscillations. Describing $K - \overline{K}$ and $B - \overline{B}$ oscillations through quark box diagrams is a reasonable ansatz a priori and a successful one a posteriori, where one finds

$$M_{12}^K \propto (\mathbf{V}_{cd}^*\mathbf{V}_{cs})^2 \frac{m_c^2}{M_W^2}; \quad M_{12}^{B_q} \propto (\mathbf{V}_{td}^*\mathbf{V}_{tq})^2 \frac{m_t^2}{M_W^2}, \quad q = d, s.$$

(14.21)

There is an analogous contribution to ΔM_D, where the internal loop contains b quarks:

$$\Delta M_D^{(b\bar{b})} \simeq -\frac{G_F^2 m_b^2}{8\pi^2} |V_{cb}^*V_{ub}|^2 \frac{\langle D^0|(\bar{u}\gamma_\mu(1-\gamma_5)c)(\bar{u}\gamma_\mu(1-\gamma_5)c)|\bar{D}^0\rangle}{2M_D};$$

(14.22)

however they are highly suppressed by the tiny CKM parameters. Using factorization to estimate the matrix element one finds: $x_D^{(b\bar{b})} \sim$ few $\times 10^{-7}$; loops with one b and one light quark likewise are suppressed.

For the light intermediate quarks $- d, s -$ a *qualitative* change occurs, since the momentum scale is set by the *external* mass m_c, not the internal quark masses as for $M_{12}^{K,B}$. Among other things this leads to a different GIM factor with a higher power, namely m_s^4/m_c^4. More specifically we find[1]

$$\Delta M_D^{(s,d)} \simeq -\frac{G_F^2 m_c^2}{8\pi^2} |V_{cs}^* V_{us}|^2 \frac{(m_s^2 - m_d^2)^2}{m_c^4} \times$$

$$\frac{\langle D^0 | (\bar{u}\gamma_\mu(1-\gamma_5)c)(\bar{u}\gamma_\mu(1-\gamma_5)c) + (\bar{u}(1+\gamma_5)c)(\bar{u}(1+\gamma_5)c) | \overline{D}^0 \rangle}{2M_D}. \quad (14.23)$$

The contribution to $\Delta\Gamma_D$ from the bare quark box is *further* reduced by an *additional* factor m_s^2. Contrary to claims in the literature the latter is *not* due to helicity suppression (the GIM factors already take care of that effect). It is of an accidental nature: it arises because the weak currents are purely $V - A$. Including radiative QCD corrections to the box diagram yields contributions $\propto m_s^4 \alpha_S/\pi$. One finds $\Delta\Gamma_D^{\text{box}} < \Delta M_D^{\text{box}} \sim$ few $\times$ 10^{-17} GeV $\hat{=}$ $x_D^{\text{box}} \sim$ few $\times 10^{-5}$.

Let us examine how the required $SU(3)_F$ breaking arises. The charm changing interaction is given by

$$\mathcal{H}_{\text{weak}}^{\Delta C} = 2\sqrt{2}G_F(\bar{c}_L\gamma_\mu s_L' \bar{d}_L' \gamma^\mu u_L + \text{h.c.}) \quad (14.24)$$

where, in terms of mass eigenstates, $d' = d\cos\theta_C + s\sin\theta_C$ and $s' = -d\sin\theta_C + s\cos\theta_C$. By performing a Fierz transformation and writing out s' and d' we arrive at

$$\mathcal{H}_{\text{weak}}^{\Delta C} = 2\sqrt{2}G_F(\bar{c}_L\gamma_\mu u_L)(\bar{d}_L'\gamma^\mu s_L'). \quad (14.25)$$

Since the strong interactions respect $SU(3)_F$ symmetry, they do not distinguish between mass eigenstates s' and d'. In the SM the latter is broken by the mass term in the Lagrangian:

$$L_{\text{Yukawa}} = -\overline{U}_L \mathcal{M}_U^{\text{diag}} - \overline{D}_L \mathcal{M}_D^{\text{diag}} U_{j,R}. \quad (14.26)$$

It thus follows

$$\langle D^0 | A(\mathcal{H}_{\text{weak}}^{\Delta C}(x)\mathcal{H}_{\text{weak}}^{\Delta C\dagger}) | \overline{D}^0 \rangle \propto (m_s - m_d)^2; \quad (14.27)$$

i.e. the $D^0 - \overline{D}^0$ amplitude is of second order in $m_s - m_d \simeq m_s$.

Now we are in a position to give a description in terms of an operator product expansion (OPE). An OPE treatment always starts from short-distance dynamics and refines it step-by-step with non-perturbative

---

[1] This contribution is saturated at the momentum scale $\sim m_c$, and thus refers to the Wilson coefficient of a dimension-six operator. Therefore they are not long-distance contributions despite being proportional to m_s^2.

corrections. The *formally* leading term in the OPE for $\Delta C = -2$ transitions comes from dimension-six four-fermion operators of the generic form $(\bar{u}c)(\bar{u}c)$; they correspond to the usual quark box diagrams. As discussed above their Wilson coefficients are tiny due to their CKM parameters and/or their GIM factors $\sim \mathcal{O}(m_s^4)$; i.e. they are unusually suppressed. The latter feature already suggests that higher order terms, which are of intrinsically non-perturbative origin, provide larger numerical contributions, since there must be terms $\sim \mathcal{O}(m_s^2)$.

This conjecture is indeed borne out by further analysis. It turns out that the $SU(3)_F$ GIM suppression per fermion line can be $\mathcal{O}(m_s/\mu_{\text{hadr}})$ if the fermion line is soft [210]; this is in full agreement with the general argument on $SU(3)_F$ breaking given above. Those contributions are shaped by quark condensates $\langle 0|\bar{q}q|0 \rangle$ introducing the hadronic scale $\mu_{\text{hadr}} \sim 1\,\text{GeV}$. The GIM factor per se amounts to m_s/μ_{hadr} yielding altogether a factor of order $(m_s/\mu_{\text{hadr}})^2 (\mu_{\text{hadr}}/m_c)^6 = (m_s^2 \mu_{\text{hadr}}^4)/m_c^6$. We should note that μ_{hadr}/m_c is not much smaller than unity. It is the presence of three mass parameters of comparable size – m_c, m_s and μ_{hadr} – that opens the door for contributions of formally higher order to become the numerically leading ones without invalidating the OPE approach. Analysing the contributions coming from higher-dimensional operators with the help of condensates one estimates [210]:

$$x_D \,,\, y_D \,\sim\, \mathcal{O}(10^{-3}), \qquad\qquad (14.28)$$

with the realization that these estimates involve high powers of the ratio of comparable scales implying considerable numerical uncertainties that are very hard to overcome.

14.3 CP violation

14.3.1 Preliminaries

Many factors favour or even mandate dedicated searches for **CP** violation in charm transitions:

$\oplus$ Baryogenesis – i.e. understanding the observed great imbalance between matter and antimatter in our universe as a dynamically generated phenomenon rather than as an arbitrary initial value – cannot be driven by CKM forces. It requires the existence of New Physics in **CP** violating dynamics.

$\oplus$ It would be unwise not to undertake dedicated searches for **CP** asymmetries in charm decays, where the 'background' from known physics is small or even absent. The Wolfenstein representation of the CKM matrix is again most revealing. Among the four CKM matrix elements controlling *tree-level* charm transitions – $\mathbf{V}_{cs}$, $\mathbf{V}_{cd}$, $\mathbf{V}_{ud}$ and

$\mathbf{V}_{us}$ – only the first one has an imaginary part through order λ^4, and it is a tiny one:

$$\text{Im}\,\mathbf{V}_{cs} = -\eta A^2 \lambda^4 \simeq -6 \cdot 10^{-4}. \tag{14.29}$$

As discussed in detail below this phase can be observed only in once Cabibbo-suppressed modes (at least in the absence of oscillations). Any direct **CP** violation in *Cabibbo allowed or doubly suppressed* channels requires the intervention of New Physics – except for final states containing a K_S (or K_L).

Reactions that involve internal quark loops – in particular $D - \overline{D}$ oscillations – are in principle sensitive also to the large phase of $\mathbf{V}_{ub}$, yet the latter's impact is basically negated, since it is accompanied by $\mathbf{V}_{cb}$ and this combination carries a tiny modulus:

$$|\mathbf{V}_{ub}\mathbf{V}_{cb}| \sim \mathcal{O}(\lambda^5). \tag{14.30}$$

While no precise KM predictions on **CP** asymmetries in the decays of charm hadrons can be given, they are expected to reach no more than the 10^{-3} level and to fall well short of that in many cases.

$\oplus$ One should keep in mind that in going from Cabibbo allowed to Cabibbo singly and doubly suppressed channels, the SM rate is *suppressed* by factors of about 20 and 400, respectively:

$$\Gamma_{SM}(H_c \to [S = -1]) : \Gamma_{SM}(H_c \to [S = 0]) : \Gamma_{SM}(H_c \to [S = +1]) \simeq$$

$$1 : 1/20 : 1/400, \tag{14.31}$$

where H_c implies any hadron carring the charm quantum number. New Physics will not necessarily follow the SM's Cabibbo pattern. The prospects for uncovering its manifestations are thus likely to improve the further down in the Cabibbo hierarchy one goes.

$\oplus$ Charm states are being produced in large numbers at e^+e^- machines – even 'parasitically' at B factories – as well as in hadroproduction. Their branching ratios into pions and kaons are large.

$\oplus$ Flavour tagging can be achieved very efficiently through $D^{*+} \to D^0 \pi^+$ versus $D^{*-} \to \overline{D}^0 \pi^-$ decays, since the slow charged pion provides a clear experimental signature.

$\oplus$ Strong phase shifts required for *direct* **CP** violation to emerge in partial widths are in general large as are the branching ratios into relevant modes; while strong final state interactions complicate the interpretation of an observed signal in terms of the microscopic parameters of the underlying dynamics, it enhances the observability of a signal.

$\oplus$ **CP** asymmetries can be linear in New Physics amplitudes thus greatly increasing sensitivity to the latter.

⊕ Decays to final states of *more than* two pseudoscalar or one pseudoscalar and one vector meson contain more dynamical information than given by their widths; their distributions, as described by Dalitz plots or **T** odd moments etc., can exhibit **CP** asymmetries that can be considerably larger than those for the width. Final state interactions while not necessary for the emergence of such effects, can fake a signal; yet that can be disentangled by comparing **T** odd moments for **CP** conjugate modes:

$$O_{\mathbf{T}}(D \to f) \neq -O_{\mathbf{T}}(\overline{D} \to \overline{f}) \implies \mathbf{CP} \text{ violation.} \qquad (14.32)$$

We view this as a very promising avenue, where we still have to develop the most effective analysis tools for small asymmetries. This will be described in Sections 14.3.2[B] and 14.3.3[B].

⊖ The 'fly in the ointment' is that $D - \overline{D}$ oscillations are on the slow side.

⊕ Nevertheless one should take on this challenge. For **CP** violation involving $D - \overline{D}$ oscillations is a reliable probe of New Physics, as explained below.

⊕ It is often overlooked that **CPT** invariance can provide non-trivial constraints on **CP** asymmetries. For it imposes equality not only on the masses and total widths of particles and antiparticles, but also on the widths for 'disjoint' subsets of channels. 'Disjoint' subsets are the decays to final states that can*not* rescatter into each other. Examples are semileptonic versus non-leptonic modes with the latter subdivided further into those with strangeness $S = -1, 0. + 1$. Observing a **CP** asymmetry in one channel one can then infer in which other channels the 'compensating' asymmetries have to arise. This has been discussed in general in Section 4.10.2.

It is true that available data show no evidence for any kind of **CP** violation in charm decays. Yet as explained below, it is only now that data have reached a sensitivity level, where one can 'realistically hope' for an effect to emerge.

In the remainder of this section we will describe the relevant phenomenology specific for charm decays, analyse the corresponding KM predictions and how New Physics can intervene. The general **CP** phenomenology discussed above for beauty decays applies here as well; yet the specifics make a big difference.

14.3.2 **CP** *asymmetries without* $\mathrm{D}^0 - \overline{\mathrm{D}}^0$ *oscillations*

In the absence of D^0 oscillations one can have only *direct* **CP** violation with the subtle exception of $D^{\pm} \to K_S \pi^{\pm}$, explained later. The following discussion applies in any case to $D^{\pm}$ decays.

Partial widths

For a **CP** asymmetry to arise in fully integrated partial widths – the only option one has for $D \rightarrow PP/PV$, where P and V denote a pseudoscalar and vector meson, respectively – two conditions have to be satisfied simultaneously, as explained in general in Section 4.10.3. Two different amplitudes have to contribute coherently with *non*-trivial weak as well as strong relative phases, see Eq. (4.140):

$$\frac{\Gamma(\overline{D} \rightarrow \overline{f}) - \Gamma(D \rightarrow f)}{\Gamma(\overline{D} \rightarrow \overline{f}) + \Gamma(D \rightarrow f)} = \frac{2\sin(\Delta\phi_W)\sin(\Delta\delta^{\text{FSI}})|\mathcal{A}_2/\mathcal{A}_1|}{1 + |\mathcal{A}_2/\mathcal{A}_1|^2 + 2|\mathcal{A}_2/\mathcal{A}_1|\cos(\Delta\phi_W)\cos(\Delta\delta^{\text{FSI}})}$$

$$(14.33)$$

where $\Delta\phi_W = \phi_1^{\text{weak}} - \phi_2^{\text{weak}}$ and $\Delta\delta^{\text{FSI}} = \delta_1^{\text{FSI}} - \delta_2^{\text{FSI}}$ denote the weak and the strong (or electromagnetic) phase shifts, respectively, and $|\mathcal{A}_{1,2}|$ the moduli of the two amplitudes. Since charm transitions proceed in an environment of virulent strong FSI, the second requirement does not pose a problem. The FSI become a theoretical headache only when one attempts to infer the size of the microscopic parameters of the underlying dynamics from observation, since strong final state interactions are not under theoretical control. On the other hand we are not completely ignorant about such strong phases. Often one can infer their size – or at least non-trivial constraints on them – from the measured branching ratios of sets of channels with the help of isospin relations. Furthermore **CPT** symmetry implies relations between **CP** asymmetries in different channels as explained in Section 4.10.2, which are controlled by re-scattering amplitudes: A **CP** asymmetry in one channel $D \rightarrow f$ has to be compensated for by a **CP** asymmetry in one or a combination of channels with final states that can rescatter with f. The strength of rescattering thus puts a bound on **CP** asymmetries. This connection is presumably of mainly academic interest in beauty hadrons due to the multitude of their decay channels. Yet for charm decays it might become of practical value.

The remarks above hold in general irrespective of the underlying forces. Applying them to CKM dynamics we can conclude the following.

- The SM provides only for *Cabibbo-suppressed channels* two different tree level amplitudes (except for final states containing a neutral kaon, see below) as the first 'conditio sine qua non': $c \rightarrow s\bar{s}u$ & $d\bar{d}u$. Of those only the first contains a weak phase *through* order λ^5 in the Wolfenstein representation, namely via $\text{Im}\mathbf{V}_{cs}\mathbf{V}_{us}^* = \lambda\text{Im}\mathbf{V}_{cs} = -\eta A^2\lambda^5$. One-loop diagrams for $c \rightarrow u + gluons$, which *look* like penguin diagrams are sensitive also to $\text{Im}\mathbf{V}_{cb}\mathbf{V}_{ub}^* \simeq A\lambda^2\text{Im}\mathbf{V}_{ub}^* = -\eta A^2\lambda^5$.

The *relative weak* phase in both cases is $-\eta A^2\lambda^4 = -\eta|\mathbf{V}_{cb}|^2 \sim 6\cdot10^{-4}$ modulo the ratio of the transition operator matrix elements. Channels like $D^0 \to K^+K^-$, $\pi^+\pi^-$, $K^{*+}\pi^-$, $\rho^+\pi^-$ or $D^+ \to \eta\pi^+$, $\rho^0\pi^+$ are prime examples.[2]

The required features have been stated above in Eq. (14.33). With charm decays proceeding in an environment shaped by many prominent hadronic resonances, there is no reason why the need for strong phase shifts should significantly suppress observable asymmetries. Therefore the typical benchmark for direct **CP** asymmetries in partial widths is about 10^{-3} or less. At present we would not rule out that in some cases they might be as high as 1% due to some fortuitous matrix element enhancements.[3]

It will be quite a while, before lattice QCD could provide useful results derived from first principles. In the meantime we see more practical value in an exercise of 'theoretical engineering', where one fits a host of charm meson decay rates to two-body or quasi-two-body nonleptonic final states in terms of a limited set of basic matrix elements allowing their moduli and strong phases to float, yet keeping their CKM parameters fixed. Some theoretical discretion has to be applied to reduce the number of the unknown matrix elements to a managable set and on how much absorption to allow for. Such an exercise has been undertaken a decade ago in Ref. [211], which found that some Cabibbo-suppressed modes indeed could exhibit direct **CP** asymmetries of up to 0.2%. With CLEO-c and BESIII about to produce more detailed and accurate measurements of branching ratios, it will be highly worthwhile to update this analysis.

Prominent Cabibbo suppressed channels are $D^0 \to K^+K^-$, $\pi^+\pi^-$, $K^{*0}\overline{K}^{*0}$, $\rho^0\pi^0$, $D^+ \to K_S K^+, \eta\pi^+, \overline{K}^{0*}K^+$, $D_s^+ \to K_S\pi^+, \overline{K}^{0*}\pi^+$. Up to small isospin violating effects no **CP** asymmetry can arise in $D^+ \to \pi^+\pi^0$ for the same reason as for $K^+ \to \pi^+\pi^0$.

On the CA and DCS level there is only one SM weak transition amplitude each thus removing the possibility of a **CP** asymmetry there. There are only two related exceptions to this general statement.

- In $D^+ \to K_S\pi^+$ (and similar modes) usually classified as CA one has also a *coherent* DCS contribution: $D^+ \to \overline{K}^0\pi^+ + K^0\pi^+ \Rightarrow K_S\pi^+$. If there were a New Physics contribution to the DCS amplitude $D^+ \to K^0\pi^+$ – quite conceivable in view of the latter's suppression by $\tan^2\theta_C \simeq 1/20$ – it could be expected to introduce a new *weak* phase.

---

[2] Even $D^0 \to K_S K_S$ is a legitimate, though experimentally unappealing candidate, while $D^+ \to \pi^0\pi^+$ described by a single isospin amplitude is not even that.

[3] Evaluating the aforementioned one-loop diagrams *literally* would yield absorptive parts and thus strong phase shifts, yet we would not trust such a computation.

With the *strong* phases between the 'exotic' $\overline{K}^0\pi^+ - \Delta I = 3/2 -$ and $K^0\pi^+ - \Delta I = 1/2 \,\& \, 3/2 -$ different we have a natural scenario for New Physics generating a direct **CP** asymmetry in $D^\pm \to K_S\pi^\pm$ [212]. Specific models suggest [213, 214] values as large as 1%.

- There is an additional twist here. While the SM unequivocally predicts $|A(D^+ \to \overline{K}^0\pi^+)|^2 = |A(D^- \to K^0\pi^-)|^2$, *known* dynamics, namely the difference observed in $K_L \to l^+\nu_l\pi^-$ versus $K_L \to l^-\overline{\nu}_l\pi^+$ as parametrized through $\mathrm{Re}\epsilon_K$ induces:

$$\left.\frac{\Gamma(D^+ \to K_S\pi^+) - \Gamma(D^- \to K_S\pi^-)}{\Gamma(D^+ \to K_S\pi^+) + \Gamma(D^- \to K_S\pi^-)}\right|_{SM} = 2\mathrm{Re}\epsilon_K = (3.32 \pm 0.06) \cdot 10^{-3}.$$

(14.34)

We want to stress again that this asymmetry, which is reliably predicted within the SM, does not reflect **CP** violation in the $\Delta C = 1$ transition, but the **CP** impurity in the K_S wave function.

Taking these two observations together means that any deviation from Eq. (14.34) would signal the intervention from New Physics – even the absence of an asymmetry.

Final state distributions

For channels with two pseudoscalar mesons or a pseudoscalar and a vector meson a **CP** asymmetry can manifest itself only in a difference between conjugate partial widths. If, however, the final state is more complex – being made up by three pseudoscalar or two vector mesons etc. – then it contains more dynamic information than expressed by its partial width, and **CP** violation can emerge also through asymmetries in final state distributions. One general comment still applies: since also such **CP** asymmetries require the interference of two *different* weak amplitudes, within the SM they can occur in Cabibbo-suppressed modes only.

(i): In the simplest such scenario one compares **CP** conjugate *Dalitz plots*. It is quite possible that different regions of a Dalitz plot exhibit **CP** asymmetries of varying signs that largely cancel each other when one integrates over the whole phase space. Subdomains of the Dalitz plot could contain considerably larger **CP** asymmetries than the integrated partial width. Once a Dalitz plot is fully understood with all its contributions, one has a powerful new probe. This is not an easy goal to achieve, though, in particular when looking for effects that presumably are not large. It might be more promising as a practical matter to start out with a more heuristic approach: one can start a search for **CP** asymmetries by just looking at conjugate Dalitz plots. One simple strategy would be to focus on an area with a resonance band and analyse the density in stripes *across* the resonance as to whether there is a difference in **CP** conjugate plots.

(ii): For more complex final states – e.g. those containing four pseudoscalar mesons – other probes have to be employed. Of all the established **CP** asymmetries only one involves final state distributions rather than partial widths, namely the forward-backward asymmetry in $K_L \to \pi^+\pi^-e^+e^-$, discussed in Section 12.1.4. There a tiny effect, namely the **CP** impurity in the K_L wave function – $|\epsilon_K| \simeq 2.3 \cdot 10^{-3}$ – induces a large forward-backward asymmetry in the orientation of the two decay planes of close to 14%; i.e. almost a factor of hundred larger! This at first sight miraculous enhancement is due to the fact that the asymmetry is generated by the interference of two amplitudes the size of which is suppressed to similar levels, albeit for very different reasons: while one is a **CP** violating $E1$ amplitude, the other is a **CP** conserving $M1$ amplitude. Of course one has to pay a price for this enhancement of the asymmetry – the branching ratio for $K_L \to \pi^+\pi^-e^+e^-$ is highly reduced down to about 3×10^{-7}: One has 'traded' branching ratio for size of the asymmetry.

A very close analogy might occur in D^0 transitions, in particular in $D_L \to K^+K^-\mu^+\mu^-$. Since it involves $D^0 - \overline{D}^0$ oscillations, we will discuss it in the next section. A more remote analogy holds for the non-leptonic mode

$$\overset{(-)}{D} \to K^+K^-\pi^+\pi^-. \tag{14.35}$$

Defining by ϕ the angle between the K^+K^- and $\pi^+\pi^-$ planes one has

$$\frac{d\Gamma}{d\phi}(D \to K^+K^-\pi^+\pi^-) = \Gamma_1 \cos^2\phi + \Gamma_2 \sin^2\phi + \Gamma_3 \cos\phi \sin\phi$$

$$\frac{d\Gamma}{d\phi}(\overline{D} \to K^+K^-\pi^+\pi^-) = \overline{\Gamma}_1 \cos^2\phi + \overline{\Gamma}_2 \sin^2\phi - \overline{\Gamma}_3 \cos\phi \sin\phi. \tag{14.36}$$

The partial width for $D[\overline{D}] \to K^+K^-\pi^+\pi^-$ is given by $\Gamma_{1,2}[\overline{\Gamma}_{1,2}]$; $\Gamma_1 \neq \overline{\Gamma}_1$ or $\Gamma_2 \neq \overline{\Gamma}_2$ represents direct **CP** violation in the *partial width*. Since we have a singly Cabibbo-suppressed mode, one expects an asymmetry of about 10^{-3} within CKM dynamics; Γ_3 and $\overline{\Gamma}_3$ constitute **T** odd correlations. By themselves they do not necessarily indicate **CP** violation, since they can be induced by strong final state interactions. However

$$\Gamma_3 \neq \overline{\Gamma}_3 \implies \mathbf{CP} \text{ violation!} \tag{14.37}$$

It is quite possible or even likely that a difference in Γ_3 versus $\overline{\Gamma}_3$ is significantly larger than in Γ_1 versus $\overline{\Gamma}_1$ or Γ_2 versus $\overline{\Gamma}_2$. Furthermore one can expect that differences in detection efficiencies can be handled by comparing Γ_3 with $\Gamma_{1,2}$ and $\overline{\Gamma}_3$ with $\overline{\Gamma}_{1,2}$. A pioneering search for such an effect has been undertaken by FOCUS [215], where an upper bound of a few percent has been placed on a **T** odd moment.

The integrated forward-backward asymmetry

$$\langle A \rangle = \frac{2\Gamma_3}{\pi(\Gamma_1 + \Gamma_2)} \tag{14.38}$$

in principle yields the full information on **CP** and **T** violation. Tracking the distribution in ϕ given in Eq. (14.36) allows us to disentangle $\Gamma_1[\overline{\Gamma}_1]$ and $\Gamma_2[\overline{\Gamma}_2]$ and – presumably more importantly – provides experimental cross checks.

(iii): Decays of *polarized* charm baryons provide us with a similar class of observables; e.g. in $\Lambda_c \rightarrow p\pi^+\pi^-$, one can analyse the T-odd correlation $\langle \vec{\sigma}_{\Lambda_c} \cdot (\vec{p}_{\pi^+} \times \vec{p}_{\pi^-}) \rangle$; here $\vec{\sigma}_{\Lambda_c}$ and $\vec{p}_{\pi^\pm}$ denote the polarization of Λ_c and the momenta of the pions, respectively. Probing $\Lambda_c^+ \rightarrow \Lambda l^+ \nu$ for $\langle \vec{\sigma}_{\Lambda_c} \cdot (\vec{p}_\Lambda \times \vec{p}_l) \rangle$ or $\langle \vec{\sigma}_\Lambda \cdot (\vec{p}_\Lambda \times \vec{p}_l) \rangle$ is a particularly intriguing case; for in this reaction there are not even electromagnetic FSI that could fake T violation. The presence of a net polarization transverse to the decay plane depends on the weak phases and Lorentz structures of the contributing transition operators. Like in the well-known case of the muon transverse polarization in $K^+ \rightarrow \mu^+\pi^0\nu$ decays the T-odd correlation is controlled by $\text{Im}(T_-/T_+)$, where T_- and T_+ denote the helicity violating and conserving amplitudes, respectively. Since the former are basically absent in the SM, a transverse polarization requires the intervention of New Physics to provide the required helicity violating amplitude. Unfortunately we do not know of an attractive model that could lead to observable effects.

14.3.3 Oscillations – the new portal to **CP** *violation*

For very personal reasons we sense a historical analogy here: In the early and mid-1980s we had been talking about searching for **CP** violation in B decays without getting much resonance – till B_d oscillations were observed! The analogy is of course not perfect: x_D is much smaller than $x(B_d)$ (about two orders of magnitude), and (re)confirming the SM's predictions for **CP** asymmetries in charm decays is not an inspiring goal. The central agenda now is to find manifestations of New Physics. Yet there are two circumstances on the positive side: as discussed below, the SM 'background' in the **CP** phenomenology is very tiny in charm transitions, unlike the case for B decays, and we can count on our experimental colleagues having become much more experienced and sophisticated in their analyses.

Oscillations per se are described by the three dimensionless quantities $x_D = \Delta M_D / \overline{\Gamma}_D$, $y_D = \Delta\Gamma_D / 2\overline{\Gamma}_D$ and q/p. Only the last one is directly sensitive to **CP** violation. Yet with oscillations on an observable level – and it seems x_D, $y_D \sim 0.005$–0.01 satisfy this requirement – the possibilities for observing **CP** asymmetries proliferate. For oscillations set the stage for an interference between a direct decay and an oscillation followed by a decay if the final state f is common to D^0 and $\overline{D}^0$ mesons. Therefore the description of neutral D decays involves also $\frac{q}{p}\overline{\rho}_f$ with the latter providing another gateway for **CP** violation to emerge.

In our discussion we can refer to the general case treated in Chapter 6 with the master equations Eq. (6.54) and Eq. (6.57). The oscillation effects here are less spectacular and more involved than for B^0 transitions, since unlike there we have $x_D \sim y_D \ll 1$. That y_D cannot be ignored is demonstrated by the experimental evidence listed above. The algebra involved is straightforward though tedious. Problem 14.1 provides some guidance to the reader.

Non-leptonic modes

If $\mathcal{H}(\Delta C = 2)$ is generated by SM dynamics alone, it produces little **CP** violation by itself. As mentioned above in discussing ΔM_D, contributions from the third family are largely decoupled and those from the first two families carry a highly diluted phase $\eta A^2 \lambda^4 \sim 6 \times 10^{-4}$. $M_{12}^D|_{\rm SM}$ and $\Gamma_{12}^D|_{\rm SM}$ thus will carry a phase not exceeding 10^{-3}. The more significant role of oscillations might then be to provide a wider stage for a **CP** asymmetry to surface where it could not occur before: it provides the second coherent amplitude, where otherwise a transition was driven by a single amplitude. One example is Cabibbo-favoured channels like $D^0 \to K_S \rho^0$ (or $D^0 \to K_S \pi^0$, $K_S \phi$): in addition to an asymmetry of $(3.32 \pm 0.06) \times 10^{-3}$ due to $|p_K| \neq |q_K|$ one obtains a time-dependent asymmetry in qualitative analogy to $B_d \to \psi K_S$ given by

$$ x_D \cdot \frac{t}{\tau_D} \cdot \mathrm{Im}\frac{q}{p}\frac{A(\overline{D}^0 \to K_S \rho^0)}{A(D^0 \to K_S \rho^0)} \simeq x_D \cdot \frac{t}{\tau_D} \cdot \mathrm{Im}\frac{\mathbf{V}^*(cs)\mathbf{V}(ud)}{\mathbf{V}(cs)\mathbf{V}^*(ud)} \simeq $$

$$ 2x_D \cdot \frac{t}{\tau_D}\eta(A\lambda^2)^2 \sim x_D \cdot \frac{t}{\tau_D} \cdot 10^{-3}. \tag{14.39} $$

With $D^0 \to K_S \phi$ representing an isoscalar mode, we have effectively a single transition operator and $\mathrm{Im}\frac{q}{p}\frac{A(\overline{D}^0 \to K_S \phi)}{A(D^0 \to K_S \phi)} \simeq 2\eta(A\lambda^2)^2$ is an accurate SM prediction with*out* much of a hadronic uncertainty. For the hadronic matrix element and its strong phase, which is the same for D^0 and $\overline{D}^0$ drop out from the ratio. Alas with $x_D \sim 0.01$ it amounts to an $\mathcal{O}(10^{-5})$ effect and is presumably too small to be observed. A more promising case is provided by doubly Cabibbo-suppressed modes like $D^0 \to K^+ \pi^-$, where

New Physics might induce a sizable weak phase in the direct amplitude. This is discussed below in general.

The more intriguing scenario arises, when New Physics contributes significantly to $\mathcal{L}(\Delta C = 2)$, which is still quite possible. We begin by drawing on analogies with two other cases, namely the retrospective one of K_L and the very topical one of B_s.

(i) Let us assume that – contrary to history – at the time of the discovery of $K - \overline{K}$ oscillations the community had already established the SM with *two* families, been aware of the possibility of **CP** violation and the need for three families to implement the latter. They would then have argued that ΔM_K could receive contributions from long-distance dynamics through *off*-shell transitions $K^0 \to "\pi^0, \eta, \eta', 2\pi" \to \overline{K}^0$ etc. Indeed roughly half the observed size of ΔM_K can be produced that way [216] with the other half due to short-distance contributions with $\bar{c}c$ intermediate states. Yet they would have realized that long-distance dynamics can*not* induce **CP** violation in $K^0 \to \overline{K}^0$, i.e. $\epsilon_K \neq 0$. The latter observable is thus controlled by short-distance dynamics. Finding a time-dependent **CP** asymmetry would then show the presence of physics beyond the SM then, namely the third family.

(ii) ΔM_{B_s} has been observed to be consistent with the SM prediction within mainly theoretical uncertainties; yet since those are still sizeable, we cannot rule out that New Physics impacts $B_s - \overline{B}_s$ oscillations significantly. This issue, which is unlikely to be resolved theoretically, can be decided experimentally by searching for a time-dependent **CP** violation in $B_s(t) \to \psi\phi$. For within the SM one predicts [119] a very small asymmetry not exceeding 4% in this transition since on the leading CKM level quarks of only the second and third family contribute; this has been discussed in detail in Section 11.5.4. Yet in general one can expect New Physics contributions to $B_s - \overline{B}_s$ oscillations to exhibit a weak phase that is not particularly suppressed. Even if New Physics affects ΔM_{B_s} only moderately, it could greatly enhance $\sin 2\phi(B_s \to \psi\phi)$, possibly even by an order of magnitude!

These examples can be seen as 'qualitative', not quantitatives analogies with $D - \overline{D}$ oscillations being (at best) quite slow. Since $y_D, x_D \ll 1$, it suffices to give the decay rate evolution to first order in those quantities [217, 218]. Since the final states in $D \to K^+K^-/\pi^+\pi^-$ are **CP** eigenstates, these transitions can be treated in analogy to $B \to \pi^+\pi^-$, albeit only a qualitative one:

$$\Gamma(D^0(t) \to h^+h^-) \propto e^{-\Gamma_1 t}|A(D^0 \to h^+h^-)|^2$$
$$\times \left[1 + y_D\frac{t}{\tau_D}\left(1 - \mathrm{Re}\,\frac{q}{p}\overline{\rho}(h^+h^-)\right) - x_D\frac{t}{\tau_D}\mathrm{Im}\,\frac{q}{p}\overline{\rho}(h^+h^-)\right] \qquad (14.40)$$

$$\Gamma(\overline{D}^0(t) \to h^+h^-) \propto e^{-\Gamma_1 t} |A(\overline{D}^0 \to h^+h^-)|^2$$

$$\times \left[1 + y_D \frac{t}{\tau_D} \left(1 - \mathrm{Re} \frac{p}{q} \rho(h^+h^-) \right) - x_D \frac{t}{\tau_D} \mathrm{Im} \frac{p}{q} \rho(h^+h^-) \right] \tag{14.41}$$

with $h = K$ or π. Some comments might elucidate Eq. (14.40) and Eq. (14.41):

- **CP** invariance implies – in addition to $|A(D^0 \to h^+h^-)| = |A(\overline{D}^0 \to h^+h^-)| - \frac{q}{p}\overline{\rho}(h^+h^-) = 1$ (and $|q| = |p|$). The transitions $D^0(t) \to h^+h^-$ and $\overline{D}^0(t) \to h^+h^-$ are then described by the same *single* lifetime. That is a consequence of the Theorem given by Eq. (6.63), since h^+h^- is a **CP** eigenstate.
- The usual three types of **CP** violation can arise, namely the direct and indirect types – $|\overline{\rho}(h^+h^-)| \neq 0$ and $|q| \neq |p|$, respectively – as well as the one involving the interference between the oscillation and direct decay amplitudes – $\mathrm{Im}\frac{q}{p}\overline{\rho}(h^+h^-) \neq 0$.
- Assuming for simplicity $|A(D^0 \to h^+h^-)| = |A(\overline{D}^0 \to h^+h^-)|$ (CKM dynamics is expected to induce an asymmetry not exceeding 0.1%) and $|q/p| = 1 - 2\epsilon_D$ one has $(q/p)\overline{\rho}(h^+h^-) = (1 - 2\epsilon_D)e^{i\phi_{h\overline{h}}}$ and thus

$$A_{\Gamma_{h\overline{h}}} = \frac{\Gamma(\overline{D}^0(t) \to h^+h^-) - \Gamma(D^0(t) \to h^+h^-)}{\Gamma(\overline{D}^0(t) \to h^+h^-) + \Gamma(D^0(t) \to h^+h^-)}$$

$$\simeq x_D \frac{t}{\tau_D} \sin\phi_{h\overline{h}} - 2y_D \frac{t}{\tau_D} \epsilon_D \cos\phi_{h\overline{h}}. \tag{14.42}$$

where $|\epsilon_D| \ll 1$ was assumed. Belle has found for $h = K$ [204]

$$A_{\Gamma_{K\overline{K}}} = (0.01 \pm 0.30 \pm 0.15)\%. \tag{14.43}$$

While there is no evidence for **CP** violation in the transition, one should also note that the asymmetry is bounded by x_D. For x_D, $y_D \leq 0.01$, as indicated by the data, A_Γ could hardly exceed the 1% range; i.e. there is no clear bound on ϕ_D or ϵ_D yet. The good news is that if x_D and/or y_D indeed fall into the 0.5–1% range, then any improvement in the experimental sensitivity for a **CP** asymmetry in $D^0(t) \to K^+K^-$ constrains New Physics scenarios – or could reveal them [219]!

Another promising channel for probing **CP** symmetry is $D^0(t) \to K^+\pi^-$ [220, 221]: being doubly Cabibbo suppressed, it should a priori exhibit a higher sensitivity to a New Physics decay amplitude. At the same time it cannot exhibit direct **CP** violation in the SM. While the final states $K^\pm\pi^\mp$

are not **CP** eigenstates, they are strictly speaking not flavour specific, since unlike semileptonic modes they can be fed by D^0 as well as $\overline{D}^0$ decays via a Cabibbo allowed or doubly suppressed transition. Thus $D^0 \to K^-\pi^+$ as well as $D^0 \to K^+\pi^-$ can both occur even in the absence of oscillations. Tracking the dependence on the time of decay allows an unambiguous separation of the two sources:

$$\frac{\Gamma(D^0(t) \to K^+\pi^-)}{|A(D^0 \to K^+\pi^-)|^2} \propto \left\{ 1 + \left(\frac{t}{\tau_D}\right)^2 \left(\frac{x_D^2 + y_D^2}{4}\right) \left|\frac{q}{p}\overline{\rho}(K^+\pi^-)\right|^2 \right.$$
$$\left. - \left(\frac{t}{\tau_D}\right) \left[y_D \mathrm{Re} \left(\frac{q}{p}\overline{\rho}(K^+\pi^-)\right) + x_D \mathrm{Im} \left(\frac{q}{p}\overline{\rho}(K^+\pi^-)\right) \right] \right\} \quad (14.44)$$

$$\frac{\Gamma(\overline{D}^0(t) \to K^-\pi^+)}{|\overline{A}(D^0 \to K^-\pi^+)|^2} \propto \left\{ 1 + \left(\frac{t}{\tau_D}\right)^2 \left(\frac{x_D^2 + y_D^2}{4}\right) \left|\frac{p}{q}\rho(K^-\pi^+)\right|^2 \right.$$
$$\left. - \left(\frac{t}{\tau_D}\right) \left[y_D \mathrm{Re} \left(\frac{p}{q}\rho(K^-\pi^+)\right) + x_D \mathrm{Im} \left(\frac{p}{q}\rho(K^-\pi^+)\right) \right] \right\}. \quad (14.45)$$

In deriving Eq. (14.44) and Eq. (14.45) we have noted as before that $\overline{\rho}(K^+\pi^-)$ and $\rho(K^-\pi^+)$ are both enhanced by $1/\tan^2\theta_C$. Therefore we have kept terms linear and quadratic in x_D or y_D and $\overline{\rho}(K^+\pi^-)$ or $\rho(K^-\pi^+)$.

In the notation of Eq. (14.7) we can express the **CP** asymmetry as follows:

$$\frac{\Gamma(\overline{D}^0(t) \to K^-\pi^+) - \Gamma(D^0(t) \to K^+\pi^-)}{\Gamma(\overline{D}^0(t) \to K^-\pi^+) + \Gamma(D^0(t) \to K^+\pi^-)}$$
$$\simeq \left(\frac{t}{\tau_D}\right) |\hat{\rho}(K^-\pi^+)| \left(\frac{2y_D'\cos\phi_{K\pi}\epsilon_D - x_D'\sin\phi_{K\pi}}{\tan\theta_C^2}\right)$$
$$+ \left(\frac{t}{\tau_D}\right)^2 |\hat{\rho}(K^-\pi^+)|^2 \frac{\epsilon_D(x_D^2 + y_D^2)}{\tan\theta_C^4}. \quad (14.46)$$

BaBar has also searched for a time-dependent **CP** asymmetry in $D^0 \to K^+\pi^-$ vs. $\overline{D}^0(t) \to K^-\pi^+$, yet so far has not found any evidence for it [205]. Yet again with x_D' and y_D' capped by about 1%, present data do not impose a non-trivial bound on the weak phase $\phi_{K\pi}$. On the other hand any further increase in experimental sensitivity could reveal a signal.

Oscillations, even in the extreme scenario of $x_D = 0$, $y_D \neq 0$, will induce a time dependence in the **T** odd moments Γ_3 and $\overline{\Gamma}_3$ of $D \to K^+K^-\pi^+\pi^-$, discussed in Eq. (14.35) and below.

Rare modes

As mentioned above, radiative transitions like $D_L \to K^+ K^- \mu^+ \mu^-$ can be treated in analogy to $K_L \to \pi^+ \pi^- e^+ e^-$. The underlying process involves an off-shell photon $D_L \to K^+ K^- \gamma^* \to K^+ K^- \mu^+ \mu^-$, which receives contributions from two different reactions

$$D_L \xrightarrow{\text{\textbf{CP} viol.}} K^+ K^- \xrightarrow{E1} K^+ K^- \gamma^* \text{ and } D_L \xrightarrow{M1} K^+ K^- \gamma^*. \quad (14.47)$$

Their interference produces a **CP** and **T** odd polarization of the virtual photon. The latter reflects itself in the orientation of the $\mu^+ \mu^-$ plane and the Γ_3 term, analogous to the discussion leading to Eq. (12.37) for $K_L \to \pi^+ \pi^- e^+ e^-$.

No explicit calculation has been performed yet on how to relate Γ_{1-3} with the strength of **CP** violation in $D \to K^+ K^-$. Yet since the **CP** conserving $M1$ amplitude will be suppressed, one can expect that the forward–backward asymmetry $\langle A \rangle = \frac{2\Gamma_3}{\pi(\Gamma_1 + \Gamma_2)}$ will provide a substantially enhanced signal. We remind the reader that it is advantageous to enhance the asymmetry at the expense of statistics, as discussed in Eq. (10.28). With an enhanced signal it should be much easier to deal with systematic uncertainties.

Semileptonic transitions

$|q/p| \neq 1$ unambiguously reflects **CP** violation in $\Delta C = 2$ dynamics. It can be probed most directly in semileptonic D^0 decays leading to 'wrong sign' leptons:

$$A_{\text{SL}}(D^0) \equiv \frac{\Gamma(D^0(t) \to l^- X) - \Gamma(\overline{D}^0 \to l^+ X)}{\Gamma(D^0(t) \to l^- X) + \Gamma(\overline{D}^0 \to l^+ X)} = \frac{|q|^4 - |p|^4}{|q|^4 + |p|^4}. \quad (14.48)$$

The corresponding observable has been studied in semileptonic decays of neutral K and B mesons. With A_{SL} being controlled by $(\Delta\Gamma/\Delta M)\sin\phi_{\text{weak}}$, it is predicted to be small in those two cases, albeit for different reasons: (i) while $(\Delta\Gamma_K/\Delta M_K) \sim 1$ one has $\sin\phi_{\text{weak}}^K \ll 1$ leading to $A_{\text{SL}}^K = \delta_l \simeq (3.32 \pm 0.06) \times 10^{-3}$ as observed; (ii) for B^0 on the other hand one has $(\Delta\Gamma_B/\Delta M_B) \ll 1$ leading to $A_{\text{SL}}^B < 10^{-3}$ (see Section 11.5.2 for details).

For D^0, on the other hand, both ΔM_D and $\Delta\Gamma_D$ are small, yet $\Delta\Gamma_D/\Delta M_D$ is not: present data indicate it is of order unity; A_{SL} is given by the smaller of $\Delta\Gamma_D/\Delta M_D$ or its inverse multiplied by $\sin\phi_{\text{weak}}^D$, which might not be that small: i.e. while the rate for 'wrong-sign' leptons is certainly small in semileptonic decays of neutral D mesons, their **CP** asymmetry might not be at all, if New Physics intervenes to induce ϕ_{weak}^D.

14.3.4 Harnessing EPR correlations

So far we have treated these decays as if occurring with a single beam of D mesons. The problem is that unlike for K_L one cannot 'wait' for the D_S component of a neutral D meson beam to decay away to obtain a D_L. So we have to resort to a by now familiar tool – EPR correlations.

Subtle quantum mechanical effects occur when a pair of neutral charm mesons is produced in a coherent state in close qualitative analogy with the situation of $e^+e^- \to B^0\overline{B}^0$ described above. Those effects are not merely intriguing and provide direct manifestations of quantum mechanics' fundamentally *nonlocal* features – they might turn out to be of practical value as well.

Consider the production of a $D^0 - \overline{D}^0$ pair in e^+e^- annihilation on the vector meson resonance $\psi(3770)$ with the subsequent decays into seemingly identical final states f_D:

$$e^+e^- \to \psi''(3770) \to D^0\overline{D}^0 \to f_D f_D. \qquad (14.49)$$

It might appear that even without oscillations such final states are possible. E.g. $f_D = K^-\pi^+$ could arise due to the Cabibbo allowed [doubly Cabibbo-suppressed] mode $D^0[\overline{D}^0] \to K^-\pi^+$; likewise for the CP conjugate channel $f_D = K^+\pi^-$. Or $f_D = K^+K^-$ would be driven by the Cabibbo-suppressed channels $D^0[\overline{D}^0] \to K^+K^-$. However in Eq. (14.49) the charm mesons are created as a **C** and **P** *odd* configuration in analogy to $e^+e^- \to \Upsilon(4S) \to B_d\overline{B}_d$. Bose–Einstein statistics then does not allow them to decay into identical final states. This is another example of exploiting EPR correlations [201].

The intervention of $D^0 - \overline{D}^0$ oscillations introduces a subtle aspect. Bose–Einstein statistics still forbid the original $D^0\overline{D}^0$ pair to evolve into a D^0D^0 or $\overline{D}^0\overline{D}^0$ configuration at one time t. The EPR correlation tells us that if one neutral D meson reveals itself as a, say, D^0, the other one has to be a $\overline{D}^0$ at *that* time. Yet at *later* times it can evolve into a D^0, since the coherence between the original D^0 and $\overline{D}^0$ has been lost. Let us consider specifically $f_D = K^-\pi^+$. We then find

$$\frac{\sigma(e^+e^- \to D^0\overline{D}^0 \to (K^\mp\pi^\pm)_D(K^\mp\pi^\pm)_D)}{\sigma(e^+e^- \to D^0\overline{D}^0 \to (K^\mp\pi^\pm)_D(K^\pm\pi^\mp)_D)} = r_D; \qquad (14.50)$$

i.e. this process can occur only through oscillation – $r_D \neq 0$ – and the EPR correlation forces the pair of charm mesons to act like a single meson.

Bose–Einstein statistics are satisfied in a subtle way, as best seen when describing the reaction in terms of the *mass* eigenstates D_H and D_L:

$$e^+e^- \to \psi''(3770) \to D_H D_L \to (K^-\pi^+)_{D_H}(K^-\pi^+)_{D_L}; \qquad (14.51)$$

i.e. the two final states $K^-\pi^+$ are *not* truly identical – their energies differ by ΔM_D. This difference is of course much too tiny to be measurable directly.

In principle the same kind of argument can be applied to more complex final states like $f_D = K^{\mp}\rho^{\pm}$, however it is less conclusive there, in particular when the ρ is identified merely by the dipion mass.

The final state $f_D = K^+K^-$ can occur in the presence of $D - \overline{D}$ oscillations, but only if **CP** invariance is violated – a point we will return to later.

The fact that the $D^0\overline{D}^0$ pair has to form a coherent **C** odd quantum state in $e^+e^- \to \psi''(3770) \to D^0\overline{D}^0$ has other subtle consequences as well. Since the EPR effect (anti)correlates the time evolutions of the two neutral D mesons as sketched above, they act like a single charm meson as far as *like*-sign dileptons are concerned; i.e.

$$\frac{\sigma(e^+e^- \to D^0\overline{D}^0 \to l^{\pm}l^{\pm} X)}{\sigma(e^+e^- \to D^0\overline{D}^0 \to ll\, X)} = \chi^D_{WS}(ll) \qquad (14.52)$$

rather than the expression found for an *incoherent* pair of $D^0\overline{D}^0$

$$\frac{\sigma(D^0\overline{D}^0|_{\text{incoh}} \to l^{\pm}l^{\pm} X)}{\sigma(D^0\overline{D}^0|_{\text{incoh}} \to ll\, X)} = 2\chi^D_{WS}(ll)[1 - \chi^D_{WS}(ll)]. \qquad (14.53)$$

An additional twist arises if one studies

$$e^+e^- \to (D^0)^*\overline{D}^0 + D^0(\overline{D}^0)^* \to D^0\overline{D}^0 + \gamma, \qquad (14.54)$$

for the $D^0\overline{D}^0$ pair now forms a **C** even configuration. Bose–Einstein statistics will now favour the original $D^0\overline{D}^0$ pair to oscillate into a D^0D^0 or a $\overline{D}^0\overline{D}^0$ pair. We then find

$$\frac{\sigma(e^+e^- \to D^0\overline{D}^0\gamma \to (K^{\pm}\pi^{\mp})_D(K^{\pm}\pi^{\mp})_D\gamma)}{\sigma(e^+e^- \to D^0\overline{D}^0\gamma \to (K^{\pm}\pi^{\mp})_D(K^{\mp}\pi^{\pm})_D\gamma)}$$

$$\simeq \frac{3}{2}\left(x_D^2 + y_D^2\right) + 4\left|\frac{A(D^0 \to K^+\pi^-)}{A(D^0 \to K^-\pi^+)}\right|^2 + 8y_D\cos\delta\left|\frac{A(D^0 \to K^+\pi^-)}{A(D^0 \to K^-\pi^+)}\right|.$$

$$(14.55)$$

Alternatively one can measure

$$\sigma(e^+e^- \to D^0\overline{D}^0 \to (K^{\pm}\pi^{\mp})_D(K^{\mp}\pi^{\pm})_D) \propto$$

$$\left(1 - \frac{x_D^2 - y_D^2}{2}\right)\left(1 - 2\cos\delta\left|\frac{A(D^0 \to K^+\pi^-)}{A(D^0 \to K^-\pi^+)}\right|^2\right). \qquad (14.56)$$

The main message is that a judicious exploitation of EPR correlations in *coherent* quantum states allows us to extract x_D and y_D (together with the strong phase δ) and thus establish unambiguously that oscillations have taken place characterized by time scales τ_D/x_D, τ_D/y_D even with data integrated over all times of decay. The practical use of these EPR correlations has been treated in detail in Ref. [222].

In close qualitative analogy to B^0 decays can one observe **CP** violation in D^0 decays also through the *existence* of a transition. The reaction

$$e^+e^- \to \psi(3770) \to D^0\overline{D}^0 \to f_\pm f'_\pm \qquad (14.57)$$

with **CP**$|f^{(\prime)}\rangle = \pm|f^{(\prime)}\rangle$ can occur *only*, if **CP** is violated, since **CP**$|\psi(3770)\rangle = +1 \neq$ **CP**$|f_\pm f'_\pm\rangle = (-1)^{l=1} = -1$. The final states f and f' can be different, as long as they possess the same **CP** parity. More explicitly one has for $x_D \ll 1$

$$\mathrm{BR}(\psi(3770) \to D^0\overline{D}^0 \to f_\pm f'_\pm) \simeq \mathrm{BR}(D \to f_\pm)\mathrm{BR}(D \to f'_\pm)$$

$$\times \left[(2 + x_D^2)\left|\frac{q}{p}\right|^2 |\overline{\rho}(f_\pm) - \overline{\rho}(f'_\pm)|^2 + x_D^2\left|1 - \frac{q}{p}\overline{\rho}(f_\pm)\frac{q}{p}\overline{\rho}(f'_\pm)\right|^2\right]. \qquad (14.58)$$

The second contribution in the square brackets can occur only due to oscillations and then also for $f'_\pm = f_\pm$; yet it is heavily suppressed by $x_D^2 \leq 10^{-4}$ making it practically unobservable. The first term arises even with $x_D = 0$, yet requires $f'_\pm \neq f_\pm$. It is possible that $|\overline{\rho}(f_\pm) - \overline{\rho}(f'_\pm)|^2$ provides a larger signal of **CP** violation than either $|1 - |\rho(f_\pm)|^2|$ or $|1 - |\rho(f'_\pm)|^2|$.

Equation (14.58) also holds, when the final states are not **CP** eigenstates, yet still modes common to D^0 and $\overline{D}^0$. Consider for example $e^+e^- \to D^0\overline{D}^0 \to f_a f_b$ with $f_a = K^+K^-$, $f_b = K^\pm\pi^\mp$. Measuring those rates will yield unique information on the strong phase shifts.

Above we have described how the forward-backward asymmetry in $D_L \to K^+K^-\mu^+\mu^-$ constitutes a powerful microscope for **CP** violation in $D \to K^+K^-$ without explaining how to obtain a D_L beam. Bose–Einstein statistics enforces

$$e^+e^- \to \gamma^* \to D_S D_L. \qquad (14.59)$$

Since **CP** invariance holds (at least) approximately in D decays, observing $D \to K^+K^-, \pi^+\pi^-$ tells us that the decaying D meson is a D_S meaning the other neutral D meson has to be a D_L.

14.4 Résumé and a call to action

The study of charm dynamics was instrumental in the acceptance of the
SM three decades ago. More recently the discovery of seemingly unconventional hadrons with open as well as hidden charm has provided fruitful
challenges to our understanding of the strong interactions. At the same
time the confluence of more realistic lattice simulations of QCD and more
precise data on the weak decays of charm hadrons could validate our
theoretical control over leptonic and semileptonic D decays.

New Physics will in general induce FCNC. We should not assume a priori
that Nature follows the SM's script for FCNC with the same faithfulness
for up-type quarks as it seems to do for down-type quarks. Charm quarks
are uniquely positioned to reveal New Physics FCNC for up-type quarks.

These qualitative considerations have been strengthened by new indica-
tions that the potential for truly seminal or even revolutionary discoveries
in charm transitions might not have been exhausted. For data presented
in the spring of 2007 show strong evidence for charm changing neutral
currents generating D^0 oscillations with x_D, $y_D \sim 0.5\text{--}1\%$. If confirmed,
this represents a seminal discovery irrespective of theory, namely the first
and probably last observation of oscillations of up-type mesons (unless
there are more than six quarks). As far as we can tell the observed size of
x_D (and y_D) could be due completely to SM dynamics; yet at the same
time x_D could contain large contributions from New Physics.

This conundrum can be resolved by a comprehensive programme of
CP studies in charm decays. Such a programme is both *mandatory* and
feasible.

  (i) The observed matter–antimatter imbalance in the Universe tells
      us there has to be New Physics in **CP** violating dynamics. Such
      new sources of **CP** violation probably affect both B and D decays,
      yet could be more easily identifiable in D than in B decays, since
      there is considerably less 'SM background' in the former than in the
      latter.
 (ii) D^0 oscillations – whatever the dynamics underlying them – provide
      a considerably wider stage where **CP** violation can emerge. We have
      seen this fact being spectacularly demonstrated in the studies of B
      decays.
(iii) This analogy is qualitative rather than quantitative. While many
      experimental features favour **CP** searches in charm processes, one
      could have wished for speedier D^0 oscillations: $x_D \sim 0.5\text{--}1\%$ is about
      a factor of hundred smaller than x_{B_d}. The fact that New Physics
      could generate **CP** asymmetries larger by orders of magnitude than
      what is predicted in the SM does not mean their absolute scale is

large. On the more positive side one should note that now we seem to have a benchmark figure.

The required program could be best pursued at a e^+e^- Super-Flavour Factory, in particular if it could also run just above charm threshold.

Problems

14.1. Derive Eq. (14.40), Eq. (14.41), Eq. (14.44) and Eq. (14.45) from the master equations Eq. (6.54) and Eq. (6.57). To do so retain at most quadratic terms of x_D, y_D when expanding $e^{\Delta\Gamma_D t}$, $\cos\Delta M_D t$ and $\sin\Delta M_D t$ and ignore the DCSD amplitude $A(D^0 \to K^+\pi^-)$ relative to $A(D^0 \to K^-\pi^+)$ in the time-dependent terms.

14.2. Show that $\text{Im}\frac{q}{p} \sim \mathcal{O}(\lambda^4)$ and $\text{Im}\bar{\rho} \sim \mathcal{O}(\lambda^4)$ in the decay given in Fig. 14.1, by drawing appropriate Feynman diagrams.

14.3. Which branching ratios do you have to measure to extract the phase shifts relevant for a non-KM difference in $D^+ \to K_S\pi^+$ vs $D^- \to K_S\pi^-$?

14.4. Derive Eq. (14.34).

14.5. Show that $O_- \Longrightarrow -O_-$ under **CP** symmetry.

14.6. Convince yourselves that direct **CP** asymmetries in once CKM suppressed decays $D \to K\bar{K}, \pi\pi$ are $\mathcal{O}(\lambda^4)$ by drawing appropriate Feynman graphs. Also discuss the presence or absence of final state interactions.

15

The strong CP problem

Old problems (like old soldiers) never die – they just fade away

15.1 The problem

It is often listed among the attractive features of QCD that it 'naturally' conserves baryon number, flavour, parity and **CP**. This had been the consensus for some time - until it was pointed out [223] that it is not quite true.[1]

Let us go back to the QCD Lagrangian given in Eq. (8.9) and note that we can add a term constructed out of the field strength tensor defined in Eq. (8.10): $\epsilon_{\mu\nu\alpha\beta}G_a^{\mu\nu}G^{a\alpha\beta}$ henceforth to be denoted by $G \cdot \tilde{G}$. It is gauge invariant and carries dimension 4; thus there is no reason why this term should be excluded from the Lagrangian. Quantum field theory teaches us to include every gauge invariant dimension-four operator in the Lagrangian: loop corrections would induce its presence even if left out from the original Lagrangian, unless it is forbidden by some other symmetry the Lagrangian has to obey. Thus we have:

$$\mathcal{L}_{\text{eff}} = \mathcal{L}_{\text{QCD}} + \frac{\theta g_S^2}{32\pi^2} G \cdot \tilde{G}. \tag{15.1}$$

The emergence of a $G \cdot \tilde{G}$ term in the Lagrangian causes a severe problem. For $G \cdot \tilde{G}$ – in contrast to $G \cdot G$ – violates both parity and time reversal

---

[1] In fact, embedding (global) SUSY into QCD even leads to flavour-changing neutral currents, see Chapter 19.

invariance! This is best seen by expressing $G_{a\mu\nu}$ and its dual through the *colour* electric and magnetic fields, $\vec{E}_a$ and $\vec{B}_a$, respectively:

$$G \cdot G \propto \sum_a |\vec{E}_a|^2 + \sum_a |\vec{B}_a|^2 \overset{\text{P,T}}{\Longrightarrow} \sum_a |\vec{E}_a|^2 + \sum_a |\vec{B}_a|^2, \qquad (15.2)$$

$$G \cdot \tilde{G} \propto \sum_a \vec{E}_a \cdot \vec{B}_a \overset{\text{P,T}}{\Longrightarrow} -\sum_a \vec{E}_a \cdot \vec{B}_a, \qquad (15.3)$$

since

$$\vec{E}_a \overset{\text{P}}{\Longrightarrow} -\vec{E}_a, \quad \vec{E}_a \overset{\text{T}}{\Longrightarrow} \vec{E}_a, \qquad (15.4)$$

$$\vec{B}_a \overset{\text{P}}{\Longrightarrow} \vec{B}_a, \quad \vec{B}_a \overset{\text{T}}{\Longrightarrow} -\vec{B}_a; \qquad (15.5)$$

i.e. with $\mathcal{L}_{\text{eff}}$ as given by Eq. (15.1) neither parity nor time reversal invariance are fully conserved by QCD. This is called – somewhat sloppily – the *strong* **CP** *problem*.

15.2 Why $G \cdot \tilde{G}$ matters and $F \cdot \tilde{F}$ does not

The analogous term $F_{\mu\nu}\tilde{F}^{\mu\nu}$ can be added to the QED Lagrangian, yet has no observable consequence there, since it amounts to a total divergence.

At first it would seem to be the same for QCD (or any other non-abelian gauge theory), since

$$G \cdot \tilde{G} = \partial_\mu K^\mu, \quad K^\mu = \epsilon^{\mu\alpha\beta\gamma} A_{i\alpha} \left[G_{i\beta\gamma} - \frac{gs}{3} f_{ijk} A_{j\beta} A_{k\gamma} \right]. \qquad (15.6)$$

If one could adopt $A_\alpha = 0$ for the gluon field at spatial infinity as one can for the photon field, then this discussion would amount to much ado about nothing. However one can*not* due to the more complex topological structure of the QCD ground or vacuum state. We will merely indicate the argument here; it is a fascinating story, and we recommend serious students of physics to look into it in greater detail [82, 224, 225, 226].

The ground state is characterized by A_μ being a *pure* gauge field: $A_\mu^{\text{vac}} = \frac{i}{gs}(\partial_\mu \Omega)\Omega^{-1}$. In the temporal gauge $A_0 = 0$ one can classify the functions Ω by their asymptotic behaviour:

$$\Omega_n \to e^{i2\pi n} \text{ as } r \to \infty \quad n = 0, \pm 1, \pm 2, \dots \qquad (15.7)$$

This suggests that the surface integral over the current K_μ, Eq. (15.6), does not vanish. Yet one can be more explicit. The phase factor $e^{i2\pi n}$ can be viewed as the topological mapping of a circle unto another circle n times; i.e. it 'wraps' or 'winds' around the circle n times. Therefore n is usually referred to as the 'winding' number. It is determined by an integral over the (pure) gauge fields corresponding to these vacuum configurations:

$$n = \frac{g_S^2}{32\pi^2} \int d^3r K^0_{(n)}, \quad K^0_{(n)} = -\frac{g_S}{3} f_{ijk}\epsilon_{abc} A^{ia}_{(n)} A^{jb}_{(n)} A^{kc}_{(n)}. \tag{15.8}$$

Transitions can occur from a configuration with n_- at $t = -\infty$ to one with n_+ at $t = +\infty$ as expressed by:[2]

$$\nu \equiv n_+ - n_- = \frac{g_S^2}{32\pi^2} \int d\sigma_\mu K^\mu|_{t=-\infty}^{t=+\infty} = \frac{g_S^2}{32\pi^2} \int d^4x \, G \cdot \tilde{G}. \tag{15.9}$$

The true vacuum is therefore a linear superposition of the n configurations:

$$|\theta\rangle = \sum_n e^{-in\theta}|n\rangle \tag{15.10}$$

with θ being a real number.

Consider a transition between two vacua at $t = \pm\infty$, $|\theta(\pm\infty)\rangle$

$$\langle\theta(+\infty)|\theta(-\infty)\rangle = \sum_\nu e^{i\nu\theta} \sum_n \langle(n+\nu)(+\infty)|n(-\infty)\rangle. \tag{15.11}$$

It can also be expressed through the path integral formalism

$$\langle\theta(+\infty)|\theta(-\infty)\rangle = \sum_\nu \int \delta A e^{iS_{\text{eff}}[A]} \delta(\nu - \frac{g_S^2}{32\pi^2} \int d^4x \, G \cdot \tilde{G}) \tag{15.12}$$

by using an effective interaction, where $\theta\frac{g_S^2}{32\pi^2} G \cdot \tilde{G}$ has been added to the customary QCD Lagrangian as in Eq. (15.1); it reproduces the phase factor $e^{i\nu\theta}$ in Eq. (15.11), see Eq. (15.9). To summarize this admittedly sketchy discussion: it is the highly complex structure of the ground state of QCD (and other nonabelian gauge theories) that transforms the surface term $G \cdot \tilde{G}$ into a dynamical agent and thus introduces **CP** violation into the strong interactions. Using the effective Lagrangian of Eq. (15.1) allows us to include these subtle non-perturbative features of QCD in our treatment.

15.3 ♠ The $U(1)_A$ problem ♠

The plots thickens further with respect to an area that is outside the subject of this book. Yet we want to include it, albeit only glancingly, in our narrative, since it provides us with an intellectually fascinating example for the interconnectedness of phenomena in quantum field theories.

---

[2] For example, a so-called instanton configuration connects vacuum states with $n-1$ and n through a process that can be viewed as a tunnelling effect.

Consider QCD with just one family made up from u and d quarks. In the limit of massless quarks – not a bad approximation, since $m_u, m_d \ll \Lambda_{QCD}$ – one might think that QCD possesses a *global* $U(2)_L \times U(2)_R$ symmetry.

The vectorial component $U(2)_{L+R}$ is indeed conserved even after quantum corrections, yet the axial part $SU(2)_{L-R}$ is seen as spontaneously broken leading to the emergence of a triplet of Goldstone bosons, the pions. Those actually acquire a mass due to $m_u, m_d \neq 0$. The puzzle arises concerning the remaining $U(1)_{L-R}$: (i) it cannot represent a symmetry, since in that case one would have 'parity doubling'; i.e. hadrons had to come in mass degenerate pairs of opposite parity, which is definitely not the case; (ii) if it were broken spontaneously, a fourth Goldstone boson had to exist. The only possible candidates are the η and η' mesons. However neither fits the role: for their masses exceed an upper bound that can be placed on their masses, if they arise as a Goldstone boson (For a modern discussion, see for example Ref. [227]). These two arguments pointing to a dead end form what is called the $U(1)$ problem of QCD.

The puzzle is solved by realizing that the flavour singlet axial current exhibits a 'quantum anomaly': a *classical* symmetry – in this case chiral invariance for massless quarks – is no longer conserved, once one-loop corrections are included:

$$\partial^\mu J_\mu^{5(0)} = \frac{g_S^2}{32\pi^2} G \cdot \tilde{G}; \tag{15.13}$$

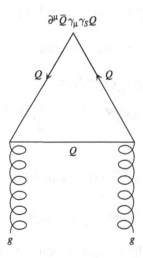

Figure 15.1 The axial current given in Eq. (15.18) is not conserved at the quantum level due to the anomaly diagram.

i.e. the assumed $U(1)_A$ symmetry was never there in the first place, even for massless quarks. Therefore only three Goldstone bosons are predicted for isospin.

The diagram which generates this chiral anomaly is shown in Fig. 15.1 and is called the 'triangle' anomaly because it is produced by a diagram with a triangular fermion loop, or the 'Adler–Bardeen–Bell–Jackiw' anomaly, named after its discoverers [228].

One sees that the resolution of the $U(1)_A$ problem due to the complex structure of the QCD vacuum appears to come with a price, namely the 'Strong **CP** Problem'. We will quantify this price below.

15.4   QCD and quark masses

The $U(1)_A$ and Strong **CP** Problems become even further intertwined, if one includes also the weak interactions.

In Chapter 8, we discussed how quark masses get generated when Higgs bosons acquire a vacuum expectation value. Starting from Eq. (8.20), when a Higgs boson ϕ^0 aquires a vacuum expectation value $\langle\phi^0\rangle = v$, the quark mass term can be written as:

$$\mathcal{L}_{\text{mass}} = v \sum_{i,j} \left((G_U)_{ij}\overline{U}_{i,L}U_{j,R} + (G_D)_{ij}\overline{D}_{i,L}D_{j,R} \right) + \text{h.c.} \qquad (15.14)$$

The mass matrices $\mathcal{M}_U = vG_U$ and $\mathcal{M}_D = vG_D$ can be diagonalized as shown in Eq. (8.22), so that Eq. (15.14) can be expressed in terms of mass eigenstates:

$$\mathcal{L}_{\text{mass}} = \overline{U}_L^m \mathcal{M}_U^{\text{diag}} U_R^m + \overline{D}_L^m \mathcal{M}_D^{\text{diag}} D_R + \text{h.c.} \qquad (15.15)$$

We are not finished yet! For example, just write the up-quark mass term:

$$\begin{aligned}
\mathcal{L}_{\text{mass}}^U &= \overline{U}_L^m \mathcal{M}_U^{\text{diag}} U_R^m + \overline{U}_R^m \mathcal{M}_U^{\text{diag}\dagger} U_L^m \\
&= \frac{1}{2}\overline{U}^m \left(\mathcal{M}_U^{\text{diag}} + \mathcal{M}_U^{\text{diag}\dagger} \right) U^m + \frac{1}{2}\overline{U}^m \left(\mathcal{M}_U^{\text{diag}} - \mathcal{M}_U^{\text{diag}\dagger} \right) \gamma_5 U^m.
\end{aligned} \qquad (15.16)$$

The terms proportional to $\overline{U}^m \gamma_5 U^m$ can be removed by performing the chiral rotation

$$U_i^m \to e^{-i\frac{1}{2}\alpha_i\gamma_5} U_i^m, \qquad (15.17)$$

where we have denoted the diagonal elements of $\mathcal{M}_U^{\text{diag}}$ by $m_i e^{i\alpha_i}$. This would be the end of the story if QCD were invariant under this chiral transformation – but it is not! In fact, the current associated with this transformation is not conserved:

$$\partial^\mu J_\mu^{5i} = \partial^\mu \overline{U}_i^m \gamma_\mu \gamma_5 U_i^m = 2m_i i \overline{U}_i^m \gamma_5 U_i^m + \frac{g_S^2}{16\pi^2} G \cdot \tilde{G} \neq 0. \quad (15.18)$$

This is the quantum anomaly stated above. For even with massless quarks the axial current is not conserved. The chiral transformation of Eq. (15.17) thus changes the action S:

$$S \longrightarrow S - \sum_i \int d^4x \, \partial^\mu J_\mu^{5,i} = S - i(\arg \det \mathcal{M}) \int d^4x \frac{g_S^2}{32\pi^2} G \cdot \tilde{G}; \quad (15.19)$$

where

$$\arg \det \mathcal{M} = \sum_i \alpha_i, \quad (15.20)$$

and the sum runs over terms arising from U and D quark masses. Equation (15.19) implies that we should have started with the modified QCD Lagrangian of Eq. (15.1). The QCD action remains unaffected by the simultaneous transformations

$$Q_i \to e^{-i\frac{1}{2}\alpha_i \gamma_5} Q_i, \quad m_i \to e^{-i\alpha_i} m_i, \quad \theta \to \theta - \sum_i \alpha_i = \theta - \arg \det \mathcal{M}.$$
$$(15.21)$$

This means that the dynamics depend on the combination

$$\overline{\theta} = \theta - \arg \det \mathcal{M}, \quad (15.22)$$

rather than θ by itself.

15.5   The neutron electric dipole moment

Since the gluonic operator $G \cdot \tilde{G}$ does not change flavour, we suspect right away that its most noticeable impact would be to generate an electric dipole moment (EDM) for neutrons. This is indeed the case, yet making this connection more concrete requires a more sophisticated argument. Let us briefly recapitulate some elementary findings on EDMs.

- They are described by an operator in the Lagrangian:

$$\mathcal{L}_{\text{EDM}} = -\frac{i}{2} d \overline{\psi} \sigma_{\mu\nu} \gamma_5 \psi F_{\mu\nu}. \quad (15.23)$$

  Since this operator's dimension is 5, its *dimensionful* coefficient d can be calculated as a *finite* quantity.

- A non-relativistic reduction (Problem 15.1) shows that the low-energy properties of an EDM are indeed satisfied by Eq. (15.23): d describes the energy shift of a system with angular momentum $\vec{j}$ when placed in an external electric field that grows linearly with the field:

$$\Delta E = \vec{d} \cdot \vec{E} + \mathcal{O}(|\vec{E}|^2) = d\vec{j} \cdot \vec{E} + \mathcal{O}(|\vec{E}|^2). \qquad (15.24)$$

- An EDM signalling **T** violation can unambiguously be differentiated against *induced* dipole moments that do not, see Chapter 3.

Here we view the neutron EDM – d_N – as due to the photon coupling to a virtual proton or pion in a fluctuation of the neutron:

$$n \implies p^* \pi^* \implies n. \qquad (15.25)$$

Of the two effective pion nucleon couplings in this one-loop process (Fig. 15.2) one is produced by ordinary QCD and conserves **P**, **T** and **CP**; the other one is induced by $G \cdot \tilde{G}$ with the help of Eq. (15.18). A rough guestimate can be gleaned from naive dimensional reasoning: the scale for a dipole moment (electric or magnetic) is set by e/M_N with M_N being the mass of the neutron; to be an EDM it obviously has to be proportional to $\bar{\theta}$ – and to the ratio of (current) quark masses m_q to M_N; this last point anticipates what will be discussed later, namely that the $\bar{\theta}$ dependence of observables can be rotated away for $m_q = 0$, i.e. in the chiral limit: $d_N \sim \mathcal{O}\left(\frac{e}{M_N}\frac{m_q}{M_N}\bar{\theta}\right) \sim \mathcal{O}\left(2 \cdot 10^{-15}\,\bar{\theta}\,\mathrm{e\,cm}\right)$. A more reliable estimate was first obtained by Baluni [229] in a nice paper using bag model computations of the transition amplitudes between the neutron and its excitations: $d_N \simeq 2.7 \cdot 10^{-16}\,\bar{\theta}\,\mathrm{e\,cm}$. In [230] chiral perturbation theory was employed instead: $d_N \simeq 5.2 \cdot 10^{-16}\,\bar{\theta}\,\mathrm{e\,cm}$. More recent estimates yield values in roughly the same range: $d_N \simeq (4 \cdot 10^{-17} \sim 2 \cdot 10^{-15})\,\bar{\theta}\mathrm{e\,cm}$ [231]. Hence

$$d_N \sim \mathcal{O}(10^{-16}\bar{\theta})\,\mathrm{e\,cm}, \qquad (15.26)$$

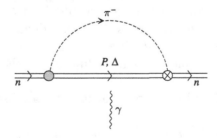

Figure 15.2 A major contribution to the neutron EDM. The blob on the left represents a strong vertex. The blob on the right represents a **CP** or **T** violating interaction.

and we infer

$$d_N \leq 2.9 \cdot 10^{-26} \, \mathrm{e\, cm}, \quad 90\% \ \mathrm{CL}, \quad \Longrightarrow \quad \bar{\theta} < 10^{-9}. \qquad (15.27)$$

Although θ_{QCD} is a QCD parameter, it might not necessarily be of order unity; nevertheless the truly tiny bound given in Eq. (15.27) begs for an explanation. The only kind of explanation that is usually accepted as 'natural' in our community is one based on symmetry. Such explanations have been put forward; the most widely discussed suggestions are based on a so-called Peccei–Quinn symmetry. Yet before we start speculating too wildly, we want to see whether there are no more mundane explanations.

15.6 Are there escape hatches?

Only a symmetry based solution of the $U(1)$ problem is seen as a natural one. Eq. (15.21) and Eq. (15.22) provide a telling example. If one of the quark masses m_i vanishes ($\det M = 0$) thus leading to a chiral symmetry, then the unphysical phase α_i can be used to dial $\bar{\theta}$ to zero! However, most authors argue that neither the up quark nor a fortiori the down quark mass can vanish [232]:

$$m_d(1 \text{ GeV}) > m_u(1 \text{ GeV}) \simeq 5 \text{ MeV}, \qquad (15.28)$$

where we have used the running mass evaluated at a scale of $1 \, \mathrm{GeV}$.

One could deny the very existence of a problem. It has been suggested that the strong **CP** problem is actually fictitious claiming that the treatment of the boundary conditions, which leads to the vacuum being characterized by the parameter θ, is incorrect [233]. We are not persuaded by this claim. Furthermore one would have to ask – at least as a theorist – what is then the solution of the $U(1)_A$ problem?

One could argue in favour of some 'engineering' solution.

Being the coefficient of a dimension- operator $\bar{\theta}$ can in general[3] be renormalized to any value, including zero. This is technically correct; however, adjusting θ to be smaller than $\mathcal{O}(10^{-9})$ by hand is viewed as highly 'unnatural'.

- A priori there is no reason why θ_{QCD} and $\Delta\theta_{\mathrm{EW}}$ should practically vanish.
- Even if $\theta_{\mathrm{QCD}} = 0 = \Delta\theta_{\mathrm{EW}}$ were set *by fiat*, quantum corrections to $\Delta\theta_{\mathrm{EW}}$ are typically much larger than 10^{-9} and ultimately actually

---

[3] Exceptions will be mentioned below.

infinite. At which order this happens depends on the electroweak dynamics, though. Within the KM ansatz $\Delta\theta_{\mathrm{ren}} \neq 0$ arises first at three loops, and it does not diverge before seven loops. In other models, though, the problem is more pressing. As described later, in models with right-handed currents or non-minimal Higgs dynamics $\delta\theta_{\mathrm{ren}} \neq 0$ arises at one loop already.

- To expect that θ_{QCD} and $\Delta\theta_{\mathrm{EW}}$ cancel so as to render $\bar{\theta}$ sufficiently tiny would require fine tuning of a kind which would have to strike even a sceptic as unnatural. For θ_{QCD} reflects dynamics of the strong sector and $\Delta\theta_{\mathrm{EW}}$ that of the electroweak sector.

15.6.1 *Soft* **CP** *violation*

A more respectable approach would be to implement **CP** symmetry *spontaneously*, which imposes $\bar{\theta} = 0$ as the leading effect, with corrections leading to a small and *calculable* deviation from zero. A priori this appears as a very attractive option.

- We have stated on several occasions the theoretical advantages of a *spontaneously* over a *manifestly* broken symmetry.
- Arranging for spontaneous **CP** breaking represents a quite manageable task. For once we go beyond the minimal structure of the SM, more VEVs emerge that can exhibit physical phases. Explicit examples will be given in our discussion of left–right symmetric models and non-minimal Higgs dynamics in Chapter 17.
- The resulting scenarios are intriguing in their own right – at least for a theorist.

Some challenges have to be met, though, chief among them two.

- The *cosmic domain wall problem* raises its unpleasant head (as it does for any *discrete* symmetry). By that we mean the following: as the universe cools down to a temperature below which **CP** invariance is broken spontaneously, domains of different **CP** phases emerge. Since it is a discrete symmetry, walls have to form to separate these domains. As shown in Ref. [234], the energy stored in such walls would greatly exceed the closure density for the universe. This cosmological disaster can be vitiated if the spontaneous **CP** breaking occurs *before* an inflationary period in our universe's past. For then we would live in a *single* domain. That means the breaking scale has to be very high, of the order of GUT scales.

- While $\bar{\theta}$ naturally emerges to be small in these scenarios – $\bar{\theta} \ll 1$ – we aim for truly tiny values: $\bar{\theta} \leq \mathcal{O}(10^{-9})$. Again, this favours a very high breaking scale.

It is important to note that **CP** being broken spontaneously does not suffice to enforce $\bar{\theta} = 0$ at tree level. For the emerging VEVs can still contribute to arg det M at tree level. Two strategies can be pursued to build viable models [235]:

- While still allowing for a complex quark mass matrix M, we impose a special form on M such that arg det $M = 0$ holds at tree level. The key ingredient here is the introduction of novel truly super-heavy quarks. Yet at low energies, accessible to experiments, we recover effectively a CKM mechanism [236, 237].
- The quark mass matrix is actually real at tree level. *No* CKM ansatz can effectively arise at low energies then, and we need alternative sources of **CP** violation to explain $K_L \rightarrow \pi\pi$. With the spectacular successes CKM theory has achieved in the last few years in its description of B decays this seems hardly a tenable option.

The last statement can actually be generalized: the ansatz of breaking **CP** invariance only softly flies in the face of the successes of CKM theory with its hard **CP** violation.

15.7 Peccei–Quinn symmetry

As just argued, $\bar{\theta} \leq \mathcal{O}(10^{-9})$ could hardly come about accidentally; an organizing principle had to arrange various contributions and corrections in such a way as to render the required cancellations. There is the general philosophy that such a principle has to involve some underlying symmetry. In the preceding section we have already sketched such an approach: a global chiral invariance allows us to rotate the dependence on $\bar{\theta}$ away; we failed however in our endeavour because this symmetry is broken by $m_q \neq 0$. Is it possible to invoke some other variant of chiral symmetry for this purpose, even if it is spontaneously broken? One particularly intriguing ansatz is to reinterpret a physical quantity that is conventionally taken to be a constant as a *dynamic degree of freedom*, that adjusts itself to a certain (desired) value in response to forces acting upon it. One early example is provided by the original Kaluza–Klein theory [238] – see Section 20.2 – invoking a six-dimensional 'space'–time manifold: two compactify dynamically and thus lead to the quantization of electric and magnetic charge.

Something similar has been suggested by Peccei and Quinn [239]. They augmented the SM by a global $U(1)_{\mathrm{PQ}}$ symmetry – now referred to as the Peccei–Quinn symmetry – which is axial with the following properties.

- It is a symmetry of the *classical* theory.
- It is subject to an axial anomaly; i.e. it is broken *explicitly* by non-perturbative effects, reflecting the complexity of the QCD ground state.
- It is broken *spontaneously* as well.

The following will happen then: the spontaneous breaking of the symmetry gives rise to Goldstone bosons called *axions*. With $U(1)_{PQ}$ being axial, it exhibits a triangle anomaly leading to a coupling of the axion field to $G \cdot \tilde{G}$. This in turn generates a mass for the axion; more importantly it transforms the quantity $\bar{\theta}$ into a dynamical one, depending on the axion field. The potential due to non-perturbative dynamics induces a vacuum expectation value for the axion such that

$$\bar{\theta} \simeq 0 \qquad (15.29)$$

emerges; i.e. $\bar{\theta}$ relaxes *dynamically* to a (practically) zero value.

To see that these words are more than just a nice yarn, let us consider the following Lagrangian:

$$\mathcal{L}_{\mathrm{PQ}} = -\frac{1}{4} G \cdot G + \sum_{j} \left[\overline{Q}_j i \gamma_\mu D^\mu Q_j - (y_j \overline{Q}_{L,j} Q_{R,j} \phi + \mathrm{h.c.}) \right]$$
$$+ \frac{\theta g_S^2}{32\pi^2} G \cdot \tilde{G} + \partial_\mu \phi^\dagger \partial^\mu \phi - V(\phi^\dagger \phi), \qquad (15.30)$$

which remains invariant *classically* under

$$\phi \xrightarrow{U_{\mathrm{PQ}}(1)} e^{i2\alpha} \phi, \quad Q_i \xrightarrow{U_{\mathrm{PQ}}(1)} e^{-i\alpha\gamma_5} Q_i. \qquad (15.31)$$

The potential $V(\phi^\dagger \phi)$ is chosen such that the axial symmetry $U_{\mathrm{PQ}}(1)$ is broken *spontaneously* by a vacuum expectation value of ϕ:

$$\langle \phi(x) \rangle = v_{\mathrm{PQ}} e^{i\langle a \rangle / v_{\mathrm{PQ}}}, \qquad (15.32)$$

with $\langle a \rangle$ denoting the vacuum expectation value of the axion field $a(x)$. With the quarks Q_i acquiring masses

$$m_i = y_i v_{\mathrm{PQ}} e^{i\langle a \rangle / v_{\mathrm{PQ}}}, \qquad (15.33)$$

we obtain

$$\bar{\theta} = \theta - \sum_i \arg y_i - N_f \langle a \rangle / v_{\text{PQ}}; \qquad (15.34)$$

i.e. the important new feature is that the quantity $\bar{\theta}$ – rather than being a mere parameter – depends on the dynamical field a through the latter's VEV. In the usual scenarios for spontaneously broken symmetries, the phase of the scalar field remains completely undetermined, which implies the masslessness of the Goldstone bosons.

Now, a second novel feature arises: the chiral anomaly is – as just sketched – implemented through a term proportional to $aG \cdot \tilde{G}$; since it is linear in the field a, $G \cdot \tilde{G}$ acts as a non-trivial effective potential for a and the resulting dynamics determine $\langle a \rangle$. To see how this comes about, let us consider the effective Lagrangian:[4]

$$\mathcal{L}_{\text{eff}} = \mathcal{L}_{\text{SM}} + \frac{g_S^2 \bar{\theta}}{32\pi^2} G \cdot \tilde{G} + \frac{g_S^2}{32\pi^2} \frac{\xi}{v_{PQ}} a G \cdot \tilde{G} - \frac{1}{2} \partial_\mu a \partial_\mu a + \mathcal{L}_{\text{int}}(\partial_\mu a, \psi).$$

$$(15.35)$$

The size of the parameters v_{PQ} and ξ and the form of $\mathcal{L}_{\text{int}}(\partial_\mu a, \psi)$ describing the (purely derivative) coupling of the axion field to other fields ψ depend on how the Peccei–Quinn symmetry is specifically realized.

Because of the chiral anomaly, the expectation value for the axion field is given by (see Problem 15.3):

$$\langle a \rangle = -\frac{\bar{\theta}}{\xi} v_{PQ}. \qquad (15.36)$$

The *physical* axion excitations are described by the shifted field

$$a_{\text{phys}}(x) = a(x) - \langle a \rangle, \qquad (15.37)$$

in terms of which Eq. (15.35) is re-written as follows:

$$\mathcal{L}_{\text{eff}} = \mathcal{L}_{\text{SM}} - \frac{1}{2} \partial_\mu a_{\text{phys}} \partial_\mu a_{\text{phys}} + \frac{g_S^2}{32\pi^2} \frac{\xi}{v_{PQ}} a_{\text{phys}} G \cdot \tilde{G} + \mathcal{L}_{\text{int}}(\partial_\mu a_{\text{phys}}, \psi);$$

$$(15.38)$$

i.e. the offending **P** and **T** violating bilinear term $\bar{\theta} G \cdot \tilde{G}$ has been traded in against a $a_{\text{phys}} G \cdot \tilde{G}$ coupling between dynamical fields:

$$\bar{\theta} = 0. \qquad (15.39)$$

Electroweak forces driving $K_L \to \pi\pi$ will actually move $\bar{\theta}$ away from zero – but only by an extremely tiny amount: $\bar{\theta} \sim \mathcal{O}(10^{-16})$!

---

[4] Here we follow the discussion given in reviews by Peccei [125, 226].

15.8   The dawn of axions – and their dusk?

The story does not end here, of course, since the breaking of $U(1)_{\text{PQ}}$ gives rise to a new dynamical entity, the physical axion field [240, 241]. Its behaviour depends on two parameters: its mass and its couplings to other fields. Since its mass is controlled by the anomaly term, we guestimate on dimensional grounds (see Problem 15.3):

$$m_a^2 \sim \mathcal{O}\left(\frac{\Lambda_{QCD}^4}{v_{PQ}^2}\right). \tag{15.40}$$

Since we expect on general grounds $v_{\text{PQ}} \gg \Lambda_{\text{QCD}}$, we are dealing with a very light boson. The question is how light would the axion be. As is shown below, e.g. Eq. (15.43), $1/v_{\text{PQ}}$ sets the scale also for the couplings of the physical axion, including its derivative couplings to fermionic fields ψ. As v_{PQ} goes up, the mass of the axion thus goes down – as does its coupling.

The electroweak scale $v_{\text{EW}} = \left(\sqrt{2}G_{\text{F}}\right)^{-\frac{1}{2}} \simeq 250\,\text{GeV}$ provides the discriminator for two scenarios, which will be discussed below.

-
$$v_{\text{PQ}} \sim v_{\text{EW}} \quad \rightarrow \quad m_a \sim \mathcal{O}(1\,\text{MeV}). \tag{15.41}$$

  In that case axions can, or even should, be seen in accelerator based experiments. Such scenarios are referred to as *visible* axions.

-
$$v_{\text{PQ}} \gg v_{\text{EW}} \quad \rightarrow \quad m_a \ll 1\,\text{MeV}. \tag{15.42}$$

  As explained below, such axions could not be found in accelerator-based experiments, because of their minute couplings; therefore they are called *invisible* scenarios. Yet that does not mean that they necessarily escape detection! They could be of great significance for the formation of stars, whole galaxies and even the universe.

15.8.1   *Visible axions*

The simplest scenario involves two $SU(2)_L$ doublet Higgs fields that possess opposite hypercharge.[5] They also carry a $U(1)$ charge *in addition* to the hypercharge; this second (and global) $U(1)$ is identified with the PQ symmetry.

---

[5] In the SM the Higgs doublet and its charge conjugate fill this role.

The couplings of the axion to fermions is purely derivative, as befits a Goldstone boson:

$$\mathcal{L}_{\text{int}} = \frac{1}{v} \partial_\mu a \left[x \sum_i \bar{u}_{R,i} \gamma_\mu u_{R,i} + \frac{1}{x} \sum_i \bar{d}_{R,i} \gamma_\mu d_{R,i} + \frac{1}{x} \sum_i \bar{l}_{R,i} \gamma_\mu l_{R,i} \right],$$

(15.43)

where

$$x = \frac{v_2}{v_1}, \quad v = \sqrt{v_1^2 + v_2^2},$$

(15.44)

with v_i denoting the VEVs of the two Higgs doublets.

The anomaly induces non-derivative couplings to the gauge fields [242]

$$\mathcal{L}_{\text{anom}} = \frac{a}{v} \left\{ \left[N_{\text{fam}} \left(x + \frac{1}{x} \right) \frac{g_S^2}{32\pi^2} G \cdot \tilde{G} + N_{\text{fam}} \left(\frac{4}{3} x + \frac{1}{3x} + \frac{1}{x} \right) \frac{g_{EW}^2}{16\pi^2} B \cdot \tilde{B} \right] \right\},$$

(15.45)

where N_{fam} denotes the number of families and $B_{\mu\nu}$ the field strength tensor of the hypercharge gauge field that couples to right-handed fermions: $B_{\mu\nu} = F_{\mu\nu}^{em} - \tan\theta_W F_{\mu\nu}^{Z^0}$. The anomaly also induces a mass for the axion [243]:

$$m_a \simeq \frac{m_\pi F_\pi}{v} N_{\text{fam}} \left(x + \frac{1}{x} \right) \frac{\sqrt{m_u m_d}}{(m_u + m_d)} \simeq 25 \, N_{\text{fam}} \left(x + \frac{1}{x} \right) \text{ keV.}$$

(15.46)

Such masses could conceivably reach the $\mathcal{O}(1\,\text{MeV})$ level, but not much beyond.

Its lifetime can be deduced from Eq. (15.43) and Eq. (15.45).

- If $m_a > 2m_e$, the axion decays very rapidly into electrons and positrons:

$$\tau(a \to e^+ e^-) \simeq 4 \cdot 10^{-9} \left(\frac{1\,\text{MeV}}{m_a} \right) \frac{x^2 \text{ or } 1/x^2}{\sqrt{1 - (4m_e^2/m_a^2)}} \text{ s.}$$

(15.47)

- If, on the other hand, $m_a < 2m_e$, then the axion decays fairly slowly into two photons, as derived from Eq. (15.45) and Eq. (15.46):

$$\tau(a \to \gamma\gamma) \sim \mathcal{O} \left(\frac{100\,\text{keV}}{m_a} \right) \text{ s.}$$

(15.48)

We have presented here a very rough sketch of scenarios with visible axions since we can confidently declare that they have been ruled out experimentally. They have been looked for in beam dump experiments – without success. Yet the more telling blows have come from searches in rare decays.

- For long-lived axions – $m_a < 2m_e$ – we expect a dominating contribution to $K^+ \rightarrow \pi^+ +$ nothing from

$$K^+ \rightarrow \pi^+ + a, \tag{15.49}$$

with the axion decaying well outside the detector. For the two-body kinematics of Eq. (15.49) we have tight bounds from published data [167]:

$$\mathrm{Br}(K^+ \rightarrow \pi^+ X^0) < 5.2 \cdot 10^{-10}, \quad 90\% \text{ CL} \tag{15.50}$$

for X^0 being a practically massless and non-interacting particle. Theoretically we would expect [244]:

$$\mathrm{Br}(K^+ \rightarrow \pi^+ a)|_{\text{theor}} \sim 3 \cdot 10^{-5} \cdot (x + 1/x)^{-2}. \tag{15.51}$$

Although Eq. (15.51) does not represent a precise prediction, the discrepancy between expectation and observation is conclusive.

- We arrive at the same conclusion that *long-lived visible* axions do not exist from the absence of *quarkonia* decay into them: neither $J/\psi \rightarrow a\gamma$ nor $\Upsilon \rightarrow a\gamma$ has been seen [244].

- The analysis is a bit more involved for *short-lived* axions – $m_a > 2m_e$. Yet again their absence has been established through a combination of experiments. Unsuccessful searches for

$$\pi^+ \rightarrow ae^+\nu, \tag{15.52}$$

figure prominently in this endeavour [245]. Likewise the *absence* of axion driven *nuclear de-excitation* has been established on a level that appears to be conclusive.

15.8.2 *Invisible axions*

The phenomenological conflicts just discussed can be avoided by separating the $SU(2)_L \times U(1)$ and $U(1)_{\mathrm{PQ}}$ breaking scales. To that purpose, we introduce a complex scalar field σ that

- is an $SU(2)_L$ *singlet*,
- yet carries a PQ charge, and
- possesses a huge VEV $v_{\mathrm{PQ}} \gg v_{\mathrm{EW}}$.

As can be inferred from Eq. (15.43), the couplings of such axions to gauge as well as fermion fields become truly tiny. These requirements can be realized in two distinct (sub-)scenarios:

(1) Only very heavy new quarks carry a PQ charge. This situation is referred to as the KSVZ axion (For prototypes see Ref. [246]). The minimal version can do with a single $SU(2)_L$ Higgs doublet.

(2) Also the known quarks and leptons carry a PQ charge. Two $SU(2)_L$ Higgs doublets are then required in addition to σ. The fermions do not couple directly to σ, yet become sensitive to PQ breaking through the Higgs potential. This is referred to as the DFSZ axion (Such models were first discussed in [247]).

From current algebra we infer for the axion mass in either case

$$m_a \simeq 0.6\,\text{eV} \cdot \frac{10^7\,\text{GeV}}{v_{\text{PQ}}}. \tag{15.53}$$

The most generic coupling of such axions is to two photons

$$\mathcal{L}(a \to \gamma\gamma) = -\tilde{g}_{a\gamma\gamma} \frac{\alpha}{\pi} \frac{a(x)}{v_{\text{PQ}}} \vec{E} \cdot \vec{B}, \tag{15.54}$$

where $\tilde{g}_{a\gamma\gamma}$ is a model-dependent coefficient of order unity. There are also generic interactions with pions and nucleons, and couplings to electrons are allowed as well [226, 248].

Axions with such tiny masses have lifetimes easily in excess of the age of the universe. Also their couplings to other fields are so minute that they would not betray their presence – hence their name *invisible* axions – under ordinary circumstances.

Yet in astrophysics and cosmology more favourable extraordinary conditions can arise, and their footprints could become visible. There are three types of such footprints.

- The two gamma coupling described by Eq. (15.54) allows stars like our sun to produce axions by transforming a photon into an axion. The solar axion flux can be searched for directly by the inverse process transforming an axion into a photon. This is sketched in Fig. 15.3: a strong magnetic field $\vec{B}$ can stimulate the conversion

$$\text{axion} \xrightarrow{\vec{B}} \text{photon} \tag{15.55}$$

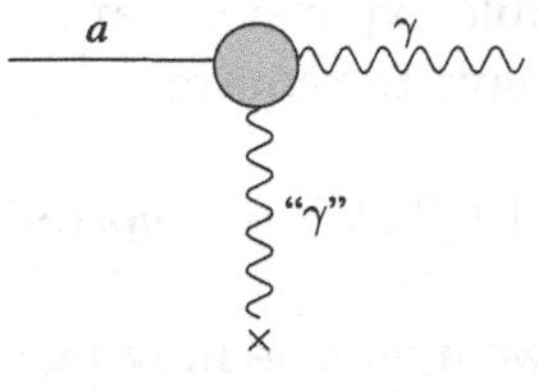

Figure 15.3 An 'invisible' axion may show itself while interacting with a magnetic field, converting itself into a photon.

where the virtual photon effects the interaction with an inhomogeneous magnetic field in a cavity. Available microwave technology allows an impressive experimental sensitivity.

No effect has been seen so far, yet the searches are on-going [248].

- There can be also indirect astrophysical signatures, since axion emission will lead to energy loss and thus provide a cooling mechanism to stellar evolution. Not surprisingly their greatest impact occurs for the lifetimes of red giants and supernovae such as SN 1987a. The actual bounds depend on the model – whether it is a KSVZ or DFSZ axion – but relatively mildly. Altogether astrophysics tells us that *if* axions exist we have [248]

$$m_a < 10^{-2}\,\text{eV}. \tag{15.56}$$

- *Cosmology*, on the other hand, provides us with a *lower* bound through a very intriguing line of reasoning. At temperatures T above Λ_{QCD} the axion is massless and all values of $\langle a(x) \rangle$ are equally likely. For $T \sim 1\,\text{GeV}$ the anomaly-induced potential turns on driving $\langle a(x) \rangle$ to a value which will yield $\bar{\theta} = 0$ at the new potential minimum. The energy stored previously as *latent heat* is then released into axions oscillating around its new VEV. Precisely because the invisible axion's couplings are so immensely suppressed, the energy cannot be dissipated into other degrees of freedom. We are then dealing with a fluid of axions. Their typical momentum is the inverse of their correlation length, which in turn cannot exceed their horizon; we find

$$p_a \sim \left(10^{-6}\,\text{s}\right)^{-1} \sim 10^{-9}\,\text{eV} \tag{15.57}$$

at $T \simeq 1\,\text{GeV}$; i.e. axions, despite their minute mass, form a very cold fluid and actually represent a candidate for cold dark matter. Their contribution to the density of the universe relative to its critical value is [249]

$$\Omega_a = \left(\frac{0.6 \cdot 10^{-5}\,\text{eV}}{m_a}\right)^{\frac{7}{6}} \cdot \left(\frac{200\,\text{MeV}}{\Lambda_{\text{QCD}}}\right)^{\frac{3}{4}} \cdot \left(\frac{75\,\text{km/s} \cdot \text{Mpc}}{H_0}\right)^2; \tag{15.58}$$

H_0 is the present Hubble expansion rate. For axions not to overclose the universe we thus have to require:

$$m_a \geq 10^{-6}\,\text{eV} \quad \leftrightarrow \quad v_{\text{PQ}} \leq 10^{12}\,\text{GeV}. \tag{15.59}$$

This means also that we might be moving or existing in a bath of cold axions making up a significant fraction of the matter of the universe.

From recent WMAP data on cold dark matter [250] somewhat stronger bounds have been inferred [226, 248, 251]:

$$m_a \geq 2.1 \cdot 10^{-5} \, \text{eV} \quad \leftrightarrow \quad v_{PQ} \leq 3 \times 10^{11} \, \text{GeV}. \qquad (15.60)$$

Ingenious suggestions have been made to search for such cosmic background axions often involving the process of Fig. 15.3.

To summarize this overview: if axions exist, their mass has probably to be in the window

$$10^{-2} \, \text{eV} \geq m_a \geq 10^{-5} \, \text{eV}. \qquad (15.61)$$

No detection strategy has been proposed yet that would allow searches in the range of 10^{-2} and 10^{-4} eV for the axion mass [248].

15.9 The pundits' judgement

The story of the strong **CP** problem is a particularly titillating one. It arises only if we go beyond mere renormalizability and insist that renormalization proceeds in a *natural* way, i.e. without fine tuning. Trying to resolve the strong **CP** problem has led to an impressive intellectual edifice that is based on an intriguing arsenal of theoretical reasoning, and has inspired fascinating experimental undertakings that are still going strong. The observation that neutrons have at best a tiny EDM has generated the suggestion that extremely light and weakly particles – axions – exist that nevertheless could make up a significant fraction of the mass of galaxies or even the universe! This is of course highly speculative, since we know of no intrinsic reason why the axion mass should lie in the cosmologically relevant range of about 10^{-5} eV.

Neither axions nor other consequences of the strong **CP** problem have been discovered so far. The allowed domain for m_a and v_{PQ} has shrunk somewhat, yet it still has a finite width - some of which we do not know how to close. Furthermore, we should keep in mind that the interpretations of the so far negative searches typically involve model assumptions.

Like many modern novels the problem – if it is indeed one – has not found a resolution. On the other hand there are still unexplored mass ranges for the axion [248, 251]. Searching for it thus still has the potential to lead the charge towards a new paradigm in high energy physics.

A cynic, however, might summarize it differently: while we certainly do not know what the solution is, we cannot be sure whether there is a problem in the first place! It thus might remind her or him of the often told story of the French officer from the period of Enlightenment who was overheard praying before a battle: 'Dear God – in case you exist – save my soul if I have got one!'.

Problems

15.1. Show that the non-relativistic reduction of Eq. (15.23) yields

$$\int d^3x \mathcal{H} = d\vec{S} \cdot \vec{E}, \quad \vec{S} = \int d^3x \psi^* \vec{\sigma} \psi, \qquad (15.62)$$

where $\psi(x)$ is the wave function of a fermion at rest. Thus, it satisfies the properties of a non-relativistic EDM described in Chap. 3.

15.2. Consider the Lagrangian of Eq. (15.35). Write down the field equation for the axion field *ignoring* the anomaly term $aG \cdot \tilde{G}$. Verify that any constant value of $\langle a \rangle$ would then satisfy this equation.

Now include the anomaly term in the field equation for the axion field and derive:

$$- \partial^2 a + \partial_\mu \frac{\partial \mathcal{L}_{\text{int}}}{\partial \partial_\mu a} = \frac{\xi}{\Lambda_{PQ}} \frac{g_S^2}{32\pi^2} G \cdot \tilde{G}. \qquad (15.63)$$

15.3. Show that the effective potential for a is minimized by $\langle G \cdot \tilde{G} \rangle = 0$. Using Eq. (15.35) and the fact that $\langle G \cdot \tilde{G} \rangle$ is periodic in $\bar{\theta}$, argue that $\langle G \cdot \tilde{G} \rangle$ is actually controlled by the combination $\bar{\theta} + \xi \langle a \rangle / v_{PQ}$. This leads to

$$\langle \bar{\theta} | a | \bar{\theta} \rangle = -\bar{\theta} \frac{v_{PQ}}{\xi}.$$

For example, we can derive

$$\langle G \cdot \tilde{G} \rangle \propto \sin(\bar{\theta} + \xi \langle a \rangle / v_{PQ}), \qquad (15.64)$$

in the one-instanton-approximation.

15.4. Show that the mass of an axion is given by:

$$m_a^2 = -\frac{\xi}{\Lambda_{PQ}} \frac{g^2}{32\pi^2} \frac{\partial}{\partial a} \langle G \cdot \tilde{G} \rangle_{\langle a \rangle = \frac{\bar{\theta}}{\xi} \Lambda_{PQ}}$$

$$\sim \mathcal{O}\left(\frac{\Lambda_{QCD}^4}{\Lambda_{PQ}^2} \right). \qquad (15.65)$$

Part III
Looking beyond the Standard Model

16

Quest for CP violation in the neutrino sector

The smoke is now curling up
From the peak of Asama.[1]

Narihara

Neutrinos have had a most remarkable career already. They were first postulated by Pauli at the VII Solvay Congress in 1933 – to less than universal acclaim – as a theoretical crutch to balance energy, momentum and angular momentum in β decay: $n \rightarrow pe^-$.... At least one theorist of towering reputation – Bohr – was more inclined to limit the validity of energy conservation in the nuclear domain. It was not until 1953 that the existence of neutrinos was verified by Reines and Cowan [252]. In 1956/57 neutrinos (through their handedness) played the central role in the unfolding drama of the demise of parity as an absolute symmetry of nature. The coupled existence of ν_L and $\bar{\nu}_R$ gave some breathing space to **CP** remaining a true symmetry – till that notion had to be abandoned in 1964. In the 1970s neutrinos moved from their role of a ghostlike oddity – albeit with an essential impact – into the mainstream as an experimental probe: tremendous advances in accelerator and detector technologies provided us with intense and well-collimated high energy beams of neutrinos and the means to track their feeble interactions. This led to the discovery of new fundamental forces – the neutral currents – on one hand, and on the other enabled us to probe the internal structure of nucleons with amazing accuracy.

Another story-line that had started out as a sideshow (at best) was gaining importance, namely neutrino astronomy and cosmology, when Davis and collaborators targeted the Sun in 1978 from the Homestake Gold Mine. It was also realized that according to 'big bang' cosmology we are

---

[1] Asama is a famous volcano in Japan.

bathing in a background radiation of neutrinos analogous to the microwave photon radiation; even tiny neutrino masses would have grave cosmological consequences. Huge instruments built to search for proton decay were taken over by neutrino astronomy; new and even bigger detectors were and are being assembled with it as primary purpose.

In 1987 neutrinos joined the All Universe Club when some of their sisters and brothers reached Earth from supernova SN 1987A that erupted long ago in a galaxy far away, specifically 160 000 years ago in the Large Magellanic Cloud.

Within the SM neutrinos are massless. As such they are mass degenerate; flavour and mass eigenstates then coincide and by definition oscillations cannot occur. Yet no fundamental reason has surfaced why neutrinos should be truly massless. In 1957, Pontecorvo indicated that if lepton numbers are not conserved, $\nu \leftrightarrow \bar{\nu}$ mixing can occur [253]. The possibility that neutrinos have non-degenerate masses and mix just like quarks had been suggested by Sakata and his collaborators [254] back in 1962.[2] They stated the potential for neutrino oscillations.

In this chapter, we sketch the events that led to the discovery of neutrino oscillations. After describing the appropriate theoretical framework we discuss the prospects for detecting **CP** violation in the neutrino sector.

16.1 Experiments

There have been many experiments hinting at neutrino oscillations. Yet since those would signal such a fundamentally new paradigm, we had to be sure. Finally after a long wait definitive proof of neutrino oscillations has been presented. We shall now go over this discovery.

16.1.1 Solar neutrinos

There is a copious source of neutrinos in our corner of the milky way, the Sun. The chain reaction producing its energy output is given by

$$\begin{pmatrix} pp \to \ ^2\mathrm{H} + e^+ + \nu_e & [p-p], \ (0 \sim 0.4)\,\mathrm{MeV} \\ p + e^- + p \to \ ^2\mathrm{H} + \nu_e & [pep], \ (1.4)\,\mathrm{MeV} \end{pmatrix}$$
$$\Rightarrow \ ^2\mathrm{H} + p \to \ ^3\mathrm{He} + \gamma$$
$$\Rightarrow \ ^3\mathrm{He} + ^3\mathrm{He} \to \ ^4\mathrm{He} + 2p$$
$$\Rightarrow \ ^3\mathrm{He} + ^4\mathrm{He} \to \ ^7\mathrm{Be} + \gamma$$

---

[2] In the same paper they also introduced what is today known as the Cabibbo angle. Unlike Cabibbo in Ref. [255], they did not check the validity of this scheme by examining the experimental data, though.

$$\Rightarrow \begin{pmatrix} e^- +{}^7\text{Be} \rightarrow {}^7\text{Li} + \nu_e & [CNO], \ (0.86, \ 0.38)\,\text{MeV} \\ p +{}^7\text{Be} \rightarrow {}^8\text{B} + \gamma \end{pmatrix}$$

$$\Rightarrow {}^8\text{B} \rightarrow {}^8\text{Be} + e^+ + \nu_e \ [B], \ (0 \sim 15)\,\text{MeV}. \tag{16.1}$$

We have stated labels in [ ] and energy ranges in ( ) for each neutrino emitted. Experiments to detect such neutrinos have been going on for many years.

The tantalizing aspect was that while the experiments find the expected order of magnitude for the neutrino flux, all of them observe significantly fewer neutrinos than predicted by the Solar Standard Model (SSM) [256]. It will be seen that they employ very distinct methods and are sensitive to different parts of the neutrino spectrum. We are impressed by the experimental achievement in finding such an elusive signal. Yet before giving in to the temptation of accepting the existence of neutrino oscillations, we have to ask ourselves conscientiously whether we truly understand the Sun – for the stakes are too high.

Below we present results from various collaborations. Obviously we cannot do justice to all the hard and ingenious work put into obtaining these results. With apologies we confine ourselves to qualitative results which give information on the global nature of neutrino oscillations.

Homestake

The pioneering Homestake experiment [257] was the first to observe a deficit in the flux of neutrinos from the Sun. Those were detected by a chemical reaction:

$$\nu_e +{}^{37}Cl \rightarrow e^- +{}^{37}Ar. \tag{16.2}$$

Since the threshold for this reaction is a mere 0.814 MeV, the detector is sensitive to all solar neutrinos except for the $p - p$ neutrino. The detector – a tank containing 615 tons of dry cleaning fluid perchloroethylene (C_2Cl_4) – was placed in a gold mine located in Lead, South Dakota. The experiment started taking data in 1970. With argon exhibiting very different chemical behaviour from chlorine one can collect even a single argon atom. The experimental result and the SSM predictions are:

$$\phi(\text{Homestake}) = 2.56 \pm 0.23 \ \text{SNU}$$
$$\phi(\text{SSM}) = 7.7^{+1.2}_{-1.0} \ \text{SNU}. \tag{16.3}$$

Here SNU – Solar Neutrino Unit – is defined as 10^{-36} interactions per target atom per second.

For those of us who are not experts in the chemistry of extracting a few atoms from 615 tons of cleaning liquid, it was not easy to decide whether one should be duly impressed that the experiment found the expected flux within an order of magnitude or take the apparent deficit seriously. We just hoped it would be the latter and that this result would be verified – showing definitively that New Physics is around the corner.

GALLEX, SAGE

Both Gallium neutrino detectors – the GALLEX ([Gall]ium [Ex]periment) [258] located in the Gran Sasso National Laboratory and the SAGE (Ru[s]sian-[A]merican [G]allium [E]xperiment) [259], located in Baksan, Russia, under the 4000 m high Mt. Andyrchi, make use of [3]

$$\nu_e + {}^{71}Ga \rightarrow e^- + {}^{71}Ge. \tag{16.4}$$

This reaction has a threshold of 0.232 MeV and is sensitive to $p - p$ neutrinos which make up about 53% of all solar neutrinos. As it can be seen in Eq. (16.1), the predicted number of $p - p$ neutrinos can be related to the number of photons produced in the core of the Sun through ${}^{2}H + p \rightarrow^{3} He + \gamma$. The SSM prediction for this flux thus has less model dependence than those for CNO and B neutrinos.

The GALLEX experiment was upgraded to the Gallium Neutrino Observatory (GNO) [260] using 101 tons of a $GaCl_3$ solution in water and HCl. The SAGE experiment employs a metalic gallium target. The data taken from 1990–2000 yield

$$\phi(\text{GNO, GALLEX}) = 77.5 \pm 6.2^{+4.3}_{-4.7} \text{ SNU},$$
$$\phi(\text{SAGE}) = 70.8^{+5.3+3.7}_{-5.2-3.2} \text{ SNU},$$
$$\phi(\text{SSM}) = 126 \pm 10 \text{ SNU}. \tag{16.5}$$

These measurements certainly strengthened the case for the neutrino flux from the Sun being considerably lower than expected.

(Super-)Kamiokande

A huge water Cherenkov detector [261, 262] had been built in the Kamioka Zinc mine in Central Japan 1000 m under ground. The original purpose of the Kamiokande detector was to look for proton decays.[4] To this day, no such decay has been detected. Instead it pioneered the new field of neutrino astronomy.

Detecting the Cherenkov electrons allows – and this is a significant advantage over the radio-chemical methods listed above – to determine the direction of the incoming neutrinos. The experimental challenge consisted

---

[3] John Ellis coined the *bon mot* of the 'Alsace–Lorraine reaction': Gallium → Germanium → Gallium ...

[4] 'Kamiokande' stands for '*Kamioka nuclear decay* experiment'.

in bringing the detector energy threshold down to 4.5 MeV, which provided access to most B neutrinos.

The Kamiokande detector has been greatly expanded to the Super-Kamiokande (SK) experiment; the measured and expected neutrino fluxes read:

$$\phi(\text{Kamiokande}) = (2.80 \pm 0.19 \pm 0.33) \times 10^6 \,\text{cm}^{-2}\text{s}^{-1}$$
$$\phi(\text{SK}) = 2.35 \pm 0.02 \pm 0.08) \times 10^6 \,\text{cm}^{-2}\text{s}^{-1}$$
$$\phi(\text{SSM}) = (5.05^{+1.01}_{-0.81}) \times 10^6 \,\text{cm}^{-2}\text{s}^{-1}. \tag{16.6}$$

Again a substantial deficit has been observed in the neutrino flux with the SK result having become a precision measurement. The new feature here is that these neutrinos have been observed to come from the direction of the Sun.

SNO

The data listed above have created a "showdown" between the SSM and the no-oscillation statement of the SM: (at least) one of them has to give. The verdict in favour of neutrino oscilations and the SSM came from the Sudbury Neutrino Observatory (SNO) [263, 264]. The SNO detector is located in the Creighton mine, 6800 feet underground near Sudbury in Ontario province, Canada. The detector contains heavy water and can register three reactions:

$$\nu_e + d \rightarrow e^- + p + p \quad (\text{CC}), \tag{16.7}$$
$$\nu_x + d \rightarrow \nu_x + p + n \quad (\text{NC}), \tag{16.8}$$
$$\nu_x + e^- \rightarrow \nu_x + e^- \quad (\text{ES}), \tag{16.9}$$

where ν_x can be ν_e, ν_μ, or ν_τ, and CC, NC, and ES stand for charged current, neutral current, and electron scattering, respectively. Note that ν_μ and ν_τ cannot undergo CC interactions, since the incoming neutrinos do not have enough energy to produce a μ or τ in the final state. This reaction can thus measure the flux of ν_e, $\phi(\nu_e)$. SNO's definitive advantage is that it can also detect neutrinos ν_μ and ν_τ through the NC interaction Eq. (16.8); i.e. they can observe neutrinos resulting from the disappearance of ν_e. Finally the ES process can go through neutral current interactions which get contributions from all three neutrinos, but also through charge current interaction for ν_e. It thus gives a cross check on $\phi(\nu_e)$. Since $\sigma(\nu_{\mu,\tau}e \rightarrow \nu_{\mu,\tau}e) \simeq \sigma(\nu_e e \rightarrow \nu_e e)/6.5$,

$$\phi^{\text{CC}}(\nu_e) = \phi(\nu_e),$$
$$\phi^{\text{ES}}(\nu_x) = \phi(\nu_e) + \frac{\phi(\nu_\mu) + \phi(\nu_\tau)}{6.5},$$
$$\phi^{\text{NC}}(\nu_x) = \phi(\nu_e) + \phi(\nu_\mu) + \phi(\nu_\tau). \tag{16.10}$$

The data yield [265]:

$$\phi^{CC}(\nu_e) = (1.76^{+0.06}_{-0.05} \pm 0.09) \times 10^6 \, \text{cm}^{-2} \, \text{s}^{-1},$$

$$\phi^{ES}(\nu_x) = (2.39^{+0.24}_{-0.23} \pm 0.12) \times 10^6 \, \text{cm}^{-2} \, \text{s}^{-1},$$

$$\phi^{NC}(\nu_x) = (5.09^{+0.44+0.46}_{-0.43-0.43}) \times 10^6 \, \text{cm}^{-2} \, \text{s}^{-1}. \qquad (16.11)$$

One should note that $\phi^{NC}(\nu_x)$ is totally consistent with the solar standard model prediction:

$$\phi(\text{SSM}) = (5.21 \pm 0.27 \pm 0.38) \times 10^6 \, \text{cm}^{-2} \, \text{s}^{-1}. \qquad (16.12)$$

Our understanding of the Sun has been vindicated – due to the steadfast efforts by theorists, in particular by the late John Bahcall! The solar neutrino deficit thus has to be due to oscillations, and this is directly borne out by the data [264]:

$$\phi(\nu_\mu) + \phi(\nu_\tau) = (3.41 \pm 0.45^{+0.48}_{-0.45}) \times 10^6 \, \text{cm}^{-2}\text{s}^{-1} \neq 0. \qquad (16.13)$$

This is a definitive proof obtained from a single experiment that the flux of neutrinos emanating from the Sun contain μ and τ neutrinos by the time they arrive on Earth – i.e. that electron neutrinos oscillate into them! This can be expressed also through the relative ν_e flux:

$$\frac{\phi(\nu_e)}{\phi(\nu_x)} = 0.340 \pm 0.023^{+0.029}_{-0.031}. \qquad (16.14)$$

This ratio is definitely below unity, yet also below 0.5 – a point we will return to below.

16.1.2   Atmospheric neutrinos

When cosmic rays hit nuclei in the Earth's atmosphere, they produce cascades of mainly pions (and some kaons). Subsequent weak decays provide a source for neutrinos and antineutrinos:

$$\pi^- \to \mu^- \, \bar{\nu}_\mu$$
$$\hookrightarrow \nu_\mu e^- \bar{\nu}_e, \qquad (16.15)$$

$$\pi^+ \to \mu^+ \, \nu_\mu$$
$$\hookrightarrow \bar{\nu}_\mu e^+ \nu_e, \qquad (16.16)$$

with twice as many muon neutrinos as electron neutrinos *produced*.

    (Super-)Kamiokande has studied ν_e and ν_μ produced in the atmosphere as a function of the zenith angle θ. The latter is defined as follows: place the origin of the coordinate system in the centre of the detector with the

z axis chosen in the radial direction away from the centre of the Earth; θ is the angle between the z axis and the momentum of the incoming neutrino. The distance between the interaction point in the atmosphere, at which the neutrino is created, and the detector is then a function of θ. For example, for $\theta = 0$, the neutrino travels about 15 km, for $\theta = 90°$, 500 km and for $\theta = 180°$, 13 000 km – the diameter of the Earth. The neutrino flux as a function of the zenith angle thus provides a measure of oscillations as a function of the distance travelled. The analysis is not completely straightforward though. For one cannot assume a priori that the cosmic ray flux is uniform from all directions. Instead one relies on Monte Carlo computations of the expected flux to compare the data with. We just mention the main conclusions important for our discussion.

- At the present level of accuracy, ν_e does not oscillate. This does not contradict the fact described above that electron neutrinos coming from the Sun oscillate; for the distance and energy scales of solar and atmospheric neutrinos differ significantly with energies of at most 15 MeV and of $\mathcal{O}(1)$ GeV, respectively.
- There is a deficit in the ν_μ flux relative to expectations, and it increases with θ in full agreement with the oscillation hypothesis.

A more reliable way to state the results is to compare the *measured* ratio between the muon and electron neutrino fluxes with the one predicted by the Monte Carlo simulations, since then uncertainties in the cosmic ray flux and the relevant cross-sections drop out:

$$R \equiv \frac{\left(\frac{(N_{\nu_\mu} + N_{\bar{\nu}_\mu})}{(N_{\nu_e} + N_{\bar{\nu}_e})} \right)_{\text{data}}}{\left(\frac{(N_{\nu_\mu} + N_{\bar{\nu}_\mu})}{(N_{\nu_e} + N_{\bar{\nu}_e})} \right)_{MC}} \qquad (16.17)$$

Without neutrino oscillations (and/or neutrino decays) we expect $R = 1$.

The Kamiokande experiment and its US counterpart IMB, which likewise had originally been constructed to search for proton decay, found this ratio R to fall below unity. Yet it was Superkamiokande with its superior statistics and systematics that provided the definitive proof for $R < 1$. At the Neutrino 98 conference held at Takayama, Japan, in June 1998, they announced their findings [265]:

$$R = 0.66 \pm 0.06 \pm 0.08, \qquad (16.18)$$

which they interpret as muon deficit rather than electron excess.

16.1.3 *Man-made neutrinos*

KamLand

The KamLand [266] experiment consists of a 1000 ton liquid scintillation detector placed in the same cavern of the Kamioka mine that had housed the Kamiokande experiment, after the latter had been removed since superseded by SuperKamiokande. KamLand is a $\bar{\nu}_e$ *disappearance* experiment with all nuclear reactors in Japan and South Korea providing the sources for the $\bar{\nu}_e$ with an average distance of about 180 km. The neutrinos are detected through inverse β decay $\bar{\nu}_e + p \rightarrow e^+ + n$, which has a threshold energy of 1.8 MeV; the $\bar{\nu}_e$ energy is inferred from the positron energy. KamLand has observed a significant deficit in the $\bar{\nu}_e$ flux in full agreement with them oscillating, as shown in Fig. 16.1.

K2K

KEK, the high energy physics laboratory within Japan's High Energy Accelerator Research Organization is located some 250 km from the SuperKamiokande detector [268]. Charged pions produced at KEK are focused such that their decay products are aimed towards the SK detector. By the time these decay products haved passed through 250 km of the

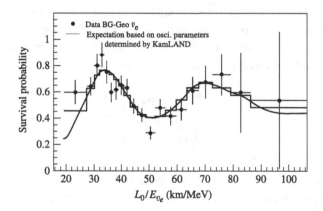

Figure 16.1 Ratio of the observed $\bar{\nu}_e$ spectrum to the expectation for no-oscillation versus L_0/E. It is compared with the oscillation hypothesis taking into account antineutrinos traveling from all sources and detector effects. The data points and expectations are plotted with $L_0 = 180$ km, as if all anti-neutrinos detected were due to a single reactor at this distance. The data are also compared with effects expected from neutrino decays, and neutrino decoherence. The second peak became apparent only recently as the background from geoneutrino (neutrino emitted by radioactivities present in rocks) has been measured and subtracted. The figure is from Ref. [267].

Earth, only the neutrinos can arrive there due to their feeble interactions. One can control the pion beam and thus the source of the neutrinos to a considerable degree; from the pion energy of $(2-3)$ GeV one infers that the decay neutrinos carry $(1.0-1.5)$ GeV.

This so-called K2K experiment has observed only 112 neutrino events at SK rather than the expected number of $158^{+9.2}_{-8.6}$. Furthermore, the L_0/E distribution for these neutrino events conforms with the one expected for oscillations, as explained in Fig. 16.1.

16.1.4   Qualitative summary

The experimental status can qualitatively be summarized in the following statements.

- Neutrinos coming from both the heavens – solar and atmospheric neutrinos – and from man-made contraptions – nuclear power stations and particle accelerators – have been shown to exhibit oscillations over different energy and distance scales.
- At the same time one should note that the Standard Solar Model has been empirically vindicated to an impressive degree.
- More specifically, electron as well as muon neutrinos oscillate into some other neutrinos with rates characterized by very different parameters. For those one finds consistent values as discussed below.
- The evidence for oscillations is not based merely on numbers – i.e. on seeing fewer neutrinos than expected – but also on more telling features like the dependence on the energies and path lengths of the neutrinos.
- The LSND experiment has reported evidence for ν_μ oscillations that does not fit into the overall pattern, at least not for three neutrino scenarios. Yet this evidence seems to be contradicted by the negative findings of the MiniBoone experiment.

16.2   Basics of neutrino oscillations

Drawing on analogies with $K^0 - \overline{K}^0$ and $B^0 - \overline{B}^0$ oscillations treated extensively before we can formulate the essence of neutrino oscillations. The neutrino mixing can be defined by

$$J_\mu = \bar{l}\gamma_\mu(1 - \gamma_5)\nu = \bar{l}^m \gamma_\mu(1 - \gamma_5)\mathbf{U}_{\text{PMNS}}\, \nu^m, \qquad (16.19)$$

where mass eigenstates of neutrinos and charged leptons are given by $(\nu^m)^{tr} = (\nu_3, \nu_2, \nu_1)$, and $(l^m)^{tr} = (\tau, \mu, e)$, respectively. $\mathbf{U}_{\text{PMNS}}$ is the mixing matrix, where PMNS stands for Pontecorvo [253], Maki, Nakagawa, and Sakata [254]. While the neutrino mixing matrix is usually labelled by MNS, we feel it is more appropriate to include P, as we shall see below that it is very likely that neutrinos are Majorana particles. However, we drop the subscript PMNS for individual matrix elements to save space. For a review on much of the subject covered below, see Ref. [269].

- The weak interaction eigenstates ν_e, ν_μ and ν_τ are non-trivial linear combinations of the mass eigenstates ν_i and their time dependences are governed by:

$$|\nu_\alpha(t)\rangle = \sum_{i=1}^{3} \mathbf{U}_{\alpha i} e^{-iE_i t} |\nu_i\rangle, \quad \alpha = e, \mu, \tau. \tag{16.20}$$

The unitary matrix $\mathbf{U_{PMNS}}$ can be parametrized as:

$$[e^{-i\phi'}] \begin{pmatrix} c_{12}c_{13} & s_{12}c_{13} & s_{13}e^{-i\delta} \\ -s_{12}c_{23} - c_{12}s_{23}s_{13}e^{i\delta} & c_{12}c_{23} - s_{12}s_{23}s_{13}e^{i\delta} & s_{23}c_{13} \\ s_{12}s_{23} - c_{12}c_{23}s_{13}e^{i\delta} & -c_{12}s_{23} - s_{12}c_{23}s_{13}e^{i\delta} & c_{23}c_{13} \end{pmatrix} [e^{i\phi}] \tag{16.21}$$

where $[e^{-i\phi^{(')}}]$ is a diagonal matrix defined by $[e^{-i\phi^{(')}}]_{ij} = \delta_{ij}e^{i\phi_i^{(')}}$.

- The transition probabilities for $\nu_\alpha \to \nu_\beta$ are easily obtained by evaluating $|\langle\nu_\beta|\nu_\alpha(t)\rangle|^2$. It is more useful to translate time into space, i.e. to express these probabilities as function of *distance* L from the production point rather than time t, since one can control L directly. For $m_i \ll E$, we have:

$$P(\nu_\alpha \to \nu_\beta; L)$$

$$= \delta_{\alpha\beta} - 4\sum_{i>j} \sin^2\left(\frac{m_i^2 - m_j^2}{4E} \cdot L\right) \cdot \text{Re}[\mathbf{U}_{\alpha i}\mathbf{U}_{\alpha j}^*\mathbf{U}_{\beta i}^*\mathbf{U}_{\beta j}]$$

$$+ 2\sum_{i>j} \sin\left(\frac{m_i^2 - m_j^2}{2E} \cdot L\right) \cdot \text{Im}[\mathbf{U}_{\alpha i}\mathbf{U}_{\alpha j}^*\mathbf{U}_{\beta i}^*\mathbf{U}_{\beta j}]. \tag{16.22}$$

Note that this expression does not depend on the phases in the diagonal matrix $[e^{i\phi}]$ nor $[e^{i\phi'}]$. So in the discussion of neutrino mixing, we shall ignore these diagonal phase matrices. For a given value of $\Delta m_{ij}^2 \equiv m_i^2 - m_j^2$, the ratio L/E characterizes the peculiar oscillation pattern. In an experiment one determines L/E independently (or even varies it in a controlled way) and then extracts the magnitude of Δm_{ij}^2 and the PMNS parameters.

The last term in Eq. (16.22) controlled by the imaginary parts of the PMNS parameters controls possible **CP** asymmetries, as will be discussed below.

- There are two classes of transitions we can probe, namely
  (1) 'appearance' reactions:

$$P(\nu_\alpha \to \nu_\beta; t) > 0 \quad \text{for} \quad \alpha \neq \beta, \tag{16.23}$$

(2) 'disappearance' reactions:[5]

$$P(\nu_\alpha \to \nu_\alpha; t) < 1. \tag{16.24}$$

16.2.1 Mass hierarchy

The expressions above assume the oscillations to occur in vacuum. This is an oversimplification. For solar neutrinos are produced mostly in the core of the Sun and thus have to traverse the Sun's bulk. Atmospheric neutrinos have to travel parts of the Earth to reach the detector. We will see below how oscillations are affected by the environment containing significant amounts of matter. For the moment we will ignore such effects.

Let us look at the results from the KamLand and K2K experiments. According to Eq. (16.22) neutrino oscillations are best observed, if

$$\frac{\Delta m_{ij}^2}{4E} \cdot L = 1.27 \Delta m_{ij}^2 [\text{in eV}^2] \frac{L[\text{in km}]}{E[\text{in GeV}]} \sim \mathcal{O}(1). \tag{16.25}$$

It reads for the two experiments as follows

$$1.27 \Delta m^2 [\text{in eV}^2] \frac{180}{(3-8) \times 10^{-3}} \sim \mathcal{O}(1) \quad KamLand$$

$$1.27 \Delta m^2 [\text{in eV}^2] \frac{250}{(0.2-3)} \sim \mathcal{O}(1) \quad K2K \tag{16.26}$$

and we infer the following sensitivity ranges for the two experiments:

$$\Delta m^2 \Big|_{KamLand} = (1.3 - 3.5) \times 10^{-5} \text{ eV},$$

$$\Delta m^2 \Big|_{K2K} = (0.63 - 9.4) \times 10^{-3} \text{ eV} ; \tag{16.27}$$

i.e. two distinct mass differences control ν_e and ν_μ oscillations.

---

[5] The disappearance of neutrinos could in principle also be caused by them decaying. This would require the intervention of New Physics as well; in any case we can distinguish the two scenarios by their dependence on t or L.

This greatly simplifies our analysis, since to a good approximation one can treat the oscillations as taking place between just two neutrinos. Yet as we will see it limits significantly the amount of **CP** violation in neutrino oscillations.

Without loss of generality – analogous to the meson case – we can define the mass eigenstate '2' to be heavier than state '1' and identify $\Delta m_{21}^2 > 0$ with the ν_e disappearance seen by KamLand and in the solar neutrino flux. This leaves two scenarios for the mass of the '3' state:

$$m_1 < m_2 \ll m_3 \quad \text{'Normal Hierarchy'}$$
$$m_3 \ll m_1 < m_2 \quad \text{'Inverted Hierarchy'}. \tag{16.28}$$

No observation made so far allows us to distinguish these two possibilities.

16.2.2   Estimating θ_{13} and θ_{12}

Let us consider the survival probability for ν_e with the values of Δm_{ij}^2 obtained above.

$$
\begin{aligned}
P(\bar{\nu}_e \to \bar{\nu}_e) = 1 &- 4|\mathbf{U}_{e2}|^2|\mathbf{U}_{e1}|^2 \sin^2(1.27\Delta m_{21}^2 L/E) \\
&- 4|\mathbf{U}_{e3}|^2|\mathbf{U}_{e1}|^2 \sin^2(1.27\Delta m_{31}^2 L/E) \\
&- 4|\mathbf{U}_{e3}|^2|\mathbf{U}_{e2}|^2 \sin^2(1.27\Delta m_{32}^2 L/E). \tag{16.29}
\end{aligned}
$$

Whenever one has a situation that can effectively be treated as a two-neutrino case – due to the smallness of PMNS parameters or the range of L/E probed – one finds for any diagonal transition

$$P(\bar{\nu}_\alpha \to \bar{\nu}_\alpha) = 1 - \sin^2(2\theta) \sin^2(1.27\Delta m^2 L/E); \tag{16.30}$$

i.e. the survival probability *averaged* over L/E can never fall below 1/2.

The CHOOZ experiment [270] was a long baseline reactor experiment where the neutrino with a typical energy of about 3 MeV was detected by a liquid scintillation counter located 1 km away. No ν_e disappearance was observed. In this kinematic region one is not sensitive to the first oscillation term $\sin^2(1.27\Delta m_{21}^2 L/E)$. Since $\Delta m_{32}^2 \simeq \Delta m_{31}^2$ and using three-family unitarity of the PMNS matrix one arrives at

$$P(\bar{\nu}_e \to \bar{\nu}_e) = 1 - 4|\mathbf{U}_{e3}|^2(1 - |\mathbf{U}_{e3}|^2) \sin^2(1.27\Delta m_{31}^2 L/E). \tag{16.31}$$

From the absence of a $\bar{\nu}_e$ deficit CHOOZ infers with $\Delta m_{31}^2 > 2 \times 10^{-3}\,\text{eV}^2$:

$$4|\mathbf{U}_{e3}|^2(1 - |\mathbf{U}_{e3}|^2) = \sin^2(2\theta_{13}) < 0.17. \tag{16.32}$$

Equation (16.31) applies also to the atmospheric ν_e flux studied by Super-Kamiokande, where no deficit is observed. $|\mathbf{U}_{e3}| \neq 0$ would generate such a deficit. All available data are consistent with

$$\mathbf{U}_{e3} = 0. \tag{16.33}$$

While such a scenario simplifies the analysis of oscillations, it is a curse in disguise. If any angle in the 3×3 PMNS matrix vanishes, then – in full analogy to the quark case – the complex phase can be rotated away, and no **CP** violation can arise in neutrino oscillations. Let us pray that nature assigned a significant value to $\mathbf{U}_{e3}$, which can be measured.

On the other hand for the KamLand experiment with $1.27\Delta m_{21}^2 L/E \sim \mathcal{O}(1)$ the last two terms in Eq. (16.29) oscillate very fast and $\sin^2(1.27\Delta m_{31}^2 L/E)$ can be averaged to $\frac{1}{2}$:

$$P(\bar{\nu}_e \to \bar{\nu}_e; \text{KamLand}) = |\mathbf{U}_{e3}|^4 + (1 - |\mathbf{U}_{e3}|^2)^2$$
$$- 4|\mathbf{U}_{e2}|^2|\mathbf{U}_{e1}|^2 \sin^2(1.27\Delta m_{21}^2 L/E). \tag{16.34}$$

From the best fit to the data one then obtains [266] (for $\theta_{13} = 0$):

$$\Delta m_{21}^2 = 6.9 \times 10^{-5} \text{ eV}^2$$
$$0.86 < \sin^2(2\theta_{12}) < 1.0. \tag{16.35}$$

16.2.3 Atmospheric neutrinos

Atmospheric neutrinos show no sign of a deficit in the electron neutrino flux or for $\nu_\mu \to \nu_e$. The present sensitivity does not allow us to probe $\nu_\mu \to \nu_\tau$ through the appearance of τ leptons in the final state. Therefore we focus on ν_μ disappearance.

In the energy (and distance) region relevant for atmospheric neutrinos, the contributions from the $\sin^2(1.27\Delta m_{32}^2 L/4E)$ term can be ignored, and we have:

$$P(\nu_\mu \to \nu_\mu) = 1 - 4|\mathbf{U}_{\mu3}|^2(1 - |\mathbf{U}_{\mu3}|^2)\sin^2(1.27\Delta m_{32}^2 L/4E). \tag{16.36}$$

From the SK measurement [262] of the ν_μ deficit one then infers

$$4|\mathbf{U}_{\mu3}|^2(1 - |\mathbf{U}_{\mu3}|^2) = \sin^2(2\theta_{23}) > 0.92. \tag{16.37}$$

16.3 Neutrino mixing parameters

Next we want to translate the experimental information into a profile of the neutrino mass eigenstates in terms to their relationship to lepton numbers. The free Schrödinger equation for the mass eigenstates ν_i^m

$$i\frac{\partial}{\partial t}\nu_i^m = \left(p + \frac{m_i^2}{2p}\right)\nu_i^m, \tag{16.38}$$

can be rewritten in terms of weak eigenstates ν_β:

$$i\frac{\partial}{\partial t}\nu_\beta = p\nu_\beta + \sum_{i,\alpha} \mathbf{U}_{\beta i}\frac{m_i^2}{2p}\mathbf{U}_{\alpha i}^*\nu_\alpha. \tag{16.39}$$

Unlike Eq. (16.38), Eq. (16.39) cannot be expressed with a *diagonal* mass matrix.

A detailed fit to all data undertaken by the PDG yields [11]:

$$\sin^2(2\theta_{12}) = 0.86^{+0.03}_{-0.04} \tag{16.40}$$

$$\Delta m_{21}^2 = (8.0^{+0.4}_{-0.3}) \times 10^{-5}\,\text{eV}^2 \tag{16.41}$$

$$\sin^2(2\theta_{23}) > 0.92 \tag{16.42}$$

$$\Delta m_{32}^2 = (1.9 - 3.0) \times 10^{-3}\,\text{eV}^2 \tag{16.43}$$

$$\sin^2(2\theta_{13}) < 0.19, \quad 90\%C.L. \tag{16.44}$$

The scenario of neutrino oscillations is thus similar to $\theta_{13} = 0$, $\sin(2\theta_{23}) = 1$. It can be shown (see Problem 16.2) that the Schrödinger equation then reads:

$$i\frac{\partial}{\partial t}\begin{pmatrix}\nu_e \\ \nu_\mu \\ \nu_\tau\end{pmatrix} = \left[\left(p + \frac{\overline{m}^2}{2p}\right)\mathbf{1} + \frac{\Delta m^2}{2p}\frac{1}{2}\begin{pmatrix}0 & 0 & 0 \\ 0 & 1 & 1 \\ 0 & 1 & 1\end{pmatrix}\right. \tag{16.45}$$

$$\left. - \frac{\delta m^2}{2p}\begin{pmatrix}c & -s/\sqrt{2} & s/\sqrt{2} \\ -s/\sqrt{2} & -c/2 & c/2 \\ s/\sqrt{2} & c/2 & -c/2\end{pmatrix}\right]\begin{pmatrix}\nu_e \\ \nu_\mu \\ \nu_\tau\end{pmatrix},$$

$$\tag{16.46}$$

where c and s denote cos and sin of $2\theta_{12}$, respectively, with mass eigenstates:

$$|\nu_3\rangle = \frac{1}{\sqrt{2}}(|\nu_\mu\rangle + |\nu_\tau\rangle) \qquad\qquad m_3^2 = \overline{m}^2 + \Delta m^2$$

$$|\nu_2\rangle = \sin\theta_{12}|\nu_e\rangle + \cos\theta_{12}\frac{1}{\sqrt{2}}(|\nu_\mu\rangle - |\nu_\tau\rangle) \quad m_2^2 = \overline{m}^2 + \delta m^2$$

$$|\nu_1\rangle = \cos\theta_{12}|\nu_e\rangle - \sin\theta_{12}\frac{1}{\sqrt{2}}(|\nu_\mu\rangle - |\nu_\tau\rangle) \quad m_1^2 = \overline{m}^2 - \delta m^2; \tag{16.47}$$

i.e. ν_μ and ν_τ mix completely, and ν_e decouples from ν_3. It seems unlikely that such a peculiar pattern arose by accident. It should therefore give us important clues towards a fundamental understanding of the lepton family structure. Alas – we have not yet decoded this message.

16.4 The MSW effect

So far we have aimed at a mainly qualitative understanding and have ignored subtleties like neutrinos oscillating differently while traveling through the earth. Without such a difference the solar neutrino flux is given by a simple expression obtained from Eq. (16.29) by replacing $\sin^2(1.27\Delta m_{21}^2 L/E)$ with $\frac{1}{2}$ due to integrating the source's position inside the Sun's core:

$$P(\bar{\nu}_e \to \bar{\nu}_e; Solar) = 1 - 2|\mathbf{U}_{e2}|^2|\mathbf{U}_{e1}|^2. \tag{16.48}$$

The formalism of neutrino oscillations in matter was first worked out by Wolfenstein [271] and applied to the solar neutrino problem by Mikheyev and Smirnov [272]; therefore it is referred to as the MSW effect. Consider a ν beam so low in energy that muons cannot be produced through scattering off nuclei and electrons.[6] While both ν_e and ν_μ undergo neutral current scattering, only ν_e can induce charged current interactions off nuclei as well as the electrons:

$$\nu_e + A \xrightarrow{\text{NC}} \nu_e + A, \ \nu_e + A \xrightarrow{\text{CC}} e + A', \ \nu_e + e \xrightarrow{\text{CC}} e + \nu_e,$$
$$\nu_\mu + A \xrightarrow{\text{NC}} \nu_\mu + A, \ \nu_\mu + A \not\xrightarrow{\text{CC}} \mu + A', \ \nu_\mu + e \not\xrightarrow{\text{CC}} \mu + \nu_e. \tag{16.49}$$

The extra force experienced by ν_e travelling through matter can be approximated by an effective interaction term

$$\mathcal{H}_{\text{eff}} = \frac{G_F}{\sqrt{2}}\bar{\nu}_e\gamma_\alpha(1-\gamma_5)\nu_e\bar{e}\gamma^\alpha(1-\gamma_5)e. \tag{16.50}$$

When the electron is nearly at rest and the neutrino relativistic, only the $\alpha = 0$ component is relevant and $\bar{e}\gamma^0(1-\gamma_5)e = n_e$ denotes the electron density:

$$\mathcal{H}_{\text{eff}} \simeq \sqrt{2}G_F n_e \approx 0.75 \times 10^{-5} \text{ eV}^2/\text{MeV}. \tag{16.51}$$

It is convenient to define

$$\xi = \frac{p}{\delta m^2}\mathcal{H}_{\text{eff}} \approx 0.2 \times p \,[\text{MeV}]. \tag{16.52}$$

The equation of motion gets modified, see Problem 16.2:

$$i\frac{\partial}{\partial t}\begin{pmatrix} \nu_e \\ \psi_- \\ \psi_+ \end{pmatrix} = \left[\left(p+\frac{\overline{m}^2}{2p}\right)\mathbf{1} + \frac{\Delta m^2}{2p}\begin{pmatrix} 0 & 0 & 0 \\ 0 & 0 & 0 \\ 0 & 0 & 1 \end{pmatrix} + \frac{\delta m^2}{2p}\begin{pmatrix} \xi & 0 & 0 \\ 0 & \xi & 0 \\ 0 & 0 & 0 \end{pmatrix} \right.$$
$$\left. + \frac{\delta m^2}{2p}\begin{pmatrix} \xi - c & s & 0 \\ s & -\xi + c & 0 \\ 0 & 0 & 0 \end{pmatrix}\right]\begin{pmatrix} \nu_e \\ \psi_- \\ \psi_+ \end{pmatrix}, \tag{16.53}$$

where $\psi_\pm = \frac{1}{\sqrt{2}}(\nu_\mu \pm \nu_\tau)$.

---

[6] This is the usual situation with ν beams from reactors and also from the Sun.

The eigenstates of the Hamiltonian shown in Eq. (16.53) are given as follows:

$$|\nu_3^M\rangle = \frac{1}{\sqrt{2}}(|\nu_\mu\rangle + |\nu_\tau\rangle) \qquad\qquad m_3^2 = \overline{m}^2 + \Delta m^2$$

$$|\nu_2^M\rangle = \sin\theta_M|\nu_e\rangle + \cos\theta_M\frac{1}{\sqrt{2}}(|\nu_\mu\rangle - |\nu_\tau\rangle) \quad m_2^2 = \overline{m}^2 + \delta m^2[\xi + \sqrt{(\xi - c)^2 + s^2}]$$

$$|\nu_1^M\rangle = \cos\theta_M|\nu_e\rangle - \sin\theta_M\frac{1}{\sqrt{2}}(|\nu_\mu\rangle - |\nu_\tau\rangle) \quad m_1^2 = \overline{m}^2 + \delta m^2[\xi - \sqrt{(\xi - c)^2 + s^2}]$$

$$(16.54)$$

where

$$\sin\theta_M = \sqrt{\frac{(\xi - c) + \sqrt{(\xi - c)^2 + s^2}}{2\sqrt{(\xi - c)^2 + s^2}}}$$

$$\cos\theta_M = \sqrt{\frac{-(\xi - c) + \sqrt{(\xi - c)^2 + s^2}}{2\sqrt{(\xi - c)^2 + s^2}}}. \qquad (16.55)$$

It is instructive to see what happens for large ξ: we then have $\sin\theta_M = 1$, and ν_e is also a mass eigenstate. After its birth the ν_e will therefore travel freely through the Sun. Yet when ν_e exits the core, ξ decreases, and the eigenvalue changes as follows:

$$\overline{m}^2 + 2\delta m^2\xi \rightarrow \overline{m}^2 + \delta m^2. \qquad (16.56)$$

The mass eigenstate changes adiabatically from ν_e to ν_2 as it exits the Sun and reaches the Earth. The probability to find again a ν_e while travelling from the Sun to Earth is $\sin^2\theta_{12}$, the value of $\sin^2\theta_M$ for $\xi = 0$. In realistic situations ξ is neither large nor small. The equation of motion Eq. (16.53) has then to be integrated numerically to get the mixing angle θ_M. It is straightforward to show that $\sin^2\theta_{12} < \sin^2\theta_M < 1$ holds; i.e. $\sin^2\theta_M$ certainly can be below 0.5. This resolves an otherwise puzzling feature in the solar neutrino flux, namely that the survival probability of ν_e falls below 0.5, see Eq. (16.14). The latter is inconsistent with neutrino oscillations *in vacuum* for an effective two-neutrino scenario.

16.5　Neutrino masses

Oscillation experiments tell us about mass *differences*. There have been intense efforts towards determining the mass of neutrinos. Since we cannot measure their energy and momentum directly, we deduce their mass from carefully analysing the kinematics of reactions where they appear. The highest sensitivity has been achieved for the electron neutrino by studying the endpoint region in tritium beta decay where the neutrino emerges almost at rest, Refs. [273, 274]. The interpretation of the data is highly non-trivial, since the sample tritium is embedded into large molecules; complex interactions then distort in particular the endpoint spectrum. The best fit is actually often obtained for negative values for the neutrino mass. For a summary of the experimental situation we refer you to Ref. [269]. The PDG upper bound reads [11]:

$$m_{\nu_e} \leq 2 \text{ eV} \quad \text{or} \quad \frac{m_{\nu_e}}{m_e} \leq 4 \cdot 10^{-6}. \tag{16.57}$$

For the other two neutrinos we find

$$\frac{m_{\nu_\mu}}{m_\mu} \leq 1.6 \cdot 10^{-3}, \quad \frac{m_{\nu_\tau}}{m_\tau} \leq 1.35 \cdot 10^{-2}. \tag{16.58}$$

Such tiny ratios call for an explanation. For Dirac neutrinos they are certainly unnatural, since there is no symmetry ensuring the smallness of the neutral to charged lepton mass. Even more generally, we know of no good a priori reason why neutrinos should be mass degenerate, let alone massless. Yet could there be a *qualitative* difference between the quark and the lepton world in this respect?

There is a mechanism that naturally leads to *tiny*, though non-zero, neutrino masses. It makes use of a property peculiar to neutral fermions which we will explain now.

Let us examine what the mass term has to look like under Lorentz symmetry. By writing:

$$\psi_R = \begin{pmatrix} \psi_+ \\ 0 \end{pmatrix}, \quad \psi_L = \begin{pmatrix} 0 \\ \psi_- \end{pmatrix}, \tag{16.59}$$

where $\psi_\pm$ are Weyl spinors with helicity $\lambda = \pm\frac{1}{2}$, the Dirac equation becomes

$$(E \mp \vec{\sigma} \cdot \vec{p})\psi_\pm = m\psi_\mp. \tag{16.60}$$

$\psi_\pm$ decouple from each other for $m = 0$. Consider general Lorentz transformations [275] for spinors which include rotations and boosts. It can be written as

$$\Lambda_\pm = \exp\left[i\frac{1}{2}\vec{\sigma} \cdot (\vec{\theta} \pm i\vec{\phi})\right] \tag{16.61}$$

and we have

$$\psi_\pm \to \Lambda_\pm \psi_\pm. \tag{16.62}$$

The physical interpretation of these operators is that $\exp[i\frac{1}{2}\vec{\sigma}\cdot\vec{\theta}]$ represents the rotation operator on $\psi_\pm$ around the direction of $\vec{\theta}$ by an angle $|\vec{\theta}|$, and

$\exp[\mp\frac{1}{2}\vec{\sigma}\cdot\vec{\phi}]$ represents a boost operator on $\psi_{\pm}$ along the direction of $\vec{\phi}$ by $|\vec{\phi}|$. Now consider another spinor $\chi_{\pm}$ which transforms as $\psi_{\pm}$. It can be shown that $\sigma_2\chi_{\pm}^*$ transforms as $\chi_{\mp}$. Noting the identity $\Lambda_{\pm}^{tr}\sigma_2\Lambda_{\pm} = \sigma_2$, we see that

$$\mathcal{L}_m = iM\chi_{\pm}^{tr}\sigma_2(\psi_{\pm}) + \text{h.c.} \tag{16.63}$$

is a Lorentz scalar [276]. The choice of the phase of the mass term will become obvious. In the general form for the mass matrix we can make two choices for χ_- which leads to two different types of the mass terms. By taking

$$\psi_R = \begin{pmatrix} i\sigma_2\chi_-^* \\ 0 \end{pmatrix}, \quad \psi_L = \begin{pmatrix} 0 \\ \psi_- \end{pmatrix}, \tag{16.64}$$

we get a Dirac mass matrix:

$$\mathcal{L}_D = iM_D\chi_-^{tr}\sigma_2\psi_- + \text{h.c.} = M_D\overline{\psi}_R\psi_L + \text{h.c.} \tag{16.65}$$

Taking the special case of Eq. (16.63), where $\chi_{\pm} = \psi_{\pm}$, and defining

$$\Psi_M = \begin{pmatrix} i\sigma_2\psi_-^* \\ \psi_- \end{pmatrix}, \tag{16.66}$$

we can write:

$$\mathcal{L}_M = iM_M\psi_-^{tr}\sigma_2\psi_- + \text{h.c.} = -M_M\overline{\Psi}_M\Psi_M. \tag{16.67}$$

Note that Ψ satisfies

$$\Psi_M^{\mathbf{C}} = \Psi_M. \tag{16.68}$$

This is called the Majorana condition, and Ψ_M is said to be a Majorana field.

By writing $\Psi_M = \frac{1}{\sqrt{2}}(\psi_L^{\mathbf{C}} + \psi_L)$, we can recast Eq. (16.67):

$$\mathcal{L}_M = M_M(\overline{\psi_L^{\mathbf{C}}}\psi_L + \overline{\psi}_L\psi_L^{\mathbf{C}}). \tag{16.69}$$

This is called a Majorana mass term.

To summarize, we have shown that there are two types of Lorentz scalars, thus two candidates for the mass term in the Lagrangian:

(1) Both left- and right-handed neutrino fields exist: $M_D\overline{\nu}_R\nu_L + \text{h.c.}$
(2) The left-handed neutrino and its charge conjugate – the right-handed antineutrino – are forced together: $M_M\overline{\nu_L^{\mathbf{C}}}\nu_L$.

The two combinations $\bar{\nu}_L \nu_R$ and $\bar{\nu}_L \nu_L^C$ in the mass term differ in one fundamental aspect: if ν carries charge q, ν^C carries charge $-q$; a term $\bar{\nu}\nu^C$ changes the charge by two units, whereas $\bar{\nu}\nu$ is neutral. A theory with left-handed neutrinos only can support neutrino masses – yet at the price of violating the corresponding lepton number. For charged fields a Majorana mass is incompatible with the conservation of electric charge.

16.6 Neutrino mixing with Majorana neutrinos

We now know that neutinos have small masses with tiny splittings. How can we understand that their pattern is so different from that among quarks? Generating neutrino masses a la quark masses appears *unnatural* in view of $\frac{m_\nu}{m_q} < \mathcal{O}(10^{-6})$.

There is an elegant way called the seesaw [277] mechanism. Consider the lepton sector to consist of three families of left-handed neutrinos and N_R families of $SU(2)_L$ singlet right-handed neutrinos. As seen in Eq. (16.65) and Eq. (16.69), the Lorentz symmetry allows for two types of mass terms. Write $\nu_L = (\nu_e, \nu_\mu, \nu_\tau)_L^{tr}$, and $\nu_R = (\nu'_1, \cdots \nu'_{N_R})_R^{tr}$. The general mass term in the Lagrangian can be written as:

$$\mathcal{L}_m = -\frac{1}{2}\begin{pmatrix} \overline{\nu_L^C} \\ \overline{\nu_R} \end{pmatrix}^{tr} \begin{pmatrix} 0 & m_D^{tr} \\ m_D & M_R \end{pmatrix} \begin{pmatrix} \nu_L \\ \nu_R^C \end{pmatrix} + \text{h.c.} \qquad (16.70)$$

where m_D is a $N_R \times 3$ Dirac mass matrix and M_R is a $N_R \times N_R$ Majorana mass matrix. Note that Majorana mass terms for ν_L cannot exist due to the $SU(2)_L \times U(1)$ symmetry of the Lagrangian. The elements of the Dirac mass matrix m_D are assumed to be comparable to quark masses arising from $SU(2)_L$ breaking in the SM; yet M_R reflects a scale (much) larger than $1\,\text{TeV}$ where some New Physics enters the stage; in the next chapter we will see that left–right gauge models provide a natural stage for that to happen. We define m_d and m_R to be the mass scales for the matrices m_D and M_R, respectively, and assume the hierarchy:

$$m_d \ll m_R. \qquad (16.71)$$

Before diagonalizing the mass matrix in Eq. (16.70), we prove that M_R is a symmetric matrix, i.e. $M_R^{tr} = M_R$. A general Majorana mass term can be written as:

$$\overline{\psi^C}\mathcal{M}\psi = \psi_i^\alpha \mathcal{M}_{ij} C^{\alpha\beta} \psi_j^\beta. \qquad (16.72)$$

Because of $C = -C^{tr}$ shown in Eq. (4.71), and Fermi statistics, $\mathcal{M}$ is a symmetric matrix. Therefore an unitary matrix can be found such that $\tilde{U}^{tr} M_R \tilde{U}$ is diagonal.[7]

By defining a unitary matrix $\mathbf{W}$ to be a matrix that relates the weak state to its mass eigenstate:

$$\begin{pmatrix} \nu_L \\ \nu_R^C \end{pmatrix} = \mathbf{W} \begin{pmatrix} \nu_L \\ \nu_R^C \end{pmatrix}^m \tag{16.73}$$

the Lagrangian becomes:

$$\mathcal{L}_m = -\frac{1}{2} \begin{pmatrix} \overline{\nu_L^C} \\ \overline{\nu_R} \end{pmatrix}^{m \ tr} \mathbf{W}^{tr} \begin{pmatrix} 0 & m_D^{tr} \\ m_D & M_R \end{pmatrix} \mathbf{W} \begin{pmatrix} \nu_L \\ \nu_R^C \end{pmatrix}^m + \text{h.c.}. \tag{16.74}$$

We shall see that the mass matrix can be diagonalized by:

$$\mathbf{W} = \begin{pmatrix} i\tilde{U}_L & m_D^{\dagger}(M_R^{-1})^{\dagger}\tilde{U}_R \\ -iM_R^{-1}m_D\tilde{U}_L & \tilde{U}_R \end{pmatrix}, \tag{16.75}$$

where $\tilde{U}_L$ and $\tilde{U}_R$ are unitary matrices, and we have neglected higher order terms in $\mathcal{O}(m_d/m_R)$. Taking advantage of the arbitrariness of the diagonal phase matrix in $\mathbf{W}$ mentioned above, we introduce the phase factor 'i' to the left column of the matrix to assure that the neutrino masses are positive. After the diagonalization, we have:

$$\mathcal{L}_m = -\frac{1}{2} \begin{pmatrix} \overline{\nu_L^C} \\ \overline{\nu_R} \end{pmatrix}^{m \ tr} \begin{pmatrix} \tilde{U}_L^{tr} m_D^{tr} M_R^{-1} m_D \tilde{U}_L & 0 \\ 0 & \tilde{U}_R^{tr} M_R \tilde{U}_R \end{pmatrix} \begin{pmatrix} \nu_L \\ \nu_R^C \end{pmatrix}^m + \text{h.c.}, \tag{16.76}$$

where we have neglected non-leading higher order terms of $\mathcal{O}(m_d/m_R)$, when appropriate. Thus we see that the flavor eigenstates are given by

$$\begin{pmatrix} \nu_e \\ \nu_\mu \\ \nu_\tau \end{pmatrix}_L = i\tilde{U}_L \begin{pmatrix} \nu_1 \\ \nu_2 \\ \nu_3 \end{pmatrix}_L \tag{16.77}$$

---

[7] There is a theorem [278] stating that a symmetric matrix $\mathcal{M}$ can be diagonalized as $\mathbf{W}^{tr} \mathcal{M} \mathbf{W} = D$, where D is diagonal with real positive diagonal elements, and $\mathbf{W}$ is a unitary matrix. To prove this theorem, first remember the Singular Value Decomposition theorem mentioned in Section 8.3.1 and Ref. [76], that we can always find unitary matrices $\mathbf{V}$ and $\mathbf{W}$ such that $\mathbf{V}^{\dagger} \mathcal{M} \mathbf{W} = D$, where D is a diagonal matrix with real positive diagonal elements. Since $\mathcal{M}^{\dagger} \mathcal{M} = \mathcal{M}^* \mathcal{M}^{tr} = (\mathcal{M} \mathcal{M}^{\dagger})^*$, we have $\mathbf{W}^{\dagger} \mathcal{M}^{\dagger} \mathcal{M} \mathbf{W} = D^{\dagger} D$ and by taking the complex conjugate, and using the fact that $D^{\dagger} D$ is real, we find $\mathbf{W}^{tr} \mathcal{M} \mathcal{M}^{\dagger} \mathbf{W}^* = D^{\dagger} D = \mathbf{V}^{\dagger} \mathcal{M} \mathcal{M}^{\dagger} \mathbf{V}$. So, we must have $\mathbf{V} = \mathbf{W}^*[e^{i\phi}]$ where $[e^{i\phi}]$ is an arbitrary diagonal phase matrix. Thus a unitary matrix $\mathbf{W}$ can be found so that $\mathbf{W}^{tr} \mathcal{M} \mathbf{W}$ is diagonal with positive diagonal elements.

where $\tilde{\mathbf{U}}_L$ is an orthogonal matrix defined by

$$\text{diag}[m_1, m_2, m_3] = \tilde{\mathbf{U}}_L^{tr} m_D^{tr} (M_R)^{-1} m_D \tilde{\mathbf{U}}_L \quad (16.78)$$

and the mass eigenstates are those that appear in Eq. (16.20).

From Eq. (16.75), we see that the resulting neutrino mass eigenstates *approximately* coincide with the chiral fields

$$\nu_L^m \simeq -i\tilde{\mathbf{U}}_L^\dagger \nu_L, \qquad (\nu_R^C)^m \simeq \tilde{\mathbf{U}}_R^\dagger \nu_R^C, \quad (16.79)$$

with masses

$$m_{\nu_L} \simeq \mathcal{O}\left(\frac{m_q^2}{m_R}\right), \qquad m_{\nu_R} \simeq \mathcal{O}(m_R), \quad (16.80)$$

where m_R is a number in the TeV range. Equation (16.79) and Eq. (16.80) show that the mass eigenstate ν_L^m is practically left handed and very light, whereas ν_R is almost right handed and very heavy. The phenomenon that $M_L \simeq 0 \ll m_D \ll M_R$ leads to *naturally* light neutrino states is called the *see-saw* mechanism; we know of no other way to achieve this goal.

16.7   Phases in the PMNS matrix

The leptonic PMNS matrix can be defined by

$$J_\mu = \bar{l}\gamma_\mu(1-\gamma_5)\nu = \bar{l}^m (\tilde{\mathbf{U}}_{\text{lep}})_L^\dagger \gamma_\mu(1-\gamma_5)(i\tilde{\mathbf{U}}_L)\nu^m, \quad (16.81)$$

where the PMNS matrix is given by:

$$\mathbf{U_{PMNS}} = (\tilde{\mathbf{U}}_{\text{lep}})_L^\dagger (i\tilde{\mathbf{U}}_L). \quad (16.82)$$

Some of the phases in $\mathbf{U_{PMNS}}$ can be removed by following the same procedure which takes advantage of the fact that quark phases are not observables. First write

$$\mathbf{U_{PMNS}} = [e^{-i\phi'}]\mathbf{V_{PMNS}}[e^{i\phi}], \quad (16.83)$$

The diagonal phase $[e^{i\phi'}]$ can be removed by redefining the phases of charged leptons. Now look what happens to the Lagrangian in Eq. (16.76), if we try to change the phases of the neutrinos. Because of the Majorana properties of the neutrino, $\nu_L \to [e^{i\phi}]\nu_L$ implies $\overline{\nu_L^C} \to \overline{\nu_L^C}[e^{i\phi}]$, and the Lagrangian is not invariant under this phase rotation. After we have taken advantage of redefining the phases of charged leptons, we obtain the leptonic PMNS matrix, which can be written as

$$\mathbf{U_{PMNS}} = \mathbf{V_{PMNS}}[e^{i\phi}]. \quad (16.84)$$

Of course one of the diagonal phases can be set to 0 as the overall phase of the interaction amplitude is not measurable.

16.8   CP and T violation in ν oscillations

Neutrino oscillations open the portal to novel searches for **CP** and/or **T** violation. Both searches can be performed independently of each other, thus allowing even for tests of **CPT** invariance. On phenomenological grounds it is conceivable that a **CPT** violation could surface there, despite previous negative searches in other systems. Of course, we would pay a heavy theoretical price for such a revolutionary discovery.

Let us denote by $P(\nu_\alpha \to \nu_\beta; t)$ the probability of finding neutrinos of type β at time t in a beam that at time $t = 0$ contained only neutrinos of type α. $P(\nu_\alpha \to \nu_\alpha; t) \neq 1$, $P(\nu_\alpha \to \nu_\beta; t) \neq 0$ for $\alpha \neq \beta$ represents manifestations of ν oscillations. **CP** invariance is then probed through comparing $P(\nu_\alpha \to \nu_\beta; t)$ with $P(\bar\nu_\alpha \to \bar\nu_\beta; t)$, whereas **T** violation is searched for in $P(\nu_\alpha \to \nu_\beta; t)$ versus $P(\nu_\beta \to \nu_\alpha; t)$. Accordingly, **CPT** symmetry is studied in $P(\nu_\alpha \to \nu_\beta; t)$ versus $P(\bar\nu_\beta \to \bar\nu_\alpha; t)$.

The appropriate formalism for going beyond these general statements is straightforward. Starting from Eq. (16.20) we note that

$$P(\nu_\alpha \to \nu_\beta; t) = \left| \sum_i e^{-i\frac{m_i^2}{2E}t} \mathbf{U}_{\alpha i} \mathbf{U}_{\beta i}^* \right|^2 = \left| \sum_i e^{-i\frac{m_i^2}{2E}t} \mathbf{V}_{\alpha i}^l \mathbf{V}_{\beta i}^{l*} \right|^2. \quad (16.85)$$

The last equality implies that the extra phases $[e^{i\phi}]$ present in **U**, when we deal with Majorana neutrinos, do not affect ν oscillations.

These transitions among light neutrinos do not exhibit a direct sensitivity to the presence of superheavy ν_R states. We have recovered the freedom to set the phases for the light neutrino mass eigenstates; i.e. decoupling is complete. For the antineutrinos we have the corresponding expression

$$\bar\nu_\alpha = \sum_i \overline{\mathbf{U}}_{\alpha i} \bar\nu_i^m, \quad (16.86)$$

where we have *not* assumed **CPT** invariance. Of course, as long as we are dealing with the mass terms discussed in Section 16.5, there is no **CPT** violation.

The transition probabilities corresponding to antineutrinos can be written in an obvious notation:

$$P(\bar\nu_\alpha \to \bar\nu_\beta; L) = \delta_{\alpha\beta}$$

$$- 4 \sum_{i>j} \sin^2 \left(\frac{\overline{m}_i^2 - \overline{m}_j^2}{4E} \cdot L \right) \cdot \mathrm{Re}[\overline{\mathbf{V}}_{\alpha i} \overline{\mathbf{V}}_{\alpha j}^* \overline{\mathbf{V}}_{\beta i}^* \overline{\mathbf{V}}_{\beta j}]$$

$$+ 2 \sum_{i>j} \sin \left(\frac{\overline{m}_i^2 - \overline{m}_j^2}{2E} \cdot L \right) \cdot \mathrm{Im}[\overline{\mathbf{V}}_{\alpha i} \overline{\mathbf{V}}_{\alpha j}^* \overline{\mathbf{V}}_{\beta i}^* \overline{\mathbf{V}}_{\beta j}].$$

$$(16.87)$$

CPT invariance implies

$$m_i = \overline{m}_i, \quad \mathbf{V}_{\alpha i}^{l*} = \overline{\mathbf{V}}_{\alpha i}^l; \tag{16.88}$$

i.e. mass-related parameters are equated for neutrinos and antineutrinos. There are actually seven physical quantities, namely the three masses m_i and the three angles, and one intrinsic complex phase by which the mixing matrix $\mathbf{V}^l$ can be described – in analogy to the CKM matrix. Symmetries impose constraints on these transition probabilities:

$$\begin{aligned} \mathbf{CP} : \quad & P(\nu_\alpha \to \nu_\beta; t) = P(\overline{\nu}_\alpha \to \overline{\nu}_\beta; t) \\ \mathbf{T} : \quad & P(\nu_\alpha \to \nu_\beta; t) = P(\nu_\beta \to \nu_\alpha; t). \end{aligned}$$

As expected on general grounds, these relations are violated only if the mixing matrix $\mathbf{V}^l$ contains irreducible complex phases.

For diagonal transitions, i.e. for $\alpha = \beta$, no **CP** asymmetry can arise since $\text{Im}|\mathbf{V}_{\alpha i}^l|^2|\mathbf{V}_{\alpha j}^l|^2 = 0$ holds. That is as expected, of course, since **CPT** invariance already implies

$$P(\nu_\alpha \to \nu_\alpha; t) = P(\overline{\nu}_\alpha \to \overline{\nu}_\alpha; t). \tag{16.89}$$

It is convenient to write

$$\begin{aligned} \Delta_{\alpha\beta}^{CP} &= P(\nu_\alpha \to \nu_\beta; L) - P(\overline{\nu}_\alpha \to \overline{\nu}_\beta; L) \\ &= 4 \sum_{i>j} \sin\left(\frac{m_i^2 - m_j^2}{2E} \cdot L\right) \cdot \text{Im}[\mathbf{V}_{\alpha i}\mathbf{V}_{\alpha j}^*\mathbf{V}_{\beta i}^*\mathbf{V}_{\beta j}]. \\ &= \pm 4J \left[\sin\left(\frac{m_3^2 - m_1^2}{2E} \cdot L\right) - \sin\left(\frac{m_3^2 - m_2^2}{2E} \cdot L\right) - \sin\left(\frac{m_2^2 - m_1^2}{2E} \cdot L\right)\right] \end{aligned}$$
$$\tag{16.90}$$

where we have used the unitarity of $\mathbf{U}$ and set

$$\begin{aligned} J &= \text{Im}[\mathbf{V}_{\alpha i}\mathbf{V}_{\alpha j}^*\mathbf{V}_{\beta i}^*\mathbf{V}_{\beta j}], \\ &= \frac{1}{8} \sin(2\theta_{12}) \sin(2\theta_{13}) \sin(2\theta_{23}) \cos(\theta_{13}) \sin\delta \\ &\le 0.05; \end{aligned} \tag{16.91}$$

the last bound is obtained from Eq. (16.44). Using Δm and δm defined in Eq. (16.47), the right-hand side can be written as:

$$\Delta_{\alpha\beta}^{CP} = \pm 4J \left[2\cos\left(\frac{\Delta m^2}{2E} \cdot L\right) \sin\left(\frac{\delta m^2}{2E} \cdot L\right) - \sin\left(\frac{\delta m^2}{E} \cdot L\right)\right]. \tag{16.92}$$

Measuring $\Delta_{\alpha\beta}^{CP}$ presents a complex challenge. Some remarks might illuminate the situation. First of all J has to be sufficiently large. At present

$J = 0$ is not ruled out, since we have only an upper bound on $\sin\theta_{13}$. There is nothing we can do about it, since the size of J is Nature's choice; yet we trust her not being malicious. Next we have to have ν_α and $\bar\nu_\alpha$ beams with known fluxes. Finally at high energies, with $\frac{\delta m^2}{2E} \cdot L << 1$, $\Delta^{CP}_{\alpha\beta}$ is suppressed by $J\frac{\delta m^2}{2E} \cdot L$. If we choose L/E such that $\cos\left(\frac{\Delta m^2}{2E} \cdot L\right)$ is rapidly oscillating, we have

$$\langle\Delta^{CP}_{\alpha\beta}\rangle \sim \pm 4J\sin\left(\frac{\delta m^2}{E} \cdot L\right). \tag{16.93}$$

Can ν_μ and $\bar\nu_\mu$ beams be obtained around the MeV reagion? Being optimists, we guess $4J \sim 0.2$ and remind experimentalists that the prize is worth the effort.

Probing lepton properties under **CP** and **T** transformations would constitute a unique opportunity, and we should make every effort to exploit it! For one it would complete the **CP** paradigm by manifesting **CP** violation also in leptodynamics. Due to the see-saw mechanism it would provide an even more intimate glimpse at dynamics at very high scales – and it could confirm an essential element of leptogenesis!

16.9   How to measure the Majorana phase?

We have seen that neutrino oscillation experiments are not sensitive to the diagonal phase $[e^{i\phi}]$ that appears in Eq. (16.84). Is there any way to get at this phase? The answer is yes – in principle. Consider double β decays($\beta\beta$ decay) of the type [279]:

$$_Z X^A \rightarrow_{Z+2} X^A + 2e^- + 2\bar\nu_R. \tag{16.94}$$

This decay can be seen only if the ordinary β decay $_Z X^A \rightarrow_{Z+1} X^A + e^- + \bar\nu_e$ is forbidden energetically (Fig. 16.2(a)). While the decay rate of this process was computed by Goeppert-Mayer [280] in 1935, it was not until 1987 that the first laboratory observation was made [281].

Neutrinoless double β decays have been considered a few years after the computation of the double β decay rate [282]. We do not derive its amplitude. The details are rather technical and given in Ref. [278]. The result is as follows:

$$T^{0\nu} = G_{0\nu}|M_{0\nu}|^2\langle m_{\beta\beta}\rangle^2, \tag{16.95}$$

where $G_{0\nu}$ is the phase space factor, $M_{0\nu}$ the nuclear matrix element, and

$$\langle m_{\beta\beta}\rangle = \left|\sum_{\nu_k^m} m_k(\mathbf{U_{PMNS}})^2_{1k}\right|. \tag{16.96}$$

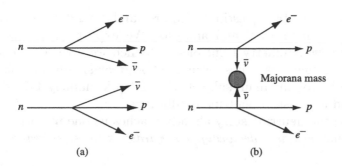

Figure 16.2   (a) A diagram responsible for double β decay where the final state contains $2\bar{\nu}_e$. As this is an ordinary β decay happening twice, certainly this will happen if the single β decay is forbidden by kinematics. (b) A diagram responsible for neutrinoless double β decay. The lepton number is violated. This violation is caused by the Majorana mass term. The neutrinoless double β decay is favored by phases space as there is not much Q values in these decays.

Using the actual representation of the $\mathbf{U_{PMNS}}$ matrix, one finds

$$|\langle m_{\beta\beta}\rangle|^2 = |m_1(c_{12}c_{13})^2 + m_2(s_{12}c_{13})^2 e^{2i\phi_2} + m_3(s_{13}e^{-1\delta})^2 e^{2i\phi_3}|^2$$

$$\sim \frac{m_1^2}{4} + \frac{m_2^2}{4} + m_3^2 s_{13}^4 + m_2 m_3 s_{13}^2 \cos 2(\delta + \phi_2 - \phi_3)$$

$$+ m_1 m_3 s_{13}^2 \cos 2(\delta - \phi_3) + \frac{1}{2}m_1 m_2 s_{13}^2 \cos 2\phi_2 \qquad (16.97)$$

where we have made use of $\sin(2\theta_{12}) \sim 1$, $c_{13} \sim 1$. We also know that $s_{13}^2 < 0.19$. It will be a while before we can even detect the presence of some Majorana phase. While it is possible in principle to obtain the value of $|\langle m_{\beta\beta}\rangle|^2$, determining the actual values of the Majorana phases ϕ_2 and ϕ_3 is another matter.

16.10   The bard's song

It has been a long-held speculation – or even suspicion – that neutrinos oscillate. This has been verified experimentally. The two mass differences that can be formed from three neutrinos provide two oscillation scales that have been observed with one being much larger than the other. In analogy to the quark sector, we have seen lepton flavor violation; the qualitative correspondence between quarks and leptons has prevailed again. It has led to a host of new questions. Why are the flavour mixing matrices so different for quarks and leptons? What is the structure of **CP** violation for leptons? Will the right-handed Majorana fields give rise to observable effects? And many more.

Such fundamental questions require and mandate dedicated and detailed experimentation and analysis. We expect such a world to be shaped by new symmetry principles, of which we give some examples in the following chapters. Those would be of great significance for a better understanding of the main subject of this book, namely **CP** violation in general and not just in neutrino oscillations.

However, the neutrino story already teaches us one important lesson of general value: *size is a decidedly poor yardstick for significance!*

Problems

16.1. Show that the survival probability in vacuum for a neutrino in a two-neutrino scenario can never fall below $1/2$ when averaged over L/E.

16.2. Show that the three states given in Eq. (16.47) are eigenstates for the equation of motion in Eq. (16.45), by first making the transformation $\psi_{\pm} = \frac{1}{\sqrt{2}}(\nu_{\mu} \pm \nu_{\tau})$, which leads to:

$$
i\frac{\partial}{\partial t} \begin{pmatrix} \nu_e \\ \psi_- \\ \psi_+ \end{pmatrix} = \left[\left(p + \frac{\overline{m}^2}{2p} \right) 1 + \frac{\Delta m^2}{2p} \begin{pmatrix} 0 & 0 & 0 \\ 0 & 0 & 0 \\ 0 & 0 & 1 \end{pmatrix} \right]
$$
$$
+ \frac{\delta m^2}{2p} \begin{pmatrix} -c & s & 0 \\ s & c & 0 \\ 0 & 0 & 0 \end{pmatrix} \begin{pmatrix} \nu_e \\ \psi_- \\ \psi_+ \end{pmatrix} ; \qquad (16.98)
$$

then diagonalize $\begin{pmatrix} -c & s \\ s & c \end{pmatrix}$.

16.3. We have shown in Section 16.4 that $\sin\theta_M = 1$ and $\sqrt{(1-c)/2}$ holds for $\xi = \infty$ and 0, respectively. Show that $\sin\theta_M$ is a monotonic function of ξ; i.e. it takes its extrema at the boundaries.

16.4. Consider a model for two lepton families coupling to charged and neutral currents. Show that if the two neutrinos are mass *degenerate*, the PMNS matrix is necessarily *diagonal!*

16.5. Does introducing a right-handed neutrino into the SM violate some fundamental principle like gauge invariance? If not, why not?

16.6. Consider, at $t = 0$, a massive $\nu_L = \frac{1}{2}(1 - \gamma_5)\nu$ with mass m moving along the z direction. Convince yourself that this is not a solution to the Dirac equation. Write this spinor as a linear combination of the solutions to the Dirac equation. Obtain the probability for $\nu_L \rightarrow \nu_R$ occurring at time t. Note that this probability is $O\left((m/E)^2\right)$ and negligible compared to transitions obtained from the mixing due to the off-diagonal term in the mass matrix.

16.7. Noting that

$$\sigma_2\sigma_i\sigma_2 = -\sigma_i^* \tag{16.99}$$

show that

$$\sigma_2\Lambda_\pm\sigma_2 = \Lambda_\mp^*$$
$$\sigma_2\Lambda_\pm^{-1}\sigma_2 = \Lambda_\pm^{tr}$$
$$\text{and} \quad \Lambda_\pm^{tr}\sigma_2\Lambda_\pm = \sigma_2, \tag{16.100}$$

where $\Lambda_\pm$ is defined in Eq. (16.61). Taking the complex conjugate of Eq. (16.62), show that

$$\sigma_2\psi_\pm^* \to \Lambda_\mp(\sigma_2\psi_\pm^*), \tag{16.101}$$

so we see that $\sigma_2\psi_\pm^*$ transforms like $\psi_\mp$.

16.8. Using Eq. (16.100), show that

$$\mathcal{L}_m = iM\chi_\pm^{tr}\sigma_2\psi_\pm \tag{16.102}$$

is a Lorentz scalar.

16.9. Given that an unitary transformation can be found so that $U^\dagger\mathcal{H}U$ is diagonal, any matrix M can be diagonalized by $V^\dagger U$. Take $V^\dagger M^\dagger MV$, and $U^\dagger MM^\dagger U$ to be diagonal. Show that for a symmetric M, $V = U^*$. Thus we can diagonalize a symmetric matrix M by $U^{tr}MU$.

16.10. In general we can write a Majorana field as

$$\chi = e^{i\alpha}\psi_L + e^{i\beta}(\psi_L)^{\mathbf{C}}, \quad \omega = e^{i\alpha'}\psi_R + e^{i\beta'}(\psi_R^{\mathbf{C}}). \tag{16.103}$$

Show that the Majorana condition becomes $\chi^c = e^{-i(\beta+\alpha)}\chi$.

17

Possible corrections to the KM ansatz: right-handed currents and non-minimal Higgs dynamics

As described before, CKM dynamics have passed their greatest challenge so far with flying colours. From a tiny **CP** asymmetry in K^0 decays it had been inferred that certain decays of B mesons exhibit **CP** asymmetries larger by two orders of magnitude. This prediction has been validated experimentally. At least the bulk of the observed effects is fully consistent with CKM dynamics as embedded within the SM. With the 'battle for supremacy' decided in favour of CKM theory, what remains to be settled is the question whether additional layers of dynamics are needed to describe **CP** violation. There is no direct evidence for the presence of New Physics in the decays of K, B and D mesons. On the other hand we know there must be a new source of **CP** violation, if baryogenesis has occurred in our universe, i.e. if the observed baryon number is not merely an arbitrary initial value of our universe.

We will discuss two generic corrections to the KM ansatz in this chapter. The first class – models with right-handed currents – has intriguing theoretical features, whereas the second class – a non-minimal Higgs sector – might be seen as an exercise in theoretical engineering. Yet the latter's redeeming feature is that it can find a natural habitat in supersymmetric extensions of the SM. Both scenarios had originally been introduced as competitors for the CKM ansatz and therefore had been calibrated to reproduce ϵ_K. We will sketch this early history, since we view it as of theoretical interest. Furthermore, some of the technical considerations might become relevant again in the future. These possible corrections tend to leave a relatively small footprint in B decays. The fact that no signal for them has been identified yet should thus not be seen as conclusive evidence for their absence. On the other hand, they are prime candidates for generating EDMs that are not only much larger than what can be produced within the SM, but also within experimental reach in the foreseeable future.

17.1 Left–right symmetric models

17.1.1 Basics

The observed parity violation is explicitely embedded into the SM with its electroweak gauge group $SU(2)_L \otimes U(1)$. It would be quite intriguing to perceive parity as a spontaneously broken symmetry[1] instead [283]. The simplest such scenario can be built on the gauge group [284]

$$SU(2)_L \otimes SU(2)_R \otimes U(1) \tag{17.1}$$

with the two $SU(2)$ couplings assumed to coincide at some (high) scale [285]

$$g_L(\mu_{LR}) = g_R(\mu_{LR}) = g. \tag{17.2}$$

At some scale below μ_{LR} the Higgs doublet field coupling to the gauge bosons of $SU(2)_R$ develops a vacuum expectation value, giving the latter a mass; at a (considerably) lower scale the process repeats itself for the gauge bosons of $SU(2)_L$. There are several attractive features to such a theory.

- We can entertain the hope of understanding the limitations of parity dynamically, rather than having to postulate them.
- The charge of the $U(1)$ group now finds a nice physical interpretation as $B - L$, the difference in baryon and lepton number.
- Such theories necessarily contain right-handed neutrino fields, and they allow naturally for large Majorana masses. Through the so-called 'seesaw' mechanism [277] we obtain the only known dynamic explanation for the long-standing mystery of why neutrinos are nearly massless.
- Closer to our main focus: also **CP** invariance can be treated as a spontaneously realized symmetry – the limitations of which can be related to those of parity. This might, however, turn out to be a 'curse in disguise'.
- The origin of **CP** violation might not reside in the family structure; this provides a counterpoint to the KM ansatz.

Since technical details can be found in excellent reviews in the literature [286, 287, 288, 289], we will merely sketch the basic structural elements of left–right symmetric models.

---

[1] We follow the customary language here, although one can argue that 'spontaneously realized symmetry' is a more fitting expression.

(1) The fields of up- and down-type quarks – U and D, neutral and charged leptons – N and E – are assigned to irreducible representations of the gauge group according to the following $(SU(2)_L, SU(2)_R, U(1))$ charges:

$$U_L, D_L: \left(\frac{1}{2}, 0, \frac{1}{3}\right) \qquad U_R, D_R: \left(0, \frac{1}{2}, \frac{1}{3}\right)$$
$$N_L, E_L: \left(\frac{1}{2}, 0, -1\right) \qquad N_R, E_R: \left(0, \frac{1}{2}, -1\right), \quad (17.3)$$

and the electric charge is given in terms of weak isospin and $B - L$ charge:

$$Q = I_{3L} + I_{3R} + \frac{B - L}{2}. \tag{17.4}$$

(2) There are three sets of gauge bosons for $SU(2)_L \otimes SU(2)_R \otimes U(1)$, namely $W_L^{\pm,0}$; $W_R^{\pm,0}$; A.

(3) The Higgs sector is considerably more complex than for the SM since it has to drive a multilayered reaction:

$$SU(2)_L \otimes SU(2)_R \otimes U(1)_{B-L} \overset{A}{\Longrightarrow} SU(2)_L \otimes U(1) \overset{B}{\Longrightarrow} U(1)_{QED}. \tag{17.5}$$

The vector bosons W_L and W_R can receive masses through the VEVs of Higgs fields Δ_L and Δ_R with quantum numbers $(1, 0, 2)$ and $(0, 1, 2)$, respectively. **P** violation is implemented spontaneously through $\langle \Delta_L \rangle \ll \langle \Delta_R \rangle$. Yet fermions cannot couple to these isovector fields. To generate their masses we introduce a third Higgs muliplet

$$\phi = \begin{pmatrix} \phi_1^0 & \phi_1^+ \\ \phi_2^- & \phi_2^0 \end{pmatrix} \tag{17.6}$$

with quantum numbers $(\frac{1}{2}, \frac{1}{2}, 0)$ and vacuum expectation values

$$\langle \phi \rangle = \begin{pmatrix} \kappa & 0 \\ 0 & \kappa' \end{pmatrix}. \tag{17.7}$$

With a hierarchy

$$\langle \Delta_L \rangle \ll \kappa, \kappa' \ll \langle \Delta_R \rangle \tag{17.8}$$

that can be achieved naturally, we can realize the sequence of Eq. (17.5). Henceforth we will set $\langle \Delta_L \rangle = 0$.

(4) The Higgs couplings induce $W_L - W_R$ mixing, leading to mass eigenstates

$$W_1 = W_R \sin\zeta + W_L \cos\zeta$$
$$W_2 = W_R \cos\zeta - W_L \sin\zeta, \qquad (17.9)$$

with

$$\tan\zeta = \frac{\kappa\kappa'}{v_R^2}, \qquad (17.10)$$

$$m_{W_1}^2 = \frac{g^2}{2}(\kappa^2 + (\kappa')^2) \ll m_{W_2}^2 = \frac{g^2}{2}(2v_R^2 + \kappa^2 + (\kappa')^2). \quad (17.11)$$

Likewise $A - W_L^0 - W_R^0$ mixing occurs, generating A_γ, Z_1 and Z_2 as fields of definite mass. Yet A, W_1 and Z_1 behave very much like the SM gauge bosons.

(5) Since natural flavour conservation has been implemented, **CP** violation can enter through the charged current couplings only. The Lagrangian for the coupling of W bosons to quark mass eigenstates reads as follows:

$$\mathcal{L}_{\mathrm{CC}} = \frac{g}{\sqrt{2}} \left[\left(\cos\zeta\, \overline{U}_L^m \gamma_\mu \mathbf{V}_L D_L^m + \sin\zeta\, \overline{U}_R^m \gamma_\mu \mathbf{V}_R D_R^m \right) W_1^\mu \right.$$
$$\left. + \left(-\sin\zeta\, \overline{U}_L^m \gamma_\mu \mathbf{V}_L D_L^m + \cos\zeta\, \overline{U}_R^m \gamma_\mu \mathbf{V}_R D_R^m \right) W_2^\mu \right].$$
$$(17.12)$$

(6) There are then three sources of **CP** violation.

- The matrix $\mathbf{V}_L$ reflecting the misalignment (in flavour space) of the quark mass matrices corresponds to the usual CKM matrix.
- Its right-handed analogue $\mathbf{V}_R$, which becomes observable due to the presence of right-handed currents.
- The fact that the two VEVs κ and κ' can turn out to be complex.

These sources are a priori independent of each other, and can yield a plethora of basic **CP** violating parameters. Rather than discussing the general case – hardly an enlightening enterprise – we will focus on two special cases with a particularly intriguing theoretical structure; common to both is that with parity being broken *spontaneously*, the Yukawa couplings form Hermitian matrices.

- *Manifest left–right symmetry*: If $\langle\phi\rangle$ remains real, the up- and down-type quark mass matrices are Hermitian; then

$$\mathbf{V}_L = \mathbf{V}_R = \mathbf{V}, \qquad (17.13)$$

and hence the name of this variant. **CP** violation enters through the CKM matrix only. The discrete symmetries **P** and **CP** are treated on a very different footing here: for **CP** violation arises in a *hard* way, namely in the Yukawa couplings described by dimension 4 operators.

- *Pseudo left–right symmetry*: Alternatively, let us assume also that **CP** is broken spontaneously, which severely constrains the Yukawa couplings. **CP** violation then arises through the VEVs of ϕ acquiring complex phases. The CKM matrices for the left- and right-handed currents are then not identical, yet closely related:

$$
\mathbf{V}_L = T_{L,U} T_{L,D}^\dagger
$$
$$
\mathbf{V}_R = J_U^* \mathbf{V}_L^* J_D, \tag{17.14}
$$

where $J_{U,D}$ are diagonal phase matrices, and T_{LQ} is defined in Eq. (8.22). While the effect enters through $W_L - W_R$ mixing, it can be parametrized in many equivalent ways. For example, we can define $W_L - W_R$ mixing as real and place the phases into $\mathbf{V}_R$. A detailed analysis shows that for the new contribution to ϵ_K from W_R, the third family can be neglected. So effectively we can set [290]

$$
\mathbf{V}^R = e^{i\gamma} \begin{pmatrix} e^{-i\delta_2} \cos\theta_C & e^{-i\delta_1} \sin\theta_C \\ -e^{-i\delta_1} \sin\theta_C & e^{-i\delta_2} \cos\theta_C \end{pmatrix}. \tag{17.15}
$$

A caveat should be noted here. With manifest left–right symmetry, it is difficult to generate baryon number when these theories are coupled to grand unified theories [291]. The left–right symmetry holds (if at all) only at high energy scales – and quite possibly only at scales that are immensely larger than the 1 TeV scale. Sizeable renormalization effects will then intervene and shift the balance between left and right handed couplings away from Eq. (17.2); this will obscure the underlying symmetry and make it non-manifest at the electroweak scale. Thus we should not take Eq. (17.2) and Eq. (17.14) as gospel. It should be kept in mind that a model with right handed currents *in general* contains a sizeable number of physical **CP** violating phases emanating from the mass matrices: for N families they number $[(N-1)(N-2) + N(N+1)]/2$, i.e. 1, 3, 6 for $N = $ 1, 2, 3, respectively.

However, one important feature of all these models should be emphasized: *the origin of* **CP** *violation does not necessarily reside in the family structure of quark (and other fermion) fields; it can enter through* W_L-W_R *mixing.*

17.1.2 The existing phenomenology in strange decays

The predictions most germane to left–right symmetry concern the existence of (1) additional weak vector bosons $W_R^{\pm,0}$ or $W_2^{\pm}$ and Z_2 and (2) non-sterile right-handed neutrinos, which also opens the gateway to the realm of neutrino masses and oscillations.

In high energy collisions we routinely search for the production of new vector bosons. None have been observed so far. The numerical bound derived from the lack of a signal depends on some assumptions, in particular concerning the strength of the relevant effective gauge coupling and whether the decay $W_R^+ \to l^+ \nu_R$ can occur; as discussed before, left- and right-handed neutrinos can possess vastly different (Majorana) masses – which turns out to be one of the most attractive features of left–right models. Studies by D0 and CDF of di-jet events yield, assuming $g_L \simeq g_R$ to hold at collider scales,

$$M_{W_R} \geq 800\,\mathrm{GeV} \qquad (17.16)$$

even for heavy right-handed neutrinos. A rather model-independent bound on $W_L - W_R$ mixing is obtained in deep inelastic $\bar{\nu}N \to \mu^+ X$ charged current reactions. The CCFR collaboration [292] gives:

$$|\zeta| \leq 0.04. \qquad (17.17)$$

One can search for right-handed currents also in $\mu^- \to e^- \bar{\nu}_e \nu_\mu$ through a detailed analysis of the generalized Michel parameters. Yet such searches are based on the assumption that right-handed neutrinos (and left-handed antineutrinos) are sufficiently light for them to be produced in μ (or π, K, etc.) decays. This is, however, far from assured and actually unnatural.

Numerically more impressive bounds on M_{W_R} can be obtained from processes that are forbidden – like $\mu \to e\gamma$ – or classically forbidden – like $K^0 - \overline{K}^0$ oscillations. Because their rates are highly suppressed in the SM, they are very sensitive to the presence of New Physics. Yet, this gain in sensitivity comes at a price: the specific features of the dynamical implementation of left–right symmetry become very important. This makes the bounds unstable; i.e. they can be weakened by adopting a particular choice of the relevant parameters and/or by allowing (or engineering) sizeable cancellations among the several possible contributions. Thus they should be viewed as typical scenarios, as benchmark figures, rather than firm bounds. The meaning of these bounds is also diluted by the fact that the SM can account at least for the bulk of **CP** violation in K and B decays. Yet even so we find the reasoning behind them instructive.

ΔS = 2 *transitions*

The effective Lagrangian is again obtained by integrating out the internal loop in box diagrams. Since $M_{W_L}^2/M_{W_R}^2 \ll 1$, we consider only two types of box diagrams: one with the exchange of two W_L and the other with one W_L and one W_R, see Fig. 17.1. Furthermore, we ignore for now the top quark contributions in the $W_L - W_R$ box: they are numerically insignificant for ΔM_K in any case; as far as ϵ is concerned we want to emphasize the impact of the new source of **CP** violation coming from $W_L - W_R$ mixing. We then obtain in a straightforward manner [293, 294]:

$$\mathcal{H}_{LR}^{\Delta S=2} \simeq \frac{G_F^2}{2\pi^2} \cdot M_{W_L}^2 x_c \text{log} x_c \frac{M_{W_L}^2}{M_{W_R}^2} (\mathbf{V}_{cd}^{L*}\mathbf{V}_{cs}^R\mathbf{V}_{cd}^{R*}\mathbf{V}_{cs}^L)$$

$$\times \left[\eta_4(\mu)\mathcal{O}_S + \eta_5(\mu) \left(\mathcal{O}_{LR} + \frac{2}{N_C}\mathcal{O}_S \right) \right], \qquad (17.18)$$

with $\eta_4 \sim 1.40$, $\eta_5 \sim 0.12$ at a renormalization scale of $\mu = 1\,\text{GeV}$,

$$\mathcal{O}_{LR} \equiv \left(\bar{d}\gamma_\mu(1-\gamma_5)s \right)\left(\bar{d}\gamma^\mu(1+\gamma_5)s \right)$$

$$\mathcal{O}_S \equiv \left(\bar{d}(1-\gamma_5)s \right)\left(\bar{d}(1+\gamma_5)s \right). \qquad (17.19)$$

The two operators $\mathcal{O}_S$ and $\mathcal{O}_{LR} + \frac{2}{N_C}\mathcal{O}_S$ possess a different chiral structure from the one encountered in the SM, which will reflect itself in their matrix elements. Factorization yields

$$\langle K^0|\mathcal{O}_S|\overline{K}^0\rangle \simeq 2\langle K^0|\bar{d}(1-\gamma_5)s|0\rangle\langle 0|\bar{d}(1+\gamma_5)s|\overline{K}^0\rangle$$

$$= -\frac{M_K^3}{m_s^2}F_K^2,$$

$$\langle K^0|\mathcal{O}_{LR} + \frac{2}{N_C}\mathcal{O}_S|\overline{K}^0\rangle \sim -F_K^2 M_K - \frac{2}{N_C}\frac{M_K^3}{m_s^2}F_K^3. \qquad (17.20)$$

This simple prescription does not reveal the μ dependence of these matrix elements, which has to cancel against the μ dependence of η_4 and η_5. On

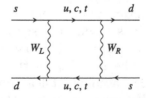

Figure 17.1   A new box diagram which generates $\Delta S = 2$ transition.

the other hand, there is no dramatic change in η_4 and η_5 when varying μ in the 'reasonable' range between 0.7 and 1.2 GeV.

The following should be kept in mind: (i) long-distance dynamics by themselves can generate about half of the observed value of ΔM_K; (ii) short-distance physics has to be called upon to generate the operators relevant for ϵ; predictions on this quantity in terms of the basic model parameters are thus more reliable; and (iii) W_L - W_R mixing can induce **CP** odd operators even without the intervention of the third family. Therefore the weight of right-handed currents could a priori be boosted considerably in ϵ since the d and s quarks couple so weakly to top quarks. Thus we can expect to deduce qualitative bounds on left–right model parameters from ΔM_K, and semiquantitative ones from ϵ. Yet even so they are instructive. For an orientation we keep just the term proportional to η_4 in Eq. (17.18) and assume manifest left–right symmetry:

$$(M_{12})_{LR} \simeq (M_{12})_{LL} \frac{6\eta_4}{B_K \eta_{cc}} e^{i(\delta_1 - \delta_2)} \cdot \log \frac{m_c^2}{M_{W_L}^2} \cdot \frac{M_{W_L}^2}{M_{W_R}^2} \cdot \frac{M_K^2}{m_s^2} ; \qquad (17.21)$$

$(M_{12})_{LL}$ is given in Eq. (9.42). Note that the contribution from the box diagram with the exchange of a W_L and W_R contains a huge enhancement factor

$$6 \left| \log \frac{m_c^2}{M_{W_L}^2} \right| \frac{M_K^2}{m_s^2} \sim 500$$

making up for its intrinsic suppression [293] by $M_{W_L}^2 / M_{W_R}^2$! This is caused by the confluence of three well understood enhancements: (i) a combinatorial factor; (ii) a much softer GIM suppression – $\log(m_c/M_{W_L})$ – which actually turns out to be an enhancement since $m_c \ll M_{W_L}$; (iii) a larger matrix element for the non-chiral operator $\mathcal{O}_S$. This means that the contribution from the LR box becomes comparable to that for the LL box for $M_{W_R} \sim 1$–2 TeV. A detailed numerical computation indeed yields [290]:

$$\frac{(\Delta M_K)_{LR}}{(\Delta M_K)_{LL}} \simeq -500 \cos(\delta_2 - \delta_1) \left(\frac{M_{W_L}}{M_{W_R}} \right)^2 ; \qquad (17.22)$$

ϵ_m, defined in Eq. (9.48), is given by

$$|\epsilon_{LR}| = \frac{1}{2\sqrt{2}} \frac{\operatorname{Im} M_{12}}{\operatorname{Re} M_{12}} = 176 \sin(\delta_2 - \delta_1) \left(\frac{M_{W_L}}{M_{W_R}} \right)^2 . \qquad (17.23)$$

Arguing that the long-distance contribution to ΔM_K cannot be significantly larger than the observed value and that therefore $\Delta M_K|_{\text{box}} > 0$ should hold, we infer $M_{W_R} \geq 1$ TeV from Eq. (17.22).

Equation (17.23) shows that in ϵ the LR contribution dominates over the LL one for a relatively light W_R. A priori it would be conceivable that a significant and even dominant fraction of ϵ is generated by the presence of right-handed currents – if W_R is not excessively heavy: $M_{W_R} \leq \mathcal{O}(20)\,\mathrm{TeV}$.

These considerations lead to the following tentative bounds: (i) to have Eq. (17.22) bounded by 1 requires $(M_{W_L}/M_{W_R})^2 \leq 1/500$. (ii) Eq. (17.23) implies $\frac{M_{W_L}^2}{M_{W_R}^2}\sin(\delta_2 - \delta_1) < 10^{-5}$.

$\Delta S = 1$ *transitions*

Since **CP** odd operators emerge without the intervention of the third family, the visibility of *direct* **CP** violation in strange decays might be boosted. Since there are several potential sources of **CP** violation, no definitive statement can be made, yet certain tendencies can be pointed out. Here we shall concentrate on those aspects of left–right symmetry that are distinct from the SM. For a detailed numerical analysis, we refer you to Refs. [290, 296, 295]. We limit ourselves to one-loop electroweak contributions without QCD radiative corrections. The relevant Feynman diagrams are given in Fig. 17.2:

$$\mathcal{H} = \mathcal{H}_{SM} + \mathcal{H}_R + \mathcal{H}_{LR} + \mathcal{H}_R^6 + \mathcal{H}_{LR}^6, \tag{17.24}$$

where $\mathcal{H}_{SM}$ is the SM Hamiltonian. The general expression for ϵ' is given in Eq. (7.34).

$$|\epsilon'| = \frac{1}{\sqrt{2}}\omega(\xi_0 - \xi_2) \ , \ \ \xi_I = arg A_I. \tag{17.25}$$

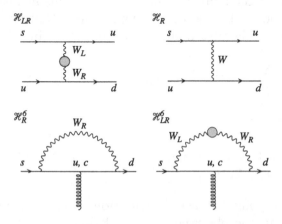

Figure 17.2   Feynman diagrams with new contributions to $\mathcal{H}(\Delta S = 1)$.

Because the $\Delta I = \frac{1}{2}$ enhancement *suppresses* the SM contribution to ϵ', New Physics can compete more favourably here, if it contributes to the $I = 2$ amplitude A_2. The following observations provide a framework for the phenomenological discussion.

(i) Since $\mathcal{H}_R^6$ and $\mathcal{H}_{LR}^6$ do not contribute to A_2, they are unlikely to cause a substantial modification to ϵ'.

(ii) From $\mathcal{H}_R$ generated by W_R exchange one obtains

$$(\xi_2)_R \sim \left(\frac{M_{W_L}}{M_{W_R}}\right)^2 \frac{\mathrm{Im}\,\mathbf{V}_{us}^{R*}\mathbf{V}_{ud}^R}{\mathbf{V}_{us}^*\mathbf{V}_{ud}} \frac{\langle(\pi\pi)_2|(\bar{s}u)_{V+A}(\bar{u}d)_{V+A}|K\rangle}{A_2}$$

$$\sim \sin(\delta_2 - \delta_1)\left(\frac{M_{W_L}}{M_{W_R}}\right)^2. \tag{17.26}$$

Using the aforementioned bound on $\frac{M_{W_L}^2}{M_{W_R}^2}\sin(\delta_2 - \delta_1) < 10^{-5}$, we obtain

$$\frac{(\epsilon')_{RR}}{\epsilon} \leq 2 \cdot 10^{-4}. \tag{17.27}$$

(iii) $W_L - W_R$ mixing yields

$$\xi_2 = \frac{\mathrm{Im}\,\langle(\pi\pi)_2|\mathcal{H}_{LR}|K^0\rangle}{A_2}$$

$$= \sin\zeta[\sin(\gamma - \delta_2) - \sin(\gamma - \delta_1)]\frac{4\langle\pi\pi;I=2|\bar{s}_L\gamma_\mu u_L\bar{u}_R\gamma^\mu d_R|K^0\rangle}{\langle\pi\pi;I=2|Q_2|K^0\rangle}. \tag{17.28}$$

The matrix elements are estimated as follows:

$$\langle(\pi\pi)_2|\bar{s}_L\gamma_\mu u_L\bar{u}_R\gamma^\mu d_R|K^0\rangle \approx \frac{2}{3}\langle\pi\pi;I=2|\bar{s}_L d_R\bar{u}_R u_L|K^0\rangle$$

$$\simeq -\frac{8}{3\sqrt{6}}\langle\pi^0|\bar{s}_L d_R|K^0\rangle\langle\pi^0|\bar{u}_R d_L|0\rangle = -i\frac{1}{6\sqrt{6}}\frac{M_K^2 - M_\pi^2}{m_s^0}\frac{M_\pi^2}{m_u^0}f_+ F_\pi \tag{17.29}$$

$$\langle(\pi\pi)_2|Q_2|K^0\rangle = \frac{8}{3\sqrt{6}}X, \tag{17.30}$$

where $X = F_\pi(M_K^2 - M_\pi^2)$. Putting these together, we have

$$\epsilon' \approx \frac{1}{5}\zeta[\sin(2\gamma - \delta_2) - \sin(2\gamma - \delta_1)]. \tag{17.31}$$

In a pseudo left–right scenario we find

$$\tan\zeta \simeq 2 \left(\frac{M_{W_L}}{M_{W_R}}\right)^2 \frac{\mathrm{Tr}\mathbf{M}_U}{\mathrm{Tr}\mathbf{M}_D}, \tag{17.32}$$

which, together with $M_{W_R} \geq 1.9\,\mathrm{TeV}$, leads to $\zeta \leq 6 \times 10^{-5}$ and $\epsilon'/\epsilon < 10^{-4}$.

With the wealth of weak phases available, no firm prediction can be made on ϵ'/ϵ [290, 295]. Yet in typical scenarios like those with *manifest* left–right symmetry contributions from right-handed currents tend to contribute to ϵ'/ϵ only on the 10^{-4} level and thus represent no more than a secondary effect. The more intriguing and novel case of having **CP** violation occur *spontaneously* only in $W_L - W_R$ mixing approximately leads to a superweak scenario with

$$\frac{\epsilon'}{\epsilon} \simeq 10^{-4}. \tag{17.33}$$

The reason for that is fairly straightforward: the concurrent enhancement factors for the W_R contribution to $\Delta S = 2$ amplitudes are basically absent in $\Delta S = 1$ amplitudes. Such a scenario is ruled out by the data.

Muon transverse polarization in $K_{\mu 3}$ decays

Muon transverse polarization in $K^+ \to \mu^+ \nu \pi^0$ can arise only if chirality conserving as well as violating couplings contribute with a relative weak phase, see Section 7.4. Yet with both W_L and W_R (or W_1 and W_2) couplings conserving chirality we need to consider box diagrams with W_L and W_R exchanges. Such a higher order effect makes $P_\perp$ unobservably tiny.

Final state asymmetries in hyperon decays

Here, too, there are several contributions of uncertain size. $\mathcal{H}_R$, $\mathcal{H}_R^6$ and $\mathcal{H}_{LR}^6$ contributions can be ignored relative to the largest one coming from $\mathcal{H}_{LR}$. The latter could generate polarization **CP** asymmetries in $\Lambda \to p\pi^-$ vs. $\bar{\Lambda} \to \bar{p}\pi^+$ that are considerably larger than in the SM, namely 10^{-4} and up to 6×10^{-4} for the S- and P-wave modes, respectively [296].

17.1.3   Electric dipole moments

EDMs represent a highly intriguing phenomenon for left–right models – particularly so when **CP** invariance is broken spontaneously, leading to the emergence of a weak phase through $W_L - W_R$ mixing:

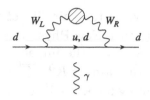

Figure 17.3 Feynman diagram which generates EDM in the left–right symmetric model.

- With **CP** breaking occurring softly, EDMs – being coefficients of dimension 5 operators – become calculable finite quantities.
- The leading contribution to the EDMs is actually coming from a one-loop diagram, as shown in Fig. 17.3. This is quite different from the SM where the EDM is produced only at the three-loop level. This means that we can expect to obtain *relatively* large effects.

Evaluating the diagram of Fig. 17.3, we find for the EDM of a neutron treated as a static combination of two d and one u quarks:

$$d_n \simeq \frac{1}{3}(4d_d - d_u) = e\frac{g_L g_R}{72\pi^2 M_{W_L}^2}\mathrm{sin}2\zeta\mathrm{sin}\theta_L\mathrm{sin}\theta_R$$
$$\times\ [5m_c\mathrm{sin}(\gamma+\delta_1) - m_s\mathrm{sin}(\gamma-\delta_1)]. \quad (17.34)$$

For $g_L \simeq g_R$, $\mathrm{sin}\theta_L \simeq \mathrm{sin}\theta_R$ and $\zeta \leq 10^{-3}$, we arrive at an order of magnitude estimate: $d_n \leq 10^{-24}\gamma$ e cm. *If* right-handed currents were to generate ϵ_K and ϵ' as observed, one would arrive at $d_n \sim \mathcal{O}\left(10^{-27}\right)$ e cm as a natural benchmark figure; i.e. at values for the neutron EDM not too far below present experimental bounds. The now more relevant comment is the following: *Even if these forces contribute little to observed* **CP** *violation in kaons, they can generate*

$$d_n \sim \mathcal{O}\left(10^{-28}\right) \text{ e cm ;} \quad (17.35)$$

i.e. values within 'striking distance' for on-going or scheduled experiments!

17.1.4 Prospects for **CP** asymmetries in beauty decays

We now know from experiments that b quarks decay predominantly via left-handed currents. This has been learnt from comparing the inclusive spectra of charged leptons and neutrinos in semileptonic beauty decays.

We do not anticipate any striking deviations from SM expectations in the usual $\Delta B = 1$ decay amplitudes for $B \rightarrow lX, \psi K, \pi\pi$ etc. if some right-handed couplings of reduced strength were added, since the rates are not particularly sensitive to the chirality of the underlying currents. There is more sensitivity to such New Physics in $B^0 - \overline{B}^0$ oscillations and radiative B decays; yet even there the impact of right-handed couplings will not be magnified, as is the case for $K^0 - \overline{K}^0$ oscillations. For two of the enhancement factors operating for $K^0 - \overline{K}^0$ oscillations cannot be enlisted here: (i) since $M_B \simeq m_b$ in contrast to $M_K \gg m_s$ there is no significant enhancement in the matrix elements of O_{LR} and O_S operators over those of O_{LL}; (ii) $B^0 - \overline{B}^0$ oscillations are driven almost exclusively by virtual top quarks with $m_t \sim \mathcal{O}(M_W)$ and no large GIM factor like $\log m_c^2/M_W^2$ arises.

The main impact – and actually a major one – could come from $\Delta B = 2$ Higgs couplings contributing to ΔM_{B_d} and ΔM_{B_s}; those quantities can be increased or decreased relative to SM expectations, and even their ratio could change. With the CKM fits allowing a self-consistent SM description there is no evidence for New Physics contributions. Yet again this is not a closed chapter due to the still sizable theoretical uncertainties in the SM value for ΔM_B. Right-handed couplings can introduce new weak phases; this would modify **CP** asymmetries observed in B^0 decays.[2] As mentioned before even a New Physics contribution that is non-leading for ΔM_{B_s} can provide the dominant one for time-dependent **CP** asymmetries in $B_s \rightarrow \psi\phi, \psi\eta, \phi\phi$. Right-handed currents might also affect Penguin-dominated transitions in an observable way, namely $B \rightarrow \gamma X$ and $B_d \rightarrow \phi K_S, \eta' K_S$ etc. A recent review [295] infers a lower bound of a few TeV on M_{W_R}, which makes it a marginal case for a direct discovery at the LHC, and of about 25 TeV for the mass of the flavour-changing neutral Higgs boson, which places it beyond the reach of the LHC. A sceptic might point out that explicit $SO(10)$ models tend to point to $M_{W_R} \gg 1$ TeV.

17.2 CP violation from Higgs dynamics

We are optimistic that the Higgs boson will be discovered at the LHC. Yet even then its sector will remain the least satisfying part of the SM. There are objections of a theoretical nature like the hierarchy problem due to the quadratic mass renormalization of scalars. Accordingly we refer to it as 'obscure' dynamics. On the other hand it can serve as a useful foil for CKM dynamics and, as discussed in the next chapter, its ills could be cured by supersymmetry.

---

[2] The relationship between the asymmetry in $B_d \rightarrow \psi K_S$ and in $B_d \rightarrow \pi\pi$ as expressed through $\phi_1 + \phi_2$ would *not* be touched, since New Physics enters through the same off-diagonal element of the B_d meson mass matrix, namely $\mathcal{M}_{12}$, rather than through a channel-specific decay amplitude.

As discussed in Section 8.2, gauge couplings can be defined to be real by rotating the phases of the gauge fields; yet in the presence of three families, quark mass matrices can introduce irreducible phases, which then enter the gauge interactions between quark mass eigenstates. Within the SM, the quark masses are generated by the quarks interacting with a Higgs doublet field which develops a vacuum expectation value. The seeds for these **CP** violating phases of the KM type are then in these Yukawa couplings and thus driven by Higgs dynamics. Nothing but simplicity keeps us from going beyond its minimal version as implemented in the SM. Once we do it, we have considerable latitude in constructing the Higgs sector.

- No bound on the number of Higgs doublets has been derived.
- A priori quarks can couple to more than one Higgs doublet.
- The self-interaction of Higgs fields can in general be rather complex.

CP breaking can then occur in a *spontaneous* as well as in a *manifest* fashion.

These models had been proposed a long time ago as the main or even only source for $K_L \rightarrow 2\pi$ [297, 298]. Such a motivation has evaporated now, yet we still see value in discussing Higgs scenarios: (i) while the huge variety of Higgs models prevents definite predictions, this can be seen also as a virtue. Since they are so rich in sources of **CP** violation, they amount to a 'teaching lab' for **CP** dynamics. (ii) Higgs dynamics could still generate non-leading, yet observable contributions to **CP** asymmetries in B decays. (iii) They can be the dominant source of EDMs and the main force behind baryogenesis. (iv) We can use **CP** studies as a tool, namely as a high sensitivity probe of Higgs dynamics. For reviews, see Refs. [299, 300, 301] and a very recent comprehensive report [302].

17.2.1 A simple example

For complex scalar fields ϕ a global phase rotation $\phi \rightarrow e^{i\beta}\phi$ leaves the theory invariant. Thus there is an arbitrary phase in the definition of **CP** transformations:

$$\mathbf{CP}\phi(t, \vec{x})\mathbf{CP}^{-1} = e^{i\delta}\phi^\dagger(t, -\vec{x}). \qquad (17.36)$$

Yet with at least two complex scalar fields ϕ_1 and ϕ_2 the relative phase cannot be removed. Let us see how **CP** violation can arise then. Define

$$\mathbf{CP}\phi_j(t, \vec{x})\mathbf{CP}^{-1} = \phi_j^\dagger(t, -\vec{x}) \qquad (17.37)$$

and rewrite it in terms of two Hermitian fields $\phi = s + ip$:

$$\mathbf{CP}s_j(t, \vec{x})\mathbf{CP}^{-1} = s_j(t, -\vec{x}), \quad \mathbf{CP}p_j(t, \vec{x})\mathbf{CP}^{-1} = -p_j(t, -\vec{x}) ; \qquad (17.38)$$

i.e. the real fields s and p represent Lorentz scalar and pseudoscalar fields, respectively. Consider the following mass terms:

$$\mathcal{L}_{\text{mass}} = \mu_1^2 \phi_1^\dagger \phi_1 + \mu_2^2 \phi_2^\dagger \phi_2 + \mu_{12}^2 \phi_1^\dagger \phi_2 + \text{h.c.},$$

where μ_{12} is assumed to be complex. The mass eigenstates obtained by diagonalizing this mass matrix are linear superpositions of the $s_{1,2}$ and $p_{1,2}$ and **CP** invariance is thus broken. In Ref. [303] an elegant, although unrealistic, model is given where such mass terms arise.

17.2.2 Sources of **CP** violation

Consider the three-family SM with a *second* Higgs doublet field added. The two Higgs doublets contain four neutral, two positively and two negatively charged fields. One neutral and two charged fields get eaten and become longitudinal Z^0, W^+ and W^- states, leaving five *physical* Higgs fields – three neutral and two charged. The most general Higgs potential is then given by [299]

$$\begin{aligned}
\mathcal{L}_H = {} & \mu_1^2 \phi_1^\dagger \phi_1 + \mu_2^2 \phi_2^\dagger \phi_2 + (\mu_{12}^2 \phi_1^\dagger \phi_2 + \text{h.c.}) \\
& - \lambda_1 (\phi_1^\dagger \phi_1)^2 - \lambda_2 (\phi_2^\dagger \phi_2)^2 - \lambda_3 (\phi_1^\dagger \phi_1 \phi_2^\dagger \phi_2) - \lambda_4 (\phi_1^\dagger \phi_2)(\phi_2^\dagger \phi_1) \\
& - \frac{1}{2} [\lambda_5 (\phi_1^\dagger \phi_2)^2 + \text{h.c.}] - (\lambda_6 \phi_1^\dagger \phi_1 + \lambda_7 \phi_2^\dagger \phi_2)(\phi_1^\dagger \phi_2 + \text{h.c.}).
\end{aligned}$$

$$(17.39)$$

The mass parameters μ_1 and μ_2 and the couplings $\lambda_1 - \lambda_4$ are necessarily real. If also μ_{12} and $\lambda_5 - \lambda_7$ happen to be real, **CP** invariance can still be broken spontaneously for $\lambda_5 > 0$ by a relative phase between the two VEVs:

$$\langle \phi_1^0 \rangle = \frac{v}{\sqrt{2}} \cos\beta \, e^{i\delta}, \qquad \langle \phi_2^0 \rangle = \frac{v}{\sqrt{2}} \sin\beta. \qquad (17.40)$$

If on the other hand μ_{12}, λ_6 and/or λ_7 are complex, $\mathcal{L}_H$ contains explicit **CP** violation as well; it is *soft* if only μ_{12} is complex.

These sources of **CP** violation, which do *not* exist for a single Higgs doublet, are in addition to the always existing possibility that the quark–Higgs couplings are complex, which represents hard **CP** violation:

$$\mathcal{L}_Y = \sum_{i,j} \left[\overline{Q}_{iL} (\Gamma_1^D)_{ij} \phi_1 D_{jR} + \overline{Q}_{iL} (\Gamma_2^D)_{ij} \phi_2 D_{jR} \right] + [D \leftrightarrow U] + \text{h.c.}$$

$$(17.41)$$

The up- and down-type quarks are denoted by $U = (u, c, t)$ and $D = (d, s, b)$, respectively; the matrices $\Gamma_{1,2}^{U,D}$ contain the Yukawa couplings.

The physical interpretation becomes more transparent when the dynamics are expressed for the quark and Higgs mass eigenstates. Once the $\phi_{1,2}$ acquire VEVs we obtain quark mass matrices

$$\mathcal{M}^D = \Gamma_1^D \langle \phi_1 \rangle + \Gamma_2^D \langle \phi_2 \rangle, \quad \mathcal{M}^U = \Gamma_1^U \langle \phi_1 \rangle + \Gamma_2^U \langle \phi_2 \rangle. \tag{17.42}$$

Upon diagonalizing, $\mathcal{M}^{D,U}$ flavour-changing neutral currents (FCNC) of the type $\bar{s}d\phi$, $\bar{c}u\phi$, $\bar{b}d\phi$, etc. will in general emerge – unless the transformations $T_{L,R}^{U,D}$ diagonalize *simultaneously*

$$\Gamma_1^{D,U} \langle \phi_1 \rangle + \Gamma_2^{D,U} \langle \phi_2 \rangle \text{ and } \Gamma_1^{D,U} + \Gamma_2^{D,U}. \tag{17.43}$$

In that case we have natural flavour conservation (NFC). This will not happen automatically, yet it can be guaranteed by imposing a discrete symmetry [304]

$$\phi_1 \to -\phi_1, \quad \phi_2 \to \phi_2, \tag{17.44}$$
$$D_R \to -D_R, \quad U_R \to U_R, \tag{17.45}$$

which prevents ϕ_1 from coupling to U_R and ϕ_2 to D_R. The same is achieved through the $U(1)$ symmetry

$$D_R \to e^{i\alpha} D_R, \quad \phi_1 \to e^{-i\alpha} \phi_1,$$
$$U_R \to U_R, \quad \phi_2 \to \phi_2. \tag{17.46}$$

The up- and down-type quark mass matrices are then controlled by a single VEV each

$$\mathcal{M}^U = \Gamma^U \langle \phi_2 \rangle, \quad \mathcal{M}^D = \Gamma^D \langle \phi_1 \rangle; \tag{17.47}$$

the $T_{L,R}^{U,D}$ diagonalize $\mathcal{M}$ and Γ simultaneously and no FCNC arise.

CP phenomenology with light fermions

There are thus two vastly different dynamical scenarios for Higgs dynamics, which have to be analysed *separately*, namely *with* or *without* FCNC. The difference between *spontaneous* and *manifest* **CP** violation – while fundamental – is of little *direct* importance for the phenomenological analysis; therefore we are going to refer to it only occasionally, mainly when considering the possible impact of higher loop effects. We begin by discussing light fermion systems.

Models with FCNC

In general FCNC will arise. Expressing the Yukawa couplings in terms of quark and Higgs *mass* eigenstates we derive from Eq. (17.41):

$$\mathcal{L}_Y^{\text{neut}}(\Delta S = 1) = \sum_{\alpha=2}^{4} \left[G_\alpha \bar{d}s + \bar{G}_\alpha \bar{d}i\gamma_5 s \right] H_\alpha, \tag{17.48}$$

where $H_1, \ldots, H_4$ are the neutral Higgs *mass* eigenstates with H_1 the Nambu–Goldstone boson, enabling Z^0 to become massive, and

$$G_\alpha = \frac{1}{2} \sum_\beta \left[(\tilde{\Gamma}^D_\beta + \tilde{\Gamma}^{D\dagger}_\beta)_{21} R^{(1)}_{\beta\alpha} + i (\tilde{\Gamma}^D_\beta - \tilde{\Gamma}^{D\dagger}_\beta)_{21} R^{(2)}_{\beta\alpha} \right]$$

$$\overline{G}_\alpha = \frac{1}{2} \sum_\beta \left[-i (\tilde{\Gamma}^D_\beta - \tilde{\Gamma}^{D\dagger}_\beta)_{21} R^{(1)}_{\beta\alpha} + (\tilde{\Gamma}^D_\beta + \tilde{\Gamma}^{D\dagger}_\beta)_{21} R^{(2)}_{\beta\alpha} \right], \quad (17.49)$$

with $\tilde{\Gamma}^D_{1,2} = T^D_L \Gamma^D_{1,2} (T^D_R)^\dagger$ and $R^{(1)}$ and $R^{(2)}$ denoting 2×4 matrices relating the H_i to the scalar and pseudoscalar components, respectively. Single exchange of such Higgs fields then generates

$$\mathcal{L}_{H-X}(\Delta S = 2) = \sum_\alpha \frac{-1}{M^2_\alpha} \left(G^2_\alpha (\bar{d}s)^2 - \overline{G}^2_\alpha (\bar{d}\gamma_5 s)^2 + 2i G_\alpha \overline{G}_\alpha (\bar{d}s)(\bar{d}\gamma_5 s) \right).$$

$$(17.50)$$

There are thus *tree* level transitions changing strangeness by two units – something that occurs in the SM only as a one-loop effect! Furthermore, since G_α and $\overline{G}_\alpha$ are in general complex, physical phases will arise in $\mathcal{L}_{H-X}(\Delta S = 2)$; i.e. in models with FCNC the presence of *two* Higgs doublets is sufficient to generate **CP** violation.

Invoking factorization for evaluating the relevant matrix elements should be quite adequate for our goal of obtaining order-of-magnitude estimates:

$$\langle K^0 | \mathcal{H}_{H-X}(\Delta S = 2) | \overline{K}^0 \rangle \simeq \frac{\tau^2}{M^2_H} \frac{F^2_K M^3_K}{(m_s + m_d)^2}, \quad \frac{\tau^2}{M^2_H} \doteq \sum_\alpha \frac{\overline{G}^2_\alpha}{M^2_\alpha}, \quad (17.51)$$

where M_H is the lightest Higgs mass. If we require that ΔM_K is not oversaturated – i.e. $2|\langle \overline{K}^0 | \mathcal{H}_{H-X}(\Delta S = 2) | K^0 \rangle| \leq \Delta M_K|_{\text{exp}}$ – we find $|\tau/M_H| \geq 1 \, \text{TeV}$, which should be interpreted as a rough benchmark figure only.

The situation becomes somewhat more definite when we consider ϵ_K: requiring $\text{Im} \langle K^0 | \mathcal{H}_{H-X}(\Delta S = 2) | \overline{K}^0 \rangle / \Delta M_K \leq \sqrt{2} |\epsilon_K|_{\text{exp}}$ we infer

$$M_H \geq (30 \cdot \text{Im} \, \tau) \, \text{TeV}. \quad (17.52)$$

The model dependence enters through the quantity τ defined in Eq. (17.51) with an a priori reasonable range of $\sim 1/30 - 1$; the observed value of ϵ_K could then be reproduced by the exchange of a very heavy Higgs field: $M_H \sim 1$–$30 \, \text{TeV}$ [305]. In this type of model $\Delta S = 1$ and $\Delta S = 2$ Higgs mediated transitions are of the same order, namely tree level; with the $\Delta S = 2$ amplitude being comparable to a second-order electroweak SM amplitude, *direct* **CP** violation in the $\Delta S = 1$ amplitude would then be greatly suppressed; for all practical purposes we would typically have a *dynamical* realization of the superweak ansatz for **CP** violation [69]. Giving up the requirement of Higgs dynamics generating ϵ_K, the mass of this non-SM neutral Higgs could of course be much heavier.

These remarks serve as general orientation rather than quantitative predictions for these models with FCNC, since the values of τ and $\mathrm{Im}\tau$ can conceivably vary by orders of magnitude. Furthermore, charged Higgs as 'light' as $\sim 100\,\mathrm{Gev}$ can exist, and through complex Yukawa couplings generate **CP** asymmetries in $\Delta S = 1$ transitions. In that case one could reproduce the observed value of ϵ_K whereas getting only a negligible contribution to ϵ'. While such contributions are not needed to reproduce the observed **CP** asymmetries in K and B decays, it is instructive to analyse different subscenarios. They center on the question of how NFC is violated.

(1) The discrete symmetry of Eq. (17.44) is violated *softly* by the μ_{12}^2 term. **CP** violation can enter the quark sector if the symmetry transformation of Eq. (17.45) for down-quark fields is modified to $\{d_{i,R}\} \rightarrow \{\eta_i d_{i,R}\}$ with η_i being $(+1)$ for some families and (-1) for others [306]. This class of models, according to [307], is of the genuinely superweak variety: it does not generate a KM phase.

(2) The invariance under Eq. (17.44) and Eq. (17.45) is violated by $\mathcal{L}_H$ as well as $\mathcal{L}_Y$ – i.e. in a hard fashion – yet by small numerical amounts. Such models contain tree level $\Delta S = 2$ transitions that violate **CP**. In addition a non-zero KM phase is induced; ϵ'/ϵ_K – while not vanishing – is expected to lie in the range 10^{-6} to 10^{-4} [308]; for all practical purposes this represents a superweak scenario as well.

(3) We abandon the idea of an approximate discrete symmetry and allow for ϕ_1 and ϕ_2 both coupling to U_R as well as D_R, as expressed in Eq. (17.41). On the other hand, we invoke some approximate $U(1)$ family symmetries to impose a special structure on the matrices $\Gamma_{1,2}^{U,D}$: they are assumed to have small off-diagonal elements, namely $\sim\mathcal{O}(0.01)$ or $\mathcal{O}(0.1)$ of the corresponding diagonal ones. This scheme has an advantage that upon diagonalization of the quark mass matrices the known form of the CKM matrix emerges and FCNC are reduced to a phenomenologically acceptable level.

$\mathcal{L}_Y$ is then re-written in terms of complex quark and Higgs mass eigenstates. The original Higgs fields ϕ_1^0 and ϕ_2^0 are expressed through neutral fields shifted by the VEVs:

$$\frac{1}{\sqrt{2}}(v + H^0 + G^0) = \cos\beta\, e^{-i\delta}\phi_1^0 + \sin\beta\,\phi_2^0$$

$$\frac{1}{\sqrt{2}}(R + iI) = \sin\beta\, e^{-i\delta}\phi_1^0 - \cos\beta\,\phi_2^0, \qquad (17.53)$$

where G^0 denotes the Goldstone boson absorbed into the Z^0; H^0, R and I are physical neutral Higgs fields. The mass eigenstates (H_1^0, H_2^0, H_3^0) are related to (R, H^0, I) by an *orthogonal* transfomation O^H.

We divide $\mathcal{L}_Y$ into two parts. (i) If the matrices $\Gamma_{1,2}^{U,D}$ were truly diagonal no FCNC would emerge and the flavour changes mediated by $H^\pm$ exchanges would be controlled by the CKM matrix; all these terms are written into $\mathcal{L}_1$. (ii) The off-diagonal elements of $\Gamma_{1,2}^{UD}$ assumed to be small induce FCNC and additional $H^\pm$ couplings; they are lumped into $\mathcal{L}_2$.

Thus we write [299]:

$$\mathcal{L}_Y = (\sqrt{2}G_F)^{1/2}(\mathcal{L}_1 + \mathcal{L}_2), \tag{17.54}$$

$$\mathcal{L}_1 = \sqrt{2}\left(H^+ \sum_{i,j}^3 \mathbf{V}_{ij}\xi_{d_j} m_{d_j} \overline{u}_L^i d_R^j - H^- \sum_{i,j}^3 \mathbf{V}_{ji}^* \xi_{u_j} m_{u_j} \overline{d}_L^i u_R^j \right)$$

$$+ H^0 \sum_i^3 \left(m_{u_i}\overline{u}_L^i u_R^i + m_{d_i}\overline{d}_L^i d_R^i \right)$$

$$+ (R + iI)\sum_i^3 \xi_{d_j} m_{d_j} \overline{d}_L^i d_R^i + (R - iI)\sum_i^3 \xi_{u_j} m_{u_j} \overline{u}_L^i u_R^i + \text{h.c.}, \tag{17.55}$$

$$\mathcal{L}_2 = \sqrt{2}\left(H^+ \sum_{i,j,k\neq j}^3 \mathbf{V}_{ik}\mu_{kj}^d \overline{u}_L^i d_R^j - H^- \sum_{i,j,k\neq j}^3 \mathbf{V}_{ki}^* \mu_{kj}^u \overline{d}_L^i u_R^j \right)$$

$$+ (R + iI)\sum_{ij\neq i}^3 \mu_{ij}^d \overline{d}_L^i d_R^j + (R - iI)\sum_{ij\neq i}^3 \mu_{ij}^u \overline{u}_L^i u_R^j + \text{h.c.}, \tag{17.56}$$

with the factors $\xi_{d_j} m_{d_j}, \xi_{u_j} m_{u_j}$ and $\mu_{ij}^{d,u}$ arising mainly from the large diagonal and small off-diagonal elements, respectively, of $\Gamma_{1,2}^{U,D}$.

Since we have postulated merely the smallness – not the absence! – of various terms, we should not be surprised by the structural complexity of the weak Lagrangian and the multitude of ways in which **CP** invariance can be broken. Those can be grouped into four major categories.

• The irreducible phase in the CKM matrix induces **CP** violation into $W^\pm$ and $H^\pm$ exchanges.

- The factors $\xi_{d,u}$ can contain additional phases; those can generate **CP** violation also in the flavour changing R and I exchanges.
- Phases in the factor $\mu_{ij}^{u,d}$ drive **CP** violation in FCNC.
- Finally, **CP** symmetry can be broken by the Higgs mixing matrix O_H. This might seem surprising at first since O_H is orthogonal, i.e. real. Yet it comes about in the following way [302, 309]. **CP** invariance requires all couplings and VEVs to be real (or that they can be all made real by adopting an appropriate phase convention for the Higgs fields). The R and I components cannot mix then and the mass eigenstates H_i^0 are **CP** eigenstates as well. However, if the VEVs develop a relative phase, then R and I can mix, as illustrated by Eq. (17.53), and the mass eigenstates are no longer **CP** eigenstates. The most intriguing consequences arise in the Higgs gauge dynamics: while Bose statistics forbids $Z^0 \to H_i^0 H_i^0$ to proceed, it is **CP** invariance that forbids $Z^0 \to H_i^0 H_j^0$ with $i \neq j$ if both H_i and H_j are either **CP** even or **CP** odd.

A priori such models could have a rich **CP** phenomenology both in $\Delta S = 1\&2$ transitions although so far none of it has surfaced. In particular, direct **CP** violation can arise in the $\Delta S = 1$ transition due to $H^\pm$ exchanges generating a value for ϵ'/ϵ_K around $10^{-5} - 10^{-4}$ for $\tan \beta \sim 1$ that can rise to 10^{-3} for $\tan \beta \gg 1$.

Models without FCNC

Imposing the transformations in Eq. (17.44) and Eq. (17.45) as symmetry leads to NFC; up-type and down-type quarks each couple to a different Higgs doublet only – a feature that is preserved by quantum corrections – and the absence of FCNC is achieved in a *natural* way. The first explicit model of this type had been put forward in Ref. [298]. It contains three Higgs doublets of which two couple to quarks in the following way:

$$\mathcal{L}_Y = \sum_{i,j} \left(\Gamma_{ij}^D \overline{Q}_{iL} \Phi_1 D_{jR} + \Gamma_{ij}^U \overline{Q}_{iL} \tilde{\Phi}_2 U_{jR} \right) + \text{h.c.} \qquad (17.57)$$

The third doublet Φ_3 remains detached from the quarks; it may be coupled to leptons. The Yukawa couplings are assumed to be real to have **CP** broken spontaneously. We postulate the Higgs potential to generate *complex* VEVs:

$$\langle \Phi_1 \rangle = e^{i\eta_1} v_1, \quad \langle \Phi_2 \rangle = e^{i\eta_2} v_2, \quad \langle \Phi_3 \rangle = v_3, \qquad (17.58)$$

where without loss of generality we have defined $\langle \Phi_3 \rangle$ as real. The *neutral* Higgs couplings can be made real by a redefinition $D_{j,R} \to e^{-i\eta_1} D_{j,R}$, $U_{j,R} \to e^{-i\eta_2} U_{j,R}$ leading to

$$\mathcal{L}_{Y,\text{neut}} = \overline{D}\Gamma^D D(v_1 + \rho_1^0) + \overline{U}\Gamma^U U(v_2 + \rho_2^0) ; \qquad (17.59)$$

the complex fields ρ_i result from expanding Φ_i around their VEVs. Thus we have for the mass matrices $\mathcal{M}^D = \Gamma^D v_1$, $\mathcal{M}^U = \Gamma^U v_2$, and they can be diagonalized, as is done in Eq. (8.22). Note that the same transformations diagonalize also the ρ_i^0 couplings to the quarks, thus banning the spectre of FCNC. Writing the shifted fields ρ_i in terms of the Higgs mass eigenstates H_i, we find for the *charged* Higgs couplings to the quarks

$$\mathcal{L}_{Y,\text{ch}} = \frac{g}{M_W}\sum_i \left[\alpha_i \overline{D}_L^m \mathbf{V}^\dagger \mathcal{M}_U^{\text{diag}} U_R^m H_i^- + \beta_i \overline{U}_L^m \mathbf{V}\mathcal{M}_D^{\text{diag}} D_R^m H_i^+ \right] ;$$

$$(17.60)$$

$\mathbf{V}$ is a CKM-like mixing matrix, yet an orthogonal one as all Yukawa couplings are real – due to the condition that $\mathbf{CP}$ be broken spontaneously. The quantities α_i and β_i contain the parameters describing the mixing among the charged Higgs fields in addition to the phase factors and the v_i, all of which depends on the details of the Higgs potential.

Since models with NFC contribute to $K^0 \to \overline{K}^0$ transitions only at the one-loop level (and likewise for $D^0 \to \overline{D}^0$ and $B^0 \to \overline{B}^0$), we do not need to invoke superheavy Higgs masses to avoid conflict with the observed value of ϵ_K. Accordingly, these models allow also for direct $\mathbf{CP}$ violation to arise on an observable level. Their phenomenology then becomes richer: we will illustrate this through the following observables [300, 310]: (i) ϵ_K and ϵ' in K_L decays, (ii) electric dipole moments for neutrons and (iii) the muon transverse polarization in $K_{\mu 3}$ decays.

ϵ_K

The expression in Eq. (7.42),

$$|\epsilon_K| \simeq \frac{1}{\sqrt{1+(\Delta\Gamma/2\Delta M_K)^2}} \left| \frac{\text{Im} M_{12}}{\Delta M_K} - \xi_0 \right| ,$$

suggests that there are two components in the dynamics driving $\Delta S = 2$ transitions.

- The short-distance contribution to $\text{Im} M_{12}$, which is given mainly by a box diagram. In the case under study it contains the exchange of a W and a charged Higgs boson H, Fig. 17.4; assuming, for definiteness, real $\mathbf{V}$ and factorization for an order of magnitude estimate, we arrive at [310]

$$\text{Im}\langle K^0|\mathcal{H}_{\text{box}}|\overline{K}^0\rangle \simeq \frac{\text{Im}(\alpha^*\beta)}{M_H^2} \frac{3G_F^2}{32\pi^2}(\sin\theta_C\cos\theta_C)^2 m_c^2 M_K^2 F_K^2$$

$$\times \frac{M_d^{\text{const}}}{m_s}\langle K^0|(\overline{d}\gamma_\mu\gamma_- s)^2|\overline{K}^0\rangle, \qquad (17.61)$$

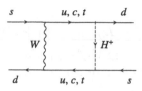

Figure 17.4 Diagram for the dominant contribution to $\mathrm{Im}M_{12}^K$.

where m_q and M_q^{const} are *current* and *constituent* quark masses, respectively.[3] As for the real part, we find that two W boson exchange dominates. Taking a ratio between $\mathrm{Im}\,M_{12}$ obtained above with the two W boson exchange diagram, we obtain

$$\frac{M_{12}|_{SD}}{\Delta M} = \frac{\mathrm{Im}\langle K^0|H|\overline{K}^0\rangle}{2\mathrm{Re}\,\langle K^0|H|\overline{K}^0\rangle} = \frac{\mathrm{Im}(\alpha^*\beta)}{M_H^2}\frac{3}{8}\frac{M_K^2 M_d}{m_s}. \tag{17.62}$$

We then find

$$\left|\frac{\mathrm{Im}M_{12}|_{\mathrm{Higgs},SD}}{\Delta M_K}\right| \leq \text{few} \times 10^{-5} \text{ for } M_H \geq 80\,\mathrm{GeV}, \ \mathrm{Im}(\alpha^*\beta) \sim 1, \tag{17.63}$$

using the experimental value for ΔM_K; i.e. the contribution from charged Higgs to ϵ_K is quite insignificant, which holds a fortiori for ΔM_K; likewise for the effects of neutral Higgs. The situation was very different at the time such models were first put forward: charged Higgs states with a mass of a few GeV were quite conceivable then.

• If ξ_0 dominates in ϵ_K, then it leads to a sizeable value for ϵ', see Eq. (7.34):

$$\left|\frac{\epsilon'}{\epsilon_K}\right| \simeq \omega \simeq 0.05, \tag{17.64}$$

which is clearly ruled out by the data.

It is still of intellectual value to see where ϵ_K then comes from. We must turn to long-distance dynamics once more and consider $K^0 \leftrightarrow \overline{K}^0$ transitions proceeding via light quark states probed off their mass shell; such a mechanism contributes to ϵ_K, yet not ϵ'. With $M_{12} = M_{12}|_{SD} + M_{12}|_{LD} \equiv M_{12}|_{SD} + DM_{12}$ we can rewrite Eq. (7.42) (see Problem 17.8) [311]:

$$\epsilon_K = \frac{1-D}{2\sqrt{2}}e^{i\pi/4} \cdot \left(\epsilon_m + 2\xi_0 + \frac{D}{1-D}\cdot\chi\right), \tag{17.65}$$

---

[3] *Constituent* quark masses – in contrast to *current* masses – are ill-defined in a quantum field theory; they are purely phenomenological constructions reflecting hadronic scales: $M_{u,d}^{\mathrm{const}} \sim 300\,\mathrm{MeV}$.

$$\left|\frac{\epsilon'}{\epsilon_K}\right| \simeq \omega \cdot \frac{2\xi_0}{(1-D)[\epsilon_m + 2\xi_0 + D\chi/(1-D)]}. \tag{17.66}$$

where $\epsilon_m = \arg M_{12}|_{SD}$, $\chi = \arg M_{12}|_{LD} + 2\xi_0$. Can *long-distance* dynamics yield $\xi_0 \ll D \cdot \chi$ such that ϵ_K is reproduced:

$$|\epsilon_K| \simeq \frac{1}{2\sqrt{2}} D \cdot \chi \,? \tag{17.67}$$

Consider the *virtual* transitions

$$K^0 \to \text{'}\pi, \eta, \eta', \pi\pi\text{'} \to \overline{K}^0. \tag{17.68}$$

$SU(3)$ symmetry, coupled with current algebra, implies that $\langle \pi^0 | \mathcal{H} | K^0 \rangle$, $\langle \eta_8 | \mathcal{H} | K^0 \rangle$ and $\langle (\pi\pi)_{I=0} | \mathcal{H} | K^0 \rangle$ all have the same phase;[4] the lightest intermediate state contributing to χ is thus $K^0 \to \text{'}\eta_0\text{'} \to \overline{K}^0$; accordingly we have to study

$$|\epsilon_K| \simeq \frac{\mathrm{Im}\langle K^0 | \mathcal{H} | \eta_0 \rangle \langle \eta_0 | \mathcal{H} | \overline{K}^0 \rangle}{2\sqrt{2}\mathrm{Re}M_{12}}. \tag{17.69}$$

In Ref. [312] we have estimated this contribution in terms of $\mathrm{Im}\alpha^* \beta / M_H^2$ using data on $\eta - \eta'$ mixing, the decay rates for $\eta \to \gamma\gamma$ and $\eta' \to \gamma\gamma$, $SU(3)$ symmetry and the bag model result for $\langle K^0 | \mathcal{H} | \pi^0 \rangle$. We found that Eq. (17.69) can be satisfied, although for $M_H \geq 40\,\mathrm{GeV}$ we have to rely on more and more extreme corners of parameter space: $vv_3/v_1v_2 \geq 50$, $v = \sqrt{v_1^2 + v_2^2 + v_3^2}$. However, this leads to a direct phenomenological problem, namely that we predict too large a value for the neutron EDM.

Electric dipole moments

A non-minimal Higgs sector provides a particularly rich dynamical substratum for EDMs. There are three mechanisms for producing an EDM:

(1) Higgs exchanges generating an EDM for individual quarks or leptons;
(2) **CP** odd gluonic operators inducing a neutron EDM;
(3) analogous electroweak operators producing an EDM for electrons.

In some of these scenarios the neutron EDM results from cooperative effects among the quarks rather than individual quarks. We should also keep in mind that most of these effects can arise also for the two Higgs

---

[4] η_8 [η_0] denotes the octet [singlet] component of the neutral η meson.

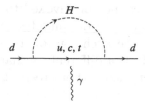

Figure 17.5   One-loop diagram for a quark EDM.

doublet models discussed before; the presence or absence of FCNC is not essential here.

(a) *Direct contributions*

A quark EDM can be generated by *charged* Higgs exchange, as shown in Fig. 17.5. With $d_N = \frac{1}{3}(4d_d - d_u)$ and $d_u \ll d_d$ in models with NFC we obtain [313]:

$$d_N = \frac{\sqrt{2}G_F e}{9\pi^2} \frac{m_c^2 m_d}{M_H^2} \mathrm{Im}(\alpha^*\beta) \cdot \left[\eta^{(c)} |V(cd)|^2 g(x_c) + \eta^{(t)} |V(td)|^2 g(x_t) \right]$$

(17.70)

$$g(x) = \frac{1}{(1-x)^2} \left[\frac{5}{4}x - \frac{1 - 3x/2}{1-x}\log x - \frac{3}{4} \right], \quad x_i = \frac{m_i^2}{M_H^2},$$

(17.71)

and $\eta^{(c,t)}$ denoting radiative QCD corrections. Imposing Eq. (17.65), we can infer a *lower* bound on d_N [312]:

$$d_N \geq 5 \cdot 10^{-25}\, e\,\mathrm{cm},$$

(17.72)

with the dependence on M_H dropping out to a large degree from this relationship of ϵ_K vs d_N. This lower bound is incompatible with the experimental upper bound quoted in Eq. (3.81).

(b) $G \cdot \tilde{G}$ *effects*

Like in the SM, we have to be concerned about the strong **CP** problem, i.e. the emergence of the **CP** odd operator $G \cdot \tilde{G}$ with coefficient $\bar{\theta}/16\pi^2$. As shown in Eq. (15.27), we infer $\bar{\theta} \leq \mathcal{O}(10^{-9})$ from the upper bound on the neutron EDM; furthermore $\bar{\theta} = \theta_{\mathrm{QCD}} + \Delta\theta_{\mathrm{EW}}$, where the second term denotes the electroweak renormalization of the quark mass matrix M: $\Delta\theta_{\mathrm{EW}} \equiv \arg \det M$. Two new aspects arise relative to the SM: (i) Higgs dynamics opens up the possibility that **CP** is broken *spontaneously*; θ_{QCD}, being the coefficient of a dimension-4 operator, has to vanish then and $\bar{\theta}$ becomes a finite and calculable quantity. (ii) Non-trivial θ renormalization occurs already on the *one*-loop level and is sizeable [314] – $\Delta\theta_{\mathrm{EW}} \sim \mathcal{O}(10^{-3})$ – making such Higgs models appear quite unnatural.

Even if Higgs dynamics cannot be invoked to generate the lion's share of ϵ_K, it could easily provide the dominant source of observable EDMs. This can happen through the exchange of heavy charged Higgs as just described – or through a host of different mechanisms. We shall present a brief discussion of such scenarios without restricting ourselves to a specific model.

(c) *Other* **CP** *odd gluonic operators*

Whatever the origins of **CP** violating electroweak forces may be, they will in general – through quantum corrections – induce a whole sequence of purely gluonic operators in increasing dimension that are *odd* under **P** and **T**:

$$\mathcal{L}_{\text{glue}}(\mathbf{CP}\ \text{odd}) = \frac{\bar{\theta}}{16\pi^2} G \cdot \tilde{G} + c_{(6)} G^2 \cdot \tilde{G} + c_{(8)} G^3 \cdot \tilde{G} + \cdots \qquad (17.73)$$

We have just commented on the first term; in the present discussion, we take $\bar{\theta} = 0$ and concentrate on possible other effects.

The coefficients $c_{(6)}, c_{(8)}, \ldots$ of the dimension 6, 8, ... operators can be calculated as *finite* numbers. $G^2 \cdot \tilde{G}$ is likely to provide the dominant contribution to d_N for large values of M_H as first discussed in Ref. [315]: coming from a dimension-6 operator it scales like $1/M_H^2$, whereas other contributions are suppressed by higher powers of $1/M_H$, by light quark masses and/or small mixing angles. The coefficient $c_{(6)}$ can be calculated in terms of the model parameters, namely the Higgs masses and Yukawa couplings with the latter expressed in terms of m_b, m_t and the VEVs v_i. Diagrams which give rise to these operators are shown in Fig. 17.6. No precise prediction for the resulting d_N has been extracted yet from these studies, since considerable theoretical uncertainties arise in evaluating the matrix element $\langle N | G^2 \cdot \tilde{G} | N \rangle$ upon which d_N depends. For $M_H \sim 80$–$100\,\text{GeV}$ we estimate [316]:

$$d_N \sim \mathcal{O}(10^{-26})\ \text{e cm}\ ; \qquad (17.74)$$

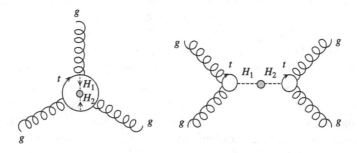

Figure 17.6   Diagrams which give rise to $G^2 \cdot \tilde{G}$ and $G^3 \cdot \tilde{G}$ operators.

i.e. a value just an order of magnitude below the present experimental upper bound. This relatively large value, despite the very heavy Higgs mass, arises due to the large Yukawa couplings of the top quarks in the loop. In any case, exchanges of intermediate mass Higgs fields with $M_H \sim (1\text{--}2) \times M_W$ are likely to induce d_N on a level that should become observable in the foreseeable future.

At first it would appear that the contribution from *neutral* Higgs exchanges can be ignored since – in the absence of FCNC – it is proportional to the cube of the u and d quark masses. Yet it induces a $G^3 \tilde{G}$ term that could conceivably generate d_N as large as $\mathcal{O}(10^{-25})$ e cm! The reason for this relatively large effect is quite interesting [317]: this contribution is related – via the anomaly – to the nucleon mass rather than the current quark masses m_q; thus it will not vanish in the chiral limit $m_q \to 0$.

It should be noted that all contributions driven by such gluonic operators involve more than one quark at a time.

(d) *Colour-electric dipole moments*

Once it was realized that diquark effects can provide the leading contribution, many more such mechanisms were found. Particularly promising seems to be the so-called quark colour magnetic moment operator shown in Fig. 17.7: it represents an electroweak two-loop effect and leads to [318, 319]

$$d_N \sim \mathcal{O}(10^{-26}) \, \text{e cm}, \tag{17.75}$$

which is close to the present upper bound.

(e) *Electric dipole moment of electrons*

Replacing the gluons in the colour-electric dipole operator by electroweak bosons and attaching it to a lepton, we obtain a two-loop contribution to the electron EDM, yielding [318, 319, 320]

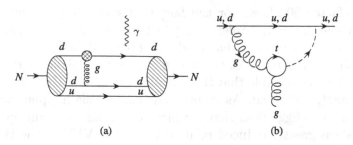

(a)   (b)

Figure 17.7 In (a), the neutron EDM is generated by a quark colour magnetic moment; in (b) the diagram giving rise to a quark colour dipole moment – the blob in (a) – is shown.

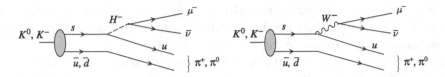

Figure 17.8   A diagram giving rise to muon transverse polarization in $K \to \pi\mu\nu$.

$$d_e \sim \text{several} \times 10^{-27}\,\text{e cm}, \tag{17.76}$$

which is just below the present experimental bound $d_e = (-0.3 \pm 0.8) \cdot 10^{-26}\,\text{e cm}$.

K → μν π

The muon transverse polarization in $K_{\mu 3}$ decay has been derived in Section 7.4. The two diagrams for $K^+ \to \mu^+\nu\pi^0$ giving rise to this **CP** violating effect are shown in Fig. 17.8. The amplitude for the charged Higgs exchange is given by

$$\mathcal{M} = -\frac{G_F}{\sqrt{2}} \sum_{i=1}^{2} \langle \pi^0 | m_s \frac{\alpha_i^* \beta_i}{M_{H_i}^2} \bar{s}(1 - \gamma_5) u \bar{\nu}(1 + \gamma_5)\mu | K^+ \rangle + \mathcal{O}(m_u). \tag{17.77}$$

Models with **CP** violation in the Higgs sector can yield a 'sizeable' $P_\perp(\mu)$ due to interference between the Higgs exchange – F_S, F_P – and the W exchange amplitude – F_V, F_A, see Eq. (7.57). 'Sizeable', however, does not mean large, since we are dealing here with *direct* **CP** violation. A rather model independent estimate on how large such an effect could be is obtained from the present bound on ϵ'/ϵ_K:

$$P_\perp(\mu) \leq 20 \cdot (\epsilon'/\epsilon_K) \cdot \epsilon_K \leq 10^{-4}, \tag{17.78}$$

where the factor 20 allows for the fact that ϵ'/ϵ_K is 'accidentally' suppressed by the $\Delta I = 1/2$ rule: $\omega \equiv (\mathrm{Re}A_2/\mathrm{Re}A_0) \simeq 1/20$. While this is much larger than anything that could be obtained in the KM ansatz, it is disappointingly small. Yet there is a relevant loophole in this generic argument: it is conceivable that the coupling of charged Higgs fields to *leptons* is strongly enhanced. As mentioned before, this happens when v_3 – the VEV of the Higgs field that couples to leptons only and gives them their mass – is greatly reduced relative to v_1, the VEV of the Higgs field coupling to the up-type quarks. In particular for $v_1/v_2 \geq \mathcal{O}(10)$ we could have [321]

$$P_\perp(\mu) \sim \mathcal{O}(10^{-3}), \tag{17.79}$$

without violating any constraints inferred from ϵ_K, ϵ' and d_N. What is even more striking: the forces leading to Eq. (17.79) are insignificant in $K_L \to \pi\pi$.

Since such a scenario is consistent with present phenomenology, it serves as a useful imagination stretcher – a commodity not to be belittled. For it illustrates that a detailed analysis of $K \to \mu\nu\pi$ could reveal a source of **CP** violation that escapes detection in $K \to 2\pi$, 3π. While present data yield $P_\perp(\mu) \leq 5 \cdot 10^{-3}$, an experiment has been proposed for the new Japanese J-Parc facility to reach the 10^{-4} level.

17.2.3 **CP** *phenomenology with heavy fermions*

Models with FCNC

Such models in general contain tree-level operators that change the flavour quantum numbers by two units:

$$\Delta C = 2 \text{ and } \Delta B = 2. \tag{17.80}$$

$\Delta B = 2$ transitions Scaling arguments suggest [322] that if Higgs induced FCNC provide a sizeable fraction of ϵ_K, they could have a significant, though hardly dominant impact on **CP** violation in B^0 decays as well. This might sound surprising since ΔM_{B_d} is a factor of 100 larger than ΔM_K, and furthermore $\text{Im}M_{12}(K) \ll \text{Re}M_{12}(K)$. However, the strength of the Higgs couplings increases with the mass of the quarks involved, as illustrated by the following generic example.

Assume tree-level FCNC to be the only source of **CP** violation in $K^0 - \overline{K}^0$ and $B^0 - \overline{B}^0$ oscillations with a more or less universal phase; the **CP** *even* parts of $(M_K)_{12}$ and $(M_B)_{12}$, on the other hand, receive SM contributions as well. The phases in $M_{12}(K)$ and $M_{12}(B)$ can then be related to each other [322]:[5] With the notation given in Eq. (7.23),

$$\epsilon_K = \frac{1}{2}\left(1 - \frac{q}{p}\right) = -\frac{i}{2}\phi\left(\frac{1}{1 - \frac{i}{2}r}\right), \tag{17.81}$$

$$\phi_K \equiv \frac{\text{Im}\langle K^0|\mathcal{H}_{\text{Higgs}}(\Delta S = 2)|\overline{K}^0\rangle}{(M_K)_{12}} \simeq 6.5 \cdot 10^{-3} \tag{17.82}$$

from ϵ_K, and thus

$$\phi_B \equiv \frac{\text{Im}\langle B^0|\mathcal{H}_{\text{Higgs}}(\Delta B = 2)|\overline{B}^0\rangle}{(M_B)_{12}}$$

---

[5] We have adopted here a phase convention such that the $\Delta S = 1$, $\Delta C = 1$ and $\Delta B = 1$ amplitudes contain *no weak* phase. This is possible in a superweak scenario.

$$\simeq \frac{\text{Im}\langle B^0|\mathcal{H}_{\text{Higgs}}(\Delta B = 2)|\overline{B}^0\rangle}{\text{Im}\langle K^0|\mathcal{H}_{\text{Higgs}}(\Delta S = 2)|\overline{K}^0\rangle} \cdot \frac{\Delta M_K}{\Delta M_B} \cdot \phi_K$$

$$= \frac{B_B F_B^2}{B_K F_K^2} \cdot \left(\frac{M_B}{M_K}\right)^3 \phi_K \sim 0.1 \, . \tag{17.83}$$

Thus we can infer that **CP** violating Higgs dynamics could affect time-dependent **CP** asymmetries in B^0 decays on the few percent level only.

$\Delta C = 2$ **transitions** Applying the scaling argument used in Eq. (17.83) to D^0 decays, we find

$$\phi_D \simeq \frac{B_D F_D^2}{B_K F_K^2} \cdot \left(\frac{M_D}{M_K}\right)^3 \cdot \phi_K \sim 0.35 \cdot \frac{\Delta M_K}{\Delta M_D}. \tag{17.84}$$

$D^0 - \overline{D}^0$ oscillations can be produced by $\Delta C = 2$ tree level FCNC. As discussed in Chapter 14, data show intriguing evidence for oscillations with $\Delta M_D, \Delta\Gamma_D/2 \sim (0.005 - 0.01) \cdot \Gamma_D$. Then time-dependent **CP** asymmetries, analogous to the one defined above for B^0 decays, could emerge in decays like $D^0 \to K^+K^-, \pi^+\pi^-$ with

$$S_{K^+K^-} \cdot \sin\Delta M_D t \sim 10^{-3} \cdot (t/\tau_D), \tag{17.85}$$

i.e. conceivably a few $\times 10^{-3}$ effect.

A considerably larger asymmetry would arise in the mode $D^0 \to K^+\pi^-$ since it is doubly Cabibbo suppressed *without* $D^0 - \overline{D}^0$ oscillations. The **CP** asymmetry in this channel has been written down explicitly in Eq. (14.44) and Eq. (14.45). Computing the magnitude of the asymmetry using Eq. (17.84) is left as an exercise.

Models without FCNC

Beauty decays The prospects are not promising that such models could have a noticeable impact on **CP** asymmetries in B decays. A priori they could generate **CP** violation both in $\Delta B = 1$ and $\Delta B = 2$ transitions. Yet unless there are unusually enhanced Yukawa couplings of charged Higgs fields to quarks, such an effect would emerge on the several percent level only for intermediate Higgs masses. Thus they would provide merely smallish corrections to the dominant KM 'background'. The most promising decay for such Higgs driven effects to emerge is in $B_s \to \psi\phi$ – within the KM ansatz we expect less than a 4% asymmetry there – and possibly rare decays like $B_d \to K\pi, K\rho, K_S\phi, B_s \to \phi\phi$, etc.

Production and decay of top quarks and τ leptons As explained in Chapter 18, production and decay of top quarks and τ leptons provide unfavourable environments for the KM mechanism to generate observable **CP** violation. There is hope, though, that Higgs dyamics with its enhanced Yukawa couplings to these heavy fermions could generate observable asymmetries there. In particular:

- a difference

$$\sigma(e^+e^- \to t_L \bar{t}_L) \neq \sigma(e^+e^- \to t_R \bar{t}_R) \qquad (17.86)$$

could arise of order 0.1 % that can be probed by comparing the energy distributions of charged 'isolated' leptons coming from W decays:

$$\frac{d}{dE_l}\Gamma(t_L \to W^+b \to l^+X) \neq \frac{d}{dE_l}\Gamma(t_R \to W^+b \to l^+X). \qquad (17.87)$$

- we can search the final state in $e^+e^- \to t\bar{t}H^0$ for **CP** odd correlations.

Details can be found in Chapter 18.

17.3 The pundits' résumé

At the first writing of this book the CKM ansatz was already seen as the leading candidate for the theory of flavour dynamics and **CP** violation. Yet for the latter there were still other legitimate contenders, namely models with right-handed currents or a non-minimal Higgs sector, which, by the way, had also been described in the KM paper [73]. With the measurements of ϵ'/ϵ_K and the large **CP** asymmetries in B_d decays the battle for supremacy has been decided in favour of CKM theory. The latter is remarkably successful in accommodating tiny asymmetries in K decays with almost maximal ones in B_d decays. Alternative theories of **CP** violation do not accommodate the whole body of data in a similarly natural way.

With the data exhibiting no need for dynamics beyond CKM theory, even the serious reader might wonder why have we given a detailed description of these New Physics models? Supremacy does not necessarily imply a monopoly. **CP** studies should thus focus on non-leading contributions from New Physics. While models with right-handed currents or an extended Higgs sector could be seen as 'theories with a great future concerning **CP** violation in their past', they still deserve discussion and even close scrutiny.

(A): With their multitude of sources for **CP** violation allowing to implement the latter *manifestly, spontaneously* or *softly* we can learn valuable theoretical lessons concerning the inner workings of **CP** dynamics. This holds also for analysing how the observed ϵ_K is accommodated within these models.

⊕ Left–right gauge models have intrinsically attractive or at least intriguing features: (i) A highly symmetric starting point gives room to the hope that parity and **CP** violation can be understood dynamically rather than being viewed as a given. This might some day allow us to calculate the basic parameters characterizing the breakings of these two symmetries and their internal relationship.

⊕ On a more practical level they can provide us through the 'see-saw' mechanism with a rather natural rationale for the extreme lightness of the observed neutrinos. It relies on two main ingredients: (i) the special ability of neutrinos to acquire Majorana masses in addition to Dirac masses; (ii) the fact that parity is broken spontaneously allows us to assign quite different Majorana masses to the left- and right-handed neutrinos.

⊕ The Higgs sector – its structure, its origin and even its mass scale – is the least understood part of the SM. No argument beyond that pragmatic one of convenience has been found in favour of limiting ourselves to the simplest possible ansatz. An elegant scheme like supersymmetry actually requires the existence of at least two different Higgs doublet fields, as discussed in the next chapter. Almost as soon as we go beyond the minimal Higgs sector we encounter additional sources of **CP** violation – whether they are 'welcome', 'needed' or neither.

Any self-respecting pundit will also point out some drawbacks.

⊖ The promise that a higher degree of initial symmetry will enable us to dynamically understand its spontaneous breaking in a quantitative fashion has remained that: a promise.

⊖ We have to deal with a proliferation of basic parameters like KM parameters for right-handed couplings, Yukawa couplings for multiple Higgs multiplets, relative phases between VEVs etc. Constraints due to the imposition of a discrete symmetry – like left–right symmetry – at presumably quite high energy scales get increasingly washed out at lower energies. By adjusting the plethora of such parameters we can – even without fine-tuning – easily allow the theory to take 'evasive action' when confronted with data. In that sense there are no benchmarks that make or break these theories.

(B): The main emphasis should of course be on the future rather than the past. Viewing these models as paradigms of New Physics is understating their potential value; they can act as an imagination stretcher – a commodity of which there can never be an overabundance! Cutting the 'umbilical cord' for these models to the size of ϵ_K can 'liberate' our imagination. For we can then see that even if these models make small contributions to the **CP** asymmetries in K and B decays, they can provide by far the dominant source of EDMs on a level that is observable in the next two rounds of experiments, namely d_N, $d_e \geq 10^{-28}$ e cm. Multi-Higgs models can produce a muon transverse polarization $P_\perp$ in $K \to \mu\nu\pi$ with $P_\perp \geq 10^{-4}$; this seems within reach of future experiments; they might generate observable **CP** asymmetries in the production and decays of top quarks and τ leptons and in D decays as discussed in Chapters 14 and 18. They could also modify significantly the SM **CP** asymmetries in B decays.

To be more explicit: a dedicated search for

- electric dipole moments for neutrons, electrons and (heavy) atoms

and detailed **CP** studies of

- the B, D and K decays;
- the muon polarization in $K \to \mu\nu\pi$; and
- the production and decay of top quarks and τ leptons;

are *meaningful even – actually in particular – if New Physics is rather irrelevant for $K_L \to \pi\pi$*. Such **CP** studies act as high sensitivity probes for New Physics.

Lastly, the following should be kept in mind: as described before, understanding the baryon number of the universe as a dynamically generated quantity rather than as an arbitrary initial value represents a fascinating intellectual challenge. As first realized by A. Sakharov, one of the essential elements of any answer to this challenge is the presence of **CP** violation. It has been suggested that **CP** violation that enters through Higgs dynamics is 'best' suited to generate the baryon number of today's universe at the electroweak scale.

The analysis of **CP** asymmetries in K and B decays – and likewise for charm decays – is clearly hampered by our failure to evaluate accurately hadronic matrix elements, since those are shaped by non-perturbative dynamics. There are however fermionic systems that are not subject to non-perturbative dynamics thus making our calculational tools more powerful. These are *leptons* – electrons, muons, τ leptons and neutrinos – and top *quarks*. The electron's EDM has been discussed in Section 3.6 and neutrinos were discussed in Chapter 16; the decays of charged leptons are addressed in Chapter 18. As pointed out in Section 10.10.3 the aforementioned gain in calculational control comes with a price – namely at best tiny **CP** asymmetries. We will see that probably final state distributions rather than partial widths have the best chance to reveal **CP** violation.

Problems

17.1. Consider a model with both right- and left-handed charged quark currents

$$\mathcal{L}_{CC} = g_L W_L^\mu \overline{U}_L \gamma_\mu D_L + g_R W_R^\mu \overline{U}_R \gamma_\mu D_R, \qquad (17.88)$$

where no restriction is placed on g_L/g_R. Diagonalizing the up- and down-type quark mass matrices will introduce the usual KM matrix $\mathbf{V}_L$, and a corresponding matrix $\mathbf{V}_R$ for the right-handed currents. For N quark families $\mathbf{V}_L$ and $\mathbf{V}_R$ are $N \times N$ unitary matrices, each described by $\frac{1}{2}N(N-1)$ angles and $\frac{1}{2}N(N+1)$ phases. Using the phase freedom in defining quark fields, show that $\mathbf{V}_L$ and $\mathbf{V}_R$ *together* contain $N^2 - N + 1$ irreducible and thus physical phases. Therefore for one family we have already a physical phase supporting **CP** violation.

17.2. Having parity broken spontaneously imposes constraints on the Yukawa couplings of the Higgs field ϕ with quantum numbers $(\frac{1}{2}, \frac{1}{2}, 0)$, see Eq. (17.7). Assume κ and κ', the VEVs of ϕ, to be *real*. Show that both the quark mixing angles and phases in the left- and right-handed sector then coincide.

17.3. Consider an $SU(2)_L \times SU(2)_R \times U(1)$ gauge theory with *spontaneous* **CP** violation realized by

$$\langle \phi \rangle = e^{i\xi} \begin{pmatrix} \kappa & 0 \\ 0 & \kappa' \end{pmatrix}, \quad \text{with } \kappa, \, \kappa' \text{ real}. \tag{17.89}$$

- Show that the quark mass matrices are symmetric.
- Derive the following relation between the left- and right-handed quark mixing matrices:

$$\mathbf{V}_R = J_U \mathbf{V}_L^* J_D^\dagger, \tag{17.90}$$

with $J_{U,D}$ being unrelated diagonal unitary matrices. (This theorem was first proved in general in Ref. [323].)

17.4. Consider an $SU(2)_L \times SU(2)_R \times U(1)$ gauge model. Show that while a Higgs field ϕ with quantum numbers $(\frac{1}{2}, \frac{1}{2}, 0)$ can produce a *Dirac* mass for neutrino fields, a Higgs field Δ_R $[\Delta_L]$ with $(0, 1, 2)$ $[(1, 0, 2)]$ can generate a *Majorana* mass for right [left]-handed neutrinos. The resulting neutrino mass matrix then reads

$$\mathcal{M} = \begin{pmatrix} \mu_L^M & m^D \\ m^D & M_R^M \end{pmatrix}, \tag{17.91}$$

with μ_L^M, M_R^M and m^D being controlled by the vacuum expectation values of Δ_L, Δ_R and ϕ, respectively.

Assuming the hierarchy

$$\mu_L^M \ll m^D \ll M_R^M, \tag{17.92}$$

find the eigenvectors and eigenvalues of this matrix.

17.5. Draw a Feynman diagram generating the muon transverse polarization in $K^+ \to \mu^+ \nu \pi^0$ decays.

17.6. Consider a Higgs potential given by

$$V = -\mu^2 \sum_i |\phi_i|^2 + \lambda \sum_i |\phi_i|^4$$
$$+ c\left[(\phi_1^*\phi_2)(\phi_1^*\phi_3) + (\phi_2^*\phi_3)(\phi_2^*\phi_1) + (\phi_3^*\phi_2)(\phi_3^*\phi_1)\right]$$
$$+ \text{h.c.} \tag{17.93}$$

If the coefficients μ, λ and c are real, then V is certainly **CP** symmetric – but what about its minima? Show that for the Higgs self-coupling $c > 0$,

$$|\alpha_2 - \alpha_1| = |\alpha_3 - \alpha_2| = |\alpha_1 - \alpha_3| = \frac{2\pi}{3} ; \tag{17.94}$$

and the vacuum is not **CP** invariant.

17.7. Starting from Eq. (17.57), show that the charged Higgs interaction to quarks can be written as:

$$\mathcal{L}_{Y,\text{ch}} = \frac{g}{M_W} \left[e^{-i\lambda_1} \overline{U}_{Li}^m V_{CKM} \mathcal{M}_D^{\text{diag}} D_{Rj}^m \rho_1^+ \right.$$
$$\left. + e^{-i\lambda_2} \overline{D}_{Li}^m V_{CKM}^\dagger \mathcal{M}_U^{\text{diag}} U_{Rj}^m \rho_2^- \right]. \tag{17.95}$$

17.8. Derive Eq. (17.65) and Eq. (17.66). First show that $M_{12} = \frac{M_{12}|_{SD}}{1-D} = \frac{1}{D} M_{12}|_{LD}$. Then show that

$$\frac{\text{Im } M_{12}}{\text{Re } M_{12}} = (1-D)\epsilon_m + D(-2\xi_0 + \chi), \tag{17.96}$$

where $\chi = \arg M_{12}|_{LD} + 2\xi_0$. This leads to Eq. (17.65).

17.9. Using the result of Section 7.4 and Eq. (17.77), compute the transversal polarization of μ in $K \to \pi\mu\nu$.

17.10. Compute the order of magnitude of **CP** asymmetry for the doubly Cabibbo suppressed decays $D \to K^+\pi^-$ and $\overline{D} \to K^-\pi^+$ using Eq. (17.84).

18

CP violation without non-perturbative dynamics – top quarks and charged leptons

The analysis of **CP** asymmetries in K and B decays – and likewise for charm – is clearly hampered by our failure to accurately evaluate hadronic matrix elements, since those are shaped by non-perturbative dynamics. There are, however, fermionic systems that are not subject to non-perturbative dynamics thus making our calculational tools more powerful. These are *leptons* – electrons, muons, τ leptons and neutrinos – and top *quarks*. The electron's EDM has been discussed in Section 3.6 and **CP** violation in neutrino oscillations in Section 16; the decays of charged leptons will be addressed here. As pointed out in Section 10.10.3 the afore-mentioned gain in calculational control comes with a price – namely at best small **CP** asymmetries. We will see that final state distributions rather than partial widths probably have the best chance to reveal **CP** violation.

18.1 Production and decay of top quarks

The existence of all members of three quark families has been established with the top quark being discovered last. Even before that time it had been realized [203] that the top quark, once it becomes sufficiently massive, will decay (semi-)weakly – $t \to bW$ – *before* it can hadronize; i.e. top states decay as quarks rather than hadrons. This transition occurs around the 110–130 GeV region (for $|V(tb)| \sim 1$), i.e. well below the mass value now observed. *Non-perturbative* dynamics thus plays hardly any role in top decays, and the *strong* forces can be treated perturbatively. While this is certainly good news for our ability to calculate observables, it carries also a negative message concerning the observability of **CP** asymmetries.

- If top *hadrons* can hardly form before top decays, $T^0 - \overline{T}^0$ oscillations become a moot point.

- Due to the huge phase space that is available in top decays, there will in general be very little interference between different sub-processes.

Thus we see that most of the **CP** sensitive observables we have discussed before do *not* work for top states, because the strong interactions are no longer sufficiently efficient in 'cooling' different transitions into a coherent state. Top quarks in many ways behave like charged leptons since the residual strong interactions can be treated perturbatively. Accordingly, we consider similar phenomenological categories when analysing possible **CP** violation in top and in τ physics: we are dealing with both the decay and the production of spin-1/2 objects whose dynamics is not prone to 'non-perturbative voodoo'; they can be polarized and their polarization is revealed in their decays.

At the same time top quarks have a mass comparable to the electroweak scale $v \sim 250\,\text{GeV}$. Thus we expect their Yukawa couplings to Higgs fields to be at least sizable. While this will complicate our computational task, we view it as mainly good news.

- Non-minimal Higgs dynamics is likely to introduce new sources of **CP** violation, as discussed in Section 17.2. Even if the latter originates from couplings among Higgs fields, it can be transmitted to fermions through their Yukawa couplings and the more so the larger those are.
- There are new classes of **CP** odd observables: in addition to **CP** asymmetries in the decays of top quarks [324] they can arise also in their *production* [325] from gluons or highly off-shell photons $- gg/\gamma^* \to t\bar{t} -$ due to the virtual exchange of Higgs fields; a recent update can be found in Section 2.8 of the comprehensive report of Ref. [302].
- Probes of **CP** symmetry in top transitions constitute searches:
  - for new **CP** breaking forces [326] that might drive baryogenesis; and
  - for some hints concerning one piece in the flavour mystery, namely why top quarks are so unexpectedly massive.

Such studies can serve also as a diagnostic *tool*: in non-minimal Higgs scenarios one typically encounters several neutral spin-zero fields, which can be scalar or pseudoscalar. Thus they can be **CP** even or odd – or be mixed like the K_L meaning **CP** breaking was involved in their creation. Whether a spin-zero Higgs field H is a **CP** eigenstate or not can discriminate between scenarios. If H is sufficiently heavy s.t. $H \to t\bar{t}$ can proceed, then one can determine the spin and **CP** characteristics of H from an analysis of correlations in the kinematics of the t and $\bar{t}$ decay products, see Section 2.7 of Ref. [302]. If not, this information can be inferred from correlations among the decay products of the associated production process $gg \rightarrow t\bar{t}H$, see Section 2.9 there.

The possibilities are vast and unexplored experimentally. We do not feel a description of a representative list of them could or at this point even should keep the reader's attention. Instead we will merely sketch two examples illustrating a qualitatively new feature, namely **CP** violation in top *production*.

18.1.1 $\sigma(t_L\bar{t}_L)$ *vs* $\sigma(t_R\bar{t}_R)$

Consider the reaction

$$e^+e^- \rightarrow \gamma^*, Z^* \rightarrow t\bar{t}. \tag{18.1}$$

The couplings of the photon and the Z boson to top quarks conserve chirality. Yet chirality and helicity coincide only for fast particles; close to threshold there is a significant component when t and $\bar{t}$ both have either helicity $+1$ or -1: $e^+e^- \rightarrow t_R\bar{t}_R, t_L\bar{t}_L$. With t_L and $\bar{t}_R$ being conjugate to each other, i.e.

$$t_R \overset{\text{CP}}{\Longrightarrow} \bar{t}_L, \tag{18.2}$$

a difference in $t_R\bar{t}_R$ vs $t_L\bar{t}_L$ production,

$$\sigma(e^+e^- \rightarrow t_R\bar{t}_R) \neq \sigma(e^+e^- \rightarrow t_L\bar{t}_L), \tag{18.3}$$

implies **CP** violation [327]; for it to arise the usual two conditions have to be satisfied: (i) two amplitudes have to contribute coherently to the production process that involve a **CP** *non*invariant interaction; and (ii) an absorptive component has to be generated.

Exchanges of Higgs fields between the quark and antiquark pair can certainly satisfy the second condition, and also the first one if the Higgs fields form part of a non-minimal Higgs sector. The leading diagrams are shown in Fig. 18.1. The asymmetry is then produced by the interference between the tree diagram and the one-loop diagram.

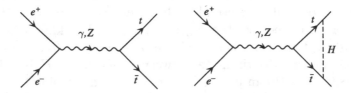

Figure 18.1   Feynman diagrams which interfere to generate **CP** violation in $e^+e^- \rightarrow t\bar{t}$ pair production.

While all of this would be true for any quark, it first becomes relevant for top quarks for two reasons:

- It is natural for top quarks to possess large Yukawa couplings to Higgs fields in general. Higgs exchanges thus can generate significant absorption and **CP** violation as well.
- Since top states decay weakly as quarks rather than *pseudoscalar* hadrons (as it would be for beauty, charm etc. production) we can employ the parity violating decay

$$t \rightarrow W^+ + b_L \tag{18.4}$$

to analyse the *polarization* of the decaying top quark and thus probe the production process where the asymmetry resides. The lack of hadronization then becomes an asset.

This scenario has been analysed in Ref. [327]. It was found that asymmetries as 'large' as $\mathcal{O}(10^{-3})$ could arise here. It remains to be seen whether an effect of this size could be observed at an e^+e^- or $\gamma\gamma$ collider or even in hadronic collisions that promise higher rates, yet present a less 'clean' environment.

18.1.2 Final state distributions in $e^+e^- \rightarrow t\bar{t}H^0$

We can probe the limits of **CP** symmetry also through analysing **T** odd correlations, as it is done in $K_{\mu 3}$ decays. The intervention of final state interactions is not required to generate a signal here although it can skew or even fake one.

One interesting possibility has been suggested in Ref. [328]. Consider

$$e^+(p_+)\, e^-(p_-) \;\rightarrow\; \gamma^*, Z^* \;\rightarrow\; t(p_t)\, \bar{t}(p_{\bar{t}})\, H^0(p_{H^0}), \tag{18.5}$$

where the inital and final state momenta have been listed. The salient points of the analysis are presented in the following list.

(1) In view of what was said before, the aim is to reconstruct the t and $\bar{t}$ momenta from their decay products and to formulate a **T** odd correlation in them:

$$O_- \equiv \langle \vec{p}_- \cdot (\vec{p}_t \times \vec{p}_{\bar{t}}) \rangle, \tag{18.6}$$

where

$$O_- \overset{\mathbf{CP}}{\Longrightarrow} -O_-. \tag{18.7}$$

We realize that the kinematics force such an observable to vanish for a two-body final state. We should also note that we search again for an asymmetry in the *production* reaction.

(2) Reconstructing the t and $\bar{t}$ momenta is not an easy task. It is then tempting to concentrate on determining the momenta of the b and $\bar{b}$ initiated jets emerging from the t and $\bar{t}$ decays, respectively:

$$O_-^{(b,\bar{b})} \equiv \langle \vec{p}_- \cdot (\vec{p}_b \times \vec{p}_{\bar{b}}) \rangle. \tag{18.8}$$

Finding $O_-^{(b,\bar{b})} \neq 0$ would manifest **CP** violation; however, the strength of a signal gets diluted in this *partial* reconstruction.

(3) So far we have mainly stated that kinematics do not force such correlations to vanish. The important dynamical point here is that the neutral Higgs H^0 can be emitted from the t (and $\bar{t}$) line through their Yukawa couplings as well as from the intermediate Z through the gauge coupling. Those three amplitudes can interfere and thus expose the presence of a relative weak phase between the gauge coupling on one hand and the t and $\bar{t}$ Yukawa couplings on the other. It has been stated that the resulting asymmetries might conceivably be of order 10% [328]!

18.2 On CP violation with leptons

There are two powerful motivations for probing **CP** symmetry in leptodynamics.

- The discovery of **CP** asymmetries in B decays close to 100% in a sense 'de-mystifies' **CP** violation. For it established that complex **CP** phases are not intrinsically small and can be close to 90° even. This demystification would be completed, if **CP** violation were found in leptodynamics as well.
- As mentioned in Section 8.3.3 we know that CKM dynamics, which is so successful in describing quark flavour transitions, is irrelevant to baryogenesis. There are intriguing arguments for baryogenesis being merely a secondary effect driven by primary leptogenesis [329], see Chapter 21. To make the latter less speculative, one has to find **CP** violation in leptodynamics.

The importance of the second motivation at least has been well recognized, as can be seen from the planned experiments to measure **CP** violation in neutrino oscillations, see Section 16, and the ongoing heroic efforts to find an electron EDM, Section 3.6. Yet there are other avenues to this goal as well that certainly are at least as challenging, which we will address in the remainder of this chapter.

No **CP** violation has been observed yet in the leptonic sector. That is not surprising since the avenues that proved most effective in the quark sector are not open here.

- No particle–antiparticle oscillations can occur for charged leptons due to the conservation of electric charge. Neutrino oscillations on the other hand could be employed, as discussed in Chapter 16.
- Within the SM neutrinos are degenerate in mass, namely massless. This additional global invariance allows us to transform the MNS matrix into a unit matrix through appropriate re-definitions of the lepton fields. Even if some relatively modest extension of the SM breaks the degeneracy in the neutrino masses, such differences are typically tiny relative to the relevant scale set by the GIM mechanism, namely the W mass:

$$\frac{\Delta M_\nu^2}{M_W^2} \ll 1. \tag{18.9}$$

Lepton flavour changing neutral currents are thus highly suppressed even on the quantum level.

- Muon decays proceed basically through a single channel – $\mu^- \rightarrow e^- \bar{\nu}_e \nu_\mu$ – with only two other modes having been seen. The only practical way to search for a **T** odd correlation involves the decay of polarized muons into electrons whose spin is measured.

18.2.1 Positronium

The best chance to find **CP** violation in dynamics involving electrons (and muons) – and an excellent one in our judgment – is to search for their EDM, as described in Section 3.6. Nevertheless one should not ignore other experimental avenues. We sketch two here, which are analogous to the observables listed above for top quarks.

In the decays of polarized ortho-positronium $[e^+e^-]_{OP} \rightarrow 3\gamma$ one can measure various **T** odd integrated moments between the polarization vector $\vec{S}_{OP}$ of ortho-positronium and the momenta $\vec{k}_i$ of two of the photons that define the decay plane:

$$A_T = \langle \vec{S}_{OP} \cdot (\vec{k}_1 \times \vec{k}_2) \rangle \tag{18.10}$$

$$A_{CP} = \langle (\vec{S}_{OP} \cdot \vec{k}_1)(\vec{S}_{OP} \cdot (\vec{k}_1 \times \vec{k}_2)) \rangle. \tag{18.11}$$

- The moment A_T is **P** and **CP** *even*, yet **T** *odd*. Rather than by **CP** or **T** violation in the underlying dynamics it is generated by higher order QED processes. It has been conjectured [330] that the leading effect is formally of order α relative to the decay width due to the exchange of a photon between the two initial lepton lines. From it one has to remove the numerically leading contribution, which has to be absorbed into the bound state wavefunction. The remaining contribution is presumably on the sub-permill level. Alternatively A_T can be generated in order α^2 – or on roughly the 10^{-5} level – through the interference of the lowest order decay amplitude with another one, where a fermion loop connects two of the photon lines.

• On the other hand the moment A_{CP} is odd under **T** as well as under **P** and **CP**. Final state interactions *cannot* generate a **CP** odd moment with **CP** invariant dynamics. Observing $A_{CP} \neq 0$ thus unambiguously establishes **CP** violation. The present experimental upper bound is around few×0.01; it seems feasible [331], to improve the sensitivity by more than three orders of magnitude down to the 10^{-5} level. The caveat arises on the theoretical level: with the 'natural' scale for *weak* interference effects in positronium given by $G_F(m_e^2) \sim 10^{-11}$, one needs a dramatic enhancement to obtain an observable effect.

18.2.2 μ decays

The most general expression for the amplitude of $\mu^- \to e^- \bar{\nu}_e \nu_\mu$ driven by a four-fermion interaction conserving lepton flavour (and without derivatives) can be written as follows [332]:

$$\mathcal{M} = \frac{4G_F}{\sqrt{2}} \sum_{i;a,b} g_{ab}^i \langle \bar{e}_a | \Gamma^i | \nu_e \rangle \langle (\bar{\nu}_\mu) | \Gamma_i | \mu_b \rangle; \tag{18.12}$$

$i = S, V, T$ denotes Lorentz scalar, vector and tensor couplings and $a, b = L, R$ the chirality of the electron or muon. The dimensionless numbers g_{ab}^i describe the relative weight of the different couplings; the ten complex g_{ab}^i represent 19 independent parameters that can be determined by experiment. No deviations from a pure 'V–A' interaction with $g_{LL}^V = 1$ and all other $g_{ab}^i = 0$ has been observed yet down to very low levels.

Of main interest in our context here is the **T** odd moment $P_T \equiv \vec{\sigma}_e \cdot (\vec{p}_e \times \vec{\sigma}_\mu)$ – the component of the decay electron polarization transverse to the electron momentum and the muon polarization. Due to the practical absence of strong or electromagnetic final state interactions here it genuinely probes **T** invariance. No such effect arises in the SM on a measurable level, and none has been seen for the energy averaged transverse polarization [332, 333]:

$$\langle P_T \rangle = (-3.7 \pm 7.7_{\text{stat.}} \pm 3.4_{\text{syst.}}) \times 10^{-3} \tag{18.13}$$

The measurement of P_T might be improved with a smaller phase space; one also needs a *pulsed* beam with high polarization (W. Fetscher, personal communication).

While the experimental effort behind these numbers is impressive, it is hard to see how even New Physics could have generated a signal on that numerical level.

18.2.3 τ decays

No **CP** violation has been observed yet in the leptonic sector. Several features enhance the chance to find that phenomenon in τ transitions [334, 335, 336, 337, 338].

(1) The **CP** phenomenology is now multi-faceted. Since there are several major decay modes, constraints from **CPT** invariance become considerably less effective. A qualitatively new feature is that hadrons can appear in the final state. Thus more types of **T** odd correlations can be constructed. Furthermore one should note that models with leptoquark bosons can contribute directly.

(2) Due to the higher mass of the charged lepton – $m_\tau \gg m_\mu$ – and of the basic objects in the final state – $m_s \gg m_{d,u} \gg m_e$ – there is a better chance for New Physics to create an observable impact, in particular in theories with a non-minimal Higgs sector with or without SUSY.

(3) The production process $e^+e^- \to \gamma^* \to \tau^+\tau^-$ provides a clean and relatively copious source of τ leptons with two additional benefits: (i) Having the electron beam longitudinally polarized leads to the τ leptons emerging with a net longitudinal polarization, which allows us to construct new types of **CP** and **T** odd correlations. (ii) Even without polarized beams one can extract information on the polarization of the τ leptons. Since they are produced off the intermediate one-photon state γ^* with their spins anti-aligned, one can infer the spin direction of one τ from the correlations among momenta or energies of the decay products of the other τ.

Partial width **CP** asymmetries in $\tau \to K\pi\nu_\tau$

The often heard statement that the SM generates no appreciable **CP** asymmetry in τ decays is incorrect. One reliably predicts [339] a difference between $\Gamma(\tau^- \to K_S\pi^-\nu_\tau)$ and $\Gamma(\tau^+ \to K_S\pi^+\bar\nu_\tau)$. This difference is actually caused by 'known' physics, namely the **CP** impurity in the K_S wave function. For

$$\left|T(\tau^- \to \bar{K}^0 \pi^- \nu_\tau)\right|^2 = \left|T(\tau^+ \to K^0 \pi^+ \bar{\nu}_\tau)\right|^2, \tag{18.14}$$

translates into

$$\frac{\Gamma(\tau^+ \to K_S \pi^+ \bar{\nu}) - \Gamma(\tau^- \to K_S \pi^- \nu)}{\Gamma(\tau^+ \to K_S \pi^+ \bar{\nu}) + \Gamma(\tau^- \to K_S \pi^- \nu)} = \frac{|p_K|^2 - |q_K|^2}{|p_K|^2 + |q_K|^2} = (3.32 \pm 0.06) \times 10^{-3} \tag{18.15}$$

It is irrelevant whether $|p_K|^2 \neq |q_K|^2$ is produced by the SM or not. An intriguing puzzle arises when calculating the **CP** asymmetry in $\tau \to K_L \pi \nu$; the puzzle is stated and resolved in Problem 18.3. Of course no such effect can occur for the other charge combination of the $K\pi$ system:

$$\frac{\Gamma(\tau^+ \to K^+ \pi^0 \bar{\nu}) - \Gamma(\tau^- \to K^- \pi^0 \nu)}{\Gamma(\tau^+ \to K^+ \pi^0 \bar{\nu}) + \Gamma(\tau^- \to K^- \pi^0 \nu)} = 0. \tag{18.16}$$

While the asymmetry in Eq. (18.15) is unrelated to leptodynamics, it is not just a curiosity, since it provides a valuable cross check. For if it is not found, then either one does not understand the experimental sensitivity – or there is New Physics interfering destructively with the 'known' physics!

CP *asymmetries in final state distributions*

Let us first illustrate such effects through simple scenarios. Let us suppose that **CP** and **T** violating τ decays exist as described by **T** odd correlations:

$$k_l \frac{d\Gamma(\tau \to \nu_\tau l(\vec{\sigma}, \vec{k}_l) \bar{\nu}_l)}{d^3 k_l} \propto C \vec{\epsilon} \cdot (\vec{\sigma}_l \times \vec{k}_l),$$

$$k_{\pi^+} k_{\pi^-} \frac{d\Gamma(\tau \to \nu_\tau \pi^+ (\vec{k}^+) \pi^- (\vec{k}^-) + X)}{d^3 k_{\pi^+} d^3 k_{\pi^-}} \propto C' \vec{\epsilon} \cdot (\vec{k}_{\pi^+} \times \vec{k}_{\pi^-}); \tag{18.17}$$

we have indicated the momentum and spin vectors for the corresponding particles; $\vec{\epsilon}$ denotes the τ polarization, and l stands for an electron or a muon. Consider the following decays of a $\tau^+ \tau^-$ pair

$$e^+ e^- \to \tau^+ \tau^- \to \nu_\tau \mu^- \bar{\nu}_\mu$$
$$\hookrightarrow \bar{\nu}_\tau e^+ \nu_e,$$
$$e^+ e^- \to \tau^+ \tau^- \to \pi^+ \pi^- + X$$
$$\hookrightarrow \bar{\nu}_\tau e^+ \nu_e. \tag{18.18}$$

How do the correlations given in Eq. (18.17) impact the actual production and decays of the $\tau^+\tau^-$ pairs, Eq. (18.18)? It can be shown [334] that the above **T** violating correlations show up as $\langle \vec{p}_+ \cdot (\vec{\sigma}_\mu \times \vec{k}_\mu)(\vec{k}_e \cdot \vec{p}_+) \rangle$ and $\langle \vec{p}_+ \cdot (\vec{k}_{\pi^+} \times \vec{k}_{\pi^-}) \rangle$, respectively, where $\vec{p}_+$ is the beam positron momentum and $\vec{k}_e$ the momentum of the positron produced in the accompanying $\tau^+ \to \nu_\tau e^+ \nu_e$ decay. Such correlations can also be used to probe for **CP** asymmetries in the production of the τ pair, as sketched below.

The $\tau \to K\pi\nu$ channels are promising candidates for revealing New Physics in final state distributions [338]:

- While they are Cabibbo suppressed in the SM, they should be most sensitive to non-minimal Higgs dynamics due to the relatively large τ and s quark masses; both features enhance the potential impact of a tree-level Higgs exchange.
- **CPT** invariance does not limit an asymmetry in partial widths: for one in $\tau^\pm \to \nu K^\pm \pi^0$ can naturally be compensated by one in $\tau^\pm \to \nu K_S \pi^\pm$.
- The final state is sufficiently complex so that **CP** asymmetries can emerge also in final state distributions. Those might well be considerably larger than a difference in the partial widths.

The requirement for **CP** violation becoming observable – to have two different amplitudes contribute – can be satisfied, since the $K\pi$ system can be produced from the (QCD) vacuum in a vector and scalar configuration with form factors F_V and F_S, respectively. Both are present in the data, with the vector component (mainly in the form of the K^*) dominant as expected [340]. Partially integrated final state distributions can in general exhibit **CP** asymmetries, since they are sensitive to the $F_V - F_S$ interference; e.g.

$$\frac{d}{dE_K}\Gamma(\tau^- \to \nu K^- \pi^0) - \frac{d}{dE_K}\Gamma(\tau^+ \to \bar{\nu}K^+ \pi^0) \propto \mathrm{Im}(F_S F_V^*)\mathrm{Im}g_S g_V^*; \tag{18.19}$$

i.e. one needs a (relative) strong phase between the form factors $F_{V,S}$ as well as a phase between their weak couplings $g_{S,V}$. The former should not represent a serious restriction, since the $K\pi$ system is produced in a mass range with several resonances. The latter does not arise within the SM on an observable level, yet it can be provided by a charged Higgs exchange in non-minimal Higgs models, which contributes to F_S. Such asymmetries might reach the 10^{-3} level. SUSY is hard pressed to reach this level, unless one invokes the 'old standby' for enhanced New Physics effects, namely SUSY with broken R parity, which could conceivably generate asymmetries even up to few percent [341].

CLEO has undertaken a pioneering search for a **CP** asymmetry in the angular distribution of the hadronic system in $\tau \to \nu K_S \pi$ placing an upper bound of a very few percent [342].

Having *polarized* τ leptons provides a powerful probe of **CP** invariance. For one can form new observables like **T** *odd* moments. In $\tau \to \nu K_S \pi$ there is one example, namely the expectation value for the τ spin vector pointing out or into the $K\pi$ plane:

$$O_T \equiv \langle \vec{\sigma}_\tau \cdot (\vec{p}_K \times \vec{p}_\pi) \rangle. \tag{18.20}$$

CP violation can surface here even with*out* a relative strong phase:

$$O_T \propto \mathrm{Re}(F_S F_V^*) \mathrm{Im} g_S g_V^*. \tag{18.21}$$

There is a caveat: final state interactions can generate **T** odd moments even from **T** invariant dynamics, when one has

$$O_T \propto \mathrm{Im}(F_S F_V^*) \mathrm{Re} g_S g_V^*. \tag{18.22}$$

Fortunately one can differentiate between the scenarios of Eq. (18.21) and Eq. (18.22) in $e^+ e^- \to \tau^+ \tau^-$, where one can compare directly the **T** odd moments for the **CP** conjugate pair τ^+ and τ^-:

$$O_T(\tau^+) \neq O_T(\tau^-) \implies \textbf{CP violation!} \tag{18.23}$$

Having at least the electron beam longitudinally polarized will lead to the τ leptons emerging in $e^+ e^- \to \tau^+ \tau^-$ with a non-zero net polarization depending on the production angle [334]. Even with unpolarized beams one can use the spin alignment of the τ pair to 'tag' the spin of the τ under study by the decay of the other τ [343].

Other τ decay modes

It appears unlikely that analogous asymmetries could be observed in the Cabibbo allowed channel $\tau \to \nu \pi \pi$, yet detailed studies of $\tau \nu 3\pi/4\pi$ look promising, also because the more complex final state allows to form **T** odd correlations with unpolarized τ leptons; yet the decays of polarized τ leptons might exhibit much larger **CP** asymmetries [344].

Particular attention should be paid to $\tau \to \nu K 2\pi$, which has potentially very significant additional advantages:

$\oplus$: One can interfere *vector* with *axial vector* $K 2\pi$ configurations.

$\oplus$: The larger number of kinematical variables and of specific channels should allow more internal cross checks of systematic uncertainties like detection efficiencies for positive versus negative particles.

18.2.4 τ production

Similar to the situation with top quarks discussed in the beginning of this chapter, **CP** asymmetries could arise also in the production process $e^+ e^- \to \tau^+ \tau^-$. Such effects are usually labeled under electric and/or weak dipole moments.

Electric and weak dipole moments of τ

Electric and weak dipole form factors $d_\tau^\gamma(s)$ and $d_\tau^Z(s)$, respectively, describing $e^+ e^- \to \tau^+ \tau^-$ are defined by [345]:

$$\mathcal{T}_{CP} = -e \left[J_\gamma^\mu \frac{d_\tau^\gamma(s)}{s} + \frac{1}{s_W c_W} J_Z^\mu \frac{d_\tau^Z(s)}{s - M_Z^2} \right] \cdot \left[\bar{u}_\tau(k_\tau) \sigma_{\mu\nu} \gamma_5 k^\nu v_\tau(k_{\bar{\tau}}) \right],$$

$$(18.24)$$

where the following notation has been used: $J_\gamma^\mu = -\bar{v}_e \gamma^\mu u_e$, $J_Z^\mu = -\bar{v}_e \gamma^\mu (1 - \gamma_5) u_e / 4 - s_W^2 J_\gamma^\mu$, $s_W = \sin\theta_W$, $c_W = \cos\theta_W$, $k = k_\tau + k_{\bar{\tau}}$ and $s = k^2$. The form factors $d_\tau^{\gamma, Z}(s)$, which in general can contain imaginary parts, are often referred to as EDM and weak dipole moment (WDM), respectively, since at threshold they are indeed dipole moments.

These form factors induce **CP** and **T** odd correlations among the decays products of the τ leptons [345]. Those have not been observed leading to the following upper bounds [11]

$$-2.2 \leq \frac{\text{Re } d_\tau^\gamma}{10^{-17} \, \text{e cm}} \leq 4.5, \qquad \frac{\text{Re } d_\tau^Z}{10^{-17} \, \text{e cm}} < 0.5 \qquad (18.25)$$

$$-2.5 \leq \frac{\text{Im } d_\tau^\gamma}{10^{-17} \, \text{e cm}} \leq 0.08, \qquad \frac{\text{Im } d_\tau^Z}{10^{-17} \, \text{e cm}} < 1.1 \qquad (18.26)$$

based on data taken at $\sqrt{s} = 10.6 \, \text{GeV}$ and at $\sqrt{s} = M_Z$ for d_τ^γ and d_τ^Z, respectively. It is impressive that experiments can infer non-trivial bounds on such subtle effects. Yet on the other hand these limits are not overly telling; for these are not significantly smaller distance scales than the 10^{-16} cm down to which leptons have exhibited pointlike behaviour.

Like for electrons and muons the SM predicts unobservably tiny values for these form factors. Yet how large could they become in conveniently chosen New Physics scenarios? The inverse Z^0 mass provides a natural yardstick:

$$d_\tau = \delta \times \frac{e}{M_Z} \simeq \delta \times 2.2 \cdot 10^{-16} \, e\,\text{cm} \qquad (18.27)$$

While the above limits read $\delta \sim 0.1$ and 0.01 for d_τ^γ and d_τ^Z, respectively, values as 'large' as $\delta_{\text{NP}} \sim \mathcal{O}(10^{-3})$ might be conceivable [345].

However there is the caveat that we cannot plead complete ignorance on the size of leptonic EDMs. For we know that the electron EDM d_e cannot be significantly larger than $10^{-27}\,e\,\text{cm}$. We can expect the τ EDM to be greatly enhanced relative to d_e due to its larger mass. The question is: how does d_τ/d_e scale with m_τ/m_e – is it the first, second or third power? Such scalings would suggest values for d_τ of about 10^{-23}, 10^{-20} and $10^{-16}\,e\,\text{cm}$, respectively. While the first range seems out of reach and the third one is already being probed experimentally, the second provides a challenge to experiment that might not be beyond hope. The first scaling scenario is clearly conservative; it depends very much on the nature of the New Physics, whether the second or third scenarios are closer to reality. Our present knowledge of leptodynamics (or lack thereof) suggest to describe the odds for finding a τ EDM as between 'not hopeless' to a definite 'maybe'.

An EDM will affect the angular distribution and polarization of the τ leptons. Those have to be inferred from the decay products. While it is not trivial to trace them due to the neutrinos that escape direct detection, it can be done [334, 346].

For proper perspective one should note that one endeavours here to identify an intrinsically weak observable competing against electromagnetic effects. Obviously one needs a considerable amount of 'fortune' – in the French sense – to succeed in finding a signal. A priori the odds are much better, when searching for **CP** violation in *weak* decays.

18.3   Résumé on top and τ transitions

The treatment of top quarks and leptons is under good theoretical control, since they do not hadronize. While this feature establishes bragging rights for theorists, it makes searches for **CP** violation in the dynamics of these states more challenging. For hadronization can provide different, yet coherent amplitudes; meson–antimeson oscillations are the most powerful example.

While SM dynamics can generate no appreciable **CP** violation, New Physics can produce effects exceeding those from the SM by orders of magnitude; i.e. the theoretical 'signal' (= New Physics) to 'noise' (= SM) ratio might be rather large. Even so signals probably tend to be on the small side – except possibly for top quarks – with the lack of hadronization being one of the major reasons. More specifically we have to acquire sensitivity to 10^{-3} level asymmetries to have a 'realistic hope' for uncovering New Physics.

In our judgment τ decays provide – next to the electron EDM and ν oscillations – the best stage to search for manifestations of **CP** breaking leptodynamics. There exists a considerable literature on the subject started by discussions on a tau-charm factory more than a decade ago [335, 336, 337, 338] and attracting renewed interest recently [339, 341, 344, 346] stressing the following points.

- There are many more channels than in muon decays making the constraints imposed by **CPT** symmetry much less restrictive.
- The τ lepton has sizable rates into multibody final states. Due to their nontrivial kinematics asymmetries can emerge also in the final state distributions, where they are likely to be significantly larger than in the integrated widths, as we have argued repeatedly in previous chapters referring to the example of $K_L \to \pi^+\pi^-e^+e^-$.
- New Physics in the form of multi-Higgs (or leptoquark) models can contribute on the tree-level like the SM W exchange.
- Having polarized τ leptons provides a powerful handle on **CP** asymmetries and control over systematics.

We have to be prepared to hunt for effects that are probably small. Even so they would have a profound impact on our paradigm of fundamental physics. We should also keep in mind that a **CP** odd observable in a SM allowed process is merely *linear* in a New Physics amplitude, since the SM provides the other amplitude. On the other hand SM forbidden transitions – say lepton flavour violation as in $\tau \to \mu\gamma$ – have to be *quadratic* in the New Physics amplitude.

$$\textbf{CP odd} \propto |T^*_{SM}T_{NP}| \text{ vs. LFV} \propto |T_{NP}|^2. \tag{18.28}$$

Probing **CP** symmetry on the 0.1% level in $\tau \to \nu K\pi$ with its branching ratio ~1% thus has roughly the same sensitivity for a New Physics amplitude as searching for BR($\tau \to \mu\gamma$) on the 10^{-8} level.

On the other hand we have not completed our 'homework' for proper evaluation of the prospects and search strategies: the required experiments are certainly challenging, but what are the limitations? How important – statistically as well as systematically – is the availability of polarized e^+e^- beams for τ (or even top) studies? Which observables are most sensitive to the intervention of which class of New Physics?

Problems

18.1. Why can we consider the $\tau^+\tau^-$ pair in the reaction shown in Eq. (18.18) to be on-shell?

18.2. Verify Eq. (18.7).

18.3. This is a problem requiring lengthy and careful calculations. Yet we think it provides a challenge to the committed student that will be seen as satisfying in the end due to the subtle insights provided. In Eq. (18.15) we have stated a **CP** asymmetry in $\Gamma(\tau^+ \to K_S\pi^+\bar{\nu})$ vs. $\Gamma(\tau^- \to K_S\pi^-\nu)$ due to the **CP** impurity in the K_S wave function. Following the same prescription, namely to determine the rate for $\tau \to K_L\pi\nu$ by forming $\langle K^0|K_L\rangle$ and $\langle\bar{K}^0|K_L\rangle$, see Eq. (6.47), we get

$$\frac{\Gamma(\tau^+ \to K_L\pi^+\bar{\nu}) - \Gamma(\tau^- \to K_L\pi^-\nu)}{\Gamma(\tau^+ \to K_L\pi^+\bar{\nu}) + \Gamma(\tau^- \to K_L\pi^-\nu)} = \frac{|p_K|^2 - |q_K|^2}{|p_K|^2 + |q_K|^2}$$

$$= (3.32 \pm 0.06) \times 10^{-3};$$

$$(18.29)$$

i.e. the same result. However straightforward this procedure seems to be, the result should raise a red flag in the mind of the attentive reader. For **CPT** invariance enforces equality in the total τ^+ and τ^- rates.[1] Which other channel(s) can compensate the asymmetries in Eq. (18.15) and Eq. (18.29)? Such channels have to rescatter with $\tau \to [\pi^+\pi^-]_K\pi\nu$.

To resolve this puzzle, we have to note that the **CPT** constraint applies to the sum over all relevant channels with their rates integrated over all times of decay and analyse more carefully the K_S and K_L classification [339]. The asymmetries of Eq. (18.15), Eq. (18.29) are measured by studying the time elapsed between the τ decay and the moment, when the 2π pair is formed. The first asymmetry is obtained by focusing on short time differences and the second one for large time differences. Exactly because $\langle K_S|K_L\rangle \neq 0$ the decay rate evolution for $\tau \to [f]_K\pi\nu$, where f is an arbitrary final state in $K_{S,L}$ decays, now contains three terms: in addition to the two contributions listed above with time dependences $\propto e^{-\Gamma_S t_K}$ and $\propto e^{-\Gamma_L t_K}$, respectively, we have an *interference* term $\propto e^{-\frac{1}{2}(\Gamma_S+\Gamma_L)t_K}$ most relevant for intermediate times $1/\Gamma_S \ll t_K \ll 1/\Gamma_L$.

---

[1] We have invoked **CPT** symmetry explicitly by using the same quantities p_K and q_K for the K_S and K_L wave functions.

Integrating over all times of decay t_K and summing over all relevant channels f we obtain (after several steps outlined in [339]):

$$\sum_f \int dt_K \Gamma(\tau^+ \rightarrow [f]_{K^0(t_K)} \pi^+ \bar{\nu}) = \sum_f \int dt_K \Gamma(\tau^- \rightarrow [\bar{f}]_{\bar{K}^0(t_K)} \pi^- \nu)$$

(18.30)

as expected from **CPT** invariance. Of course it has to be so. Yet nature achieves it in a remarkable way by a savvy use of $K^0 - \bar{K}^0$ oscillations.

19

SUSY-providing shelter for Higgs dynamics

Supersymmetry – a theory with a great future in its past?

Our discussion of **CP** violation based on left–right models was driven by two complementary goals, namely to implement **CP** *non*-invariance in a spontaneous fashion and to have the dynamics subjected to a higher degree of symmetry. The motivation for analysing non-minimal Higgs dynamics was much less profound: since no rationale more compelling than simplicity has emerged for limiting ourselves to minimal Higgs dynamics, we should be obliged to look beyond a minimalistic version – even if it served only as an imagination stretcher.

The Higgs sector is quite commonly perceived as the product of some effective, yet ultimately unsatisfactory, theoretical engineering. Two types of scenario have been suggested to provide a more appealing framework.

(A): Higgs fields are composites rather than elementary and represent an effective description of some unknown underlying dynamics. Techni-colour models are one implementation of this scenario that used to be quite popular. Few definite statements can be made in such models. Yet it would be miraculous if a *minimal* Higgs sector emerged, and extra sources of **CP** violation are likely to surface following the classification given in Chapter 17.

(B): There is one very elegant theoretical scheme that provides a natural habitat for scalars – namely supersymmetry (SUSY). Since we consider it so attractive, we will describe and analyse this scenario explicitly.

At the same time we note that over the last few years there have been two developments of great significance in this context.

- As stated before, detailed analyses of B decays and probes of neutron and electron EDMs have not revealed (yet) any source of **CP** violation beyond that of the SM.
- It is often stated that *generic* SUSY models should have affected B decays in an observable way, *if* characterized by mass scales on the 1 TeV level. This feature, which holds also for other New Physics scenarios is often referred to as the 'New Physics Flavour Problem' .

We are *not* convinced that such a problem truly exists at present. Below we will address this issue in the specific context of SUSY and present the numbers on which such a statement is based. Our general perspective can be expressed as follows: while there is one feature of SUSY that is beyond any doubt, namely that it is broken by considerable mass gaps between ordinary fields and their SUSY partners, we are quite ignorant about the mechanism underlying it and the ensuing pattern. This is certainly a blow to the professional pride of us theorists, yet we should accept this lesson in humility and keep working on SUSY rather than abandon it. Some recent arguments suggest that the present absence for a SUSY signal in B decays is rather natural even with SUSY quanta existing with masses below 1 TeV or so, at least if one takes a more liberal view of what 'generic' means [347]. We will sketch that line of reasoning after introducing SUSY and discussing its phenomenology relevant for our topic.

We will not give a detailed description of SUSY models, since it would clutter the narrative of our book. Even worse – most of it might soon become obsolete: for we see an excellent chance that the LHC (scheduled to start up in 2009) will provide direct evidence for SUSY. In that case we will seriously consider writing a third edition of this book. We hope the reader considers it a promise, not a threat.

19.1 The virtues of SUSY

SUSY is the ultimate symmetry, it forms a bridge to gravity and it alleviates several vexing theoretical problems.

- The Coleman–Mandula theorem [348] proves, on very general grounds, that all possible symmetry groups S in quantum field theories can be expressed as the direct product of the Poincaré group $\mathcal{P}$ and an internal symmetry group G:

$$S = \mathcal{P} \otimes G. \tag{19.1}$$

This means that no symmetry can connect states of different spins; i.e. each irreducible representation of S contains only states of the same spin. There is a loophole in the proof of this theorem, though: it admits only *bosonic* operators as generators of a symmetry. It was realized later that *fermionic* operators can also generate a symmetry group and that they relate fermions and bosons to each other. It thus represents the 'ultimate' symmetry of a quantum field theory and was aptly named supersymmetry.

- Once SUSY is implemented as a *local* gauge symmetry, coordinate covariance has become local as well – i.e. general relativity has to emerge. That is why local SUSY is referred to as *supergravity*. SUSY also forms an integral part of superstring theories.
- Higgs dynamics play an essential role in theoretical engineering, since they allow us to break (gauge) symmetries *spontaneously*. This happens through neutral Higgs fields ϕ acquiring vacuum expectation values (VEVs): minimizing a potential of the type $V_H = M_\phi^2 |\phi|^2 + \lambda |\phi|^4$ with $M_\phi^2 < 0$ leads to $\langle \phi \rangle = \sqrt{-M_\phi^2 / 2\lambda}$. From the size of the Fermi constant G_F we infer $\langle \phi \rangle \simeq 174$ GeV for the neutral component of the $SU(2)_L$ doublet scalar; very roughly, we then expect $|M_\phi| \sim \mathcal{O}(100)$ GeV as well. Yet already the one-loop correction to *scalar* masses depends on Λ_{UV}^2, the square of the UV cut-off:

$$\Delta M_\phi^2 \propto \Lambda_{UV}^2 + \cdots . \tag{19.2}$$

i.e. it introduces a quadratic infinity in the scalar mass, which has to be removed through the renormalization process, a procedure referred to as quadratic mass renormalization. Mathematically this can can be arranged for by introducing counterterms into the Lagrangian. Yet physicists tend to view such a situation as very contrived requiring extreme *fine tuning*. In their perspective Λ_{UV} is not merely a mathematical 'place holder' for an infinity. It parametrizes the sensitivity of an observable – in this case the scalar mass – to the dynamics being relevant at (very) high energy scales. In a quantum field theory there is always some sensitivity to Λ_{UV}, usually a very mild one expressed by a logarithmic dependence. Yet a quadratic dependence means that the potential sensitivity to new dynamics operating at, say, GUT scales $\sim \mathcal{O}(10^{15})$ GeV or the Planck scale $\sim \mathcal{O}(10^{19})$ GeV is very high. Having contributions of order $(10^{15}$ GeV$)^2$ or more that are generated by quantum corrections cancelled by a term introduced by hand into the Lagrangian such that a physical scalar mass with $M_\phi^2 \sim \mathcal{O}(10^4)$ GeV^2 arises – while mathematically feasible – appears as highly 'unnatural'. For it would imply that the dynamics being effective at the electroweak scale of about 100 GeV had to be highly cognizant of whether new layers of dynamics enter at much higher scales, to achieve the necessary cancellation. This is referred to as the 'gauge hierarchy problem' of the SM.

SUSY per se does not solve this problem, but makes it more tractable: for once a Higgs potential is chosen such that it yields the required large ratio to tree-level – i.e. classically – we can invoke SUSY to stabilize this ratio against *quantum* corrections. This is referred to as one of the *non-renormalization theorems of SUSY* . Light scalars then become natural by riding piggy-back on the shoulders of light fermions. Technically this is achieved by cancelling the $\Lambda^2_{\rm UV}$ terms in the scalar mass quantum corrections due to loops of bosons against those due to fermions. The intrinsic connection between fermions and bosons unique to SUSY is essential here.

- With the *quadratic* Higgs mass renormalization removed by SUSY, electroweak symmetry breaking can be induced *radiatively* if the top quark is sufficiently massive, since its Yukawa coupling g_t^Y then dominates the renormalization of the Higgs mass: $\partial M_\phi^2 / \partial {\rm log}\mu = 3(g_t^Y)^2 m_t^2 / 8\pi^2 + \cdots$ This happens for $m_t \geq 160\,{\rm GeV}$ or so. This feature had been noted [349] when such a mass was perceived as extravagantly high.

For SUSY to exist each state has to possess a superpartner, i.e. a state whose spin differs by half a unit.[1] For *manifest* SUSY the superpartner must be mass degenerate. Not a single such superpartner has been observed, i.e. SUSY breaking has to be implemented. This can be achieved by arranging for the superpartners to have sufficiently heavy masses so that they would have escaped direct experimental detection, like well in excess of 100 GeV. On the other hand, not all of their masses can exceed 1 TeV if SUSY is called upon, as sketched above, to stabilize the huge hierarchy in the GUT scales relative to the weak scales.

Intellectual honesty compels us to concede that at the moment of writing the second edition of this book there is still no direct evidence for SUSY being realized at all in nature: no two observed states can be considered superpartners. Even the common *bon mot* that half the SUSY states have been found is not quite true. For the Higgs sector in SUSY is considerably more complex than in the SM, and even the SM Higgs state has not been observed. Yet all of that can change very quickly, in particular since various indirect lines of reasoning suggest that SUSY partners enter below the TeV scale or so. Real SUSY quanta could then be produced at the FNAL and LHC colliders and high energy e^+e^- linear colliders, at present on the drawing board. Yet in the spirit of the whole book, it is reasonable to

---

[1] We consider neither *non*linear realizations of SUSY where the superpartners can be composites nor extended SUSY combining states with a wider range in the spin quantum number.

ask whether the existence of the SUSY degrees of freedom could not be 'felt' through the impact of their quantum corrections, in particular, in transitions that are highly suppressed like $K^0 - \overline{K}^0$ and $B^0 - \overline{B}^0$ oscillations or $B \to \gamma X$, $l^+ l^- X$ transitions. We will discuss in particular how *virtual* SUSY states will affect the **CP** phenomenology and rare decays.

With the increased number of layers in the dynamical structure there are many additional gateways through which **CP** violation can enter, in particular since SUSY has to be broken 'softly', i.e. through terms in the Lagrangian of operator dimension less than 4. Three *qualitatively* new features emerge here.

- **CP** odd couplings can now arise both in chirality conserving and changing couplings.
- **CP** violation can enter flavour-diagonal couplings.
- Intriguing connections between **CP** violation and the flavour problem can be formulated.

To be able to go beyond generalities we will sketch the minimal version of SUSY still consistent with nature – the minimal supersymmetric Standard Model (MSSM) – and use it as a reference point to make generic comments on more general cases.

19.2   Low-energy SUSY

We can be brief in outlining the general structure of SUSY and the MSSM, since the details can be found in several excellent reviews like Refs. [350, 351]. We introduce R parity – a multiplicative quantum number – assigning it a value $+1$ for ordinary fields (quarks, leptons, gluons, gauge bosons, Higgs, graviton etc.) and -1 for their superpartners (squarks, sleptons, gluinos, gauginos, higgsinos, gravitinos, etc.). The theory is most concisely formulated in terms of *superfields* which (for $N = 1$ SUSY) combine fields differing by half a unit in their spin, namely $J = 0$ and $J = 1/2$ – *chiral* or *matter* superfields – or $J = 1/2$ and $J = 1$ – *gauge* or *vector* superfields; there is also a superfield combining the $J = 2$ graviton with the $J = 3/2$ gravitino. These superfields are denoted by their $R = +1$ component. *Non*-gauge interactions are introduced through a superpotential G coupling up to three superfields together. Constructing a SUSY model thus requires five steps:

- adopt a gauge group;
- choose superfields with the appropriate gauge quantum numbers;
- formulate the gauge interactions;

- construct the superpotential with them; and, finally,
- implement SUSY breaking.

The last step provides the greatest challenge. It is also the one which obscures the intrinsic elegance of SUSY. One can adopt a pragmatic 'engineering' approach and add terms to the Lagrangian that break SUSY explicitly, yet softly. That is the approach we are going to follow mostly. Alternatively and more ambitiously one can construct theories that generate SUSY breaking dynamically, which turns out to be a rather involved affair. One typically starts from a so-called hidden sector of dynamical fields that do not carry SM quantum numbers. SUSY can be broken there by, say, fields acquiring a vacuum expectation value. This breaking is then 'mediated' through a sector of 'messenger' fields that carry non-trivial SM quantum numbers to produce a supersymmetric SM with the required breakings. This 'mediation' has to occur in a 'flavour blind' way – otherwise it would be even harder to understand why SUSY has not manifested itself in heavy flavour transitions. There are several ways known for achieving this goal such as gauge mediation or anomaly mediation [352, 353, 354]. Yet they seem to require a great deal of theoretical finesse and thus are not particularly robust; we will return to this point.

19.2.1 The MSSM

The superfield content of MSSM is given in Table 19.1. Each field of the SM is extended into a superfield. There is one non-trivial element: at least two *distinct* Higgs superfields H_1 and H_2 are needed, since otherwise chiral anomalies arise due to higgsino loops.

Table 19.1 *Superfields of MSSM with* $i = 1, 2, 3$ *as family index.*

Superfield	Colour	Isospin	Hypercharge	Lepton number	Baryon number
Q_i	3	$\frac{1}{2}$	$\frac{1}{6}$	0	$\frac{1}{3}$
U_i^c	3	0	$-\frac{2}{3}$	0	$\frac{1}{3}$
D_i^c	3	0	$\frac{1}{3}$	0	$\frac{1}{3}$
L_i	0	$\frac{1}{2}$	$-\frac{1}{2}$	1	0
E_i^c	0	0	1	-1	0
H_1	0	$\frac{1}{2}$	$-\frac{1}{2}$	0	0
H_2	0	$\frac{1}{2}$	$+\frac{1}{2}$	0	0

Non-gauge interactions are introduced through the superpotential G:

$$G = \mu H_1 H_2 + Y_{ij}^u Q_i H_2 U_j^c + Y_{ij}^d Q_i H_1 D_j^c + Y_{ij}^l L_i H_1 E_j^c \,. \tag{19.3}$$

The dimensionless numbers $Y_{ij}^{u,d,l}$ contain – among other things – the Yukawa couplings of the ordinary fermion fields. With superfields carrying dimension 1 (in mass units) and G thus dimension 3, we see that the mixing parameter μ has dimension 1 as well.

The interactions among the scalar fields can be extracted from G in a straightforward way and expressed through two types of ordinary potentials (of dimension 4):

$$\begin{aligned}
V_F &= |\mu h_1 + Y_{ij}^u \tilde{Q}_i \tilde{U}_j^c|^2 + |\mu h_2 + Y_{ij}^d \tilde{Q}_i \tilde{D}_j^c + Y_{ij}^l \tilde{L}_i \tilde{E}_j^c|^2 \\
&\quad + \sum_j |Y_{ij}^u \tilde{Q}_i h_2|^2 + |Y_{ij}^d \tilde{Q}_i h_1|^2 + |Y_{ij}^l \tilde{L}_i h_1|^2 \\
&\quad + \sum_i |Y_{ij}^u h_2 \tilde{U}_j^c|^2 + |Y_{ij}^d h_1 \tilde{D}_j^c|^2 + |Y_{ij}^l h_1 \tilde{E}_j^c|^2
\end{aligned}$$

$$\begin{aligned}
V_D &= \frac{g'^2}{2} \left(\frac{1}{6} \tilde{Q}_i^\dagger \tilde{Q}_i - \frac{2}{3} \tilde{U}_i^{c*} \tilde{U}_i^c + \frac{1}{3} \tilde{D}_i^{c*} \tilde{D}_i^c \right. \\
&\quad \left. - \frac{1}{2} \tilde{L}_i^\dagger \tilde{L}_i + \tilde{E}_i^{c*} \tilde{E}_i^c + \frac{1}{2} h_1^* h_1 - \frac{1}{2} h_2^* h_2 \right)^2 \\
&\quad + \frac{g^2}{8} \left(\tilde{Q}_i^\dagger \vec{\tau} \tilde{Q}_i + \tilde{L}_i^\dagger \vec{\tau} \tilde{L}_i + h_1^\dagger \vec{\tau} h_1 + h_2^\dagger \vec{\tau} h_2 \right)^2 \\
&\quad + \frac{g_s^2}{8} \left(\tilde{Q}_i^\dagger \vec{t} \tilde{Q}_i - \tilde{U}_i^{c*} \vec{t} \tilde{U}_i^c - \tilde{D}_i^{c*} \vec{t} \tilde{D}_i^c \right)^2, \tag{19.4}
\end{aligned}$$

with $\tilde{Q}$, $\tilde{U}$, $\tilde{D}$, $\tilde{L}$ and $\tilde{E}$ now representing the squark and slepton fields and h_1 and h_2 the Higgs fields rather than the full superfields; the $\vec{\tau}$ and $\vec{t}$ denote the Pauli and Gell-Mann matrices, respectively.

SUSY obviously cannot be a manifestly realized symmetry. *If* we want to invoke SUSY to shield Higgs masses against getting *quadratically* renormalized, as mentioned before, we infer that the spectrum of SUSY partners should start below the TeV scale. A *spontaneous* breaking of SUSY could be achieved through a conventional Higgs mechanism in V_D and V_F. However, that would lead to a phenomenological conflict in the relationship between the quark and squark masses. For a sum rule given by the supertrace

$$\text{Str } M^2 = \sum_J (-1)^{2J} M_J^2 = 0, \tag{19.5}$$

which holds trivially in *manifest* SUSY, is *not* modified when the spontaneous breaking of SUSY is driven by V_D or V_F (see Problem 19.1). As the sum of all fermion masses squared has to equal that of all boson masses squared, it predicts that some squarks and sleptons must be light,[2] superpartners should then have been found by now. Equation (19.5) holds in all models in which SUSY is broken at the *tree* level.

To avoid this phenomenological conflict we introduce *soft* breaking terms into the MSSM Lagrangian, where all dimension-4 operators obey SUSY:

$$\mathcal{L} = \mathcal{L}_{\text{SUSY}}(d \leq 4) + \mathcal{L}_{\text{soft}}(d \leq 3). \tag{19.6}$$

This is seen merely as an 'engineering' device to generate higher masses for all SUSY partners. We have actually in mind that New Physics *beyond* MSSM dynamically generates such soft SUSY breaking. More specifically, in addition to the *visible* or low-energy sector described by MSSM, we envision a *hidden* sector characterized by very high scales where SUSY breaking originates, which then is transmitted to the MSSM through *flavour-blind* interactions. Such an ansatz effectively disassociates the scale at which the intrinsic SUSY breaking occurs – Λ_{SSB} – from the one that gives rise to the different flavours – Λ_{Flav} – with $\Lambda_{SSB} \ll \Lambda_{\text{Flav}}$. SUSY breaking masses emerge as flavour singlets; squark masses are then degenerate up to small corrections of the order of the corresponding quark masses. As described below, this severely suppresses flavour-changing neutral currents (FCNC) in the SUSY sector as well; i.e. a restrictive super-GIM mechanism arises naturally. Two basic scenarios are:

(1) SUSY breaking is *gravity* mediated and enters $\mathcal{L}_{\text{soft}}(d \leq 3)$ through soft mass terms, for which we find

$$M_{\text{soft}} \sim \mathcal{O}\left(\frac{\Lambda_{SSB}^2}{M_{\text{Planck}}}\right), \tag{19.7}$$

implying $\Lambda_{SSB} \sim \mathcal{O}(10^{11})\,\text{GeV}$ to yield $M_{\text{soft}} \sim 1\,\text{TeV}$.

(2) It can be *gauge* mediated instead, with the soft mass terms being generated by virtual exchanges of messenger fields from the hidden sector which possess $SU(3)_C \times SU(2)_L \times U(1)$ couplings:

$$M_{\text{soft}} \sim \mathcal{O}\left(\frac{\alpha_{\text{gauge}}}{4\pi}\frac{\Lambda_{SSB}^2}{M_{\text{mess}}}\right), \tag{19.8}$$

where α_{gauge} denotes the gauge coupling of the messenger fields and M_{mess} their typical mass scale. It turns out that $M_{\text{mess}} \sim \Lambda_{SSB} \sim$

---

[2] We might entertain the idea that gauginos, being much heavier than the corresponding gauge bosons, would allow some sfermions to be lighter than their fermions without violating Eq. (19.5). Yet it turns out that such a trade-off is not possible: Eq. (19.5) has to be satisfied sector by sector.

$\mathcal{O}(10^5)\,\mathrm{GeV}$ or so is quite a natural value; i.e. the SUSY breaking occurs at much lower scales than in the gravity-mediated case.

The phenomenological problem expressed by Eq. (19.5) is resolved in these schemes – at least for the time being – through the emergence of a sizeable term on the right-hand side. For example, when SUSY is implemented as a *local* symmetry, it requires a spin-3/2 fermion – called the gravitino – as the superpartner for the spin-2 graviton. One manifestation of SUSY breaking is the emergence of a finite gravitino mass leading to

$$\mathrm{Str}\ M^2 = \sum_J (-1)^{2J} M_J^2 \sim m_{3/2}^2 \,, \qquad (19.9)$$

since the graviton remains massless. Equation (19.9) implies that superpartners of leptons and quarks are, in general, heavier than quarks and leptons characterized by $m_{3/2} > 100\,\mathrm{GeV}$ or so.

Let us illustrate these general remarks by one example for a flavour-*blind* scenario. We make the following ansatz for soft SUSY breaking being generated at some high scale:

$$
\begin{aligned}
\mathcal{L}_{soft} = m_{3/2}^2 \sum_i |A_i|^2 \ & + A m_{3/2} [Y_{ij}^u \tilde{Q}_i h_2^* \tilde{U}_j^c + Y_{ij}^d \tilde{Q}_i h_1 \tilde{D}_j^c \\
& + Y_{ij}^l \tilde{L}_i h_1 \tilde{E}_j^c + \mathrm{h.c.}] + B m_{3/2} \mu h_1 h_2 \\
& + \frac{1}{2} M (\lambda_1 \lambda_1 + \lambda_2 \lambda_2 + \lambda_3 \lambda_3),
\end{aligned}
\qquad (19.10)
$$

where A_i denote scalar fields in general, $\tilde{Q}$, $\tilde{U}$, $\tilde{D}$, $\tilde{L}$ and $\tilde{E}$ fermion fields, $h_{1,2}$ Higgs fields and $\lambda_{1,2,3}$ gaugino fields for the gauge groups $U(1)$, $SU(2)_L$ and $SU(3)_C$, respectively. The low energy effective Lagrangian is then obtained by running the parameters down to the electroweak scale, as described below.

In addition to the SM parameters, there are *five new* classes of dimensional parameters in MSSM: μ, describing the mixing between the two Higgs superfields, plus four more representing SUSY breaking, namely the gravitino and gaugino masses, $m_{3/2}$ and M, respectively, a higgsino mixing term, $B m_{3/2}$, and quark–squark–higgsino couplings, $A m_{3/2}$. For simple cases $B = A - 1$ and the number of parameters is reduced to four. In addition, we must require that the electroweak symmetry breaking occurs at the right energy scale: $4 M_W^2 / g^2 = v_1^2 + v_2^2$, with

$$\frac{v_2}{v_1} \equiv \tan\beta \qquad (19.11)$$

being a new parameter. One might think that $v_1 \sim v_2$ is a reasonable guestimate, and 'small $\tan\beta$' scenarios have been examined extensively.

On the other hand one can also argue that the Yukawa couplings of the Higgs fields $h_{1,2}$ should be of comparable strength. To obtain $m_b \ll m_t$ then requires $v_1 \ll v_2$, i.e. a 'large tanβ' scenario. Those have become rather popular with tanβ ranging as high as 50 or even higher [355].

To summarize this lightning review.

- There is a well defined minimal extension of the SM, the MSSM. It contains at least two Higgs doublet superfields, H_1 and H_2.
- SUSY cannot be a manifest symmetry. Realizing it spontaneously *within* MSSM leads to phenomenological disaster. Instead, we introduce *explicit soft* SUSY breaking terms. In the simplest versions they are flavour singlets; yet – as described below – they could contain a non-trivial flavour structure.
- It is understood that these explicit breaking terms put in by hand are the product of dynamics *beyond* MSSM operating at much higher scales and handed down to the low energy sector in different ways.

19.3   Gateways for CP violation

SUSY models in general introduce a host of possible sources for **CP** violation, entering through

- the superpotential[3] and/or
- the soft breaking terms.

They generate phases that at low energies can surface in charged as well as neutral current couplings and in Higgs interactions. Their actual pattern depends on the specifics of the physics driving SUSY breaking.

This can be expressed more concisely and bluntly: the fact that realistic SUSY models have many potential sources of **CP** violation can be viewed as intriguing – and as frustrating at the same time.

19.3.1   A first glance at CP phases in MSSM

Inspecting Eq. (19.10) as derived from a gravity-induced breaking pattern allows us to arrive at the following observations.

(1) Hermiticity of the potential requires $m_{3/2}^2$ to be real; $m_{3/2}$ can be made real by absorbing its remaining phase into the definition of A and B.

---

[3] The irreducible phase contained in the Yukawa couplings of quarks that is present already in the SM is embedded here.

(2) The gaugino mass term M can be chosen real by adjusting the phases of the gaugino fields λ_i.

(3) The squark–quark–gluino coupling will be affected by such a choice of the gaugino phase.

(4) We will also show that the phase associated with a linear combination of A and μ can be detected by experiments.

Next we will analyse how such phases can affect the low-energy couplings.

19.3.2   Squark mass matrices

The superpartners of left- and right-handed quarks are referred to as left- and right-handed squarks although the latter – being scalars – possess no chiral structure. Squark masses are greatly affected by the soft SUSY breaking. In particular the term in $\mathcal{L}_{\text{soft}}$ proportional to the trilinear scalar coupling A mixes left- and right-handed squarks. From Eq. (19.10) we can read off the squark mass matrices. For down-type squarks we find

$$
V_{D-\text{mass}} = \begin{pmatrix} \tilde{D} \\ \tilde{D}^{c*} \end{pmatrix}^{tr} \begin{pmatrix} \tilde{\mathbf{M}}^2_{DLL} & \tilde{\mathbf{M}}^2_{DLR} \\ \tilde{\mathbf{M}}^{2\dagger}_{DLR} & \tilde{\mathbf{M}}^2_{DRR} \end{pmatrix} \begin{pmatrix} \tilde{D}^* \\ \tilde{D}^c \end{pmatrix} \tag{19.12}
$$

$$
\tilde{\mathbf{M}}^2_{DLL} = \left(m^2_{3/2} + (v^2_1 - v^2_2)\left(\frac{g'^2}{12} - \frac{g^2}{4}\right) \right)\mathbf{1} + \mathcal{M}_D \mathcal{M}^\dagger_D
$$

$$
\tilde{\mathbf{M}}^2_{DLR} = (A^* m_{3/2} + \mu^* \tan\beta)\mathcal{M}_D
$$

$$
\tilde{\mathbf{M}}^2_{DRR} = \left(m^2_{3/2} + (v^2_1 - v^2_2)\frac{g'^2}{6} \right)\mathbf{1} + \mathcal{M}^\dagger_D \mathcal{M}_D, \tag{19.13}
$$

with $\mathcal{M}_D$ representing the mass matrix for down-type quarks. We have written the three generations of down squarks as six component vectors with $\tilde{q}^c_L = \tilde{q}^*_R$:

$$
(\tilde{D}^{\text{tr}}, \tilde{D}^{c\,\text{tr}}) = (\tilde{d}_L, \tilde{s}_L, \tilde{b}_L, \tilde{d}^*_R, \tilde{s}^*_R, \tilde{b}^*_R). \tag{19.14}
$$

Equation (19.13) and its analogue for the up squark mass matrix constitute the super-GIM mechanism: the leading contribution to the squark masses is flavour independent, namely the gravitino mass, with the flavour *splittings* given by the masses for the corresponding quarks, namely the down-type quarks here. Let us say that $\mathcal{M}_D$ is diagonalized by

$$
\mathcal{M}^{\text{diag}}_D = \mathbf{U}^D_L \mathcal{M}_D \mathbf{U}^{D,\dagger}_R. \tag{19.15}
$$

Making a unitary transformation

$$\begin{pmatrix} U_L^D e^{i\phi_A} & 0 \\ 0 & U_R^D e^{-i\phi_A} \end{pmatrix} \tilde{M}_D^2 \begin{pmatrix} U_L^{D\dagger} e^{-i\phi_A} & 0 \\ 0 & U_R^{D\dagger} e^{i\phi_A} \end{pmatrix} = \begin{pmatrix} \hat{M}_{DLL}^2 & \hat{M}_{DLR}^2 \\ \hat{M}_{DLR}^{2\dagger} & \hat{M}_{DRR}^2 \end{pmatrix},$$

$$(19.16)$$

where the phase factor

$$\phi_A = \arg(Am_{3/2} + \mu \tan\beta) \tag{19.17}$$

is introduced to absorb the phase in the off-diagonal matrix elements, we obtain:

$$\hat{M}_{DLL}^2 = \left(m_{3/2}^2 + (v_1^2 - v_2^2)\left(\frac{g'^2}{12} - \frac{g^2}{4}\right)\right)1 + (\mathcal{M}_D^{\text{diag}})^2$$

$$\hat{M}_{DRR}^2 = \left(m_{3/2}^2 + (v_1^2 - v_2^2)\frac{g'^2}{6}\right)1 + (\mathcal{M}_D^{\text{diag}})^2$$

$$\hat{M}_{DLR}^2 = \left(|A|m_{3/2} + \mu^*\frac{v_1}{v_2}\right)\mathcal{M}_D^{\text{diag}}. \tag{19.18}$$

While this mass matrix mixes left- and right-handed squarks, it still is *diagonal* in flavour space. Equation (19.18) holds at some large unification scale. In the much lower energy region $\sim M_W$ and below, coupling constants acquire radiative corrections [356] shown in Fig. 19.1. They induce corrections to $\tilde{M}_D^2$ proportional to $Y^U Y^{U\dagger}$, with Y^U denoting the Yukawa couplings of up-type quarks. While the precise effects depend on details

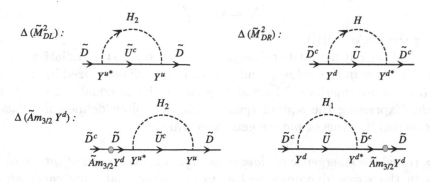

Figure 19.1 Example of diagrams which drive the renormalization group equation for M_U and M_D.

of the SUSY model under study, we can draw some generic conclusions. At scale M_W the mass matrix becomes for *small* $\tan\beta$:

$$\tilde{\mathbf{M}}_D^2 = \begin{pmatrix} \tilde{\mathbf{M}}_{DLL}^{\text{tree}}\tilde{\mathbf{M}}_{DLL}^{\text{tree}\dagger} + c_1\mathcal{M}_U\mathcal{M}_U^\dagger & Am_{3/2}\mathcal{M}_D(1 + \frac{c_2}{M_W^2}\mathcal{M}_U^\dagger\mathcal{M}_U) \\ A^*m_{3/2}(1 + \frac{c_2}{M_W^2}\mathcal{M}_U^\dagger\mathcal{M}_U)\mathcal{M}_D^\dagger & \tilde{\mathbf{M}}_{DRR}^{\text{tree}\dagger}\tilde{\mathbf{M}}_{DRR}^{\text{tree}} \end{pmatrix}$$

$$(19.19)$$

where we have kept only those contributions which give non-trivial modifications. The renormalization coefficients c_1 and c_2 can be computed from Ref. [356] within a given theory as functions of $t = \log(M_P/M_W)$ with M_P denoting the large scale where Eq. (19.10) holds; here we leave them as parameters.

Using Eq. (19.15) and Eq. (19.17) and the equivalent for up-type quarks, we perform a unitary transformation to find

$$\begin{pmatrix} \mathbf{U}_L^{D\dagger}e^{i\phi_A} & 0 \\ 0 & \mathbf{U}_R^{D\dagger}e^{-i\phi_A} \end{pmatrix} \tilde{\mathbf{M}}_D^2 \begin{pmatrix} \mathbf{U}_L^D e^{-i\phi_A} & 0 \\ 0 & \mathbf{U}_R^D e^{i\phi_A} \end{pmatrix} = \begin{pmatrix} \mathbf{W} & \mathbf{X} \\ \mathbf{Y} & \mathbf{Z} \end{pmatrix},$$

$$(19.20)$$

where

$$\mathbf{W} = (\hat{\mathbf{M}}_{DLL}^{\text{tree}})^2 + c_1\mathbf{V}^\dagger(\mathcal{M}^{\text{diag}})_U^2\mathbf{V}$$

$$\mathbf{X} = |A|m_{3/2}\left(1 + \frac{c_2}{M_W^2}\mathbf{V}^\dagger(\mathcal{M}_U^{\text{diag}})^2\mathbf{V}\right)\mathcal{M}_D^{\text{diag}}$$

$$\mathbf{Y} = |A|m_{3/2}\mathcal{M}_D^{\text{diag}}\left(1 + \frac{c_2}{M_W^2}\mathbf{V}^\dagger(\mathcal{M}_U^{\text{diag}})^2\mathbf{V}\right)$$

$$\mathbf{Z} = (\hat{\mathbf{M}}_{DRR}^{\text{tree}})^2,$$

$$(19.21)$$

with

$$\mathbf{V} = \mathbf{U}_L^U\mathbf{U}_L^{D\dagger}$$

$$(19.22)$$

being the CKM matrix.

A qualitatively new feature has emerged here due to the radiative corrections: because *in general* $\mathcal{M}_D$ and $\mathcal{M}_U$ *cannot* be diagonalized by the same unitary transformations, $\tilde{\mathbf{M}}_D^2$ and $\mathcal{M}_D$ cannot be diagonalized *simultaneously*. Expressing the squark–quark–gluino coupling defined for *flavour* eigenstates in terms of *mass* eigenstates will then

- reveal the emergence of *flavour changing neutral current interactions*;
- in the *strong* dynamics sector, in the sense that they carry *strong* coupling strength α_S;
- the flavour structure of which is controlled by the *CKM matrix*!

To analyse the experimental consequences of the new flavour changing interactions induced by Eq. (19.19) it is convenient to work with a basis [357]

$$\begin{pmatrix} \mathbf{U}_L^D e^{i\phi_A} & 0 \\ 0 & \mathbf{U}_R^D e^{-i\phi_A} \end{pmatrix} \begin{pmatrix} \tilde{D}^* \\ \tilde{D}^c \end{pmatrix}, \qquad (19.23)$$

leading to squark propagators being *not* diagonal: there is left–right squark mixing. These can be treated perturbatively using mass insertions:

$$\Delta_{LL}^2 = c_1 \mathcal{M}_U \mathcal{M}_U^\dagger$$

$$\Delta_{LR}^2 = A m_{3/2} \mathcal{M}_D \left(1 + \frac{c_2}{M_W^2} \mathcal{M}_U^\dagger \mathcal{M}_U \right)$$

$$\Delta_{RL}^2 = A^* m_{3/2} \left(1 + \frac{c_2}{M_W^2} \mathcal{M}_U^\dagger \mathcal{M}_U \right) \mathcal{M}_D^\dagger. \qquad (19.24)$$

Thus we have learnt the following lesson: in MSSM we usually envision flavour-*blind* SUSY breaking to be generated at high energy scales. Evolving the effective dynamics down to low energies we find that:

- FCNC emerge even in the *strong* sector described by gluino couplings;
- FCNC arise *radiatively* and thus are *reduced* considerably in strength;
- in general they do *not* conserve **CP**;
- **CP** violation occurs even in flavour *diagonal* transitions.

19.3.3 Beyond MSSM

New Physics even beyond MSSM operating in a hidden sector is invoked to generate the soft breaking terms of MSSM. In this situation minimality is not necessarily a virtue or even unambiguously defined. It is intriguing to consider scenarios with $\Lambda_{\text{Flav}} \sim \Lambda_{SSB}$, as suggested if there is a connection between SUSY breaking and the physics of flavour generation. The communication between the high energy and low energy domains no longer proceeds in a flavour blind manner even before radiative corrections are included, and the soft breaking terms can be expected to exhibit a non-trivial flavour structure. For example, the mass terms for the third family squarks might be different than for the first two families; or the mass matrices for the up- and down-type squarks might follow different patterns.

In any case FCNC will arise that in general are not suppressed by a super-GIM mechanism; the resulting phenomenology is less conservative and potentially much richer than in the previous scenario. More sources of **CP** violation become relevant and thus there is the potential for conflict with existing or future data – unless some *approximate symmetry* is invoked to suppress the size of **CP** phases. This opens the door to a very rich and multi-layered phenomenology; yet there is little we can say concretely beyond these generalities.

In Eq. (19.24) we introduced a parametrization of the FCNC and left–right mixing within MSSM in terms of an expansion in the matrices Δ_{LL}^2, Δ_{LR}^2 and Δ_{RL}^2. This procedure can be generalized for non-minimal models by expanding the squark propagators in powers of

$$(\delta_{ij})_{LL}, \qquad (\delta_{ij})_{LR}, \qquad (\delta_{ij})_{RL}, \qquad (\delta_{ij})_{RR}, \qquad (19.25)$$

where $\delta \equiv \Delta^2/\tilde{m}^2$ with $\tilde{m}$ being an appropriate mass scale like the average squark mass. In other words: while in this treatment the quark–squark–gluino vertex conserves flavour, there is a mass insertion in the squark propagator that is *not* flavour diagonal – $\delta_{ij} \neq 0$ for $i \neq j$ – and can also mix left- and right-handed squarks. This is commonly referred to as Mass Insertion (MI) approximation.

On one hand, such representation allows us to express phenomenological bounds in a model-independent way, while on the other, the parameters $(\delta_{ij})_{AB}$ are readily evaluated in a given model.

Within MSSM the quantities $(\delta_{ij})_{AB}$ contain an effective super-GIM filter reducing them below the 10^{-2} level or so without fine tuning. In non-minimal SUSY models, though, we have merely $|\delta| < 1$.

The story repeats itself for other quark–squark–gaugino couplings, albeit on a reduced numerical level.

19.4   Confronting experiments

The biggest *qualitative* change relative to the KM ansatz within SM is the emergence of **CP** violation in flavour-*diagonal* transitions, allowing electric dipole moments (and **CP** odd nuclear effects in general) to arise on an observable level.

On the experimental side the most important development since the turn of the millenium has been the harvest of high quality data on B transitions mostly from the B factories with the BaBar and Belle experiments, yet with some essential assistance from the Fermilab experiments CDF and D0.

19.4.1   Electric dipole moments

A tight experimental constraint can be placed on ϕ_A, the phase of the trilinear coupling A in $\mathcal{L}_{\text{soft}}$, as it generates a neutron electric dipole moment

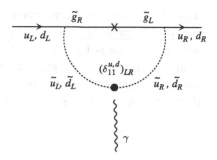

Figure 19.2   Feynman diagram which gives the neutron electric dipole moment.

already on the one-loop level. The diagram shown in Fig. 19.2 gives (using, as before, the static approximation $d_N \simeq 2d_d + d_u$)

$$d_N = -\frac{4}{27}\frac{e\alpha_s}{\pi}\frac{M_{\tilde{g}}}{M_{\tilde{q}}^2}\mathrm{Im}\left[2(\delta_{11}^D)_{LR} + (\delta_{11}^U)_{LR}\right]F_1(x), \qquad (19.26)$$

with

$$F_1(x) = \frac{1}{(1-x)^3}\left(\frac{1+5x}{2} + \frac{2+x}{1-x}\ln x\right), \quad x = \frac{M_{\tilde{g}}^2}{M_{\tilde{q}}^2}, \qquad (19.27)$$

where $M_{\tilde{g}}$ and $M_{\tilde{q}}$ denote the gluino and average squark mass, respectively. Since the dipole operator is chirality changing, a complex mixing between left- and right-handed squarks is needed to produce an effect: $\mathrm{Im}\delta_{LR} \neq 0$. The experimental bound on d_N yields for $M_{\tilde{g}} \simeq M_{\tilde{q}}$

$$\mathrm{Im}\,(\delta_{11}^D)_{LR}, \quad \mathrm{Im}\,(\delta_{11}^U)_{LR} \leq \text{few} \times 10^{-6}. \qquad (19.28)$$

Within MSSM $(\delta_{11}^D)_{LR} = (\delta_{11}^U)_{LR} = A \cdot m_{3/2}m_d/M_{\tilde{q}}^2$; i.e. assuming $m_{3/2} \sim M_{\tilde{q}} \sim 300\,\mathrm{GeV}$ as a benchmark, we infer from Eq. (19.28)

$$\arg(A) \leq \mathcal{O}(10^{-2}). \qquad (19.29)$$

While the super-GIM mechanism goes a long way towards explaining the minute size of $\mathrm{Im}(\delta_{11})_{LR}$, the phase of A is small for $M_{\tilde{q}}$ well below $1\,\mathrm{TeV}$. It is then tempting to conjecture the intervention of some symmetry to yield $\arg(A) \simeq 0$. In that case there is basically only one source for **CP** violation as in the SM, namely the irreducible KM phase (in addition to $\bar{\theta}_{QCD}$), yet it can enter through a new gateway – the renormalized squark mass matrices.

In non-minimal models *without* an effective super-GIM filter, the need for a symmetry suppressing new phases in the low energy sector appears compelling.

Of course you can also say that any improvement in experimental sensitivity for EDMs just might reveal a non-vanishing result!

Another view-point we might take is to say that the extra phases that appear in MSSM are maximal, and thus the experimental value on EDM leads to bounds on masses of supersymmetric partners [358]. For these considerations, other diagrams may also give significant contributions to EDM.

19.4.2   *SUSY contributions to* $\Delta S \neq 0 \neq \Delta B$ *transitions*

The wealth of experimental information on K and B decays is well described by CKM dynamics alone. Such an outcome cannot be *counted on a priori* if New Physics enters at the 1 TeV scale. This is true in particular, if the New Physics is of the SUSY variety with the latter's host of potential sources for flavour and **CP** violation. We will list those sources relevant for K and B decays, then summarize the resulting constraints as expressed through the MI approximation and comment on them in the next section.

• *The $K^0 - \overline{K}^0$ mass matrix*

An example of new diagrams contributing to the $K^0 - \overline{K}^0$ mass matrix is given in Fig. 19.3. For a complete set, we refer you to Ref. [359]. The resulting contributions are complicated functions of the various δ, the gluino mass $M_{\tilde{g}}$, and the average squark mass $\tilde{m}$. Large cancellations can occur among the various terms.

• *Direct **CP** violation in $\Delta S = 1$ decays*

There are several new diagrams for $\Delta S = 1$ decays, see Fig. 19.4. They can make significant contributions to ϵ'/ϵ_K that could a priori enhance or reduce it.

• *$b \to s\gamma$ (and other one-loop decays)*

Gluino exchanges mediate $b \to s\gamma$ decay through the diagrams shown in Fig. 19.5 [357];[4] they yield

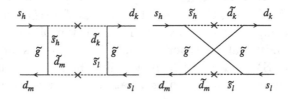

Figure 19.3   Example of diagrams which contribute to the $K - \overline{K}$ mass matrix.

---

[4] Similar diagrams contribute to $b \to se^+e^-$ and $b \to sg^{(*)}$.

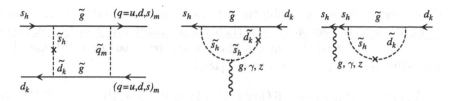

Figure 19.4 Example of diagrams which contribute to $\Delta S = 1$ transitions.

$$\text{Br}(b \to s\gamma) = \frac{\alpha_s \alpha}{81\pi^2 M_{\tilde{q}}^4} m_b^3 \tau_b \left[|m_b F_3(x)(\delta_{23}^D)_{LL} + M_{\tilde{g}} F_1(x)(\delta_{23}^D)_{LR}|^2 \right.$$
$$\left. + (L \leftrightarrow R) \right]. \tag{19.30}$$

where $F_1(x)$ is defined in Eq. (19.27) and

$$F_3(x) = \frac{1 - 8x - 17x}{12(x-1)^4} + \frac{x^2(3+x)\ln x}{2(x-1)^5}, \quad x = \frac{M_{\tilde{g}}^2}{m_{\tilde{q}}^2}. \tag{19.31}$$

The LR mixing term, being enhanced relative to the LL term by $M_{\tilde{g}}/m_b$, which is a large number, could play a dominant role in these rare decays. In practice, however, the same LR squark mixing reduces the mass of the lightest stop. Requiring that the light stop is heavy enough to evade experimental detection forces us to adjust other parameters. Down-type squarks then have to be nearly degenerate, which in turn reduces the contribution of gluino–squark loops to K and B transitions. MSSM contributions to the rare decays can still be sizeable and significant, but they do not modify the SM predictions in any dramatic way: at most a factor of two difference in rate can arise. However, quite dramatic modifications can arise in non-minimal implementations of SUSY.

The inclusive transition $B \to \gamma X$ has been first observed by Cleo at CESR. Belle [181] and BaBar [179] have now achieved rather accurate measurements; the average reads:

$$\text{Br}(B \to \gamma X_s; E_\gamma > 1.6\,\text{GeV}) = \left(3.55 \pm 0.24|_{\text{exp}} \, {}^{+0.09}_{-0.10}\big|_{\text{th}}\right) \times 10^{-4}; \tag{19.32}$$

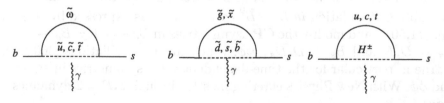

Figure 19.5 Feynman diagram which gives the SUSY contribution to the $b \to s\gamma$ amplitude.

the first error reflects the combined statistical and systematic experimental uncertainty and the second one a theoretical uncertainty in the extrapolation of the measured slice of the photon *spectrum* to the total one. The state-of-the-art SM prediction yields [180]:

$$\text{Br}(B \to \gamma X_s;\ E_\gamma > 1.6\,\text{GeV}) = (3.15 \pm 0.23) \times 10^{-4}\ ;\qquad (19.33)$$

the quoted uncertainty is largely of different origin than the theoretical one in Eq. (19.32): it refers mostly to the integrated photon spectrum rather than its shape as in Eq. (19.32). The prediction agrees quite well with the measurement. There is little space left for contributions from New Physics irrespective of the latter's features (unless the New Physics transition operator produces photons with a different helicity).

- $B^0 - \overline{B}^0$ *oscillations and* **CP** *violation*

With $\arg(A)$, shown in Eq. (19.29), too small to be significant in beauty decays, the phases driving **CP** asymmetries in B transitions within MSSM still reside in the KM matrix:

$$\phi_1|_{\text{KM}} = \phi_1|_{\text{MSSM}},\quad \phi_2|_{\text{KM}} = \phi_2|_{\text{MSSM}},\quad \phi_3|_{\text{KM}} = \phi_3|_{\text{MSSM}},$$

$$\text{Im}\frac{q}{p}\overline{\rho}_{B_s \to J/\psi\phi}|_{\text{MSSM}} \simeq 0, \qquad (19.34)$$

$$\left.\frac{\Delta M_{B_d}}{\Delta M_{B_s}}\right|_{\text{SM}} \simeq \left.\frac{\Delta M_{B_d}}{\Delta M_{B_s}}\right|_{\text{MSSM}}. \qquad (19.35)$$

As discussed in Section 10.1.4 and 10.2.1 the accurately measured ratio $\Delta M_{B_d}/\Delta M_{B_s}$ is fully consistent with the SM prediction. The latter contains a sizable uncertainty, which should be reduced in the future. It might happen that the ratio $\Delta M_{B_d}/\Delta M_{B_s}$ still agrees with the refined SM prediction, yet the *absolute* values of ΔM_{B_d} and ΔM_{B_s} do not. MSSM would be a good candidate for explaining such a discrepancy.

Non-minimal SUSY, on the other hand, can make significant contributions to ΔM_{B_d} and ΔM_{B_s} without having to satisfy Eq. (19.35). Furthermore, it can introduce large new weak phases thus enhancing considerably **CP** violation *in* $B^0 - \overline{B}^0$ oscillations – as expressed by A_{SL}, see Eq. (11.50) – and modify the **CP** asymmetries in $B_d \to \psi K_S$, $B_d \to \pi^+\pi^-$, $B_s \to D_s K$ and $B_s \to D_s\overline{D}_s$, $\psi\phi$, which involve oscillations [360]. This is true in particular for the time-dependent **CP** asymmetries in $B_s \to \psi\phi$ and $\phi\phi$. With New Physics entering mostly through $\Delta B = 2$ dynamics we have:

$$\phi_1 \to \phi_1 - \phi_{\text{NP}}(B_d),\quad \phi_2 \to \phi_2 + \phi_{\text{NP}}(B_d),\quad \phi_3 \to \phi_3 + \phi_{\text{NP}}(B_s), \quad (19.36)$$

i.e.:

$$\phi_1 + \phi_2|_{\text{meas}} = \phi_1 + \phi_2|_{KM}, \quad \phi_1 + \phi_2 + \phi_3|_{\text{meas}} = \pi + \phi_{\text{NP}}(B_s). \quad (19.37)$$

Transitions that proceed in the SM only (or predominantly) on the loop-level and thus are pure quantum effects have a priori the best chance to reveal the intervention of New Physics. This applies in particular to $B_d \to K^+K^-K_S$, $\eta'K_S$, $B \to K\pi$ and $B_s \to J/\psi\phi$, $\phi\phi$. The experimental landscape has been surveyed with increasing sensitivity, as summarized below.

19.4.3 Bounds on MI SUSY parameters

The SUSY contributions to K and B decays depend on the MI parameters δ_{ij}^D, the gluino mass $m_{\tilde{g}}$ and various squark masses $m_{\tilde{q}}$; the latter is assumed to be a practically universal quantity to implement the Super-GIM mechanism. From the observed values of ϵ_K, ϵ'/ϵ_K, BR$(B \to \gamma X_s)$, $\Delta M_{B_{d,s}}$ and $\sin 2\phi_1$ one infers the upper bounds listed in Table 19.2 [355]. A few caveats and comments are in order for proper perspective.

- These bounds have to be taken with quite a grain or two of salt: on one hand they are conservative, since one allows for the SUSY contribution to saturate the observable; on the other hand one ignores cancellations between different classes of contributions.
- All MI parameters are small compared to unity, which is a necessary condition for applying the MI approximation.
- The fact that these quantities are so tiny has led to the statement that there is a 'SUSY Flavour Problem' or more generally a 'New Physics Flavour Problem', as mentioned in the beginning of this chapter. It implies that if New Physics entered at the 1 TeV scale, it should already have made its presence felt in K and B decays. Such a claim should, however, not be accepted uncritically; we remain unconvinced at present.
  - The lowest bounds arise for $\sqrt{|\text{Im}(\delta_{12}^D)_{LR}^2|}$. Yet those numbers might be less telling than it appears at first sight. For this MI quantity is proportional to fermion masses in realistic models and thus contains a suppression factor $m_s/M_{\text{SUSY}} \leq 10^{-4}$ [355].
  - One should not expect δ_{12}^D, δ_{13}^D and δ_{23}^D to be of similar size. It is quite possible or even likely that in particular δ_{23}^D is significantly larger than δ_{12}^D and thus could reveal itself in future studies of, say, $B \to l^+l^-X_s$, $B_d \to \phi K_S$ or $B_s \to J/\psi\phi$, $\phi\phi$.
- In any case there might be some natural mechanism for generating MI parameters such that they do not exceed the bounds significantly, yet at the same time are not too small to be observable. Such a 'generic'

Table 19.2   *Upper bounds on the MI parameters inferred from ϵ_K, ϵ'/ϵ_K, $BR(B \to \gamma X_s)$, $\Delta M(B_{d,s})$ and $sin 2\phi_1$ for $m_{\tilde{q}} = 500\,GeV$ and $x = m_{\tilde{g}}^2/m_{\tilde{q}}^2$; bounds on $(\delta_{12}^D)_{LR}$ and $(\delta_{13}^D)_{LR}$ scale as $(m_{\tilde{q}}/500\,GeV)^2$ and the others as $m_{\tilde{q}}/500\,GeV$; from [355].*

x	$\sqrt{\left\lvert \mathrm{Im}(\delta_{12}^D)_{LL}^2 \right\rvert}$	$\sqrt{\left\lvert \mathrm{Im}(\delta_{12}^D)_{LR}^2 \right\rvert}$	$\sqrt{\left\lvert \mathrm{Im}(\delta_{12}^D)_{LL}(\delta_{12}^D)_{RR} \right\rvert}$
0.3	$2.9 \cdot 10^{-3}$	$1.1 \cdot 10^{-5}$	$1.1 \cdot 10^{-4}$
1.0	$6.1 \cdot 10^{-3}$	$2.0 \cdot 10^{-5}$	$1.3 \cdot 10^{-4}$
4.0	$1.4 \cdot 10^{-2}$	$6.3 \cdot 10^{-5}$	$1.8 \cdot 10^{-4}$
x	$\sqrt{\left\lvert \mathrm{Re}(\delta_{13}^D)_{LL}^2 \right\rvert}$	$\sqrt{\left\lvert \mathrm{Re}(\delta_{13}^D)_{LR}^2 \right\rvert}$	$\sqrt{\left\lvert \mathrm{Re}(\delta_{13}^D)_{LL}(\delta_{13}^D)_{RR} \right\rvert}$
0.3	$4.6 \cdot 10^{-2}$	$2.6 \cdot 10^{-2}$	$1.6 \cdot 10^{-2}$
1.0	$9.8 \cdot 10^{-2}$	$3.3 \cdot 10^{-2}$	$1.8 \cdot 10^{-2}$
4.0	$2.3 \cdot 10^{-1}$	$3.6 \cdot 10^{-2}$	$2.5 \cdot 10^{-2}$
x	$\left\lvert \mathrm{Im}(\delta_{13}^D)_{LL} \right\rvert$	$\left\lvert \mathrm{Im}(\delta_{13}^D)_{LL} \right\rvert$	
0.25	$1.3 \cdot 10^{-1}$	$6.6 \cdot 10^{-2}$	
1.0	$3.0 \cdot 10^{-1}$	$7.4 \cdot 10^{-2}$	
4.0	$3.4 \cdot 10^{-1}$	$1.0 \cdot 10^{-1}$	
x	$\sqrt{\left\lvert \mathrm{Re}(\delta_{23}^D)_{LL}^2 \right\rvert}$	$\left\lvert (\delta_{23}^D)_{LR}^2 \right\rvert$	$\sqrt{\left\lvert \mathrm{Re}(\delta_{23}^D)_{LL}(\delta_{23}^D)_{RR} \right\rvert}$
0.3	0.21	$1.3 \cdot 10^{-2}$	$7.4 \cdot 10^{-2}$
1.0	0.45	$1.6 \cdot 10^{-2}$	$8.3 \cdot 10^{-2}$
4.0	1	$3.0 \cdot 10^{-2}$	$1.2 \cdot 10^{-1}$

scenario is sketched below. It depends on details of the SUSY model, whether measurements of K or of B decays carry a higher promise of revealing the existence of SUSY.

Very intriguing relations arise in SUSY extensions of Grand Unified Theories (GUT) that connect quark and lepton parameters. Examples are lepton flavour violating transition rates like $\tau \to \mu\gamma$, 3μ versus **CP** asymmetries in $B_d \to \phi K_S$ or $B_s \to \phi\phi$ [355]. Such relations depend on details of the SUSY GUT model. Thus no reliable predictions can be made. On the other hand, scenarios can be sketched that gain significance, once a non-SM signal has emerged.

19.4.4   *Can SUSY be generic?*

A host of mostly theoretical arguments points to the presence of New Physics (beyond the SM Higgs state) at the 1 TeV scale. This led to the expectation that experiments at the LHC will find *direct* evidence for new degrees of freedom. This confidence has been moderated recently by the fact that neither precision measurements of electroweak observables (like the masses of the weak bosons) nor the detailed data on B decays coming out of the B factories has so far shown any evidence for such New Physics entering there through quantum corrections. One potential conclusion is that the mass scale of the new SUSY states is significantly higher than the 1 TeV scale and might be beyond the reach of the LHC for direct production.

Alternatively one can search for a symmetry principle to constrain sufficiently the size of the mass insertion parameters δ_{ij} while retaining squark and gluino masses below 1 TeV or so. It has even be suggested [347] that the required SUSY breaking can be generated in a hidden sector consisting of a new supersymmetric 'QCD' (or other non-chiral gauge theory) in a 'generic' way and be transmitted to the SUSY SM in such a way that SUSY masses are retained below the 1 TeV scale without violating bounds from heavy flavour decays.

These are stimulating ideas. Even more importantly nature has often shown an unexpected inventiveness in coming up with novel solutions or reformulations. Therefore we firmly believe in the primacy of experiment and conclude the following: the fact that SUSY has so far not revealed its presence in heavy flavour transitions does not imply that there are no specific SUSY effects there – it only sets the bar higher for the required experimental scrutiny.

19.5 The pundits' résumé

After this crash course in SUSY, we can evaluate **CP** violation in these models in a slightly more informed and considerably more refined manner.

(1) SUSY is truly a symmetry 'sans-pareille' – like no other. To realize all its attractive features we have to formulate it as a local theory – SUGRA – embedded in a GUT scheme.

(2) The credentials of SUSY as a symmetry that nature cannot have overlooked are further enhanced by the realization that it forms an integral part of string theories – maybe our best bet to bring gravity into the quantum world.

(3) SUSY's many new dynamical layers can support a plethora of additional sources of **CP** violation in the form of complex Yukawa couplings and other Higgs parameters, including those driving SUSY breaking.

(4) Through quark–squark–gluino couplings, flavour changing neutral currents mediating even **CP** violation enter the (nominally) strong dynamics.

(5) Remarkably enough, in MSSM only two observable phases emerge: one is the usual KM phase having its origin in the misalignment of the mass matrices for up- and down-type quarks; it also migrates into the squark mass matrices and controls the **CP** properties of the quark–squark–gluino couplings. The other one is ϕ_A, see Eq. (19.17), reflecting soft SUSY breaking; it is severely restricted by the experimental bound on the neutron EDM, which makes it irrelevant for $K_L \to \pi\pi$ decays.

(6) Yet once we enter the vast regime of non-minimal SUSY models – even by an apparently slight modification of MSSM – the floodgates open for additional sources of **CP** violation:

- observable effects are likely or at least quite conceivable for the EDMs of neutrons and electrons.

Beyond that, completely different scenarios can occur.

(a) MSSM: the same **CP** asymmetries arise in beauty decays as with the SM implementation of the KM ansatz, although the $B^0 - \overline{B}^0$ oscillation rates are different.

(b) Non-minimal SUSY: while large effects emerge in beauty decays there are – due to $\Delta B = 2$ dynamics being modified by SUSY – sizeable deviations from the KM expectations.

(7) What was just stated does not mean that an interpretative chaos of 'everything goes and nobody knows' will rule. For within each scenario there are numerous non-trivial correlations among the **CP** observables, rare decay rates, the KM parameters and gross features of the sparticle spectrum.

(8) Last, but certainly not least concerning SUSY: it would be highly premature to conclude from the present lack of experimental evidence for a SUSY impact in heavy flavour transitions that squarks and gluinos are significantly heavier than usually anticipated; or that such evidence cannot be derived from the next round of experiments on B, D and K decays or EDMs. It is our considered judgement that reaching the percent level sensitivity in B decay studies will be required for a realistic chance of finding SUSY's impact there.

Problems

19.1. Consider spontaneous breaking of SUSY through a conventional Higgs mechanism being implemented in V_D and V_F, as defined in Eq. (19.4). Let h_1 and h_2 – the scalar components of the superfields H_1 and H_2 – acquire VEVs v_1 and v_2, respectively.

Verify that the fermion and sfermion masses are then described by the following expressions:

$$\mathcal{L}_{\text{fermion masses}} = -v_1 Y^u_{ij} u_i u^c_j - v_2 Y^d_{ij} d_i d^c_j - v_2 Y^l_{ij} l_i l^c_j, \quad (19.38)$$

$$\begin{aligned}
\mathcal{L}_{\text{sfermion masses}} = {} & v_1^2 (\tilde{u} Y^u Y^{u\dagger} \tilde{u}^* + \tilde{u}^{c*} Y^{u\dagger} Y^u \tilde{u}^c) \\
& + v_2^2 (\tilde{d} Y^d Y^{d\dagger} \tilde{d}^* + \tilde{d}^{c*} Y^{d\dagger} Y^d \tilde{d}^c \\
& + \tilde{e} Y^l Y^{l\dagger} \tilde{e}^* + \tilde{e}^{c*} Y^{l\dagger} Y^l \tilde{e}^c) \\
& + \mu^* v_2 \tilde{U} Y^u \tilde{U}^c + \mu^* v_1 (\tilde{d} Y^d \tilde{d}^c + \mu^* v_1 \tilde{e} Y^l \tilde{e}^c) + \text{c.c.} \\
& + \frac{g'^2}{2} (v_1^2 - v_2^2) \left[\frac{1}{6} (\tilde{u}^* \tilde{u} + \tilde{d}^* \tilde{d}) - \frac{2}{3} \tilde{u}^{c*} \tilde{u}^c + \frac{1}{3} \tilde{d}^{c*} \tilde{d}^c \right. \\
& \left. \qquad - \frac{1}{2} \tilde{e}^* \tilde{e} - \frac{1}{2} \tilde{\nu}^* \tilde{\nu} + \tilde{e}^{c*} \tilde{e}^c \right] \\
& + \frac{g^2}{4} (v_1^2 - v_2^2) \left[\tilde{u}^* \tilde{u} - \tilde{d}^* \tilde{d} + \frac{1}{2} \tilde{e}^* \tilde{e} - \frac{1}{2} \tilde{\nu}^* \tilde{\nu} \right];
\end{aligned}$$

$$(19.39)$$

here we have used the following notation: u, u^c, d, d^c, l, l^c are left-handed quark and charged lepton fields with their charge conjugates; $\tilde{u}$, $\tilde{u}^C$, $\tilde{d}$, $\tilde{d}^C$, $\tilde{e}$, $\tilde{e}^C$, $\tilde{\nu}$, $\tilde{\nu}^C$ denote squark, charged and neutral slepton fields, respectively, together with their charge conjugates; the $Y^{U,D,l}$ represent the Yukawa couplings.

Do Eq. (19.38) and Eq. (19.39) exhibit SUSY breaking? Evaluate the supertrace of Eq. (19.5).

19.2. Evaluate the Feynman diagram in Fig. 19.2 to reproduce Eq. (19.26).

19.3. Verify Eq. (19.37); i.e. show that if New Physics contributes to $B_d - \overline{B}_d$ oscillations such that $\left.\frac{q}{p}\right|_{B_d} = \left.\frac{q}{p}\right|_{B_d,KM} e^{-i\phi_{NP}(B_d)}$, then the new phase $\phi_{NP}(B_d)$ drops out from the sum of the two angles ϕ_1 and ϕ_2 of the KM triangle.

20
Minimal flavour violation and extra dimensions

A host of mostly theoretical arguments points to the presence of New Physics (beyond the SM Higgs state) at the 1 TeV scale. This led to the expectation that experiments at the LHC will find *direct* evidence for new degrees of freedom. This confidence has been moderated recently for some people by the fact that neither precision measurements of electroweak observables (like the masses of the weak bosons) nor the detailed data on B decays have so far shown any evidence for such New Physics entering there through quantum corrections. One potential conclusion is that the mass scale of the New Physics states – like SUSY quanta – is significantly higher than the 1 TeV scale and might be beyond the reach of the LHC for direct production. In any case, we view it as a challenge to increase our efforts in B studies.

Alternatively one can search for a *principle* to suppress sufficiently the impact of New Physics on B and K decays, even when the New Physics quanta enter with masses below 1 TeV. In the SUSY framework described in Chapter 19, that amounts to requiring the mass insertion parameters δ_{ij} to satisfy the bounds listed in Table 19.2 (and potentially tighter ones in the future) while retaining squark and gluino masses below 1 TeV or so. This might be feasible along the lines sketched in Section 19.4.4.

20.1 On minimal flavour violation

As discussed in the preceding Chapter minimal SUSY extensions of the SM can be constructed in such a way that *no additional* sources of **CP** violation arise. This and of course the fact that no signal for New Physics has been found in K, D and B decays has given rise to a new classification scheme for theories of flavour dynamics, namely minimal flavour violation (MFV), which we will sketch here.

When searching for New Physics *indirectly*, one treats the SM as an *effective* theory valid at energies below a scale Λ, where only the SM fields ϕ_{SM} provide *dynamical* degrees of freedom. This is best expressed through an operator product expansion:

$$\mathcal{L}_{\text{eff}}(\phi_{SM}) = \mathcal{L}_{SM}(\phi_{SM}) + \sum_{n=1}^{\infty} \frac{1}{\Lambda^n} \mathcal{O}^{(n)}(\phi_{SM}); \qquad (20.1)$$

the $\mathcal{O}^{(n)}(\phi_{SM})$ denote all polynomials in the SM fields of operator dimension $4+n$ that are consistent with the gauge symmetries of the SM and of course with Lorentz symmetry as well. The scale Λ calibrates the observable impact of the anticipated New Physics at energies $E < \Lambda$ through an expansion in powers of E/Λ. Thus they disappear for $\Lambda \to \infty$.[1] It will suffice to retain only the leading contribution in the sum, which is $n = 2$ for flavour-changing transitions.

As discussed in detail in the preceding chapters, no clear manifestation for New Physics has so far surfaced in quark flavour dynamics. This has lead to the speculation that the anticipated New Physics by itself might be 'flavour neutral'; i.e. the dynamics driving all flavour-*non*diagonal transitions – including **CP** breaking effects – are related to the known structure of the SM Yukawa couplings. This conjecture can be labelled as 'minimal flavour violation' (MFV), since we infer from the successes of the SM that its dynamical elements do exist in nature. We will give a more specific definition for MFV to transform it into a quantitative criterion. Then we will sketch, how the MFV ansatz is implemented and its findings. We do not intend to give a detailed discussion of the technicalities involved; for those we refer to the ample literature that can be tracked through the two references [361, 362].

20.1.1 Defining, implementing and probing MFV

Without a Higgs sector the SM is a perfectly self-consistent theory (at least in the ultraviolet domain), albeit one with profoundly puzzling features like family replication. It is also out of touch with reality; for all of its gauge bosons and fermions are massless then. Coupling Higgs fields to gauge bosons in the required manner takes care of the former omission – at the expense of the gauge hierarchy problem between the electroweak, GUT and Planck scales. The need for New Physics has been inferred from it: SUSY (as described above) can provide an attractive solution;

---

[1] Since the operators $\mathcal{O}^{(n)}(\phi_{SM})$ carry dimensions higher than 4, they represent non-renormalizable terms with couplings proportional to $1/\Lambda^n$. This does not cause problems, since we apply this ansatz only at energies (well) below Λ.

postulating extra dimensions (as sketched later) represents an intriguing one. Obtaining a *natural* solution to the aforementioned hierarchy problem typically implies Λ not to exceed (very) few TeV.

The 'Yukawa-less' SM introduced above with three families of $SU(2)_L$ quark and lepton doublets Q_L^i and L_L^i, respectively, right-handed quark and lepton singlets U_R^i, D_R^i and E_R^i, respectively, – all with $i = 1, 2, 3$ – has a large *global flavour* symmetry group G_{Fl} [370]:

$$G_{Fl} \equiv SU(3)_q^3 \otimes SU(3)_l^2$$

$$SU(3)_q^3 \equiv SU(3)_Q \otimes SU(3)_U \otimes SU(3)_D, \ \ SU(3)_l^2 \equiv SU(3)_L \otimes SU(3)_E$$
$$(20.2)$$

with the obvious transformation laws

$$Q_L^i \sim (3, 1, 1), \ U_R^i \sim (1, 3, 1), \ D_R^i \sim (1, 1, 3) \qquad (20.3)$$

under $SU(3)_q^3$ and

$$L_L^i \sim (3, 1), \ E_R^i \sim (1, 3) \qquad (20.4)$$

under $SU(3)_l^2$.

'Switching on' the needed Yukawa couplings of the Higgs fields to the fermions breaks G_{Fl} massively – pun intended – and creates non-diagonal flavour transitions. A priori New Physics can induce two types of transition operators, namely those with the same flavour and Lorentz structure as the SM and those with novel structures; i.e. the former can produce merely a re-calibration of SM rates, while the latter can lead to qualitatively new effects, in particular new sources of **CP** violation beyond the KM phase. The former scenario is labeled MFV. It is *not* a theory per se – it is a classification scheme similar to the 'superweak model' of **CP** violation. When one considers a specific model, one has to analyse whether this model represents a dynamical implementation of MFV and to which degree. We do not consider it very likely that MFV holds *exactly*. Yet even then it proves to be useful, in particular in the context of SUSY theories, where MFV arises naturally in minimal versions at least. It has little *predictive* power per se; yet once deviations from the SM are observed, it provides a valuable diagnostic tool. For by comparing the pattern in different transitions it allows to decide whether the emerging New Physics is of the MFV type or requires novel sources of flavour and in particular **CP** violations.

The above given qualitative discussion has to be made more precise to be of real use. The global flavour symmetry group G_{Fl} gets broken by the SM Yukawa couplings. This symmetry can *formally* be recovered by introducing matrices Y_U, Y_D and Y_E of dimensionless *auxiliary* fields that transform non-trivially under $SU(3)_q^3$ and $SU(3)_l^2$, respectively:

$$Y_U \sim (3, \bar{3}, 1), \ Y_D \sim (3, 1, \bar{3}), \ Y_E \sim (3, \bar{3}). \tag{20.5}$$

The Y_U, Y_D and Y_E are *not* physical fields; they are merely book-keeping devices convenient for tracking the G_{Fl} transformation properties of the interactions. Thus they are also referred to as 'spurions'. With them one can express the SM Yukawa couplings contained in $\mathcal{L}_{SM}$ formally as G_{Fl} invariant:

$$\mathcal{L}_Y = \overline{Q}_L Y_D D_R H + \overline{Q}_L Y_U U_R H_C + \overline{L}_L Y_E E_R H + h.c. \tag{20.6}$$

with H and H_C describing the Higgs field and its charge conjugate, respectively. Using the (formal) $SU(3)_q^3$ and $SU(3)_l^2$ symmetry we can transform the matrices of the auxiliary fields:

$$Y_D \to \lambda_d, \ Y_L \to \lambda_l, \ Y_U \to V^\dagger \lambda_u \tag{20.7}$$

with the λ denoting the diagonal matrices of Yukawa couplings and V the CKM matrix.

A theory satisfies *Minimal Flavour Violation*, if all transition operators constructed from the SM fields and the spurions Y are (formally) invariant under G_{Fl}. In MFV the dynamics of flavour transitions is then completely controlled by the pattern of the SM Yukawa couplings; in particular all **CP** violation originates from the KM phase (ignoring here the possibility of $\bar{\theta} \neq 0$ in QCD).

The effective transition operators for beauty, charm and strange decays are obtained by iterating the Lagrangian of Eq. (20.1) using Eq. (20.6); the leading terms arise on the dimension-6 level. Let us focus on $\Delta F = 2$ transitions for mesons made up from down-type quarks, namely the **CP** odd contribution to $K^0 - \overline{K}^0$ and all aspects of $B^0 - \overline{B}^0$ oscillations. Those are driven by four quark operators involving the quark bilinears

$$\overline{Q}_L Y_U Y_U^\dagger Q_L. \tag{20.8}$$

For most practical purposes we can use an approximate expression for the Y. Noting that all SM Yukawa couplings are small except for the top quarks, the leading non-diagonal element arises from the contraction of two Y_U, which transforms as $(8, 1, 1)$ under $SU(3)_q^3$. The only sizable elements of the matrix $Y_U Y_U^\dagger$ are [361]:

$$(\lambda_{FC})_{ij} = \left(Y_U Y_U^\dagger \right)_{ij} \simeq \lambda_t^2 V_{3i}^* V_{3j}, \ i \neq j \tag{20.9}$$

In this approximation the ΔS & $\Delta B = 2$ operators are controlled by a single term, namely $(\overline{Q}_L \lambda_{FC} \gamma_\mu Q_L)$ with an coefficient proportional to $1/\Lambda^2$.

From the measured values of ϵ_K, ΔM_{B_d} and ΔM_{B_s} one infers [361, 362]

$$\Lambda > 5\,\text{TeV} \quad 95\% \text{ C.L.} \tag{20.10}$$

with similar bounds of a few TeV from other processes like $B \rightarrow \gamma X_s$, $l^+ l^- X_s$.

For scenarios that are not of the MFV variety one can actually infer bounds higher by an order of magnitude, namely several $10\,\text{TeV}$.

In all these exercises one simplifying assumption has been made out of necessity, namely there is only a single New Physics contribution to an observable; i.e. no allowance is made for cancellations between different possible contributions. Even with that caveat one cannot simply equate the scale Λ with the mass scale of new states. For the coupling g_X of such a state X to the quarks (or the SM fields in general) might be suppressed in a flavour universal way

$$\frac{1}{\Lambda^2} \simeq \frac{g_X^2}{M_X^2} \tag{20.11}$$

thus opening the door for even $M_X \ll \Lambda$.

20.2 Extra (space) dimensions

A dynamical implementation of MFV could arise in an unusual setting, namely in models with extra (space) dimensions. Those lend themselves readily to a treatment with the formalism of effective field theory.

The speculation that there might be more than the obvious three space and one time dimensions has played a very marginal role till recently. At first sight such a speculation might seem to be a 'non-starter', since the theories that successfully describe nature are based on four dimensions. How come extra dimensions have been overlooked over the centuries? Kaluza and Klein addressed this question in the 1920s [227, 363] in an intriguing way. They considered general relativity in five dimensions and suggested that the fifth dimension has been compactified or 'curled' up – possibly due to quantum corrections – into a closed circle with a radius so tiny that it could not be resolved experimentally, that even its existence had not been inferred. Considering only fields that do not depend on coordinates in the fifth dimension allows to decompose the metric tensor in five dimensions into one in four dimensions plus a four-dimensional vector field that acts like the vector potential of electrodynamics. In other words, 'general relativity' in five dimensions leads to the usual photon field coupled to gravity in four dimensions: general coordinate invariance in five dimensions – a purely geometric concept in the higher-dimensional space – induces general coordinate invariance in four dimensions plus local gauge invariance; the latter is usually viewed as an internal rather than geometric symmetry. Such a unification of Maxwell's and Einstein's theories leaves some other potential footprints: With the fifth dimension having the topology of a circle, fields must satisfy periodic boundary conditions in the fifth dimension: $\psi(x_5) = \psi(x_5 + 2\pi R)$ with R denoting the radius of the fifth dimension. With Newton's constant the only dimensional quantity available, we have $1/R \sim 1/\sqrt{G_N} = M_{\text{Pl}}$. Such fields can be expanded as follows:

$$\psi(x_5) = \sum_{n=0} \psi_n e^{inx_5/R}. \tag{20.12}$$

While the zero modes ψ_0 can be identified with the known fields, there is a whole 'Kaluza–Klein' (KK) tower for such a field with their masses given in units of $M_{\text{Pl}} \simeq 10^{19}$ GeV.

This bold proposal soon ran into some road blocks that could not be overcome [227]. Thus it turned into a 'sleeping beauty' till re-awakened by the emergence of super-string theories as a candidate for a 'theory of everything'. Those theories combine the concept of higher dimensions with that of internal symmetries to have anomalies cancelled that otherwise would destroy the viability of the theory; i.e. they have abandoned the program of reducing all internal symmetries to a general coordinate invariance in higher dimensions. In exchange super-string theory gives an almost unique answer to the question how many dimensions are realized in nature.

• Subsequently speculations about extra dimensions morphed into different versions not necessarily connected with super-string theory [364]. The idea of 'large' extra dimensions – 'large' compared to the Planck length $\sim 10^{-33}$ cm – was invoked to 'solve' the gauge hierarchy problem. It was argued that the huge ratio between the Planck scale $\sim 10^{19}$ GeV and the electroweak scale is an artefact of our basic misconception about why gravity is so weak. The tiny value of $G_N = 1/M_{\text{Pl}}^2$ can be seen as a geometric effect: gravity's strength is diluted, since gravitational fields unlike all other SM fields get dissipated into extra space dimensions. Denoting by M_{QG} the scale induced by quantizing gravity and by n the number of compactified extra dimensions, one obtains for the Planck mass as inferred from measuring the classical gravitational potential:

$$M_{\text{Pl}}^2 = M_{\text{QG}}^{n+2} (2\pi R)^n. \tag{20.13}$$

If there were one or two extra dimensions – $n = 1, 2$ – then the observed validity of Newtonian gravity rules out a scale of $M_{QG} \sim 1\,\mathrm{TeV}$ by a long shot. Yet for $n \geq 3$, $M_{QG} \sim 1\,\mathrm{TeV}$ becomes a tenable scenario and one that for practical reasons cannot be probed through searching for deviations from Newtonian gravity.[2] The ADD [365] and RS models [366] differ in the topology assumed for the extra dimensions, namely 'flat' in the former and 'warped' in the latter case. The New Physics effects are due to the exchanges of the graviton's KK tower. While its members are only gravitationally coupled, they are so densely populated in mass that the overall production rate for KK gravitons and black holes can be sizable as can be contributions to particle production due to virtual KK graviton exchanges [364]. Yet no new sources to flavour violation arise.

One should note that a priori there is nothing special about the $1\,\mathrm{TeV}$ scale in this context. It is picked as a reference scale, since we know that M_{QG} cannot be significantly lower; for otherwise one should have seen the footprints of such extra dimensions in, say, hadronic collider experiments at Fermilab's Tevatron.

Then the interpretive 'floodgates' opened.

• The ansatz of 'universal' extra dimensions (UED), as exemplified by the ACD model [367], opens up those dimensions to the 'masses'; i.e. they allow all SM fields access to the new dimensions. In such an ansatz one no longer undertakes to explain the feebleness of gravity as a geometric effect as sketched above, and the inverse radius R is related to the mass scale of the KK towers of the various SM fields. In return one can obtain many new sources of flavour and **CP** violation. UED models have another qualitatively new feature: 'KK parity' can be defined as a conserved quantity allowing KK states to be produced only in pairs; i.e. at low energies they can contribute only through loop effects. This lowers the bounds on their masses as inferred from phenomenological constraints on KK contributions. A value of $1/R$ as low as $250\,\mathrm{GeV}$ is still allowed leading to a potentially exciting phenomenology at the LHC. The ACD version of UED models will affect rare K and B decays in a sizable and noticable way, albeit in a MFV fashion [368].

• Rather than the very exclusive access to the extra dimensions of the ADD and RS models or the 'democratic' way of the ACD model one can organize the 'commonwealth' of fields also according to 'estates'.[3] This is the principle behind the 'split-fermion' approach [369]: different fermion fields – quark versus lepton fields, left-handed versus right-handed

---

[2] For $n = 3$ one has $M_{QG} \sim 1\,\mathrm{TeV}$ and $R \sim 10^{-6}\,\mathrm{cm}$. While the intrinsic scales of quantum gravity and of the weak forces no longer exhibit an 'unnatural' hierarchy, no good answer is offered to why R exceeds its 'obvious' scale $1/M_{QG} \sim 10^{-17}\,\mathrm{cm}$ by so many orders of magnitude.

[3] This manner of organizing the public domain was typical for feudal Europe.

fields – are described by Gaussian wave functions centred at different locations in the extra dimension(s). The overlap between such Gaussians is very small, unless their locations are very close to each other. Splitting quark and lepton fields provides a very efficient way to suppress proton decay. Separating the chiral components of the quark flavours by different distances can naturally lead to vastly different mass terms.

Yet at the same time the KK towers of the gauge fields can generate FCNC in K, B and D decays that are not particularly suppressed and thus could lead to sizable deviations from SM predictions with the resulting phenomenology *not* of the MFV variety. Constraints derived from present data depend very much on details of such models like whether the extra dimension is 'flat' or 'warped'; the latter is more promising in the sense that it allows for a New Physics scale as low as a few TeV [370].

20.3 The pundits' call

'Generic' versions of the New Physics anticipated for the 1 TeV scale should have impacted B and K decays in a discernible way. Yet past experience should have taught us that nature has a tendency not to follow our notions of what is 'natural'. It is thus premature to argue that New Physics quanta only of the MFV variety can be found at the LHC.

The principle of MFV represents a classification scheme rather than a theory or even class of theories. It is more likely to be realized approximately than exactly. Future heavy flavour studies could actually reveal manifestations of New Physics *beyond* the *direct* reach of the LHC – i.e. with masses well above 10 TeV – in particular if it is not of the MFV variety.

The realm of models with extra dimensions does not look like a unitarian state. It rather resembles a loose federation between several 'tribes' held together by little more than just the symbol of extra dimensions. It is not surprising then that those 'tribes' at times co-operate and at other times compete with each other. On the other hand postulating extra dimensions is a radical departure from conventional model building. As such it might lead us out of the 'dead end' in our efforts to decode the presumably profound message that nature has given us through the family structure.

In the language of our colleague L. Sehgal we view the ansatz of extra dimensions as an 'imagination stretcher' rather than 'credulity generator'. Some see SUSY as a solution in search of a problem. We think this quote is a more apt characterization of the concept of extra dimensions. The term 'extra dimension' is usually meant as extra *space* dimension; it is intriguing, in particular in the context of this book, to speculate about extra *time* dimensions.

21

Baryogenesis in the universe

21.1 The challenge

One of the most intriguing aspects of big bang cosmology is to 'understand' nucleosynthesis, i.e. to reproduce the abundances observed for the nuclei in the universe as *dynamically* generated rather than merely dialled as input values. This challenge has been met successfully, in particular for the light nuclei, and actually so much so that it is used to obtain information on dark matter in the universe, the number of neutrinos, etc. It is natural to ask whether such a success could be repeated for an even more basic quantity, namely the baryon number density of the universe, which is defined as the difference in the abundances of baryons and antibaryons:

$$\Delta n_{\text{Bar}} \equiv n_{\text{Bar}} - n_{\overline{\text{Bar}}}. \tag{21.1}$$

Qualitatively, we can summarize the observations through two statements.

- The universe is not empty.
- The universe is almost empty.

More quantitatively, we find

$$r_{\text{Bar}} \equiv \frac{\Delta n_{\text{Bar}}}{n_\gamma} \sim \text{few} \times 10^{-10}, \tag{21.2}$$

where n_γ denotes the number density of photons in the cosmic background radiation. We know more: at least in our corner of the universe, there are practically no primary antibaryons:

$$n_{\overline{\text{Bar}}} \ll n_{\text{Bar}} \ll n_\gamma. \tag{21.3}$$

444

It is conceivable that in other neighbourhoods antimatter dominates, and that the universe is formed by a patchwork quilt of matter- and antimatter-dominated regions, with the whole being matter–antimatter symmetric. Yet it is widely held to be quite unlikely – primarily because no mechanism has been found by which a matter–antimatter symmetric universe following a big bang evolution can develop sufficiently large regions with non-vanishing baryon number. While there will be statistical fluctuations, they can be nowhere near large enough. Likewise for dynamical effects: baryon–antibaryon annihilation is by far not sufficiently effective to create pockets with the observed baryon number, Eq. (21.2). For the number density of *surviving* baryons can be estimated as [371]

$$n_{\text{Bar}} \sim \frac{n_\gamma}{\sigma_{\text{annih}} m_N M_{\text{Pl}}} \simeq 10^{-19} n_\gamma, \tag{21.4}$$

where σ_{annih} denotes the cross-section of nucleon annihilation, and m_N and M_{Pl} the nucleon and Planck mass, respectively. Hence we conclude, for the universe as a whole,

$$0 \neq \frac{n_{\text{Bar}}}{n_\gamma} \simeq \frac{\Delta n_{\text{Bar}}}{n_\gamma} \sim \mathcal{O}(10^{-10}), \tag{21.5}$$

which makes more explicit the meaning of the statement quoted above that the universe has been observed to be almost empty, but not quite. Understanding this double observation is the challenge we are going to address now.

21.2 The ingredients

The question is: under what conditions can we have a situation where the baryon number of the universe that vanishes at the initial time – which for all practical purposes is the Planck time – develops a non-zero value later on?

$$\Delta n_{\text{Bar}}(t = t_{\text{Pl}} \simeq 0) = 0 \overset{?}{\implies} \Delta n_{\text{Bar}}(t = \text{`today'}) \neq 0. \tag{21.6}$$

We can and should actually go one step further in the task we are setting for ourselves: explaining the observed baryon number as dynamically generated *no matter what its initial value was.*

In a seminal paper that appeared in 1967, Sakharov listed the three ingredients that are essential for the feasibility of such a program [372, 373].

(1) Since the final and initial baryon numbers differ, there have to be baryon number violating transitions:

$$\mathcal{L}(\Delta n_{\text{Bar}} \neq 0) \neq 0. \tag{21.7}$$

(2) **CP** invariance has to be broken. Otherwise for every baryon number changing transition, $N \to f$, there is its **CP** conjugate one, $\overline{N} \to \overline{f}$, and no net baryon number can be generated, i.e. we need

$$\Gamma\left(N \xrightarrow{\mathcal{L}(\Delta n_{\text{Bar}} \neq 0)} f\right) \neq \Gamma\left(\overline{N} \xrightarrow{\mathcal{L}(\Delta n_{\text{Bar}} \neq 0)} \overline{f}\right). \tag{21.8}$$

(3) Unless we are willing to allow for **CPT** violations, the baryon number and **CP** violating transitions have to proceed out of thermal equilibrium. For in thermal equilibrium time becomes irrelevant globally and **CPT** invariance reduces to **CP** symmetry, which has to be avoided, see above:

$$\textbf{CPT}\text{ invariance} \xrightarrow{\text{thermal equilibrium}} \textbf{CP}\text{ invariance.} \tag{21.9}$$

It is important to keep in mind that these three conditions have to be satisfied *simultaneously*. The other side of the coin is, however, the following: once a baryon number has been generated through the concurrence of these three effects, it can be washed out again by these same effects.

21.3   GUT baryogenesis

Sakharov's paper was not noticed (except by [374]) for several years until the concept of Grand Unified Theories (=GUTs) emerged, starting in 1974 [375]; for those naturally provide all three necessary ingredients [376, 377]:

(1) Baryon number changing reactions have to exist in GUTs. For placing quarks and leptons into common representations of the underlying gauge groups – the hallmark of GUTs – means that gauge interactions exist changing baryon and lepton numbers. Those gauge bosons are generically referred to as X bosons and have two couplings to fermions that violate baryon and/or lepton number:

$$X \leftrightarrow qq, \quad q\overline{l}. \tag{21.10}$$

(2) Those models are sufficiently complex to allow for several potential sources of **CP** violation. Since X bosons have (at least) two decay channels open **CP** asymmetries can arise

$$\Gamma(X \to qq) = (1 + \Delta_q)\Gamma_q, \quad \Gamma(X \to q\overline{l}) = (1 - \Delta_l)\Gamma_l,$$
$$\Gamma(\overline{X} \to \overline{qq}) = (1 - \Delta_q)\Gamma_q, \quad \Gamma(\overline{X} \to \overline{q}l) = (1 + \Delta_l)\Gamma_l, \tag{21.11}$$

where

$$\mathbf{CPT} \implies \Delta_q \Gamma_q = \Delta_l \Gamma_l$$
$$\mathbf{CP} \implies \Delta_q = 0 = \Delta_l$$
$$\mathbf{C} \implies \Delta_q = 0 = \Delta_l. \tag{21.12}$$

(3) Grand Unification means that a phase transition takes place around an energy scale M_{GUT}. For temperatures T well above the transition point – $T \gg M_{\mathrm{GUT}}$ – all quanta are relativistic with a number density

$$n(T) \propto T^3. \tag{21.13}$$

For temperatures around the phase transition – $T \sim M_{\mathrm{GUT}}$ – some of the quanta, in particular those gauge bosons generically referred to as X bosons, aquire a mass $M_X \sim \mathcal{O}(M_{\mathrm{GUT}})$ and their equilibrium number density becomes Boltzmann suppressed:

$$n_X(T) \propto (M_X T)^{\frac{3}{2}} \exp\left(-\frac{M_X}{T}\right). \tag{21.14}$$

More X bosons will decay according to Eq. (21.10) than are regenerated from qq and $q\bar{l}$ collisions, ultimately bringing the number of X bosons down to the level described by Eq. (21.14). Yet that will take some time; the expansion in the big bang cosmology leads to a cooling rate that is so rapid that thermal equilibrium cannot be maintained through the phase transition. That means that X bosons decay – and in general interactions – drop out of thermal equilibrium [373].

To the degree that the back production of X bosons in qq and $q\bar{l}$ collisions can be ignored, we find as an order-of-magnitude estimate

$$r_{\mathrm{Bar}} \sim \frac{\frac{4}{3}\Delta_q \Gamma_q - \frac{2}{3}\Delta_l \Gamma_l}{\Gamma_{\mathrm{tot}}} \frac{n_X}{n_0} = \frac{\frac{2}{3}\Delta_q \Gamma_q}{\Gamma_{\mathrm{tot}}} \frac{n_X}{n_0}, \tag{21.15}$$

with n_X denoting the initial number density of X bosons and n_0 the number density of the light decay products.[1] The three essential conditions for baryogenesis are thus naturally realized around the GUT scale in big bang cosmologies, as can be read off from Eq. (21.15):

- $\Gamma_q \neq 0$ representing baryon number violation;
- $\Delta_q \neq 0$ reflecting **CP** violation; and
- the absence of the back reaction due to an absence of thermal equilibrium.

---

[1] Due to thermalization effects we can have $n_0 \gg 2n_X$.

The fact that this problem can be formulated in GUT models and answers obtained that are very roughly in the right ballpark is a highly attractive feature of GUTs, in particular since this was *not* among the original motivations for constructing such theories.

On the other hand, it would be highly misleading to claim that baryogenesis has been understood. There are serious problems in any attempt to have baryogenesis occur at a GUT scale.

- A baryon number generated at such high temperatures is in grave danger of being washed out or diluted in the subsequent evolution of the universe.
- Very little is known about the dynamical actors operating at GUT scales – and that is putting it mildly. Actually even in the future we can only hope to obtain some slices of indirect information on them.

Of course it would be premature to write-off baryogenesis at GUT scales, yet it might turn out that it is best characterized as a proof of principle – that the baryon number of the universe can be understood as dynamically generated – rather than as a semi-quantitative realization.

21.4 Electroweak baryogenesis

Baryogenesis at the electroweak scale [378] is the most actively analysed scenario at present. For it possesses several highly attractive features.

- We know that dynamical landscape fairly well.
  - In particular **CP** violation has been found to exist there.
  - A well-studied phase transition, namely the spontaneous breaking

$$SU(2)_L \times U(1) \implies U(1)_{QED} \tag{21.16}$$

  takes place.
- Future experiments will certainly probe that dynamical regime with ever increasing sensitivity, both by searching for the on-shell production of new quanta – like SUSY and/or Higgs states – and the indirect impact through quantum corrections on rare decays and **CP** violation.

However, the reader might wonder: 'What about the third required ingredient, baryon number violation? At the electroweak scale?' It is often not appreciated that the electroweak forces of the SM by themselves violate baryon number, though in a very subtle way. We find here what is called an anomaly: the baryon number current is conserved on the classical, yet *not* the quantum level:

$$\partial_\mu J_\mu^{\text{Bar}} = \partial_\mu \sum_q (\bar{q}_L \gamma_\mu q_L) = \frac{g^2}{16\pi^2} \text{Tr} G_{\mu\nu} \tilde{G}_{\mu\nu} \neq 0, \tag{21.17}$$

where g denotes the $SU(2)_L$ gauge coupling, $G_{\mu\nu}$ the electroweak field strength tensor

$$G_{\mu\nu} = \tau_a \left(\partial_\mu A_\nu^a - \partial_\nu A_\mu^a + g\epsilon_{abc} A_\mu^b A_\nu^c \right) \tag{21.18}$$

(with the τ_a being the $SU(2)$ generators) and $\tilde{G}_{\mu\nu}$ its dual:

$$\tilde{G}_{\mu\nu} = \frac{1}{2} \epsilon_{\mu\nu\alpha\beta} G_{\alpha\beta}. \tag{21.19}$$

The right-hand side of Eq. (21.17) can be written as the divergence of a current

$$\text{Tr} G_{\mu\nu} \tilde{G}_{\mu\nu} = \partial_\mu K_\mu, \quad K_\mu = 2\epsilon_{\mu\nu\alpha\beta} \text{Tr} \left(A_\nu \partial_\alpha A_\beta - \frac{2}{3} ig A_\nu A_\alpha A_\beta \right). \tag{21.20}$$

We have encountered this situation before, in our discussion of the strong **CP** problem in Chapter 15: a conservation law is vitiated through a triangle anomaly on the quantum level; although the offending term can be written as a total divergence, it still affects the physics of non-abelian gauge theories in a non-trivial way. For there is an infinity of inequivalent ground states differentiated by the value of their K charge, which is the space integral of K_0, the zeroth component of the current K_μ constructed from their gauge field configuration. This integral reflects differences in the gauge topology of the ground states and therefore is called the *topological charge*.

In the present context the triangle anomaly induces *baryon number violation* because of the chiral nature of the weak interactions. Equations (21.17) and (21.20) show that the difference $J_\mu^{\text{Bar}} - K_\mu$ is conserved. The transition from one ground state to another, which represents a tunnelling phenomenon, is thus accompanied by a change in baryon number. Elementary quantum mechanics tells us that this baryon number violation is described as a barrier penetration and exponentially suppressed at low temperatures or energies [379]: $\text{Prob}(\Delta n_{\text{Bar}} \neq 0) \propto \exp(-16\pi^2/g^2) \sim \mathcal{O}(10^{-160})$ – a suppression that reflects the tiny size of the weak coupling.

There is a corresponding anomaly for the lepton number current, implying that lepton number is violated as well, with the selection rule

$$\Delta n_{\text{Bar}} - \Delta n_{\text{lept}} = 0. \tag{21.21}$$

This is usually referred to by saying that $B - L$, the difference between baryon and lepton number, is still conserved.

At sufficiently high energies this huge suppression of baryon number changing transition rates will evaporate, since the transition between different ground states can be achieved classically through a motion *over* the barrier. The question then is at which energy scale this will happen and how quickly baryon number violation will become operative. Some semi-quantitative observations can be offered and answers given [373, 380].

There are special field configurations – called sphalerons – that carry the topological K charge. In the SM they induce effective multistate interactions among left handed fermions that change baryon and lepton number by three units each:

$$\Delta n_{\text{Bar}} = \Delta n_{\text{lept}} = 3. \tag{21.22}$$

At high energies where the weak bosons W and Z are massless, the height of the transition barrier between different ground states vanishes likewise, and the change of baryon number can proceed in an unimpeded way and presumably faster than the universe expands. Thermal equilibrium is then maintained, and any baryon asymmetry existing before this era is actually washed out![2] Rather than generating a baryon number, sphalerons act to drive the universe back to matter–antimatter symmetry at this point in its evolution.

At energies below the phase transition, i.e. in the broken phase of $SU(2)_L \times U(1)$ baryon number is conserved for all practical purposes, as pointed out above. The value of Δn_{Bar} as observed today can thus be generated only in the transition from the unbroken high energy to the broken low energy phase. With $\Delta n_{\text{Bar}} \neq 0$ processes operating there, the issue now turns to the strength of the phase transition: is it relatively smooth like a second-order phase transition or violent like a first-order one? Only the latter scenario can support baryogenesis.

A large amount of interesting theoretical work has been done on the thermodynamics of the SM in an expanding universe. Employing perturbation theory and lattice studies, we have arrived at the following result: for light Higgs masses up to around 70 GeV, the phase transition is first order, for larger masses it is second order [381]. Since no such light Higgs states have been observed at LEP, we infer that the phase transition is

---

[2] To be more precise, only $B + L$ is erased within the SM, whereas $B - L$ remains unchanged.

second order, thus apparently foreclosing baryogenesis occurring at the electroweak scale.

We have concentrated here on the questions of thermal equilibrium and baryon number, while taking **CP** violation for granted, since it is known to operate at the electroweak scale. Yet most authors – with the exception of some notable heretics – agree that KM theory is not at all up to *this* task: it fails by several orders of magnitude. On the other hand, New Physics scenarios of **CP** violation – in particular of the Higgs variety – can reasonably be called upon to perform the task.

21.5 Leptogenesis driving baryogenesis

If the electroweak phase transition is indeed a second-order one, sphaleron mediated reactions cannot drive baryogenesis, as just discussed, and they will wipe out any pre-existing $B + L$ number. Yet if at some high energy scales a lepton number is generated, the very efficiency of these sphaleron processes can *communicate* this asymmetry to the baryon sector through their maintaining conservation of $B - L$.

There are various ways in which such scenarios can be realized. The simplest one is to just add *heavy right-handed Majorana* neutrinos to the SM. This is highly attractive in any case, as described in Section 16.5, since it enables us to implement the see-saw mechanism for explaining why the observed neutrinos are (practically) massless; it is also easily embedded into $SO(10)$ GUTs.

The basic idea is the following [382].

- A primordial lepton asymmetry is generated at high energies well above the electroweak phase transition:
  - Since a Majorana neutrino N is its own **CPT** mirror image, its dynamics necessarily violate lepton number. It will possess at least the following classes of decay channels:

$$N \to l\overline{H}, \ \overline{l}H, \qquad (21.23)$$

  with l and $\overline{l}$ denoting a light-charged or neutral lepton or antilepton and H and $\overline{H}$ a Higgs or anti-Higgs field, respectively.
  - A **CP** asymmetry will in general arise:

$$\Gamma(N \to l\overline{H}) \neq \Gamma(\overline{N} \to \overline{l}H). \qquad (21.24)$$

For the observation of neutrino oscillations has shown us that they carry *non*-universal masses, as discussed in Chapter 16. The resulting MNS matrix can be expected to contain an irreducible complex

phase and thus induce **CP** violation in *qualitative* analogy to the
KM mechanism in the quark sector.

- These neutrino decays are sufficiently slow to occur out of thermal
equilibrium around the energy scale where the Majorana masses
emerge.

• The resulting lepton asymmetry is transferred into a baryon number
through sphaleron mediated processes in the unbroken high energy
phase of $SU(2)_L \times U(1)$:

$$\langle \Delta n_{\text{lept}} \rangle = \frac{1}{2} \langle \Delta n_{\text{lept}} + \Delta n_{\text{Bar}} \rangle + \frac{1}{2} \langle \Delta n_{\text{lept}} - \Delta n_{\text{Bar}} \rangle$$

$$\Longrightarrow \frac{1}{2} \langle \Delta n_{\text{lept}} - \Delta n_{\text{Bar}} \rangle. \tag{21.25}$$

• The baryon number thus generated survives through the subsequent
evolution of the universe.

21.6 Wisdom – conventional and otherwise

We understand how nuclei were formed in the universe, given protons and
neutrons. Obviously it would be even more fascinating if we could understand how these baryons were generated in the first place. We do not
possess a specific and quantitative theory successfully describing baryogenesis. However, leaving it at that statement would – we believe – miss
the main point. We have learnt which kinds of dynamical ingredients are
neccessary for baryogenesis to occur in the universe. We have seen that
these ingredients can be realized naturally.

• GUT scenarios for baryogenesis provide us with a proof of principle
that such a programme can be realized. In practical terms, however,
they suffer from various shortcomings.

- Since the baryon number is generated at the GUT scales, very little
is and not much more might ever be known about that dynamics.

- It appears quite likely that a baryon number produced at such high
scales is subsequently washed out.

• The highly fascinating proposal of baryogenesis at the electroweak
phase transition has attracted a large degree of attention – and
deservedly so.

- A baryon number emerging from this phase transition would be in
no danger of being diluted substantially.

- The dynamics involved here is known to a considerable degree and
will be probed even more with ever increasing sensitivity over the
coming years.

However, it seems that the electroweak phase transition is of second order and thus not sufficiently violent.

- A very intriguing variant turns some of the vices of sphaleron dynamics into virtues, by attempting to understand the baryon number of the universe as a reflection of a *primary* lepton asymmetry. The required new dynamical entities – Majorana neutrinos and their decays – obviously would impact on the universe in other ways as well.
- The observation of neutrino oscillations strongly hints at the existence of heavy Majorana neutrinos, as explained in Chapter 16. This makes it very likely that lepton dynamics play a crucial role in the generation of the baryon number of the universe.

The challenge to understand baryogenesis has already inspired our imagination, prompted the development of some very intriguing scenarios and thus has initiated many fruitful studies – and in the end we might even be successful in meeting it!

Part IV
Summary

22

Summary and perspectives

The discovery that the weak forces break previously unquestioned discrete symmetries – first parity **P** and charge conjugation **C**, then **CP** and **T** – had a revolutionizing impact on our perception of nature and how we analyse the elements of its grand design. We realized that symmetries should not be taken for granted; some even began questioning that sacrosanct fruit of quantum field theory, **CPT** invariance. We learnt from the violation of **CP** symmetry – *not* from that of **P** and **C** separately – that *left* and *right* or *positive* and *negative* charge are dynamically distinct rather than being mere labels based on a convention; furthermore that nature distinguishes between *past* and *future* even on the *microscopic* level. From 1964 to 2001 **CP** violation had been observed only in a single system – the decays of K_L mesons – as a seemingly unobtrusive phenomenon. Yet even so we had come to understand that it represents not only a profound intellectual insight, but has also many and far-reaching concrete consequences.

- The huge predominance of matter over antimatter apparently observed in our universe requires **CP** violation if it is to be understood as *dynamically generated* rather than merely reflecting the initial conditions.
- Once the dynamics are sufficiently complex to support **CP** violation, the latter can manifest itself in numerous different ways; we can even say the floodgates open.
- The three-family SM can implement **CP** violation through the KM mechanism without requiring so-far unobserved degrees of freedom. It is already highly non-trivial that it can accommodate the data on ϵ_K and ϵ' within the uncertainties.
- Despite this phenomenological success, it is incorrect to claim the KM ansatz provides us with an understanding. Since **CP** violation enters

457

through the quark mass matrices, its source is related to three central mysteries of the SM.
    − How are fermion masses generated?[1]
    − Why is there a family structure?
    − Why are there three families rather than one?
 • **CP** studies can be employed as a high sensitivity probe for New Physics in an indirect fashion; i.e. when new dynamical degrees of freedom enter through quantum corrections only.

These insights have been strengthened and become more profound since 2001 from the studies of B decays in particular.

 • The 'KM Paradigm of Large **CP** Violation in B Decays' has been validated experimentally to an impressive degree even on the quantitative level. The data are consistent with CKM dynamics providing at least the dominant source of the observed **CP** violation in B and K decays.
 • While the issue is no longer to search for *alternatives* to KM theory, it makes eminent sense to probe its completeness; i.e. to look for *corrections* to it. This is no 'wild goose chase': for baryo- or lepto-genesis to occur we need other sources of **CP** violation as the dominant engine; thus we can be confident that New Physics exists in the **CP** odd sector.
 • Furthermore New Physics scenarios suggested for various mainly theoretical reasons − like the gauge hierarchy problem − contain a set of new potential sources of **CP** violation, unless one limits oneself to the simplest version of such models.
 • The large **CP** asymmetries observed in B decays have shown us that if the dynamics are complex enough to allow for **CP** violation, there is no intrinsic reason why the relevant weak phase should be small. Thus we can conjecture that new sources of **CP** violation − relevant, say, for baryogenesis or EDMs − can make their presence felt with sizable or even large weak phases.

CP violation is both a fundamental and a mysterious phenomenon, which exists in nature, yet contains a message that has not been decoded yet. In our judgement it would be unrealistic to expect progress in answering these questions through pure thinking. We strongly believe we have to appeal to nature through experimental efforts to provide us with more

---

[1] Or more generally: how are masses produced in general? For in New Physics scenarios **CP** violation enters through the mass matrices for gauge bosons, Higgs bosons and/or other scalar fields.

pieces still missing from the puzzle. **CP** studies are essential for obtaining the full dynamic information concerning the dynamics driving the generation and differentiation of fermion and other masses – and there is a wide realm open for them!

22.1 The cathedral builder's paradigm

The dynamical ingredients for numerous and multi-layered manifestations of **CP** and **T** violations do exist or are likely to exist. Accordingly, those have been, are being and will be searched for in many phenomena, namely in

- the electric dipole moments of neutrons, deuterons and other nuclei in numerous laboratories in particular in Europe and North America; this is an area that had been pioneered at the Institute Laue-Langevin in Grenoble, France;
- the electric dipole moments of electrons and atoms the study of which had been begun at Seattle, Berkeley and Amherst in the US and has spread to many other places now;
- the transverse polarization of muons in $K^- \to \mu^- \bar{\nu} \pi^0$ at KEK in Japan;
- ϵ'/ϵ_K as obtained from K_L decays at FNAL and CERN;
- decay distributions of hyperons at FNAL;
- likewise for τ leptons at LEP, the beauty factories and BES in Beijing;
- **CP** violation in the decays of charm hadrons produced at FNAL and the beauty factories;
- last, and certainly not least, **CP** asymmetries in beauty decays studied at LEP, the B factories at KEK and SLAC, at the FNAL collider and soon at the LHC.

This list makes it clear that frontline research on this topic is pursued at high-energy labs all over the world – and then some; techniques from several different branches of physics – atomic, nuclear and high-energy physics – are harnessed in this endeavour, together with a wide range of set-ups; lastly, experiments are performed at the lowest temperatures that can be realized on Earth – ultracold neutrons – and at the highest – in collisions produced at the LHC. All of that is dedicated to one profound goal. At this point we can explain what we mean by the term *cathedral builder's paradigm*. The building of cathedrals required inter-regional collaborations, front line technology (for the period) from many different fields, and commitment; it had to be based on solid foundations – and it took time. The analogy to the ways and needs of high energy physics is obvious – but it goes deeper than that. At first sight a cathedral looks like a very complicated and confusing structure with something here and something there. Yet further scrutiny reveals that a cathedral is more appropriately characterized as a complex rather than a complicated structure, one that is multi-faceted and multi-layered – yet with a coherent theme! We cannot (at least for first rate cathedrals) remove any of its elements without diluting (or even destroying) its architectural soundness and intellectual message. Neither can we in our efforts to come to grips with **CP** violation!

22.1.1 *Present status and general expectations*

What has happened over the last few years can be characterized as the 'expected' triumph of a 'peculiar' theory. There would not have been plausible deniability, if KM theory's paradigm of large **CP** violation in B decays had failed to materialize in the data. It is a 'peculiar' theory, since it is based on the existence of three families with a seemingly unusual pattern of fermion masses and CKM parameters. The triumph could be seen as 'expected', since even the semi-quantitative success of the CKM description before 2001 was quite non-trivial. The following paragraphs identify the specific items.

- Based on a tiny effect in K_L decays – a few $\times 10^{-3}$ **CP** asymmetry – CKM theory had led to the bold prediction that some channels of the ten times heavier B mesons have to exhibit asymmetries two orders of magnitude larger and thus close to the maximal value mathematically possible. The predicted and measured values for the **CP** asymmetry in $B_d \to J/\psi K_S$ agree within a few percent uncertainty.
- The impressive success of the CKM description can be illustrated in a nutshell as follows: the measured values of two **CP** *in*sensitve ratios of observables, namely $\Delta M_{B_d}/\Delta M_{B_s}$ and $|V_{ub}/V_{cb}|$, allow us to construct the CKM triangle. One finds unequivocally that the resulting triangle can no longer be flat; i.e. it implies the existence of **CP** violation and actually in a way that is fully in agreement with the measured values of the **CP** sensitive observables $\sin 2\phi_1$ and ϵ_K.
- We see no need to have ϵ_K and ϵ' measured with higher accuracy. While we are optimistic that the theoretical description of ϵ_K will become more precise in the near future, we are much less sanguine about ϵ'.
- The arguments for the SM being incomplete used to be based purely on theoretical grounds. Now we have found experimental proofs – mainly of celestial origin – for the presence of New Physics: the SM cannot support the observed neutrino oscillations; it has no

candidates for 'Dark Matter' (deduced from the observed rotation curves of stars and galaxies) or for the even more mysterious phenomenon of 'Dark Energy' (inferred from the red shifts of distant supernovae).

- It is, however, also true that generic versions of New Physics entering around the 1 TeV scale should have caused the electroweak parameters to deviate from SM predictions and likewise for transition rates of B mesons in particular. No such deviation has been observed so far. Yet it would be highly premature to conclude that therefore one will not find any impact from New Physics on heavy flavour transitions or only from scenarios strictly obeying MFV. The lack so far of clear cut deviations from SM predictions is actually in line with our conjecture expressed in the first edition of this book that such deviations will typically be of at most moderate size except for some special cases like a time dependent **CP** asymmetry in $B_s \to J/\psi\phi$.

- There are many models of New Physics in the wings eagerly awaiting their turn that contain novel sources of **CP** violation. Even when they are quite insignificant for K and B decays they could be the leading source for observable EDMs. In some of these models significant **CP** asymmetries could arise in the charm, τ and top transitions. The relevant phenomenology is far from fully developed even theoretically.

22.1.2 A look back

A look back can provide us with a proper perspective. *The comprehensive study of kaon and hyperon physics has been instrumental in guiding us to the SM.*

- The $\tau - \theta$ puzzle led to the realization that parity is not conserved in nature.
- The observation that the production rate exceeded the decay rate by many orders of magnitude – this was the origin of the name 'strange particles' – was explained through postulating a new quantum number – 'strangeness' – conserved by the strong, though not the weak forces. This was the beginning of the second quark family.
- The absence of flavour-changing neutral currents was incorporated through the introduction of the quantum number 'charm', which completed the second quark family.
- **CP** violation finally led to the postulation of yet another, the third family.

All of these elements, which are now essential pillars of the SM, were New Physics at *that* time!

We take this historical precedent as a clue that a detailed, comprehensive and thus neccessarily long-term programme on heavy flavour physics will lead to a new paradigm, a *new* SM!

22.2 Agenda for the future

CKM theory and the SM have been impressively successful in describing heavy flavour physics, and the latter represents a mature field now. 'Mature', however, does not mean 'finished' – more needs to be done; there are four main reasons.

- To complete the 'demystification' of **CP** violation we have to find it also in the decays of charged hadrons – a goal we are on the brink of achieving – and even more importantly in leptodynamics, be it in the neutral sector in neutrino oscillations or in the charged sector through EDMs or decays.
- We have made little progress towards realizing maybe the most ambitious goal, namely to explain the observed baryon number of the Universe as a dynamically generated quantity. So far we have failed to identify the **CP** violating forces that are essential for baryo- or lepto-genesis. The latter provides another powerful motivation for probing **CP** invariance in the lepton sector with the highest sensitivity possible.
- The CKM as well as PMNS parameters in the quark and lepton sectors, respectively, represent fundamental elements in 'Nature's Grand Design'. We have reasons to believe that their values reflect basic forces operating at very high scales rather than being arbitrary inputs, but we have hardly a clue what those forces are. The values of CKM and PMNS parameters might help us in the future to sort out different ideas; yet for that purpose one has to know these values with great accuracy, for the differences in effects that arise at high scales tend to get 'washed out' significantly at much lower scales due to quantum corrections.
- We share the optimism that experiments at the LHC will uncover a new paradigm for nature's fundamental dynamics. This anticipated shift might actually be of a revolutionary scope, if it reveals SUSY or extra dimensions or something we have not even thought of. At the same time it is our considered judgment that to identify the salient features of this new dynamics we have to carefully analyse its impact on flavour dynamics – even if it has none. Therefore a comprehensive program of flavour physics is not a luxury – i.e. a noble undertaking of marginal consequence – but an essential complement to the study of high $p_\perp$ collisions at the LHC. Thus it is central to our ability to

answer the generational challenge in front of us, namely to identify the forces behind electroweak symmetry breaking.

To put it differently: the main justification for continuing a comprehensive programme of flavour physics is *not* the hope for enlightenment about the origin of the family structure of the SM – although one should not rule it out – but on one hand to identify a central element in baryogensis and on the other to discriminate between different scenarios of the anticipated TeV scale New Physics. It might even be possible to catch a glimpse of a dynamical layer that is beyond the direct reach of the LHC.

In such an endeavour we will be greatly helped by the considerable progress achieved over the last 20 years in theoretical engineering and by the many innovative analysis techniques developed by our experimental colleagues.

B decays constitute an almost ideal, certainly optimal and unique lab, due to the dynamic interplay of all three quark families, which is greatly enhanced by the 'long' lifetimes and speedy oscillations. It will be possible in the near future to make predictions with an uncertainty not exceeding very few percent and make measurements with commensurate accuracy. It will be of the greatest value to harness the statistical muscle of LHC experiments for such measurements. We anticipate a body of exciting measurements to be produced by the LHCb collaboration in particular. Even so we hope for the realization of a Super-B factory with the unique capabilities it would bring.

In addition, the following activities must be pursued.

- A vigorous research programme must be continued for nuclear and atomic EDMs, for kaon decays, in particular the super-rare $K \rightarrow \pi \nu \bar{\nu}$ modes and for the muon transverse polarization in $K_{\mu 3}$ decays.
- A comprehensive analysis of charm decays with special emphasis on $D^0 - \overline{D}^0$ oscillations and **CP** violation is a moral imperative – in particular after the recent evidence for such oscillations! Likewise for **CP** studies with τ leptons and top quarks. We cannot expect numerically large effects there; yet the redeeming feature is that SM dynamics can hardly produce a significant 'background' there.
- Close feedback between experiment and theory will be essential.

22.3   Final words

In the book on what **CP** and **T** invariance and their limitations can teach us about Nature's Grand Design, we are still closer to the beginning than the end.

(1) There is still unfinished business in our analysis of the weak decays of strange hadrons.

(2) The quest for EDMs has to be continued with renewed vigour.

(3) A determined effort has to be mounted for studying the **CP** properties of τ leptons, charm hadrons and top quarks.

(4) A comprehensive, detailed and high-statistics analysis of beauty decays is bound to provide us with essential information on fundamental dynamics. It represents a unique opportunity where data can be compared with predictions that are, or will be, numerically accurate by then.

Such a program of inquiry will be intriguing, exciting – but neither easy nor quick! Yet we have to keep the following in mind: insights into Nature's Grand Design which can be obtained from a comprehensive and detailed programme of **CP** studies:

- are of essential and fundamental importance;
- cannot be obtained any other way; and
- cannot become obsolete!

If we want to understand mass generation, we should endeavour to acquire all information concerning it, and that includes the observable content of the non-diagonal mass matrices. No direct observation of new fields – like SUSY partners – can supersede that information. On the contrary, it would be of great help by providing us with essential input for our predictions.

References

[1] R. G. Sachs, *The Physics of Time Reversal,* University of Chicago Press, Chicago, 1987.

[2] T. D. Lee, *Particle Physics and Introduction to Field Theory.* New York: Harwood Academic Publishers GmbH, 1988.

[3] E. P. Wigner, *Nachr. Ges. Wiss. Göttingen, Math.-Physik. Kl.,* **32** 546 (1932).

[4] A. Messiah, *Quantum Mechanics Vol. II.* Amsterdam: North-Holland Publishing Co., 1965.

[5] H. A. Kramers, *Proc. Acad. Sci. Amsterdam* **33** 959 (1930); see also: F. J. Dyson, *J. Math. Phys.* **3** 140 (1962).

[6] A. Abragam and B. Bleaney, *Electron Paramagnetic Resonance of Transition Ions.* Oxford: Clarendon Press, 1970.

[7] J. J. Sakurai, *Invariance Principles and Elementary Particles,* Princeton, NJ: Princeton University Press, 1964.

[8] I. B. Khriplovich and S. K. Lamoreaux, *CP Violation without Strangeness,* Berlin: Springer, 1997.

[9] N. F. Ramsey, 'Earliest Criticisms of Assumed P and T Symmetries', M. Skalsey *et al.* (eds.), *Time Reversal – The Arthur Rich Memorial Symposium,* AIP Conf. Proc. 270. New York: American Institute of Physics, 1993.

[10] B. Heckel, *Fourty Years of Neutron Electric Dipole Moment,* M. Skalsey *et al.* (eds.), Time Reversal – The Arthur Rich Memorial Symposium, AIP Conference Proceedings 270. New York: American Institute of Physics, 1993.

[11] Particle Data Group: W.-M. Yao *et al., J. Phys.* **G 33**, 1 (2006).

[12] L. I. Schiff, *Quantum Mechanics.* New York: McGraw-Hill Inc., 1968.

[13] W. Bernreuther and M. Suzuki, *Rev. Mod. Phys.* **63** 313 (1991).

[14] S. A. Murthy, D. Krause, Z. L. Li and L. R. Hunter, *Phys. Rev. Lett.* **63** 965 (1989).

[15] J. D. Bjorken and S. D. Drell, *Relativistic Quantum Mechanics.* New York: McGraw-Hill Book Company, 1964.

J. D. Bjorken and S. D. Drell, *Relativistic Quantum Fields*. New York: McGraw-Hill Book Company, 1964.

[16] J. D. Jackson, *Classical Electrodynamics*. New York: John Wiley & Sons, Inc., 1963.

[17] G. Lüders, *Ann, Phys.* **2** 1 (57).

[18] K. M. Watson, *Phys. Rev.* **D95** 228 (1954).

[19] A. Pais, 'CP violation: the first 25 years', in: *CP Violation in Particle and Astrophysics*, J. Tran Than Van (ed.). Gif-sur-Yvette, France: Edition Frontières, 1989.

[20] A. Pais, *Phys. Rev.* **D86** 663 (1952).

[21] M. Gell-Mann, *Phys. Rev.* **D92** 833 (1953).

[22] T. Nakano and K. Nishijima, *Prog. Theor. Phys.* **10** 580 (1953).

[23] T. D. Lee and C. N. Yang, *Phys. Rev.* **D104** 254 (1956).

[24] C. S. Wu, E. Ambler, R. W. Hayward, D. D. Hoppes and R. P. Hudson, *Phys. Rev.* **D105** 1423 (1957).

[25] M. Gell-Mann and A. Pais, *Proceedings of the Glasgow Conference Nuclear and Meson Physics*, E. H. Bellamy and R. G. Moorhouse (eds.). Oxford: Pergamon Press, 1955.

[26] M. Gell-Mann and A. Pais, *Phys. Rev.* **D97** 1387 (1955).

[27] K. Lande *et al.*, *Phys. Rev.* **D103** 1901 (1956).

[28] A. Pais and O. Piccioni, *Phys. Rev.* **D100** 1487 (1955).

[29] L. B. Okun, *Weak Interactions of Elementary Particles*, Pergamon, 1965. [The Russian original had appeared in 1963, i.e. clearly *before* the discovery of CP violation.]

[30] R. K. Adair, 'CP-Nonconservation – the Early Experiments', *CP Violation in Particle and Astrophysics*, J. Tran Thanh Van (ed.). Gif-sur-Yvette, France: Editions Frontières, 1989.

[31] F. Muller *et al.*, *Phys. Rev. Lett.* **4** 418 (1960); R. H. Good *et al.*, *Phys. Rev.* **D124** 1221 (1961).

[32] L. B. Leipuner *et al.*, *Phys. Rev.* **132** 2285 (1963).

[33] J. H. Christensen *et al.*, *Phys. Rev. Lett.* **13** 138 (1964).

[34] J. Cronin, Nishina Memorial Lecture (1997).

[35] B. Laurent and M. Roos, *Phys. Lett.* **13**, 269 (1964); *ibid.* **15**, 104.

[36] T. D. Lee and C. N. Yang, *Phys. Rev.* **98** 1501 (1955).

[37] T. D. Lee, R. Oehme and C. N. Yang, *Phys. Rev.* **106** 340 (1957).

[38] R. G. Sachs, *Ann, Phys.* **22** 239 (1963).

[39] V. F. Weisskopf and E. P. Wigner, *Z. Phys.* **63** 54 (1930); *Z. Phys.* **65** 18 (1930).

[40] P. K. Kabir, Appendix A of: *The CP Puzzle*, Academic Press, 1968.

[41] T. D. Lee and L. Wolfenstein, *Phys. Rev.* **D138** B1490 (1965).

[42] T. D. Lee and C. S. Wu, *Ann. Rev. Nucl. Sci.* **16** 471 (1966).

[43] Y. Nir and H. R. Quinn, in: *B decays*, S. Stone (ed.). Singapore: World Scientific, 1994.

[44] I. I. Bigi, V. Khoze, A. Sanda and N. Uraltsev, in: [126].

[45] J. S. Bell and J. Steinberger, *Proc. 1965 Oxford Int. Conf. on Elementary Particles*, Rutherford High Energy Laboratory, Didcot.

[46] NA31 collaboration: G. D. Barr, *et al.*, *Phys. Lett.* **B317** 233 (1993);

[47] E731 collaboration: L. K. Gibbons *et al.*, *Phys. Rev. Lett.* **70** 1203 (1993).

[48] NA48 Collaboration: V. Fanti *et al.*, *Phys. Lett.* **B465** 335 (1999).

[49] KTeV Collaboration: A. Alavi-Harati *et al.*, *Phys. Rev. Lett.* **83** 22 (1999).

[50] F. Ambrosino *et al.*, *Phys. Lett.* **B642** 315 (2006).

[51] CPLEAR Collaboration: R. LeGac, *Proceedings of a Workshop on K Physics*, L. Fayard (ed.). Gif-sur-Yvette, France: Editions Frontières.

[52] P. K. Kabir, *Phys. Rev.* **D2** 540 (1970).

[53] A. Angelopoulos *et al.*, *Phys. Rep.* **374**, 165, (2003).

[54] C. Geweniger *et al.*, *Phys. Lett.* **B48** 487 (1974).

[55] A. Bohm *et al.*, *Nucl. Phys.* **B9** 605 (1969).

[56] N. W. Tanner and R. H. Dalitz, *Ann. Phys. (N.Y.)* **171**, 463 (1986).

[57] A. R. Zhitnitskii, *Sov. J. Nucl. Phys.* **31**, 529 (1980).

[58] L. B. Okun and I. B. Khriplovich, *Sov. J. Nucl. Phys.* **6**, 598 (1968); E. S. Ginsberg and J. Smith, *Phys. Rev.* **D8** 3887 (1973);

[59] KEK experiment E246: M. Abe *et al.* *Phys. Rev. Lett.* **93** 131601 (2004).

[60] S. W. MacDowell, *Nuov. Cim.*, **9** 258 (1958).

[61] N. Cabibbo and A. Maksymowicz, *Phys. Lett.* **9** 352 (1964); *Phys. Lett.* **11** 360 (1964); *Phys. Lett.* **14** 72 (1966).

[62] G. Belanger and C. Q. Geng, *Phys. Rev.* **D44** 2789 (1991).

[63] S. Weinberg, *Phys. Rev. Lett.* **4** 87 (1960).

[64] T. D. Lee and C. N. Yang, *Phys. Rev.* **D108** 1645 (1957); R. Gatto, *Nucl. Phys.* **5** 183 (1958).

[65] L. Roper *et al.*, *Phys. Rev.* **D138** 190 (1965).

[66] J. F. Donoghue, X.-G. He and S. Pakvasa, *Phys. Rev.* **D34** 833 (1986).

[67] HyperCP Collab. (E871), T. Holmstrom *et al.*, *Phys. Rev. Lett.* **93** 262001 (2004).

[68] HyperCP Collab. (E871), L. C. Lu *et al.*, *Phys. Rev. Lett.* **96** 242001 (2006).

[69] L. Wolfenstein, *Phys. Rev. Lett.* **13** 562 (1964).

[70] M. Gell-Mann and Y. Ne'eman, *The Eightfold Way*. New York: W. A. Benjamin Inc., 1964.

[71] S. Glashow, J. Illiopolous and L. Maiani, *Phys. Rev.* **D2** 1285 (1970).

[72] K. Niu, E. Mikumo and Y. Maeda, *Prog. Theor. Phys.* **46** 1644 (1971).

[73] M. Kobayashi and T. Maskawa, *Prog. Theor. Phys.* **49** 652 (1973).

[74] N. Cabibbo, *Proc. XIIth Int. Conf. in High Energy Physics*, C. E. Mauk (ed.). Berkeley, CA: University of California Press, 1967.

[75] For example: M. E. Peskin and D. V. Schroeder, *An Introduction to Quantum Field Theory*, Reading, MA: Perseus Books, 1995.

[76] For example, *Matrix Analysis*, R. A. Horn and C. R. Johnson, Cambridge: Cambridge University Press, 1990.

[77] C. Jarlskog, *Proceedings of the IVth LEAR Workshop*, Villars-sur-Ollon, Switzerland, 1987.

[78] L. Wolfenstein, *Phys. Rev. Lett.* **51** 1945 (1983).

[79] L.-L. Chau, in: *CP Violation*, C. Jarlskog (ed.). Singapore: World Scientific, 1988.

[80]   J. L. Rosner, A. I. Sanda, and M. Schmidt *Proc. Fermilab Workshop on High Sensitivity Beauty Physics at Fermilab*, A. J. Slaughter, N. Lockyer and M. Schmidt (eds.), 1987. See also C. Hamzaoui, J. L. Rosner and A. I. Sanda, same Proceedings.

[81]   G. Buchalla, A. J. Buras and M. E. Lautenbacher, *Nucl. Phys.* **B370** 69 (1992); A. J. Buras, M. Jamin and M. E. Lautenbacher, *Nucl. Phys.* **B408** 209 (1993).

[82]   T.-P. Cheng and L.-F. Lee, *Gauge Theory of Elementary Particles.* Oxford: Oxford University Press, 1982.

[83]   A. J. Buras, in: *CP Violation*, C. Jarlskog (ed.). World Scientific, Singapore, 1988.

[84]   M. K. Gaillard and B. W. Lee, *Phys. Rev. Lett.* **33** 108 (1974).

[85]   G. Altarelli and L. Maiani, *Phys. Lett.* **B52** 351 (1974).

[86]   A. I. Vainshtein, M. B. Voloshin, V. I. Zakharov and M. A. Shifman, *Sov. Phys. JETP* **45** 670 (1977).

[87]   F. J. Gilman and M. B. Wise, *Phys. Rev.* **D20** 2392 (1979).

[88]   J. M. Flynn and L. Randall, *Phys. Lett.* **B224** 224 (1989).

[89]   H. Georgi, *Weak Interactions and Modern Particle Theory.* Menlo Park, CA: The Benjamin/Cummings Publishing Co., USA.

[90]   R. S. Chivukula, J. M. Flynn and H. Georgi, *Phys. Lett.* **B171** 453 (1986).

[91]   T. Morozumi, C. S. Lim and A. I. Sanda, *Phys. Rev. Lett.* **65** 404 (1990).

[92]   P. Franzini, in: *Les Rencontres de Physique de la Vallée d'Aoste, Results and Perspectives in Particle Physics*, M. Greco (ed.). Gif-sur-Yvette, France: Editions Frontières, 1991.

[93]   J. F. Donoghu, E. Golowich, B. R. Holstein, and J. Tranpetic, *Phys. Lett.* **179B** (1986)A. J. Buras and J.-M. Gerard, *Phys. Lett.* **B192** 156 (1987), G. Ecker, G. Muller, H. Neufeld and A. Pich, *Phys. Lett.* **B477** 88 (2000); C. E. Wolfe and K. Maltman, *Phys. Rev.* **D63** 014008 (2001).

[94]   G. Buchalla, A. J. Buras and M. E. Lautenbacher, *Rev. Mod. Phys.* **68** 1125 (1996) .

[95]   A. J. Buras and M. Jamin, JHEP **0401**0482004.

[96]   T. Inami and C. S. Lim, *Prog. Theor. Phys.* **65** 297 (1981).

[97]   A. Soni, *Nucl. Phys.* **B47** (Proc. Suppl.) 43 (1996).

[98]   J. F. Donoghue, E. Golowich, B. R. Holstein and J. Trampetic, *Phys. Lett.* **179B** 361 (1986); A. J. Buras and J.-M. Gérard, *Phys. Lett.* **192** 156 (1987); M. Lusignoli, *Nucl. Phys.* **B325** 33 (1989).

[99]   A. I. Sanda, in Proceedings of *b20: Twenty Beautiful Years of Bottom Physics*, D. Kaplan (ed.), Illinois Institute of Technology, 1997.

[100] J. H. Christenson *et al.*, *Phys. Rev. Lett.* **25** 1524 (1970).

[101] J. J. Aubert *et al.*, *Phys. Rev. Lett.* **33** 1404 (1974).

[102] J. E. Augustin, *Phys. Rev. Lett.* **33** 1406 (1974).

[103] J. W. Herb *et al.*, *Phys. Rev. Lett.* **39** 252 (1977).

[104] C. W. Darden *et al.*, *Phys. Lett.* **76B** 246 (1978).

[105] D. Besson *et al.*, *Phys. Rev. Lett.* **54** 382 (1985).

[106] D. M. J. Lovelock *et al.*, *Phys. Rev. Lett.* **54** 377 (1985).

[107] E. Fernandez *et al.*, *Phys. Rev. Lett.* **51** 1022 (1983).

[108] N. Lockyer *et al.*, *Phys. Rev. Lett.* **51** 1316 (1983).

[109] G. Bellini, I. Bigi and P. Dornan, *Phys. Rep.* **289** 1 (1997).

[110] H. Albrecht *et al.*, *Phys. Lett.* **B192** 245 (1987).

[111] L. B. Okun, V. I. Zakharov and M. P. Pontecorvo, *Nuov. Cim. Lett.* **13** 218 (1975).

[112] HFAG, E. Barberio *et al.*, arXiv:0704.3575 [hep-ex] and online update at http://www.slac.stanford.edu/xorg/hfag.

[113] A. Abulencia *et al.*, *Phys. Rev. Lett.* **97** 242003 (2006).

[114] A. J. Buras, M. Jamin and P. Weisz, *Nucl. Phys.* **B347** 491 (1990).

[115] CLEO Collaboration: M. Artuso *et al.*, *Phys. Rev. Lett.* **62** 2233 (1989).

[116] A. Pais and S. B. Treiman, *Phys. Rev.* **D12** 2744 (1975).

[117] M. Bander, D. Silverman and A. Soni, *Phys. Rev. Lett.* **43** 242 (1979).

[118] A. B. Carter and A. I. Sanda, *Phys. Rev.* **D23** 1567 (1981).

[119] I. I. Bigi and A. I. Sanda, *Nucl. Phys.* **B193** 85 (1981).

[120] K. Abe *et al.*, *Phys. Rev. Lett.* **87** 091802 (2001).

[121] B. Aubert *et al.*, *Phys. Rev. Lett.* **87** 091801 (2001).

[122] K. F. Chen *et al.*, *Phys. Rev. Lett.* **98** 031802 (2007).

[123] B. Aubert *et al.*, hep-ex/0703021.

[124] BaBar Collaboration: B. Aubert *et al.*, *Phys. Rev. Lett.* **99** 021603 (2007).

[125] Belle Collaboration: H. Ishino *et al.*, *Phys. Rev. Lett.* **98** 211801 (2007).

[126] C. Jarlskog (ed.) *CP Violation*, Singapore: World Scientific, 1994.

[127] BaBar Collaboration: B. Aubert *et al.*, *Phys. Rev. Lett.* **99** 171803 (2007).

[128] M. Gronau, *Phys. Lett.* **B300** 163 (1993).

[129] A. I. Sanda and Z.-Z. Xing, *Proceedings of 7th Int. Symp. on Heavy Flavor Physics*, Santa Barbara (1997).

[130] B. Kayser, M. Kuroda, R. D. Peccei and A. I. Sanda, *Phys. Lett.* **237** 508 (1990).

[131] H. J. Lipkin, *Proc. SLAC Workshop on Physics and Detector Issues for the High-Luminosity Asymmetric B Factory*, D. Hitlin (ed.), SLAC 373; I. Dunietz, H. R. Quinn, A. Snyder, W. Toki and H. J. Lipkin, *Phys. Rev.* **D43** 2193 (1991).

[132] N. Sinha and R. Sinha, *Phys. Rev. Lett.* **80** 3706 (1998); G. Krammer and W. F. Palmer, *Phys. Rev.* **D45** 193 (192); *ibid.* **D46** 3197 (1992).

[133] A. S. Dighe, I. Dunietz, H. J. Lipkin and J. Rosner, *Phys. Lett.* **B369** 144 (1996); *Phys. Lett.* **165B** 429 (1985); *Phys. Rev. Lett.* **59** 958 (1987).

[134] I. Dunietz, H. R. Quinn, A. Snyder, W. Toki and H. J. Lipkin, *Phys. Rev.* **D43** 2193 (1991).

[135] BaBar Physics Book Editors P. F. Harrison and H. R. Quinn.

[136] Belle Collaboration: R. Itoh *et al.*, *Phys. Rev. Lett.* **95** 091601 (2005).

[137] BaBar Collaboration: B. Aubert *et al.*, *Phys. Rev.* **D71** 032005 (2005).

[138] L.-L. Chau and H.-Y. Cheng, *Phys. Rev. Lett.* **53** 1037 (1984); *Phys. Lett.* **165B** 429 (1985); *Phys. Rev. Lett.* **59** 958 (1987).

[139] D. Zeppenfeld, *Z. Phys.* **C8** 77 (1981).

[140] M. Gronau, O. F. Hernandes, D. London and J. L. Rosner, *Phys. Rev.* **D52** 6356 (1995).

[141] M. Gronau and D. London, *Phys. Rev. Lett.* **27** 3381 (1990).

[142] H. Lipkin, Y. Nir, H. R. Quinn and A. E. Snyder, *Phys. Rev.* **D44** 1454 (1991).

[143] A. E. Snyder and H. R. Quinn, *Phys. Rev.* **D48** 2139 (1993).

[144] A. Deandrea and A. D. Polosa, *Phys. Rev. Lett.* **86** 216 (2001).

[145] S. Gardner and U. G. Meissner, *Phys. Rev.* **D 65**, 094004 (2002).

[146] I. I. Bigi and A. I. Sanda, *Phys. Lett.* **B221**, 213 (1988).

[147] M. Gronau and D. Wyler, *Phys. Lett.* **B265** 172 (1991).

[148] D. Atwood, I. Dunietz and A. Soni, *Phys. Rev. Lett.* **78** 3257 (1997).

[149] A. Poluektov *et al.*, Belle Collaboration *Phys. Rev.* **D70** 072003 (2004).

[150] A. Giri, Y. Grossman, A. Soffer and J. Zupan *Phys. Rev.* **D68** 054018 (2003).

[151] M. Voloshin *et al.*, *Sov. J. Nucl. Phys.* **46** 112 (1987); see also Ref. [109].

[152] J. S. Hagelin, *Nucl. Phys.* **B193** 123 (1981).

[153] A. Lenz, arXiv:hep-ph/0710.0940; A. Lenz and U. Nierste, arXiv:hep-ph/0612167.

[154] CKMfitter Collab., http://ckmfitter.in2p3.fr

[155] UT$_{\text{fit}}$ Collab., http://utfit.roma1.infn.it/

[156] M. K. Gaillard and B. W. Lee, *Phys. Rev.* **D10** 897 (1974).

[157] F. J. Gilman and M. B. Wise, *Phys. Rev.* **D21** 3150 (1980).

[158] M. Vysotskii *Sov. J. Nucl. Phys.* **31** 797 (1980).

[159] A. J. Buras and R. Fleischer, *Heavy Flavors II*, A. J. Buras and M. Lindner (eds.). Singapore: World Scientific, 1997.

[160] W. J. Marciano and A. I. Sanda, *Phys. Lett.* **67B** 303 (1977).

[161] A. J. Buras, F. Schwab, and S. Uhlig hep-ph/0405132.

[162] G. D. Barr *et al.*, *Phys. Lett.* **B284** 440 (1992).

[163] V. Papadimitriou *et al.*, *Phys. Rev.* **D44** 573 (1991).

[164] L. M. Sehgal, *Phys. Rev.* **D38** 808 (1988).

[165] G. Isidori, *et al.*, *Nucl. Phys.* **B718** 319 (2005).

[166] A. Buras *et al.* *JHEP* **0611** 002 (2006).

[167] E787 S. Adler *et al.*, *Phys. Rev. Lett.* **79** 2204 (1997); E949 A. V. Artamonov *et al.*, *Phys. Rev.* **D72** 091102 (2005).

[168] Exp. E391a, J. K. Ahn, *et al.*, *Phys. Rev.* **D74** 051105 (2006).

[169] Y. Grossman and Y. Nir, *Phys. Lett.* **B398** 163 (1997).

[170] M. McGuigan and A. I. Sanda, *Phys. Rev.* **D36** 1413 (1987).

[171] E. J. Ramberg *et al.* (E731 Collab.), *Phys. Rev. Lett.* **70** 2525 (1993).

[172] L. M. Sehgal and M. Wanninger, *Phys. Rev.* **D46** 1035 (1992); *Phys. Rev.* **D46** 5209 (1992) (E).

[173] See also the earlier papers: A. D. Dolgov and L. A. Ponomarev, *Sov. J. Nucl. Phys.* **4** 262 (1967); D. P. Majumdar and J. Smith; *Phys. Rev.* **187** 2039 (1969).

[174] J. K. Elwood, M. B. Wise and M. J. Savage, *Phys. Rev.* **D52** 5095 (1995); **D53** 2855(E) (1996); J. K. Elwood, M. B. Wise, M. J. Savage and J. M. Walden, *Phys. Rev.* **D53** 4078 (1996).

[175] M. Arenton, Invited Talk given at *HQ98, Workshop on Heavy Quarks at Fixed Target*, Fermilab, Oct. 10–12, 1998, to appear in the proceedings; see also: KTeV Collab., J. Adams *et al.*, *Phys. Rev. Lett.* **80** 4123 (1998).

[176] P. Heiliger and L. M. Sehgal, *Phys. Rev.* **D48** 4146 (1993).

[177] CLEOII, *28th Int. Conf. on High Energy Physics*, Warsaw, Poland, 1996.

[178] I. I. Bigi, M. Shifman, and N. Uraltsev, *Ann. Rev. Nucl. Part. Sci.* **47** 591 (1997), with reference to earlier work.

[179] M. Misiak *et al.*, *Phys. Rev. Lett.* **98**, 022002 (2007).

[180] Belle Collaboration: P. Koppenburg *et al.*, *Phys. Rev. Lett.* **93**, 061803 (2004).

[181] BaBar Collaboration: B. Aubert *et al.*, *Phys. Rev. Lett.* **97**, 171803 (2006).

[182] ALEPH Collaboration, *Proceedings of 28th Int. Conf. on High Energy Physics*, Warsaw, Poland, 1996.

[183] C. S. Lim, T. Morozumi and A. I. Sanda, *Phys. Lett.* **B218** 343 (1989).

[184] N. G. Deshpande, J. Trampetic and K. Panose, *Phys. Rev.* **D39** 1461 (1989); P. J. O'Donnel and H. K. K. Tung, *Phys. Rev.* **D43** R2067 (1991).

[185] A. J. Buras and M. Munz, *Phys. Rev.* **D52** 186 (1995).

[186] A. Ali, T. Mannel and T. Morozumi, *Phys. Lett.* **B273** 505 (1991).

[187] R. F. Streater and A. S. Wightmann, *PCT, Spin and Statistics, and All That*. New York: Benjamin, 1964.

[188] S. W. Hawking, *Phys. Rev.* **D14** 2460 (1975); *Commun. Math. Phys.* **87** 395 (1982).

[189] J. Ellis, J. S. Hagelin, D. V. Nanopoulos and M. Srednicki, *Nucl. Phys.* **B241** 381 (1984).

[190] J. D. Bjorken, *Ann. Phys.* **24** 174 (1963).

[191] F. A. Bais and J.-M. Frere, *Phys. Lett.* **98** 431 (1981).

[192] T. Banks and A. Zaks, *Nucl. Phys.* **B184** 303 (1981).

[193] I. I. Bigi, *Z. Phys.* **C12** 235 (1982).

[194] H. B. Nielsen, in: *Fundamentals of Quark Models*, J. M. Barbour and A. T. Davies (eds.), University of Glasgow, Glasgow, 1977, p. 528 ff.

[195] V. A. Kostelecky and R. Potting, *Nucl. Phys.* **B359** 545 (1991); *Phys. Rev.* **D51** 3923 (1995); *Phys. Lett.* **B381** 89 (1996); D. Colladay and V. A. Kostelecky, *Phys. Lett.* **B344** 259 (1995).

[196] M. Hayakawa and A. I. Sanda, *Phys. Rev.* **D48** 1150 (1993).

[197] L. Lavoura, *Ann. Phys.* **207** 428 (1991).

[198] I. Dunietz, J. Hauser and J. L. Rosner, *Phys. Rev.* **D35** 2166 (1987).

[199] P. Heiliger, L. M. Sehgal *Phys. Lett.* **B265** 410 (1991); M. E. Luke *Phys. Lett.* **B256** 265 (1991); C. O. Dib and B. Guberina *Phys. Lett.* **B255** 113 (1991).

[200] E. Shabalin, *Phys. Lett.* **B396** 335 (1996).

[201] A. Einstein, B. Podolsky and N. Rosen, *Phys. Rev.* **D47** 777 (1935).

[202] M. Kobayashi and A. I. Sanda, *Phys. Rev. Lett.* **69** 3139 (1992).

[203] K. Fujikawa *Prog. Theor. Phys.* **61** 1186 (1979). I. I. Bigi, Y. Diakonov, V. Khoze, J. Kühn and P. M. Zerwas, *Phys. Lett.* **B181** 157 (1986).

[204] Belle Collaboration: M. Staric *et al.*, *Phys. Rev. Lett.* **98** 211803 (2007).

[205] BaBar Collaboration: B. Aubert *et al.*, hep-ex/0703020.

[206] Belle Collaboration: L. M. Zhang *et al.*, *Phys. Rev. Lett.* **99** 131803 (2007).

[207] E. Golowich, S. Pakvasa and A. Petrov, *Phys. Rev. Lett.* **98** 181801 (2007).

[208] A. Falk *et al.*, *Phys. Rev.* **D65** 054034 (2002).

[209] A. Falk *et al.*, *Phys. Rev.* **D69** 114021(2004).

472 *References*

[210] I. I. Bigi and N. G. Uraltsev, *Nucl. Phys.* **B592** 92 (2001).

[211] F. Buccella, M. Lusignoli and A. Pugliese, *Phys. Lett.* **B379** 249 (1996).

[212] I. I. Bigi and H. Yamamoto, *Phys. Lett.* **B 349** 363 (1995).

[213] H. Lipkin and Z.-Z. Xing, *Phys. Lett.* **B 450** 405 (1999).

[214] G. D'Ambrosio and D.-N. Gao, *Phys. Lett.* **B 513** 123 (2001).

[215] FOCUS Collaboration: J. M. Link *et al.*, *Phys. Lett.* **B622** 239 (2005).

[216] I. I. Bigi and A. I. Sanda, *Phys. Lett.* **B148** 205 (1984).

[217] I. I. Bigi and A. I. Sanda, *Phys. Lett.* **B171** 320 (1986)

[218] S. Bianco *et al.*, *La Rivista del Nuov. Cim.* **26**, # 7–8 (2003).

[219] Y. Grossman, A. Kagan and Y. Nir, *Phys.Rev.* **D75** 036008 (2007).

[220] I. I. Bigi, in: *Proc. XIII Int. Conf. on High Energy Physics*, S. C. Loken (ed.), World Scientific, Singapore, 1986, p. 857.

[221] G. Blaylock, A. Seiden and Y. Nir, *Phys. Lett.* **B355** 555 (1995).

[222] CLEO Collaboration: D. Asner *et al.*, *Int. J. Mod. Phys.* **A21** 5456 (2006); W. M. Sun, hep-ex/0603031.

[223] G. t'Hooft, *Phys. Rev. Lett.* **37** 8 (1976); *Phys. Rev.* **D14** 3432 (1976).

[224] C. G. Callan, R. Dashen and D. Gross, *Phys. Lett.* **63B** 334 (1976).

[225] R. Jackiw and C. Rebbi, *Phys. Rev. Lett.* **37** 172 (1976).

[226] R. D. Peccei, arXiv:hep-ph/0607268.

[227] M. Kaku, *Quantum Field Theory – A Modern Introduction*. Oxford: Oxford University Press, (1993).

[228] S. L. Adler, *Phys. Rev.* **177** 2426 (1969); J. S. Bell and R. Jackiw, *Nuov. Cim.* **60** 47 (1969); W. A. Bardeen, *Phys. Rev.* **184** 1848 (1969).

[229] V. Baluni, *Phys. Rev.* **D19** 2227 (1979).

[230] R. J. Crewther *et al.*, *Phys. Lett.* **88B** 123 (1979); Errata, **91B** 487 (1980).

[231] R. D. Peccei, *CP Violation*, C. Jarlskog (ed.). Singapore: World Scientific, 1988.

[232] J. Gasser and H. Leutwyler, *Phys. Report* **87** 77 (1982).

[233] S. Khlebnikov and M. Shaposhnikov, *Phys. Lett.* **B203**, 121 (1988).

[234] Y. B. Zeldovich, I. B. Kobzarev and L. B. Okun, *Sov. Phys. JETP* **40** 1 (1975).

[235] S. Barr and A. Zee, *Phys. Rev. Lett.* **55** 2253 (1985).

[236] A. Nelson, *Phys. Lett.* **136B** 387 (1983).

[237] S. Barr, *Phys. Rev. Lett.* **53** 329 (1984); *Phys. Rev.* **D30** 1805 (1984).

[238] Th. Kaluza, *Sitz. Preuss. Akad. Wiss.* **Kl**, 966 (1921); O. Klein, *Z. Phys.* **37**, 895 (1926).

[239] R. Peccei and H. Quinn, *Phys. Rev. Lett.* **38** 1440 (1977); *Phys. Rev.* **D16** 1791 (1977).

[240] S. Weinberg, *Phys. Rev. Lett.* **40** 223 (1978).

[241] F. Wilczek, *Phys. Rev. Lett.* **40** 279 (1978).

[242] N. A. Bardeen, R. D. Peccei and T. Yanagita *Nucl. Phys.* **B279** 401 (1987).

[243] N. A. Bardeen and S. H. H. Tye, *Phys. Lett.* **74B** 580 (1979).

[244] S. Yamada, *Proc. 1983 Int. Symp. on Lepton and Photon Interations at High Energies*, D. G. Cassel and D. L. Kreinick (eds.). Cornell, 1983.

[245] R. Eichler *et al.*, *Phys. Lett.* **175B** 101 (1986).

[246] J. Kim, *Phys. Rev. Lett.* **43** 103 (1979); M. A. Shifman, A. I. Vainshtein and V. I. Zakharov, *Nucl. Phys.* **166** 493 (1980).

[247] M. Dine, W. Fischler and M. Srednicki, *Phys. Lett.* **104** 199 (1981); A. P. Zhitnitskii, *Sov. J. Nucl. Phys.* **31** 260 (1980).

[248] G. G. Raffelt, arXiv:hep-ph/0611118.

[249] P. Sikivie, *Dark Matter Axions and Caustic Rings*, preprint hep-ph/9709477, 1997.

[250] D. N. Spergel *et al.*, *ApJS* **170**, 377 (2007).

[251] P. Sikivie, arXiv:hep-ph/0610440.

[252] F. Reines and C. L. Cowan, Jr., *Nature* **178**, 446 (1956); C. L. Cowan *et al.*, *Science* **124**, 103 (1956).

[253] B. Pontecorvo, *J. Exp. Theor. Phys.* **33** 549 (1957).

[254] Z. Maki, M. Nakagawa and S. Sakata, *Prog. Theor. Phys.* **30** 727 (1963).

[255] N. Cabibbo, *Phys. Rev. Lett.* **10** 531 (1963).

[256] J. N. Bahcall, M. H. Pinsonneault, and S. Basu, *Astrophys. J.* **555** 990 (2001); A. S. Brun, S. Turck-Chi'eze, and J. P. Zahn, *Astrophys. J.* **525** 1032 (2001).

[257] R. Davis, Jr., *Prog. Part. Nucl. Phys.* **32**, 13 (1994).

[258] P. Anselmann *et al.*, *Phys. Lett.* **B342** 440 (1995).

[259] J. N. Abdurashitov *et al.*, *Phys. Lett.* **B328** 234 (1994).

[260] GNO Collaboration: M. Altman *et al.*, *Phys. Lett.* **B616** 174 (2005).

[261] Kamiokande Collaboration: Y. Fukuda *et al.*, *Phys. Rev. Lett.* **77** 1683 (1996).

[262] Superkamiokande Collaboration: J. Hosaka *et al.*, *Phys. Rev.* **D73** 112001 (2006).

[263] SNO Collaboration: Q. R. Ahmad *et al.*, *Phys. Rev. Lett.* **89** 011301 (2002).

[264] SNO Collaboration: B. Aharmim *et al.*, *Phys. Rev.* **DC75** 045502 (2007).

[265] The Super-Kamiokande Collaboration: Y. Fukuda *et al.*, *Phys. Rev. Lett.* **81** 1562 (1998).

[266] KamLand Collaboration: K. Eguchi *et al.* *Phys. Rev. Lett.* **90** 021802 (2003); T. Arali *et al.*, *Phys. Rev. Lett.* **94** 081801 (2005).

[267] KamLand Collaboration: S. Abe *et al.*, arXiv:0801.4589v2 [hep-ex].

[268] K2K Collaboration: M. H. Ahn *et al.* *Phys. Rev.* **D74** 072003 (2006).

[269] M. Fukugita and T. Yanagida, *Physics and Astrophysics of Neutrinos*, Springer-Verlag, Tokyo, 1994.

[270] CHOOZ Collaboration: M. Apollonio *et al.*, *Eur. Phys. J.* **C27** 331 (2003).

[271] L. Wolfenstein, *Phys. Rev.* **D17** 2369 (1978).

[272] S. P. Mikheyev and A. Yu. Smirnov, *Sov. J. Nucl. Phys.* **42** 913 (1985); *Nuov. Cim.* **C9** 17 (1986).

[273] H. Backe *et al.*, in *Proc. Neutrino 92*, Granada, 1992, A. Morales (ed.), NPB (Proc. Suppl.) **31** 46 (1993).

[274] C. Weinheimer, *Proc. Neutrino 98, XVIII Int. Conf. on Neutrino Physics and Astrophysics*, Takayama, Japan, 4–9 June 1998, Y. Suzuki and Y. Totsuka (eds.).

[275] L. H. Ryder, *Quantum Field Theory*. Cambridge: Cambridge University Press, 1985.

[276] P. H. Frampton and P. Vogel, *Phys. Rep.* **82** 339 (1982).

[277] M. Gell-Mann, R. Slansky and P. Ramond, in: *Supergravity*, North Holland, 1979, p. 315; T. Yanagita, in: *Proc. Workshop on Unified Theory and Baryon Number in the Universe*, KEK, Japan, 1979.

[278] M. Doi, T. Kotani, E. Takasugi, *Prog. Theor. Phys.* **Suppl. 83** 1 (1985).

[279] For a recent review of double β decay, see F. T. Avignone III, S. Elliot, and J. Engel, arXiv 0708.1033.

[280] M. Goeppert-Mayer, *Phys. Rev.* **D48** 512 (1935).

[281] S. R. Elliot, A. Hahn, and M. K. Moe, *Phys. Rev. Lett.* **59** 2020 (1987).

[282] *Nuov. Cim.* **14** 322 (1937).

[283] J. C. Pati and A. Salam, *Phys. Rev.* **D10** 275 (1974).

[284] R. N. Mohapatra and J. C. Pati, *Phys. Rev.* **D11** 566 (1975); R. N. Mohapatra, *Phys. Rev.* **D11** 2558 (1975); R. N. Mohapatra and G. Senjanovic, *Phys. Rev.* **D12** 1502 (1975).

[285] M. Beg, *Phys. Rev. Lett.* **38** 1252 (1977); G. Senjanovic *Nucl. Phys.* **B153** 334 (1979).

[286] R. N. Mohapatra, in: CP Violation, C. Jarlskog (ed.). Singapore: World Scientific, 1988.

[287] W. Grimus, *Fortschr. Phys. Berlin* **36** 201 (1988).

[288] G. Ecker, W. Grimus and H. Neufeld, *Nucl. Phys.* **B247** 70 (1984).

[289] P. Langacker and S. U. Sankar, *Phys. Rev.* **D40** 1569 (1989).

[290] J.-M. Frere *et al.*, *Phys. Rev.* **D46** 337 (1992).

[291] P. Langacker, in: CP Violation, C. Jarlskog (ed.). Singapore: World Scientific, 1988.

[292] S. R. Mishra *et al.*, *Phys. Rev. Lett.* **68** 3499 (1992).

[293] G. Beall, M. Bander and A. Soni, *Phys. Rev. Lett.* **48** 848 (1982).

[294] See also G. Barenboim, J. Bernabeu, J. Prades and M. Raidal, *Phys. Rev.* **D55** 4213 (1997).

[295] Yue Zhang *et al.*, arXiv:hep-ph/0712.4218.

[296] D. Chang, X.-G. He and S. Pakvasa *Phys. Rev. Lett.* **74** 3927 (1995).

[297] T. D. Lee, *Phys. Rev.* **D8** 1226 (1973); *Phys. Rep.* **96** (1979).

[298] S. Weinberg, *Phys. Rev. Lett.* **37** 657 (1976).

[299] A concise review is found in L. Wolfenstein and Y. L. Wu, *Phys. Rev. Lett.* **73** 1762 (1994).

[300] For an update, see: H.-Y. Cheng, *Phys. Rev.* **D42** 2329 (2990); *Int. J. Mod. Phys.* **A7** 1059 (1992).

[301] I. I. Bigi and A. I. Sanda, in: CP Violation, C. Jarlskog (ed.). Singapore: World Scientific, 1988.

[302] S. Kraml *et al.*, *Report of the CPNSH Workshop, May 2004–Dec 2005*, [arXiv:hep-ph/0608079].

[303] G. C. Branco, *Phys. Rev. Lett.* **44** 504 (1980); *Phys. Rev. Lett.* **73** 1762 (1994).

[304] S. Glashow and S. Weinberg, *Phys. Rev.* **D15** 1958 (1977).

[305] G. C. Branco and A. I. Sanda, *Phys. Rev.* **D26** 3176 (1982).

[306] H. Georgi, *Had. J.* **1**, 155 (1978).

[307] L. Lavoura, *Int. J. Mod. Phys.*, **A9**, 1873 (1994).

[308] J. Liu and L. Wolfenstein, *Nucl. Phys.* **B289** 1 (1987).

[309] A. Mendez and A. Pomarol, *Phys. Lett.* **B272** 313 (1991).

[310] A. I. Sanda, *Phys. Rev.* **D23** 2647 (1981); N. G. Deshpande, *ibid.* **23**, 2654 (1981). J. F. Donoghue and B. R. Holstein, *Phys. Rev.* **D32** 1152 (1983).

[311] J.-M. Frere, J. Hagelin and A. I. Sanda, *Phys. Lett.* **151B** 161 (1985).

[312] I. I. Bigi and A. I. Sanda, *Phys. Rev. Lett.* **58** 1604 (1987).

[313] G. Beall and N. G. Deshpande, *Phys. Lett.* **132B** 427 (1983).

[314] R. Akhoury and I. I. Bigi, *Nucl. Phys.* **B234** 459 (1984).

[315] S. Weinberg, *Phys. Rev. Lett.* **63** 2333 (1989).

[316] D. Chang, W.-Y. Keung and T. C. Yuan, *Phys. Lett.* **251** 608 (1990); I. I. Bigi and N. G. Uraltsev, *Nucl. Phys.* **B353** 321 (1991).

[317] A. Anselm *et al.*, *Phys. Lett.* **152B** 116 (1985).

[318] M. S. Barr and A. Zee, *Phys. Rev. Lett.* **65** 21 (1990).

[319] J. Gunion and D. Wyler, *Phys. Lett.* **248B** 170 (1990); D. W. Chang, W. Y. Keung and T. C. Yuan, *Phys. Lett.* **251B** 608 (1990).

[320] D. Bowser-Chao, D. Chang and W. -Y. Keung, preprint hep-ph/9703435.

[321] R. Garisto and G. Kane, *Phys. Rev.* **D44** 2038 (1991).

[322] T. Nakada, *Phys. Lett.* **B261** 474 (1991).

[323] G. Branco, J. M. Frere and J. M. Gerard, *Nucl. Phys.* **B221** 317 (1983).

[324] W. Bernreuther and P. Overmann, *Z. Phys.* **C61** 599 (1994).

[325] W. Bernreuther and A. Brandenburg, *Phys.Rev.* **D49** 4481 (1994); W. Bernreuther, A. Brandenburg and M. Flesh, hep-ph/9812387.

[326] D. Atwood *et al.*, *Phys.Rep.* **347** 1 (2001) [arXiv:hep-ph/0006032].

[327] C. R. Schmidt and M. Peskin, *Phys. Rev. Lett.* **69** 410 (1992).

[328] S. Bar-Shalom, D. Atwood, G. Eilam and A. Soni, *Z. Phys.* **C72** 79 (1996).

[329] W. Buchmüller, R. D. Peccei, T. Yanagida, *Ann. Rev. Nucl. Part. Sci.* **55** (2005)311.

[330] W. Bernreuther, U. Low, J. P. Ma and O. Nachtmann, *Z. Phys.* **C41** (1988) 143; W. Bernreuther, private comm.

[331] M. Felcini, *Int. J. Mod. Phys.* **A19**, 3853 (2004) [arXiv:hep-ex/0404041].

[332] W. Fetscher and H.-J. Gerber, *Eur. Phys. J.* **C15** (2000) 316.

[333] I. C. Barnett *et al.*, *Nucl. Instrum. Meth.* **A455** (2000) 329.

[334] S. Y. Pi and A. I. Sanda, *Ann. Phys.* **106** 171 (1977).

[335] Y. S. Tsai, in: Stanford Tau Charm 1989:0387-393 (QCD183:T3:1989).

[336] Y. S. Tsai, *Phys.Rev.* **D51** 3172 (1995).

[337] J. H. Kühn and E. Mirkes, *Z.Phys.* **C56** (1992); Erratum, *ibid.* **C67** 364.

[338] J. H. Kühn and E. Mirkes, *Phys. Lett.* **B398** 407 (1997).

[339] I. I. Bigi and A. I. Sanda, *Phys. Lett.* **B625** 47 (2005).

[340] A. Pich, *Int. J. Mod. Phys.* **A21** 5652 (2006).

[341] D. Delepine *et al.*, *Phys. Rev.* **D74** 056004 (2006) [hep-ph/0503090]; D. Delepine, G. Faisel and S. Khalil, *Phys. Rev.* **D77** 016003 (2008) [arXiv:hep-ph/0710.1441].

[342] CLEO Collaboration: S. Anderson *et al.* *Phys. Rev. Lett.* **81** 3823 (1998).

[343] J. Kühn, *Nucl. Phys. Proc. Suppl.* **76** 21 (1999) [arXiv:hep-ph/9812399].

[344] A. Datta *et al.*, arXiv:hep-ph/0610162.

[345] W. Bernreuther, A. Brandenburg and P. Overmann, *Phys. Lett.* **B391** 413 (1997).

[346] J. Bernabeu *et al.*, *Nucl. Phys.* **B763** 283 (2007).

[347] H. Murayama, arXiv:0709.3041[hep-ph]; H. Murayama, Y. Nomura and D. Poland, arXiv:0709.0775 [hep-ph].

[348] S. Coleman and J. Mandula, *Phys. Rev.* **D159** 1251 (1967).

[349] K. Inoue *et al.*, *Prog. Theor. Phys.* **C 68**, 927 (1982); *ibid.* **C 71**, 413 (1984); L. E. Ibanez and G. G. Ross, *Phys. Lett.* **B 110** 215 (1982).

[350] H. P. Nilles, *Phys. Rep.* **110** 1 (1982); S. P. Martin, preprint hep-ph/9709356.

[351] S. P. Martin, preprint hep-ph/9709356.

[352] L. Alvarez-Gaume, M. Claudson and M. B. Wise, *Nucl. Phys.* **B 207**, 96 (1982); M. Dine and W. Fischler, *Phys. Lett.* **B 110**, 227 (1982).

[353] D. E. Kaplan, G. D. Kribs and M. Schmaltz, *Phys. Rev.* **D 62**, 035010 (2000); Z. Chacko *et al.*, *JHEP* **0001**, 003 (2000).

[354] L. Randall and R. Sundrum, *Nucl. Phys.* **B 557** 79 (1999); G. F. Giudice *et al.*, *JHEP* **9812**, 027 (1998).

[355] A. Masiero and O. Vives, in: *Proceedings of the International School of Physics "Enrico Fermi"*, M. Giorgi *et al.* (eds.), Course CLXIII. IOS Press, Amsterdam; SIF, Bologna, 2006, p. 133; J. L. Hewett *et al.*, arXiv:hep-ph/0503261, p. 391ff.

[356] S. Bertolini *et al.*, *Nucl. Phys.* **B353** 591 (1991).

[357] F. Gabbiani and A. Masiero, *Nucl. Phys.* **B322** 235 (1989).

[358] Y. Kizukuri and N. Oshimo, *Phys. Rev.* **D46** 3025 (1992).

[359] F. Gabbiani, E. Gabrielli, A. Masiero and L. Silvestrini, *Nucl. Phys.* **B477** 321 (1996).

[360] I. I. Bigi and F. Gabbiani, *Nucl. Phys.* **B352** 309 (1991); M. Ciuchini *et al.*, *Phys. Rev. Lett.* **79** 978 (1997).

[361] G. D'Ambrosio *et al.*, *Nucl. Phys.* **B645**, 155 (2002).

[362] UT*fit* Collab., M. Bona *et al.*, arXiv:0707.0636 [hep-ph].

[363] Th. Kaluza, *Sitz. Preuss. Akad. Wiss.* **K1**, 966 (1921); O. Klein, *Z. Phys.* **37**, 895 (1926).

[364] Nice introductions to the theories with extra dimensions can be found in three TASI lecture series: C. Csaki, arXiv:hep-ph/0404096; R. Sundrum, arXiv:hep-ph/0508134; G. Kribs, arXiv:hep-ph/0605325.

[365] N. Arkani-Hamed, S. Dimopoulos and G. R. Dvali, *Phys. Lett.* **B429**, 263 (1998) [arXiv:hep-ph/9803315]; I. Antoniades *et al.*, *Phys. Lett.* **B436**, 257 (1998) [arXiv:hep-ph/9804398]; N. Arkani-Hamed, S. Dimopoulos and G. R. Dvali, *Phys. Rev.* **D59**, 086004 (1999) [arXiv:hep-ph/9807344].

[366] L. Randall and R. Sundrum, *Phys. Rev. Lett.* **83**, 3370 (1999) [arXiv:hep-ph/9905221].

[367] T. Appelquist, H. C. Cheng and B. A. Dobrescu, *Phys. Rev.* **D64**, 035002 (2001) [arXiv:hep-ph/0012100]; A. J. Buras, M. Spranger and A. Weiler, *Nucl. Phys.* **B 660**, 225 (2003) [arXiv:hep-ph/0212143].

[368] A. Buras *et al.*, *Nucl. Phys.* **B678**, 455 (2004)[arXiv: hep-ph/0306158].

[369] N. Arkani-Hamed and M. Schmaltz, *Phys. Rev.* **D61**, 033005 (2000) [arXiv:hep-ph/9903417].

[370] B. Lillie in: *The Discovery Potential of a Super B Factory*, J. Hewett, D. Hitlin (eds.), arXiv:hep-ph/0503261.

[371] A. D. Dolgov and Ya. B. Zel'dovich, *Rev. Mod. Phys.* **53** 1 (1981).

[372] A. D. Sakharov, *JETP Lett.* **5** 24 (1967).

[373] A. D. Dolgov, in: *Proceedings of the XXVth ITEP Winterschool of Physics*, Feb. 18–27, Moscow, Russia.

[374] V. A. Kuzmin, *Pis'ma Z. Eksp. Teor. Fiz.* **12** 335 (1970).

[375] J. C. Pati and A. Salam, *Phys. Rev.* **D10** 275 (1974); H. Georgi and S. L. Glashow, *Phys. Rev. Lett.* **32** 438 (1974).

[376] M. Yoshimura, *Phys. Rev. Lett.* **41** 281 (1978).

[377] D. Touissaint, S. B. Treiman, F. Wilczek and A. Zee, *Phys. Rev.* **D19** 1036 (1979); S. Weinberg, *Phys. Rev. Lett.* **42** 850 (1979); M. Yoshimura, *Phys. Lett.* **88B** 294 (1979).

[378] V. A. Kuzmin, V. A. Rubakov and M. E. Shaposhnikov, *Phys. Lett.* **B155** 36 (1985).

[379] G. t'Hooft, *Phys. Rev. Lett.* **37** 8 (1976); *Phys. Rev.* **D14** 3432 (1976).

[380] V. A. Rubakov and M. E. Shaposhnikov, *Usp. Fiz. Nauk* **166** 493 (1996).

[381] K. Kajantie *et al.*, *Nucl. Phys.* **B466** 189 (1996).

[382] M. Fukugita and Y. Yanagida, *Phys.Rev.* **D42** 1285 (1990).

Index

Printed in the United States
By Bookmasters